Chemistry in the Laboratory

Seventh Edition

James M. Postma
California State University, Chico

Julian L. Roberts
University of Redlands

J. Leland Hollenberg
University of Redlands

W. H. FREEMAN AND COMPANY

NEW YORK

Publisher: Clancy Marshall

Sponsoring Editor: Kathryn Treadway

Media and Supplements Editor: Dave Quinn

Project Editor: Vivien Weiss

Marketing Manager: John Britch

Design Manager and Cover Designer: Victoria Tomaselli

Illustrations: Fine Line Illustrations

Senior Illustrations Coordinator: Bill Page

Production Coordinator: Paul W. Rohloff

Composition: Prepare, Inc.

Manufacturing: RR Donnelley

Library of Congress Control Number: 2009925431

W. H. Freeman and Company
41 Madison Avenue
New York, NY 10010
Houndmills, Basingstoke RG21 6XS, England
www.whfreeman.com

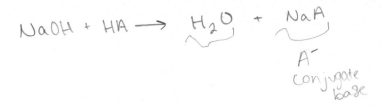

NaOH + HA $\longrightarrow$ H$_2$O + NaA

A$^-$
conjugate
base

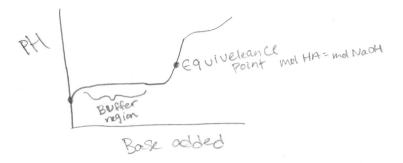

PH

Equiveleance Point   mol HA = mol NaOH

Buffer region

Base added

To our colleagues whose stimulating writing in the Journal of Chemical Education *and now on the Web has inspired us.*

CONTENTS

EXPERIMENT 42
Ultraviolet Light, Sunscreens, and DNA

EXPERIMENT 43
Natural Radioactivity
The Half-Life of Potassium-40 in
Potassium Chloride

Note to Users of Atkins and Jones, **Chemical Principles: The Quest for Insight**

The table below shows the Experiments and Study Assignments in this manual that are suggested as companions to the chapters of *Chemical Principles: The Quest for Insight,* by Atkins and Jones

Chapter Number	Experiment Number
A, B, C: Matter and Energy; Elements and Atoms; Compounds	1, 2
D: The Nomenclature of Compounds	Study Assignment A
E, F: Moles and Molar Masses; Determination of Chemical Formulas	6, 8, 9, 10
G: Mixtures and Solutions	3
H: Chemical Equations	4, 5, 8, 9
I, J, K: Aqueous Solutions and Precipitation; Acids and Bases; Redox Reactions	5, 7, Study Assignment A, 18, 34, 35
L, M: Reaction Stoichiometry; Limiting Reactants	5, 8, 9
1. Atoms: The Quantum World	17
2. Chemical Bonds	18, Study Assignment B
3. Molecular Shape and Structure	19
4. The Properties of Gases	11, 12, 13
5. Liquids and Solids	20
6. Inorganic Materials	42
7. Thermodynamics: The First Law	14, 15, 16
8. Thermodynamics: The Second and Third Laws	
9. Physical Equilibria	21, 22
10. Chemical Equilibria	23, 24
11. Acids and Bases	27, 28
12. Aqueous Equilibria	27, 28, 29, 30, 31, 32, 33
13. Electrochemistry	34, 36, 37
14. Chemical Kinetics and the Elements: The First Four Main Groups	9, 13, 25, 26, 31
15. The Elements: The Last Four Main Groups	7, 38, 40, 41
16. The *d* Block: Metals in Transition	24, 39
17. Nuclear Chemistry	43
18. Organic Chemistry I: The Hydrocarbons	
19. Organic Chemistry II: Functional Groups	20, 42

PREFACE

When we revised *Chemistry in the Laboratory* for the fourth edition in 1997, we thought we were being very "green" by shrinking the scale of many experiments to save materials (and costs). In the fifth and sixth editions, we continued this green trend by transitioning to chemistries with fewer hazardous materials as reactants and products.

In this exciting seventh edition, we add a new experiment altogether, *An Introduction to Green Chemistry: Making Biodiesel and Soap from Vegetable Oils and Fats* (Experiment 33), that specifically targets environmental topics. We have also redirected the content of Experiment 29 (formerly *The Chemistry of Natural Waters*) to that of a current environmental issue, *What's in Your Drinking Water? Comparing Tap Water and Bottled Water.*

Another topic of current interest is highlighted in Experiment 3, *Forensic Chemistry: Crime Scene Investigation.* We hope that you can couple the creativity and versatility of our text with your own ingenuity to create an interesting and productive learning environment in the laboratory. We understand that some instructors prefer that their students use laboratory notebooks rather than report forms; we believe our text works well with either format. Those who desire an atmosphere of discovery and open inquiry in the laboratory will find our *Consider This* feature an asset, while those who want the experiments to work well and illustrate a concept clearly will be equally satisfied. We believe that we have stayed true to our reputation for thorough introductions, well-tested experiments, and attention to detail.

We've maintained many of the new features of the sixth edition, such as briefer introductions, a Web-based activity, and applied topics like atmospheric ozone, natural radioactivity measurements, and explorations of ultraviolet light, DNA, and sunscreens. We've also continued our emphases on safety and waste disposal, the use of reduced scale where appropriate, and the *Consider This* feature.

Always Striving to Improve

Each time we gather to plan a new edition, a bit of neurosis settles in with us—the wish to please all chemistry instructors and their students. Our confidence renewed, we have composed a revised edition that maintains the traditions of our past texts while piquing the curiosity of new users.

Timeless perfection is our goal with every edition of *Chemistry in the Laboratory.* Each time we take it into the laboratory with our students, or read the

monthly *Journal of Chemical Education*, or discuss a topic with a colleague, we find ourselves making notes on improvements we'd like to add to the manual. We hope this edition is a step closer to your idea of perfect, whether you are an instructor or a student of chemistry.

Similar in structure to past editions, *Chemistry in the Laboratory*, Seventh Edition, contains 43 experiments and is sufficient for a one-year course in general chemistry. It serves as a companion to the general chemistry text *Chemical Principles: The Quest for Insight* by Peter Atkins and Loretta Jones, as well as to Zumdahl's *Chemical Principles*, Sixth Edition, and Brown and LeMay's *Chemistry: The Central Science*, Eleventh Edition. But this manual can be used with any modern general chemistry text.

Our goals for this edition included the tightening of introductory material. Some argue that you cannot have too much of a good thing, others say that brevity is the soul of wit; we've tried to be a bit more witty.

The new format of Experiment 41, the Web-based activity on atmospheric ozone, was well received in the sixth edition and so we've kept it for this revision. Experiment 42, which explores the interactions of ultraviolet light, sunscreens, and DNA, continues to be of contemporary interest. And measurements of radioactivity from a common laboratory material, potassium chloride (Experiment 43), should improve students' perspective on a little-understood phenomenon.

We all envision an ideal educational laboratory setting where, at an unhurried pace, a student and an instructor explore the physical world together, discovering the important laws and theories of science through systematic observation, challenging discussion, and insightful deduction. But the realities of limited budgets, demanding schedules, and inadequate preparation, coupled with the need to "just get it done" or the desire to "make it work right," leave us all unsettled about the laboratory experience and its place in a modern education system.

Some have suggested abandoning the lab altogether or using ever-advancing technology to simulate lab experiences. We reject the former out of hand and are very cautious of the latter. Computer simulations augment the educational experience wonderfully, providing models that students can manipulate and thereby explore arenas of the physical world that were previously inaccessible. But the finest models cannot connect us to natural processes and their accompanying colors, textures, and smells as well as the startling realities of the imperfections of our idealized models.

Thus, we have retained many of the popular features in previous editions of *Chemistry in the Laboratory* and have carried through the improvements of each edition. The *Consider This* sections, first introduced in the fifth edition, continue to generate enthusiasm for the lab experience. We have shared many instructors' desire to open vistas for students, confining them less to a procedure (or recipe, as some would describe it) and allowing for more exploration and discovery in the general chemistry laboratory. These discovery/exploration models of lab experience, however, are very demanding of instructors' time and expertise, requiring patience and support systems such as stockroom help, budgetary support, and equipment availability.

After evaluating our own experiences and our experience with students, we decided that a combination of the two extremes would create a healthy, positive laboratory experience. We have retained the student-tested, directive laboratory exercises that have characterized our laboratory manual (and most other successful texts) to create a foundation of positive laboratory experiences and techniques to build on. The somewhat expansive background discussions (not without some paring down) also remain; they create an interesting context for the experiment.

The *Consider This* feature encourages users to expand the principles of the experiment into interesting applications, open-ended experiments, or unexplored corners. Some of these can serve as a jumping-off point for undergraduate research projects and perhaps even publications. Eager to learn how this feature is applied and the insights gleaned for pursuing these explorations, we hope you will take advantage of the W. H. Freeman Web site, e-mail, or traditional communication to let us know of your explorations.

Expanded Use of Reduced Scale Techniques

We have continued to emphasize reduced-scale experimentation (e.g., titrations involving 1-mL burets), but we have also expanded and added detail for experimental procedures for those who prefer to use macroscale experimentation using 25- or 50-mL burets. In general, we see advantages to the smaller scale manipulations. These advantages include the following:

• Student exposure to toxic and corrosive chemicals is reduced.

• Laboratory air quality is improved.

• The amount of hazardous waste is reduced by a factor of 10, decreasing disposal costs and the burden on the environment.

• Purchase costs for apparatus and chemicals are significantly lower.

• Breakage is reduced, as are replacement costs for apparatus.

• Microscale experiments often take less time than their macroscale counterparts.

• Students gain an introduction to microscale experimentation, which is now widely used in the second-year organic chemistry course.

On the minus side, we note that:

• Working on a smaller scale may demand more attention to development of laboratory skills (although many students are now being introduced to reduced scale laboratory work in their high school chemistry courses.)

• The precision of volumetric titrations is generally reduced. This is partially offset by the increase in the sharpness of the end point as the titration volume is reduced. (These points are discussed more fully in the Introduction.)

• Balances with at least milligram sensitivity are required for decent quantitative work.

Many experiments in this manual have not been reduced in scale tenfold. For example, a chemical synthesis on a milligram scale doesn't produce a satisfying aesthetic experience—there's not enough product to see and handle. So although we have generally reduced the scale of synthesis experiments, they are still at the gram level. Experiments to measure the enthalpy of a reaction also suffer when scaled too small. The larger surface-area-to-volume ratio means that a greater proportion of energy will leak through the walls of the container. Furthermore, the thermal mass of thermometers and stirrers will become a larger proportion of the heat capacity of the system, necessitating corrections that complicate the calculation of the enthalpy change. For these reasons, we have not reduced the scale of the reaction enthalpy experiments.

Although the term *microscale* is by now firmly entrenched, it is something of a misnomer. The scale of almost all experiments described in the current chemical education literature as *microscale* would more appropriately be called *milliscale*, because sample sizes are in the milligram-to-gram range and reaction volumes and titration volumes are in the range of 1–10 mL. In coining the term *microscale* to describe the reduced scale approach, modern practitioners appear to be ignoring the priority of the already well-established fields of microchemistry and microanalysis that work with truly "micro" samples in the microgram and microliter range.

These microchemical techniques began to be developed as early as 1910, the journal *Mikrochemie* made its first appearance in 1923, and the use of microchemical techniques was well established in the 1930s. In short, microscale is new only to undergraduate laboratories and, in the form presented in contemporary "microscale" lab manuals, is not truly microscale. That said, we think the trend to reduced scale experimentation in undergraduate laboratories is laudable and will prove not to be simply a fad.

Most of us who have been teaching chemistry for more than a few years are familiar with passing fads in chemical pedagogy, and we all have a tendency to stick with methods that are tried and true. So there is a natural reluctance to change, particularly when institutional inertia is a barrier or when the idea is not our own. Reduced scale experimentation is not a panacea, but it is worth trying simply because it's safer for students and because small messes are easier and less costly to clean up than big ones. So we urge you to begin by looking for ways to reduce the scale of your experiments wherever feasible. This lab manual is offered in that spirit. Once you begin performing reduced scale experiments, we believe you will come to prefer them in many instances.

We have tried to give the manual a modern look by incorporating several new or redrawn illustrations and by making the writing clearer and simpler. Where possible, we have updated the references and Bibliography sections of the experiments.

We believe quite strongly in the traditional format of our laboratory manual. Each exercise contains a substantial introduction that provides a context for the experiment and discussion of the theory that we believe can set the stage for important student insights. We provide all the relevant equations so that each experiment can stand on its own without the need to consult other texts for lab preparation. We provide an option, where appropriate, of reduced scale and macroscale directions with sufficient detail to allow success in either format.

Safety and Waste Collection Procedures

We have maintained from previous editions our emphasis on safety and proper handling of labora-

tory wastes. Using our own experience, our reading of the chemical literature, and comments from readers, we have made improvements in the experimental procedures, safety precautions, and waste collection procedures. All these changes are designed to make the laboratory a safer place to work, to raise the safety consciousness of students, and to protect students and the environment from exposure to hazardous materials.

Safety

No instructor should assume that students will read carefully and practice the precautions contained in the experiments. Safety consciousness is not inborn; it must be developed through training and practice. Consequently, general chemistry laboratory instructors must take a few minutes each laboratory period to emphasize the safety precautions and waste collection procedures.

Instructors must also rigorously enforce good safety practices, particularly the rule that students must wear safety goggles at all times. Here students will take their cue from the instructor. If the instructor is lax in observing good safety practices, students will be lax as well.

Good safety practices are in the best interests not only of students but also the instructor and the institution, which may be liable if a student is injured because of negligence or ignorance on the part of the instructor.

A number of experiments in this manual require the use of a fume hood. Where one is not available, we suggest that the experiment not be performed.

Waste Collection

Good waste collection and waste management procedures require that wastes not be mixed together indiscriminately. For each hazardous waste, a properly labeled waste receptacle should be provided. The segregated and properly identified wastes can then be combined and packed for proper disposal by someone who is trained in waste management.

Most sizable research institutions have already put waste management programs in place, but many smaller institutions continue to grapple with the problem of maintaining proper procedures for handling and disposing of laboratory wastes. They are also discovering that the costs of disposing of chemical wastes can easily surpass the costs of purchasing the chemicals in the first place. Thus, for economic reasons, as well as for protection of the students and the environment, we have tried to devise experiments that produce a minimum amount of hazardous waste.

Instructor and Student Resources

The *Instructor's Manual*

We believe that our *Instructor's Manual* can be a most useful tool for the instructor, the grader of laboratory results, and stockroom personnel. It includes equipment and materials lists, directions for chemical preparations, and filled-in report forms for each of the experiments. We hope that these features will save time and energy and facilitate the grading process.

To access and download the *Instructor's Manual*, please visit our password-protected instructor's Web site, which can be accessed through www.whfreeman.com/postma.

The Book Companion Web Site

The Book Companion Web site (www.whfreeman.com/postma) includes a number of electronic resources for students and instructors alike. Students can access *pre-lab quizzes* to prepare for class, as well as a periodic table and calculators to use for assignments. The password-protected *Instructor Resources* can also be accessed through this Web site; features include *answers to pre-lab quizzes*, the *Instructor's Manual, correlation guides*, and much more.

Acknowledgments

We wish to acknowledge with thanks the assistance of several staff at W. H. Freeman and Company who were instrumental in the production of our book: Publisher Clancy Marshall, Sponsoring Editor Kathryn Treadway, Marketing Manager John Britch, Media and Supplements Editor Dave Quinn, Project Editor Vivien Weiss, Assistant Editor Anthony Petrites, Design Manager Victoria Tomaselli, Senior Illustration Coordinator Bill Page, Production Coordinator Paul W. Rohloff, and Copy Editor David Roemer.

For this and past editions, special thanks are due to Fred E. Wood, now the Associate Dean of Undergraduate Education at the University of California, Davis. Margaret Korte and Ron Cooke of the Chemistry Department at California State University, Chico, have provided detailed critiques of each experiment and suggestions for further improvement.

Reviewers of the seventh edition have helped us sharpen our focus and improve the text before you. They include:

Hafed Bascal, University of Findlay, Findlay, Ohio

Edward Chikwana, California State University Chico, Chico, California

Michael Garlick, Delta College, Bay City, Michigan

Chad Gatlin, Indian Hills Community College, Centerville, Iowa

Cynthia Peck, Delta College, Bay City, Michigan

Jess Vickery, Adirondack Community College, Queensbury, New York.

We are grateful to them and to all of you who have made suggestions and offered comments over the years.

Welcome to the Chemistry Lab

Your laboratory work is the core of your chemistry course. Everything we know about atoms and molecules is almost entirely the result of laboratory experimentation. You have a challenging opportunity to make many observations of chemical reactions under controlled laboratory conditions. You also will experience firsthand the method of inquiry that is the foundation of all experimental sciences. Planning experiments and making and interpreting observations are at the heart of the scientific approach to understanding the interactions of atoms and molecules.

Here are some brief points that will give you a good start in this course.

1. Gain self-confidence by working individually, unless the experiments call for teamwork. However, don't hesitate to ask questions of the instructor if you are uncertain about a procedure or the interpretation of a result.

2. Use your ingenuity and common sense. Laboratory directions, while quite specific, leave ample opportunity for clear-cut, logical, original, and imaginative thinking. This attitude is a prerequisite in any scientific endeavor.

3. Don't waste time by coming unprepared to the laboratory. Prepare for each experiment by studying it (both the Pre-Lab Preparation and, briefly, the Experimental Procedure) before you come to the laboratory.

4. Note beforehand any extra equipment required from the stockroom, and obtain all of it at the same time.

5. Prepare your Laboratory Reports on each experiment with care. Use a permanently bound laboratory notebook, as prescribed by your instructor, to record your data. All data should be recorded directly into your notebook, not on loose sheets or scraps of paper. Where calculations of data are involved, show an orderly calculation for the first set of data, but do not clutter the calculation section with arithmetic details. Likewise, *as you perform the experiment,* think through and answer important questions that are intended to give you an understanding of the principles on which the experimental procedure is based.

6. Scientists test their own ideas and knowledge by discussing them with one another, even arguing about them. You likewise may profit by discussion with your classmates, but don't blindly copy their work. Do your own thinking and work and remember that integrity is the keystone of scientific work. You may also find it useful to refer to your text while working in the laboratory. (Books are generally even more reliable and complete sources of information than are your classmates!)

7. For tabular data on the properties of substances, you may wish to consult handbooks such as the *Handbook of Chemistry and Physics* (CRC Press, Inc., Boca Raton, Florida) or *Lange's Handbook of Chemistry* (McGraw-Hill, New York).

Safety Rules

Familiarize yourself with the safety rules given in the Introduction. Observance of these rules, as modified or added to by your instructor, is essential for the sake of your safety and that of others in the laboratory.

Your instructor will indicate the location and show you the proper use of the fire extinguishers, fire blanket, eyewash fountain, safety shower, and first-aid cabinet and supplies. The instructor will also tell you where to obtain safety glasses or goggles.

Good Laboratory Practices

Familiarize yourself with the good laboratory practices in the Introduction. It is essential that you carefully follow these regulations, as modified or added to by your instructor.

Basic Laboratory Equipment and Procedures

Check the equipment in the laboratory locker assigned to you, following the procedure described in the Introduction or as directed by your instructor. If time remains, or if your instructor so directs, read over the procedures for the handling of chemicals, the use of laboratory burners, the use of microscale equipment, the care of laboratory glassware, and volumetric measurements of liquids.

Reduced Scale (Microscale) Experimentation

Several of the experiments in this lab manual that involve titrations provide a dual set of instructions—either for microscale titrations, using 1-mL burets, or for the use of conventional 50-mL burets. The description of several devices used as microburets and instructions for the use of conventional burets can be found in the Introduction. Some of the advantages of microscale experimentation are discussed in the Preface.

Introduction

The chemistry laboratory can be a place of enjoyment, discovery, and learning. It can also be a place of frustration—and danger.

Although every effort has been made in these experiments to minimize exposure to hazardous, toxic, or carcinogenic substances, there is an unavoidable hazard involved in using a variety of chemicals and glass apparatus. In experiments where a potential danger exists, you will find a Safety Precautions section at the beginning of the experimental procedure. Read this section very carefully before proceeding with the experiment.

Your skin and the delicate membranes of your eyes, mouth, and lungs are largely made of protein. We hope that you do not experience firsthand the adverse effect that concentrated solutions of acids and bases have on protein. The eyes are especially sensitive to acids, bases, and oxidizing agents and must be protected at all times by wearing safety goggles. In addition, the open flame of a Bunsen burner presents a constant hazard to loose clothing and hair.

It's likely that you will become very frustrated if you come unprepared to the laboratory, neglect to record important data, or frantically try to write up reports an hour before they are due. You can minimize these problems by reading the experiments carefully beforehand, noting the critical data that must be recorded, and thoughtfully considering the data as you collect it in order to avoid careless blunders.

We strongly advise you to learn and observe at all times the following laboratory rules and good laboratory practices. By doing so, you will minimize the potential dangers and frustrations of laboratory work and maximize the enjoyment.

Safety Rules

These rules are designed to ensure that all work done in the laboratory will be safe for you and your fellow students. In addition to the rules listed here, your institution may have a set of rules that you will be asked to read and to sign as evidence that you have read them.

Your laboratory instructor should also point out the location of various pieces of safety equipment—such as fire extinguishers, fire blankets, safety show-

ers, eyewash fountains, and equipment for handling spills—and should demonstrate how to use these items.

Finally, you should know where first-aid supplies are kept and where a telephone is located for emergencies that require paramedical, fire, or police assistance. The telephone numbers of these on- or off-campus services should be posted in a prominent place.

1. The most important rule is that *safety goggles* or *safety glasses with side panels* must be worn at all times in the laboratory. Goggles provide the best protection. Ordinary prescription glasses sometimes cover only parts of the eyes and lack side panels that protect the wearer from chemical splashes that might get into the eyes from the side. For this reason, eyeglasses should be covered by safety goggles. Contact lenses, if worn in the laboratory, should be covered with safety goggles. By themselves, they offer no protection from splashes, and they may be considered unsafe even under safety goggles because various fumes—for instance, hydrogen chloride gas—may accumulate under the lens and cause injury. If you do not own a pair of safety goggles, obtain one from the stockroom or your campus bookstore.

If any chemical comes in contact with the eye, the most effective first aid is to flush the eye immediately with a steady stream of tap water. You are seldom more than a few seconds from a faucet or eyewash fountain. Continue flushing for at least five minutes and then consult a physician as soon as possible. If your laboratory is equipped with eyewash fountains, familiarize yourself with their use and their location.

2. Fire is always a danger. Learn where the nearest fire extinguisher is and how to use it. Your laboratory should also be equipped with a fire blanket and safety shower. *If your hair or clothing should catch fire, smother the fire with a blanket or douse yourself in the shower.*

Because some synthetic fabrics (polyester) melt and stick to the skin when they catch fire, we recommend that you wear old clothes made of cotton, or a cotton labcoat or laboratory apron. Cotton shirts are more fire resistant and a pair of cotton jeans affords some protection to the legs from minor spills.

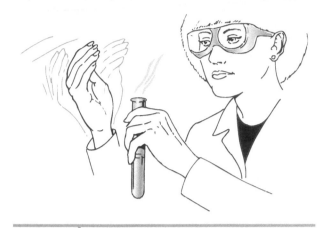

FIGURE I-1 Chemical substances should not be inhaled or smelled unless you are instructed to do so. Even in such cases, you must exercise great care in noting the odor of a substance, using your hand to waft its vapor gently toward your face. Whenever possible, avoid breathing fumes of any kind.

3. Bare feet are not allowed in the chemistry laboratory. Broken glass and spilled chemicals, such as concentrated acids, are all too common on the floors of chemistry labs. Wear shoes that protect the tops of the feet from spills; don't wear sandals or flip-flops.

4. Minor burns, cuts, and scratches are fairly common injuries. However, you must report every such injury to your instructor, who will determine what first aid is appropriate. If you or another student must report to the infirmary or hospital, be certain that someone accompanies the injured person.

5. The vapors of a number of solutions are quite potent and can irritate or damage the mucous membranes of the nasal passages and the throat. Use the technique displayed in Figure I-1 when you need to note the odor of a gas or vapor.

6. In many experiments, it is necessary to heat solutions in test tubes. Never apply heat to the *bottom* of the tube; always apply it to the point at which the solution is highest in the tube, working downward if necessary. Be extremely careful about the direction in which you point a tube; a suddenly formed bubble of vapor may eject the contents violently (an occurrence called *bumping*). Indeed, a test tube can become a miniature cannon (see Figure I-2).

7. Avoid tasting anything in the laboratory. (Poisonous substances are not always labeled as such in the laboratory.) Do not use the laboratory as an eating place, and do not eat or drink from laboratory glassware.

8. Perform no unauthorized experiments.

9. Never work in the laboratory alone.

10. Beware of hot glass tubing; it *looks* cool long before it can be handled safely.

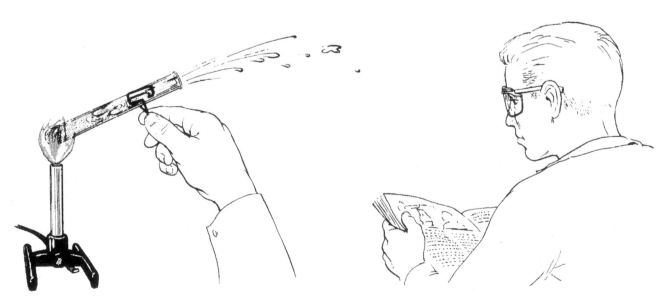

NEVER point a test tube that is being heated at your neighbor—it may bump and eject its contents.
SAFETY GOGGLES worn regularly in the laboratory will protect your eyesight.

FIGURE I-2 Two important safety precautions.

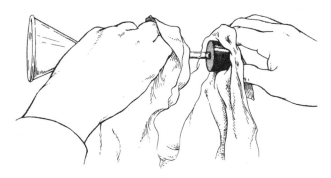

FIGURE I-3 **The procedure for inserting a glass tube into a stopper. NEVER force a thistle tube or funnel into a stopper by grasping the large end. Grasp the stem and twist as you push. ALWAYS wrap your hands in a towel when putting a glass tube into a stopper. Moisten the tube with water or glycerol and insert with a twisting motion.**

11. For reactions involving poisonous gases, *use the hood,* which draws gases or vapors out of the lab and into the exhaust system.

12. Neutralize a spilled acid or base as follows: (a) Acid on clothing, use dilute sodium bicarbonate solution; (b) base on clothing, use boric acid solution (50 g/L); (c) acid or base on the desk, use solid sodium bicarbonate for either, followed by water.

13. To insert glass tubing (including thermometers, long-stemmed funnels, thistle tubes, etc.) through a rubber stopper, first lubricate the tube and stopper with water or *glycerol.* Hold the tubing with a cloth near *the end to be inserted,* and insert with a twisting motion. (If you twist a long-stemmed funnel by the funnel end, it is easily broken. See Figure I-3.)

Good Laboratory Practices (Do's and Don'ts)

These practices are designed to guide you in developing efficient laboratory techniques and in making your laboratory a pleasant place to work.

1. Read each experiment thoroughly before entering the lab. If you do not, you will waste a great deal of time (both your own and your instructor's), you may expose yourself and others to unnecessary hazards, and you will probably not obtain reliable, useful data. (You will also routinely fail all pre-lab quizzes if your instructor chooses to use them.)

2. *Don't throw matches, litmus, or any insoluble solids into the sink.* Labeled waste containers should be provided to collect hazardous solid or liquid wastes.

3. Leave reagent bottles at the side shelves. Bring test tubes or beakers *to the shelf* to obtain chemicals.

4. Read the label *twice* before taking anything from a bottle.

5. Avoid using excessive amounts of reagent—1 to 3 mL is usually ample for test-tube reactions.

6. *Never* return unused chemicals to the stock bottle. You may make a mistake that later will cause other students' experiments to suffer.

7. Don't insert your own pipets or medicine droppers into the reagent bottles. Avoid contamination of the stock solution by pouring only what you need into a small, clean beaker or test tube.

8. Don't lay down the stopper of a bottle. The stopper may pick up impurities and thus contaminate the solution when the stopper is returned. (The proper way to pour liquids from a glass-stoppered bottle is shown later.)

9. Don't heat thick glassware such as volumetric flasks, graduated cylinders, or bottles; they break easily, and heating distorts the glass so that the calibrations are no longer valid (see Figure I-4). If test tubes are heated above the liquid level, they may break and then splash liquid over the hot glass. Evaporating dishes and crucibles may be heated red hot. Avoid heating any apparatus too suddenly; apply the flame intermittently at first.

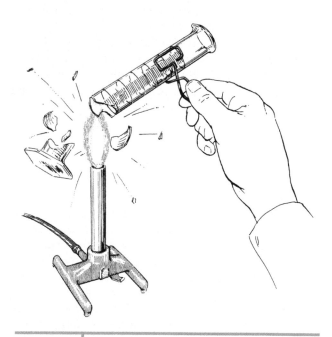

FIGURE I-4 **If heat is applied to the wrong type of laboratory apparatus, the outcome can be disastrous. NEVER heat a graduated cylinder or bottle.**

Basic Laboratory Equipment and Procedures

The laboratory locker

Check the equipment in the locker assigned to you. Refer to Figure I-5 for the identification of any unfamiliar items. Ascertain that all items are present and examine them carefully to be sure they are in acceptable condition. You are responsible for this equipment and may be charged for any breakage or shortage at the conclusion of the course.

The handling of chemicals

Some suggestions bear repeating: Be considerate of others by always bringing your container to the reagent shelf to obtain a chemical. Don't take the bottle to your desk. Maintain the purity of the chemicals in the reagent bottles. Don't withdraw more than you need, and never return any chemical to the bottle. Never contaminate the stopper by laying it down; hold it by your fingers. Don't insert your own transfer pipet into a reagent bottle or the transfer pipet from a reagent bottle down into your own test tube or solutions (Figure I-6). If necessary, clean the outside of the reagent bottle of accumulated dust, ammonium chloride, or other contaminant; rinse the neck and stopper with distilled water; and wipe dry before removing the stopper.

Some of these simple suggestions on the proper handling of solid and liquid chemicals are illustrated in Figures I-7 and I-8.

Observing these suggestions carefully will prevent the contamination of the stock bottles. If you do spill any chemical, clean it up completely, at once. A dirty laboratory prevents good work from being done.

Laboratory burners

The Bunsen burner, used for most laboratory heating, produces a cone-shaped flame, as illustrated in Figure I-9. Ordinary beakers, crucibles, and other objects to be heated are placed just above the hottest portion of the flame. This allows the most heat to spread about them. If placed down in the cold inner cone of the flame, which consists of unburned gas, the objects are not heated effectively.

The modern Fischer burner is designed to give a concentrated, very hot flame (Figure I-10). For the maximum temperature, have the gas on full pressure and, with the air vents open, adjust the needle valve at the base (or the air valve if the compressed air type of burner is used) to produce a short blue flame of many short cones that are about 0.5 cm high. The object to be heated is placed about 1 cm above the grid.

Operations with glass tubing

Glass is not a true crystalline solid and therefore does not have a sharp melting point. (In this respect, glass more nearly resembles a solid solution or an extremely viscous liquid that gradually softens when heated.) It is this property that makes glassworking possible.

Soda-lime glass, made by heating a mixture of soda (Na_2CO_3), limestone ($CaCO_3$), and silica (SiO_2), softens readily at about 300 °C to 400 °C in the burner flame. Tubing of this glass is easily bent, but because of its high temperature coefficient of expansion, it must be heated and cooled gradually to avoid undue strain or breakage. Annealing by mild reheating and uniform cooling helps reduce breakage. Such glass must not be laid on a cold surface while it is hot, since this introduces strains and causes breakage.

Borosilicate glass (such as Pyrex or Kimax) does not soften much below 700 °C to 800 °C and must be worked in an oxygen–natural-gas flame, preferably using a glassworking torch. Because it has a low temperature coefficient of expansion, objects made of it can withstand sudden temperature changes.

Figure I-11 shows the proper way to cut glass tubing and fire polish the ends. Fire polishing smooths the sharp edges so that the glass tubing can be inserted easily into a rubber stopper without cutting your fingers. Figure I-12 illustrates the way to make a bend or a constricted tip.

Care of laboratory glassware

Examine all glassware for cracks and chips. When heated, flasks or beakers with cracks may break and cause injury. Small chips in borosilicate glassware can sometimes be eliminated by fire polishing; otherwise, chipped glassware should be discarded because it is easy to cut oneself on sharp glass edges.

The recommended procedure for cleaning glassware is to wash the object carefully with a brush in hot water and detergent, then rinse thoroughly with tap water, and finally rinse once again with a small quantity of distilled or deionized water. Then allow the glassware to drain dry overnight in your locker. If you must use a piece of glassware while it is still wet, rinse it with the solution to be used.

In a general chemistry laboratory, we don't recommend the use of "cleaning solution" (a solution of CrO_3 or $K_2Cr_2O_7$ in concentrated sulfuric acid). Such hazardous solutions should be employed only under

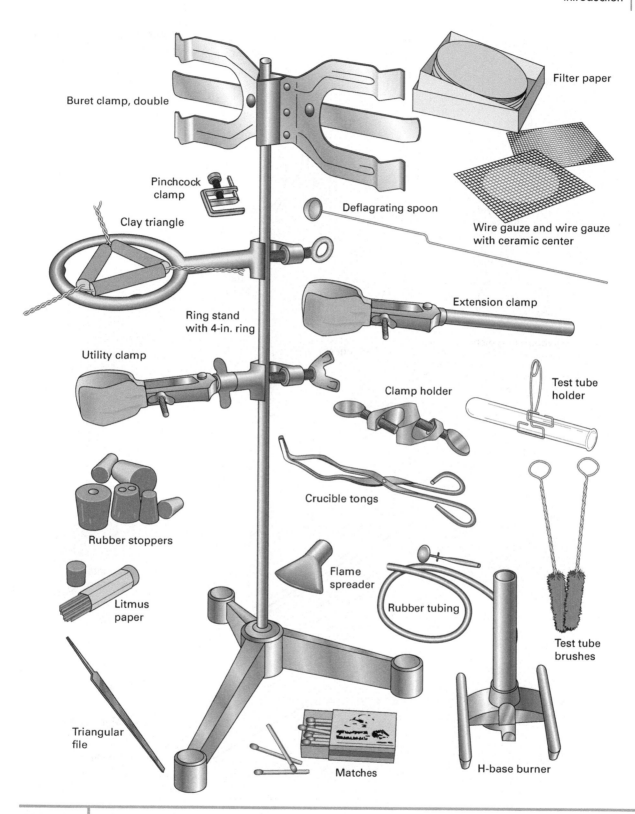

Buret clamp, double

Filter paper

Pinchcock clamp

Clay triangle

Deflagrating spoon

Wire gauze and wire gauze with ceramic center

Ring stand with 4-in. ring

Extension clamp

Utility clamp

Clamp holder

Test tube holder

Crucible tongs

Rubber stoppers

Litmus paper

Flame spreader

Rubber tubing

Test tube brushes

Triangular file

Matches

H-base burner

FIGURE I-5 | Common laboratory equipment.

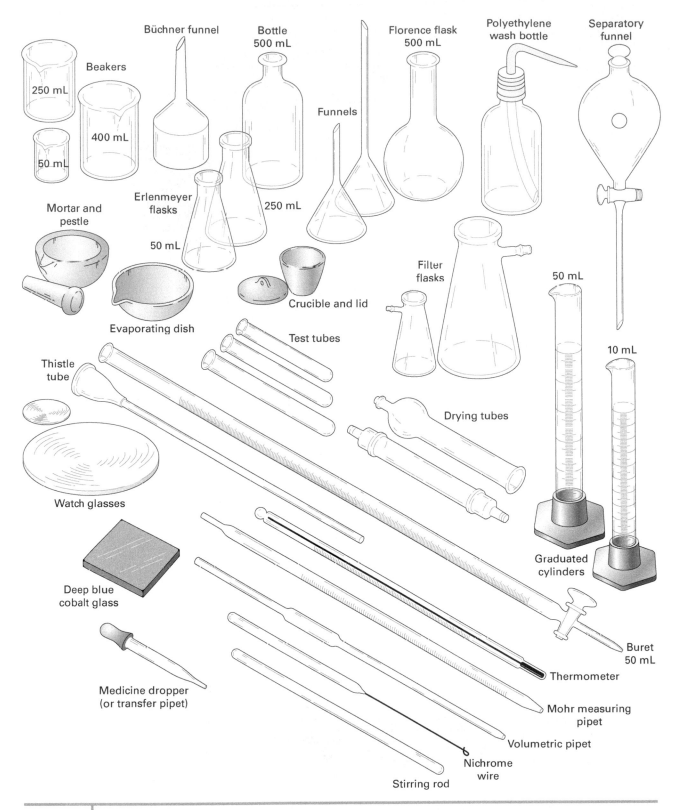

FIGURE I-5 Continued

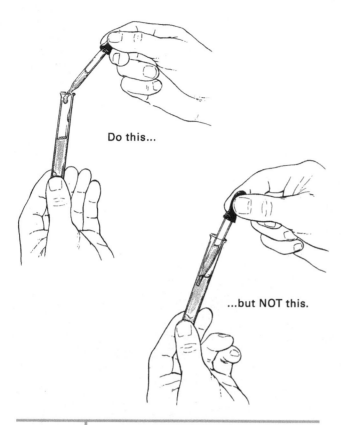

Do this...

...but NOT this.

FIGURE I-6 **The proper use of a medicine dropper or transfer pipet.**

the *direct* supervision of the instructor, because the solution is very corrosive and under certain conditions concentrated acids can produce noxious gases, endangering you and others nearby. Using such cleaning solutions also poses a considerable hazardous waste disposal problem.

Volumetric measurements of liquids

Volumetric measurements of liquids are made with graduated cylinders, burets, volumetric transfer pipets, and volumetric flasks.

Graduated Cylinders The graduated cylinder (Figure I-13) is usually used to measure approximate volumes of liquids. Aqueous solutions wet the glass walls, forming a concave meniscus (curved surface of a liquid); the bottom of the meniscus is used to indicate the volume of liquid. To avoid parallax error (a misreading caused by an improper location of the observer's eye level), your eye should always be level with the meniscus when you are making a reading. The volume is estimated to one-fifth of the smallest division. Graduated cylinders may be calibrated to contain (TC) or to deliver (TD). If it is calibrated to deliver, the cylinder actually contains slightly more than the volume read, thus compensating for the

thin film of liquid left on the walls when the contents are poured out.

Burets for Reduced Scale Titration (Microburets)

The benefits of reduced scale experimentation are firmly established, as described in greater detail in the Preface. The generic term *microscale* has become the established term for reduced scale experimentation in undergraduate laboratories. We will conform to this usage, although it would be more appropriate to call these techniques *milliscale* because sample sizes are typically greater than 100 mg, and titration volumes are in the milliliter range.

In general, we prefer a 10–25-fold scaling down of macroscale titration volumes, using 1-mL burets in place of 25- or 50-mL burets. Typically, microscale titration procedures in this manual are scaled to require about 0.8 mL of titrant. (However, we continue to provide instructions for traditional macroscale titrations using 25- or 50-mL burets.) When 1-mL burets are used, students can put into a 13×100-mm test tube enough reagent for three or four titrations. This markedly reduces consumption and incidental spilling and waste of reagent.

We have used three different devices for microscale titrations, which are illustrated in Figure I-14. All give satisfactory results when properly used. The conventional 50-mL buret is more fragile and costs 300 to 1000 times more than the devices described here. Although the devices can be reused and will last a semester, they are inexpensive enough to be considered disposable.

The first microburet, shown in Figure I-14(A), uses a $1 \times 1/100$ mL Pyrex glass disposable serological pipet.[1] (A polystyrene pipet can be substituted, although there is some sacrifice in chemical resistance and ease of reading.) Added to the bottom of the serological pipet is a fine tip, cut from a polyethylene transfer pipet [shown in Figure I-14(C)]. Addition of the fine tip allows 0.02-mL (1 drop) volume resolution.

At the top, to control the flow, is a 2-cm length of $1/8$-in. ID $\times$ $1/32$-in. wall thickness silicone rubber tubing containing a 4-mm soda-lime glass ball. The ball is located close to the top of the serological pipet to give maximum control of delivery by minimizing the space between the bead and the top of the pipet. Otherwise, there is a tendency to pinch the tubing below the bead; this causes a rebound effect that can

---

[1] $1 \times 1/100$ mL Pyrex disposable serological pipet (Corning 7078D-1 or equivalent, cost approx. $0.20 each in quantities of 500).

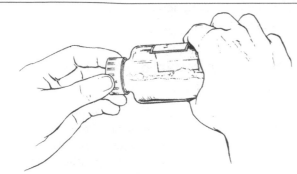

1. Roll and tilt the bottle until some of the contents enters the inside of the plastic cap.

2. Carefully remove the cap so that some of the contents remains in it.

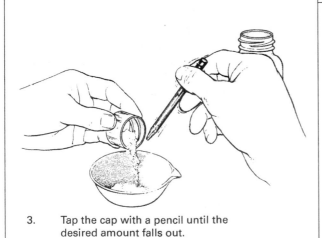

3. Tap the cap with a pencil until the desired amount falls out.

First method

Scoop out a little of the material with the spatula provided.

Tap the spatula until the desired amount falls off.

Second method

Roll and tilt the jar until the desired amount falls out.

Third method

FIGURE I-7 Methods for transferring powders and crystals.

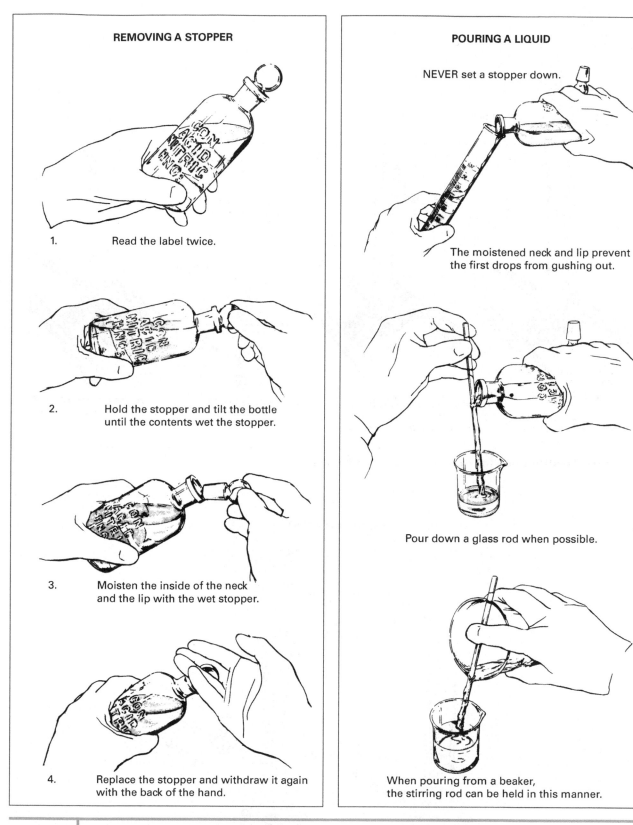

REMOVING A STOPPER

1. Read the label twice.

2. Hold the stopper and tilt the bottle until the contents wet the stopper.

3. Moisten the inside of the neck and the lip with the wet stopper.

4. Replace the stopper and withdraw it again with the back of the hand.

POURING A LIQUID

NEVER set a stopper down.

The moistened neck and lip prevent the first drops from gushing out.

Pour down a glass rod when possible.

When pouring from a beaker, the stirring rod can be held in this manner.

FIGURE I-8 | **Methods for transferring liquids. When handling corrosive materials, latex gloves should be worn to protect the skin.**

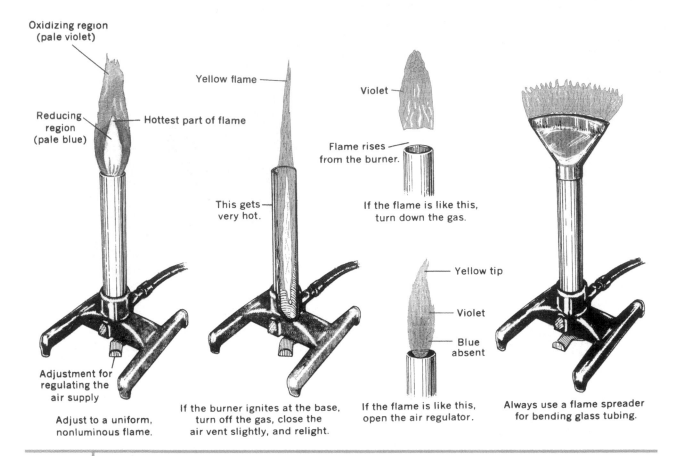

Oxidizing region (pale violet)

Reducing region (pale blue)

Hottest part of flame

Adjustment for regulating the air supply

Adjust to a uniform, nonluminous flame.

Yellow flame

This gets very hot.

If the burner ignites at the base, turn off the gas, close the air vent slightly, and relight.

Violet

Flame rises from the burner.

If the flame is like this, turn down the gas.

Yellow tip

Violet

Blue absent

If the flame is like this, open the air regulator.

Always use a flame spreader for bending glass tubing.

FIGURE I-9 | **Instructions for operating a Bunsen burner.**

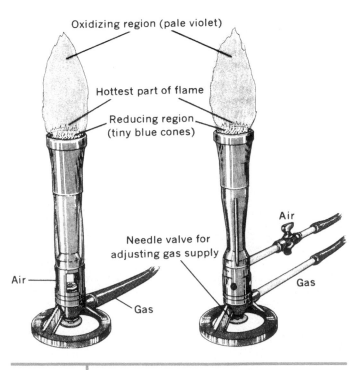

Oxidizing region (pale violet)

Hottest part of flame

Reducing region (tiny blue cones)

Needle valve for adjusting gas supply

Air

Air

Gas

Gas

FIGURE I-10 | **Fischer burners.**

BREAKING A TUBE

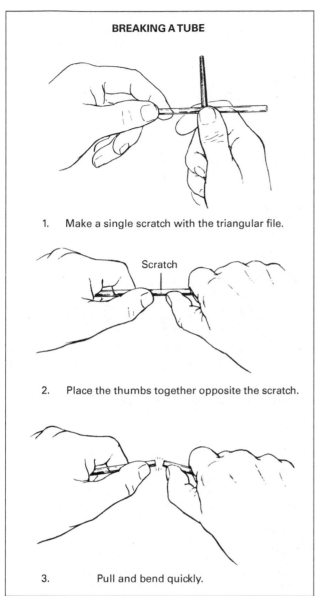

1. Make a single scratch with the triangular file.

2. Place the thumbs together opposite the scratch.

3. Pull and bend quickly.

FIRE POLISHING THE END OF A TUBE

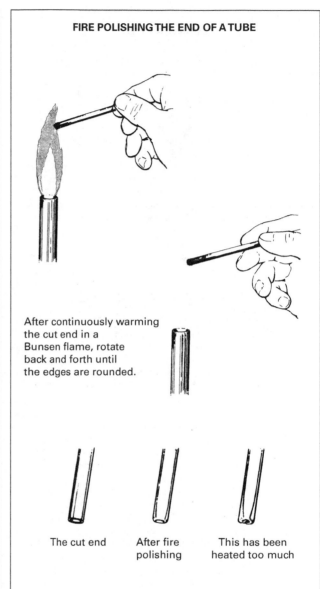

After continuously warming the cut end in a Bunsen flame, rotate back and forth until the edges are rounded.

The cut end   After fire polishing   This has been heated too much

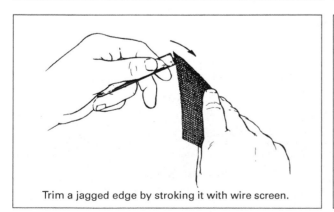

Trim a jagged edge by stroking it with wire screen.

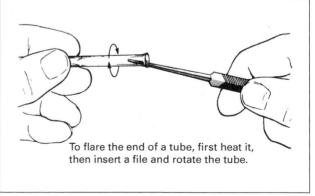

To flare the end of a tube, first heat it, then insert a file and rotate the tube.

FIGURE I-11 | **Some elementary manipulations of glass tubing.**

MAKING A BEND

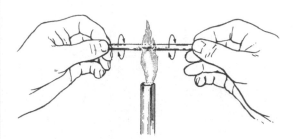

Roll the tube back and forth in the high part of a flat flame (use a flame spreader) until it has become quite soft.

Remove from the flame and hold for a couple of seconds to let the heat become more uniform.

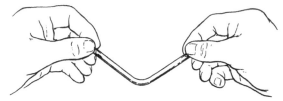

Bend quickly to the desired shape and hold until it hardens.

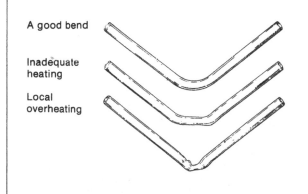

A good bend

Inadequate heating

Local overheating

MAKING A CONSTRICTED TIP

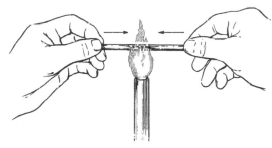

Roll the tube in a Bunsen flame until it softens. Don't use a flame spreader.

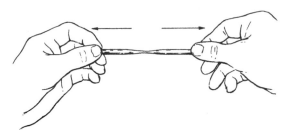

Allow the tube to become shorter as the walls thicken to about twice their original thickness.

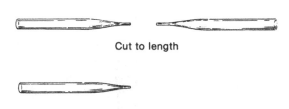

Remove from the flame and after a moment pull until the softened region is as small as desired.

Cut to length

Fire polish, or file the tip

FIGURE I-12 Additional manipulations of glass tubing.

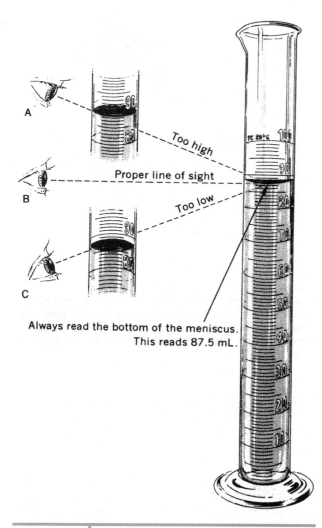

FIGURE I-13 | The proper way to read a meniscus, to avoid parallax error.

Always read the bottom of the meniscus. This reads 87.5 mL.

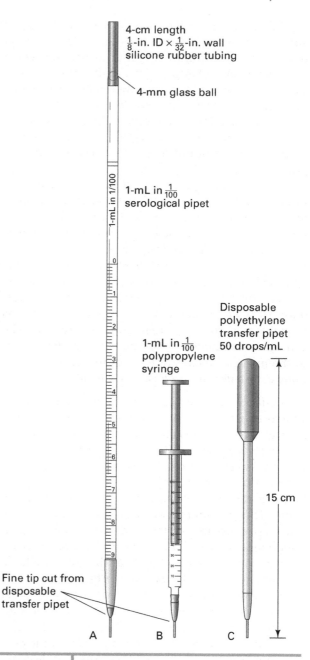

4-cm length
$\frac{1}{8}$-in. ID × $\frac{1}{32}$-in. wall
silicone rubber tubing

4-mm glass ball

1-mL in $\frac{1}{100}$
serological pipet

1-mL in $\frac{1}{100}$
polypropylene
syringe

Disposable
polyethylene
transfer pipet
50 drops/mL

15 cm

Fine tip cut from
disposable
transfer pipet

A   B   C

FIGURE I-14 | Three devices that can be used as simple and inexpensive microburets.

draw air into the tip when the pressure on the silicone tubing is relaxed. Silicone rubber tubing is preferred to polyvinyl chloride (Tygon) tubing because silicone rubber is more flexible and allows easier control of the flow.

The second device is a 1 × 0.01 mL all polypropylene tuberculin syringe,[2] shown in Figure I-14(B). A fine tip, cut from a polyethylene transfer pipet, makes a snug fit over the Luer tip of the syringe and allows 0.02-mL (1 drop) volume resolution.

The simplest and least expensive device is a fine-tipped polyethylene transfer pipet,[3] producing about 50 drops per mL, shown in Figure I-14(C). The pipet is calibrated by counting the number of drops per mL

of solution delivered, then using the drop count method to determine the volume of titrant added (volume added = no. of drops × 1 mL/50 drops).

Advantages and Disadvantages of the Microburet Devices Described We consider that any of the three inexpensive devices described works satisfactorily; the choice of which is best is somewhat a matter of personal preference. (At the University of Redlands, we use the 1-mL tuberculin syringe.)

Transfer Pipet. The polyethylene transfer pipet has an appealing simplicity and is the least expensive, but the drop count method, in our view, has some

---

[2]1 × 0.01 mL Luer tip syringe (Air-Tite Norm-Ject Luer Slip syringe; cost approx. $0.15 each in quantities of 100; Air-Tite Products Co., Inc., www.air-tite-shop.com).
[3]Fine-tip polyethylene transfer pipet, 50 drops/mL (Samco Scientific catalog no. 232; cost approx. $0.13 each in quantities of 500; www.samcosci.com).

drawbacks. It's relatively easy to become distracted and lose count. It's difficult to interrupt and then resume the addition of reagent, particularly when air gets introduced into the pipet tube after the pressure on the bulb is temporarily relaxed. Some who like this type of buret have improved it by constructing some device (operating like a tubing screw clamp) to replace finger pressure. This allows you to interrupt the titration without getting air into the pipet tube.

The transfer pipet is well suited to weight titrations; one simply weighs the transfer pipet, bulb down, in a small beaker before and after the titration. The quantity of reagent delivered can be determined by the simple relation *mmol of titrant = g of titrant solution × mol of reagent/kg of solution*. This approach requires that solutions be made up by mass rather than by volume, which is a simple procedure. The concentration unit *mol/kg solution* also has the advantage of being temperature invariant. The concentration in mol/kg solution is essentially equal to the molar concentration for dilute aqueous solutions whose density is close to 1.0 g/mL.

Tuberculin Syringe. We generally prefer the tuberculin syringe or serological pipet versions of the buret. The syringe is robust and compact. It can be laid down on the bench top. With a little practice, it can be operated with one hand, pressing the thumb on the plunger and applying pressure on the shaft of the plunger held between the first two fingers and the palm of the hand. This technique avoids "stick-slip" friction, which can cause inadvertent addition of more reagent than intended (see Figure I-15). It's easy to interrupt the titration and not lose track of the volume added. The syringe can also be used to stir the solution.

Filling the syringe and expelling air bubbles must be done with care. Reagent may dribble over the outside of the syringe and get on the fingers, or worse, squirt out the end as students work to expel the bubbles. Students with vision problems may find the graduations too closely spaced to easily read the volume scale printed on the side of the syringe. Instructors may also be concerned about the possible theft of syringes and the possibility that they could be put to illicit uses.

Serological Pipet. Using a serological pipet to make a buret counteracts some of the disadvantages of the syringe. It is easily filled, using as a pipet filler a 3- or 5-mL disposable plastic syringe with a Luer tip that can be inserted into the tubing at the top of the buret. The graduations are three times farther apart, so they are easier to read than those on the syringe. The serological pipet, like the syringe version, can be operated with one hand and can also be used to stir

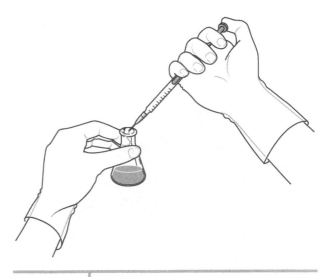

FIGURE I-15 Using a 1-mL syringe as a microburet. The thumb applies pressure to the plunger, while the first two fingers lightly grip the plunger, and the last two fingers hold the barrel of the syringe. The left hand swirls the solution in the flask.

the solution. Although we prefer glass, because of its chemical resistance and the clarity and durability of its volume markings, polystyrene serological pipets are also available and have the advantage that they won't break if dropped. Serological pipets generally come with a cotton plug. This plug should be removed to eliminate the possibility of cross-contamination when different solutions are put in the buret, since it's almost inevitable that the cotton plug will become wet with the titrant solution.

Precision in Microscale Titration When the end point of a titration is approached in a dropwise manner with the buret tip not immersed in the sample solution, the minimum uncertainty in locating the end point is determined by the volume resolution of the buret, ΔV, the smallest volume increment of the titrant that can be delivered in a controlled and reproducible fashion (~1 drop volume).

For a sharp end point, where the volume interval of the color change is smaller than the volume resolution, a good predictor of the relative uncertainty (or relative standard deviation) is the ratio *volume resolution/buret capacity*, $\Delta V/V$. We show in Table I-1 the *percentage relative uncertainty*, which we define as the *volume resolution* (mL)/*buret capacity* (mL) × 100.

Compared to a 1-mL syringe or pipet, a 2-mL pipet used as a microburet decreases the relative uncertainty by a factor of 2, a significant improvement. For this reason, we recommend that you use a 2-mL pipet rather than a 1-mL pipet if you choose to use a serological pipet as a microburet. Using a pipet

TABLE I-1 Relative Uncertainty of Different Kinds of Burets

Type of Buret	Total Capacity (mL)	Volume Resolution (mL)	Relative Uncertainty (%)
Conventional buret	50	0.05 (1 drop)	0.1
Conventional buret	25	0.05 (1 drop)	0.2
Gilmont syringe buret[a]	2.0	0.002 (tip immersed)	0.1
Serological pipet	1.0	0.02 (1 drop)	2.0
Serological pipet	2.0	0.02 (1 drop)	1.0
Tuberculin syringe	1.0	0.02 (1 drop)	2.0
Fine-tipped polyethylene transfer pipet (3-mL bulb)	1.0	0.02 (1 drop)	2.0

[a]The Gilmont syringe buret provides 2-mL capacity with the precision of a 50-mL buret, when operated so that the tip is immersed in the sample solution. It has a polypropylene barrel that rotates like a micrometer, driving a Teflon plunger that displaces the titrant solution. It is much more expensive than the simple devices we have described, the cost being about the same as a Class A 50-mL buret with Teflon stopcock.

with 2-mL capacity requires that the titrant concentration specified in the experiment instructions be reduced by a factor of 2, because the titrant concentrations are scaled for 1-mL microburets.

The best (smallest) relative standard deviation reported for a 2-mL microburet is about 0.4%.[4,5] This is in reasonable agreement with the predicted relative uncertainty, assuming a volume resolution of 0.01 mL (~0.5 drop):

percentage relative uncertainty =
$$0.01 \text{ mL}/2 \text{ mL} \times 100\% = 0.5\%$$

This is consistent with an uncertainty in the third digit, for example, (2.00 ± 0.01 mL). However, the claim that the titration volume can be determined to four significant figures (that is, 2.000 ± 0.001 mL) is overstated because the limiting uncertainty in locating the end point is the volume resolution of the buret (~0.01 mL), not the ability to read the volume to the nearest 0.001 mL.[6]

[4]Singh, M. M.; McGowan, C.; Szafran, Z.; Pike, R. M. "A Comparative Study of Microscale and Standard Burets," *J. Chem. Educ.* **2000,** *77*, 625.
[5]Singh, M. M.; McGowan, C.; Szafran, Z.; Pike, R. M. "A Modified Microburet for Microscale Titration," *J. Chem. Educ.* **1998,** *75*, 371.
[6]Roberts, J. L. "Precision in Microscale Titration" (Letter), *J. Chem. Educ.* **2002,** *79*, 941.

The typical volume resolution of burets of the type shown in Figure I-14 is about 0.02 mL (~1 drop). So the expected relative uncertainty $\Delta V/V$ is about 0.02 mL/1 mL × 100 = 2%. This is improved to 1% using a 2-mL buret.

End Point Sharpness—The Microscale Advantage
A smaller relative uncertainty means that one can, in principle, achieve a more precise titration. Superficially, it would appear from Table I-1 that the relative uncertainty achievable with a 1-mL buret is a factor of 10 or 20 times larger (larger is worse) than can be attained with a conventional 25- or 50-mL buret, which typically has a volume resolution of about 0.05 mL (~1 drop). In practice, we have observed that the relative precision is typically a factor of 2–4 times larger using 1-mL burets. Part of this has to do with the level of training, practice, skill, and motivation of general chemistry students. Using conventional macroscale burets, general chemistry students typically achieve a relative precision of about 1–2%, rather than the theoretical 0.1% that can be achieved in very careful work:

volume resolution/buret capacity =
$$0.05 \text{ mL}/50 \text{ mL} \times 100\% = 0.1\%$$

The second factor is the increase in the sharpness of the end point as the volume of the titration is reduced. This effect is often overlooked because it is so counterintuitive that it has been called *paradoxical*.[7] Yet on closer examination, it may be explained in simple terms—adding a fixed volume increment of titrant produces a larger effect as the sample volume is made smaller. So, in a titration where the response variable R is a logarithmic function of concentration, the rate of change of system response (R) with volume (V) of added titrant, $\Delta R/\Delta V$, be it pH reading or change of indicator color, is greater as the sample volume is made smaller *even when the solution concentrations (titrant and solution titrated) remain constant*. If the sample volume is reduced tenfold, the end point sharpness increases tenfold. This effect makes it easier to locate the end point in a small scale titration. The sharper end point obtained in a smaller sample volume partially offsets the tendency of the relative uncertainty to get larger (worse) as you make the buret volume smaller.

Using a Conventional (Macro) Buret The use of a 25-mL buret is shown in Figure I-16. If it is clean, the solution will leave an unbroken film when it drains; if

[7]Rossi, M.; Iha, M. E. V. S.; Neves, E. A. "A Paradoxical Effect of the Titrated Volume in Potentiometric Titrations," *J. Chem. Educ.* **1990,** *67*, 65–66.

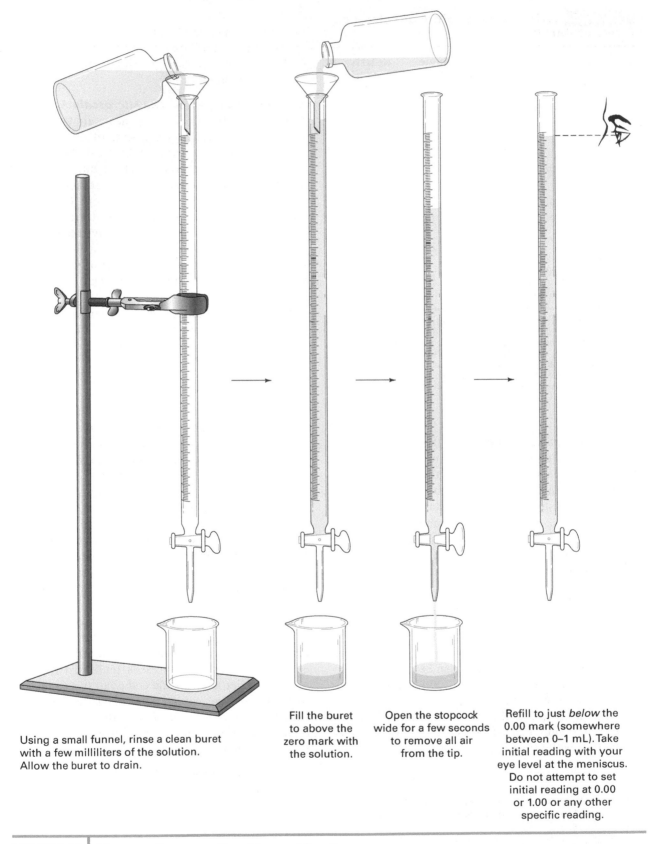

Using a small funnel, rinse a clean buret with a few milliliters of the solution. Allow the buret to drain.

Fill the buret to above the zero mark with the solution.

Open the stopcock wide for a few seconds to remove all air from the tip.

Refill to just *below* the 0.00 mark (somewhere between 0–1 mL). Take initial reading with your eye level at the meniscus. Do not attempt to set initial reading at 0.00 or 1.00 or any other specific reading.

FIGURE I-16 | **The use of a conventional (macro) buret.**

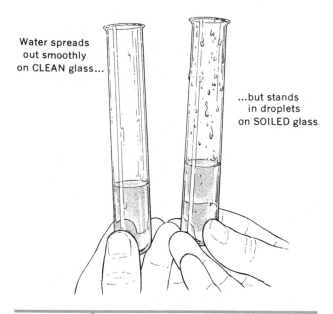

Water spreads out smoothly on CLEAN glass...

...but stands in droplets on SOILED glass.

FIGURE I-17 | **Clean and dirty glassware.**

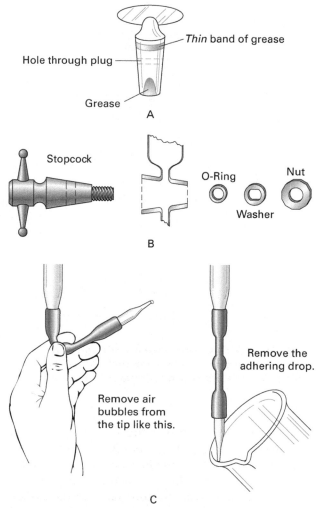

Hole through plug — *Thin* band of grease

Grease

A

Stopcock

O-Ring — Nut

Washer

B

Remove air bubbles from the tip like this.

Remove the adhering drop.

C

FIGURE I-18 | **Three varieties of stopcocks.
(A) Glass stopcock: To grease a glass stopcock, remove old grease from both parts with organic solvent such as cyclohexane; wipe dry; then apply a thin film of stopcock grease as shown. (B) Teflon stopcock: No grease is used on a Teflon stopcock. (C) Rubber-tubing, glass-bead stopcock: Pinching the tube near the bead allows the solution to drain from the buret.**

drops of solution adhere to the inside of the buret, it should be cleaned with a brush, hot water, and detergent until it drains properly (Figure I-17). Cleanliness is important because the volume of a 25- or 50-mL buret can ordinarily be estimated to the nearest 0.02 mL, and the error caused by a single large drop adhering to the inside of a buret causes an error of about 0.05 mL. The presence of several drops would obviously cause an error in reading the volume delivered.

When filling a freshly cleaned buret with solution, add a 5- to 10-mL portion of the solution, being sure the stopcock is turned off, and tip and rotate the buret so that the solution rinses the walls of the buret completely. Repeat the procedure with at least two more fresh portions of solution; then fill the buret above the zero mark and clamp in a buret holder. Open the stopcock wide to flush out any air bubbles between the stopcock and the tip of the buret. Next, drain the buret below the zero mark and take the initial reading, being careful to avoid parallax error. The smallest division on a 25- or 50-mL buret is ordinarily 0.1 mL; estimate the volume to the nearest fifth of the smallest division, or the nearest 0.02 mL.

Burets may have stopcocks made of glass, which must be periodically cleaned and lubricated, as shown in Figure I-18(A). Teflon stopcocks [Figure I-18(B)] ordinarily require no lubrication, but they may have to be cleaned if they are plugged or adjusted if the tension nut is too tight or too loose. A drawn-out glass tip that is connected to the buret by a

short length of rubber tubing containing a round glass ball makes the simple yet effective stopcock shown in Figure I-18(C).

The buret is calibrated to deliver (TD) and is capable of a precision of approximately ±0.02 mL when carefully used. The tip should be small enough that the delivery time is not less than 90 s for a 50-mL buret. This allows adequate time for drainage so that you can obtain a proper reading. If the delivery time of your buret is faster, wait a few seconds before taking a reading so that you allow the buret to drain.

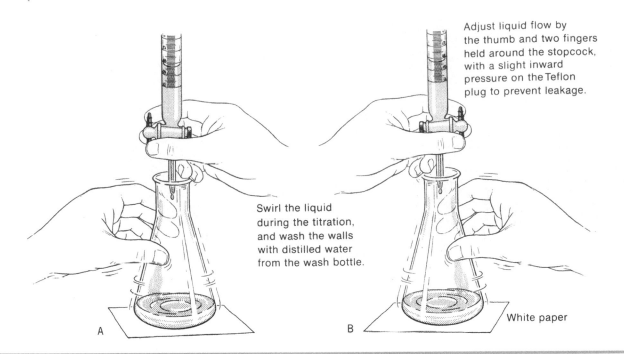

Adjust liquid flow by the thumb and two fingers held around the stopcock, with a slight inward pressure on the Teflon plug to prevent leakage.

Swirl the liquid during the titration, and wash the walls with distilled water from the wash bottle.

White paper

A   B

FIGURE I-19 | **Recommended technique for manipulation of a buret stopcock. Most left-handed students will manipulate the stopcock** **with the right hand (A), whereas most right-handed students will prefer to manipulate it with the left hand (B).**

Figure I-19 illustrates the recommended technique for manipulation of a buret stopcock. You may add the solution from the buret quite rapidly until it is close to the end point, but then you should reduce the flow until individual drops fall into the flask. As you add the last few drops slowly, swirl the flask to obtain thorough mixing.

The Use of Volumetric Pipets Volumetric transfer pipets are designed to deliver a single fixed volume of liquid. The graduation mark is located on the narrow part of the pipet to assure precision. These pipets come in sizes ranging from less than 1 mL to 100 mL. They are calibrated to deliver (TD) the specified volume if they are handled in the prescribed manner.

Pipets are ordinarily calibrated at 20 °C. In very careful work, temperature corrections are necessary if the solution temperature is markedly different from the calibration temperature of the pipet. Fill the pipet by placing the tip in a flask of the solution and using a suction bulb to draw the liquid up past the calibration mark (Figure I-20). Then slip off the bulb and, with the forefinger, quickly seal the top of the pipet before the solution drops below the calibration mark. Wipe the outside of the pipet with tissue or a clean towel; then allow the liquid to flow out until the bottom of the meniscus is just at the calibration ring. To pick off the last drop adhering to the outside of the tip, touch the tip to the side of the flask.

Next, withdraw the pipet from the flask and hold it over the vessel into which the liquid is to be transferred. Allow the pipet to drain in a vertical position, with the tip against the side of the vessel. After it appears that most of the liquid has drained out, allow an additional 15 to 20 s for drainage. The tip of the pipet will still contain some liquid. Ordinarily, this has been accounted for in the calibration of volumetric pipets (marked TD) and should not be blown out. (Some pipets, such as calibrated serological pipets, are calibrated to be blown out. Most of these will have a ring, or double ring, at the top of the pipet.) Like the buret, the pipet must be scrupulously clean if precise results are to be obtained. If the pipet is still wet from cleaning, rinse with several portions of the solution to be pipetted.

The Use of Volumetric Flasks Volumetric flasks come in a variety of common sizes from 5 mL to 2 L. They are calibrated to contain the specified volume at a given temperature, usually 20 °C. They are used in the preparation of accurate solution concentrations by weighing a pure substance, quantitatively transferring the substance to the volumetric flask, adding water to dissolve the substance, and diluting the solution precisely to the mark inscribed on the neck of the volumetric flask. (The bottom of the curved portion of the meniscus should coincide with the mark.)

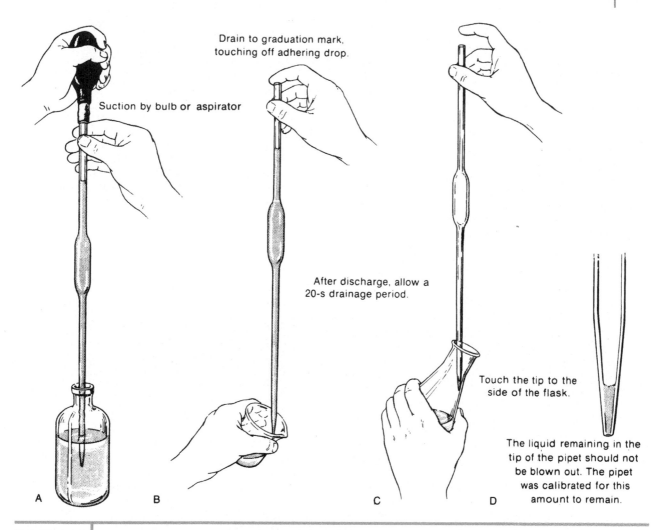

Suction by bulb or aspirator

Drain to graduation mark, touching off adhering drop.

After discharge, allow a 20-s drainage period.

Touch the tip to the side of the flask.

The liquid remaining in the tip of the pipet should not be blown out. The pipet was calibrated for this amount to remain.

A   B   C   D

FIGURE I-20 | **The procedure for using a transfer pipet.**

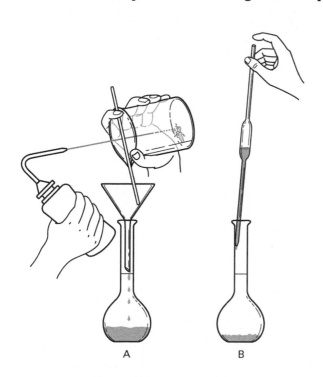

A   B

Small amounts of solid (1–2 g) can be weighed on a creased sheet of weighing paper and transferred directly into the mouth of a volumetric flask. A small spatula is used to push the solid into the mouth of the volumetric flask. Larger amounts of solid are usually first dissolved in a small amount of water in a beaker, then quantitatively transferred to a volumetric flask using a ribbed funnel in the mouth of the volumetric flask, a glass stirring rod, and a wash bottle, as shown in Figure I-21.

Volumetric flasks are also used to produce accurate dilutions of solutions, by pipetting a known volume of solution into the volumetric flask, then diluting the solution to the mark (see Figure I-21). Mix the solution thoroughly by inverting the stoppered flask five or six times.

FIGURE I-21 | **Using volumetric flasks. (A) Method for quantitatively transferring a dissolved solid to a volumetric flask. (B) Accurate dilution using a volumetric pipet and a volumetric flask.**

Scientific Measurements

Purpose

• Understand how measurements are made and the utility of basic and derived units of measurement.

• Identify the different kinds of errors that can be present in measurements.

• Display random measurement variations in a histogram that approximates the Gaussian normal distribution.

• Calculate the standard deviation for a set of randomly distributed measurements.

• Express the uncertainty of a measured quantity by using the appropriate number of significant figures.

• Learn how errors propagate when quantitites are derived from measurements that are multiplied and divided or added and subtracted.

• Present the data obtained from measurements visually, by using graphs.

• Practice making measurements of mass, length, and volume and graphing the results.

Pre-Lab Preparation

The introduction of *quantitative* measurement brought chemistry from the magical, mystical alchemist's laboratory to the practical science it is today. In the 1600s, experimenters began to measure the properties of gases in a quantitative way; the relationship between the pressure and volume of a fixed quantity of gas was discovered and published by Robert Boyle.

It wasn't until the late 1700s, when chemists began to measure the masses of the starting materials and of the products of a reaction, that some of the fundamental laws of chemistry were recognized and stated. These include the laws of mass conservation and of definite composition (the discovery that pure compounds have constant fixed ratios of the masses of their elements).

Such quantitative observations provided supporting evidence for the concept that matter was composed of atoms with definite masses. The ability to determine accurately the relative masses of the elements and to systematize their chemical properties eventually led to the creation of the periodic table of elements. If you look carefully at the periodic table you will note that, with few exceptions, the elements are arranged in order of increasing atomic mass.

When quantitative measurements were made of living things, it became apparent that the processes of life were themselves chemical reactions subject to the same laws as other chemical processes.

How Measurements Are Made

Most quantitative measurements are operations in which two quantities are compared. For example, to measure the length of an object, you compare it to a ruler with calibrated marks on it. As the ruler itself was designed and manufactured, it was compared to other standards of length that conform to the standard unit of length agreed upon by international conventions.

The double-pan equal-arm analytical balance is another example of measurement by direct comparison. With this apparatus, the mass of an object is determined by comparing it directly with a set of masses of known value.

The single-pan analytical balance also involves a comparison of masses, using the principle of substitution—an unknown mass on the pan substitutes for known masses on the beam above the pan. The point of balance occurs when the sum of the known masses that have been removed is exactly equal to the unknown mass.

Figure 1-1 shows how unknown and known masses are compared by means of double-pan, single-pan, and electronic balances. Even a modern electronic balance involves an indirect comparison because standard masses are used to calibrate the force coil of the balance.

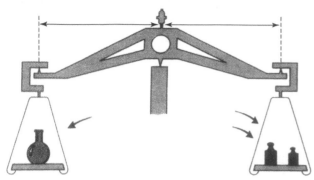

Equal-arm double-pan balance with three knife edges

A

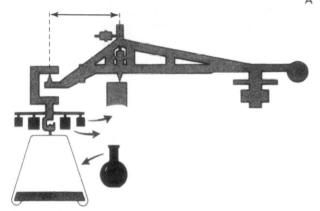

Substitution-type single-pan balance with only 2 knife edges
Constant load = constant sensitivity
Lever arm error impossible

B

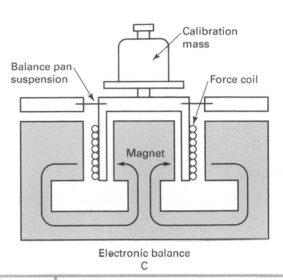

Electronic balance
C

FIGURE 1-1 (A) How masses are compared on an equal-arm balance. The lever arms of the balance must have equal lengths. (B) How masses are compared on a substitution-type single-pan balance. Masses whose sum is equal to the unknown mass are removed from the hanger to keep the beam in equilibrium. (C) Calibration of an electronic balance with a known mass. A sensitive detector/DC current amplifier (not shown) keeps the balance pan precisely at the rest (or null) position by adjusting the electric current in the force coil. The current, proportional to the mass on the pan, is transformed into a digital readout of the mass.

Time is usually measured by counting the number of periods of a system that oscillates at some natural frequency. The pendulum of a grandfather clock swings at a natural frequency; an escapement mechanism turns a toothed wheel that is connected by gears to the hands of the clock. The hands of an electric clock are driven by a synchronous electric motor supplied with alternating current of a stable frequency. A quartz digital watch measures time by counting (and subdividing) the oscillations of a tiny quartz crystal. The most accurate clocks, which serve as our fundamental time standard, are based on counting the frequency of an oscillator that is continuously compared with the frequency associated with a transition between two energy states of the electrons in gaseous cesium atoms. From these descriptions of various timekeeping devices, we see that the measurement of time is an operation in which the natural periods of various oscillating systems are compared.

The two most common kinds of measurements made in a general chemistry laboratory are measurements of mass, using a balance, and measurements of volume, using graduated cylinders, pipets, and burets. When measuring volume, we generally use our eyes to note the position of a meniscus relative to calibration marks and sometimes estimate the position of the meniscus between two marks, as shown in Figure I-13 of the Introduction.

We also make measurements of temperature using a thermometer, measurements of time using a timer or stopwatch, measurements of length using a meter stick or ruler, and measurements of voltage or current using a multimeter.

Basic and Derived Units

By an international agreement reached in 1960, certain basic metric units and units derived from them are to be preferred in scientific use. The preferred units are known as the International System units (commonly called SI units, from the French, *Système Internationale*). The basic units of the SI system are given in the Appendix, Table 1.

The National Institute of Standards and Technology (NIST) Internet site (http://physics.nist.gov) provides a full discussion of the fundamental physical constants, including historical background information and the best values of the constants and their uncertainties.

Units obtained from the basic units are called *derived* units. The volume of a cube or rectangular box can be determined by measuring the lengths of its three dimensions and computing the volume as the product of those lengths. Thus, volume is derived from length measurements and has the dimensions of length cubed—for example, cubic centimeters (cm^3). Cylindrical containers are easier to manufacture than rectangular containers, so volume is more often measured in containers such as graduated cylinders and burets. The volume of a cylinder is equal to the product of its cross-sectional area times its length. This formula allows us to use a length scale, often inscribed along the length of the cylinder, to determine volume.

The Assessment of Experimental Errors

Every measurement involves some measurement uncertainty (often referred to as "error," but *not* in the normal sense of this term as a synonym for "mistake"). Because many generalizations or laws of science are based on experimental observations involving quantitative measurements, it is important for a scientist (or a student of a quantitative science) to take into account any limitations in the reliability of the data from which conclusions are drawn. In the following section, we will discuss different kinds of errors—systematic, random, and personal errors—and assess the quality of a measurement by considering its accuracy and precision.

Accuracy and precision

The limitations of both accuracy and precision will contribute to uncertainty in the measurement. The error in a measurement, or better, the average of several measurements, is the difference between the measured value and the true value of the quantity measured. *The smaller the error, the closer the measured value is to the true value and the more accurate is the result.* The true value is sometimes difficult to establish. Generally, it involves applying the same measurement technique to an unknown sample and to a carefully prepared or analyzed standard sample that resembles the unknown as much as possible. Preparing these kinds of standards requires trained specialists; standards for many kinds of measurements can be purchased from the National Institute of Standards and Technology (NIST).

Typically, when we make a series of several measurements on the same sample or replicate samples,

we do not get exactly the same value for each measurement. There will be a dispersion (or spread) of the measured values. *The smaller the spread of values, the more precise will be the average (or mean) value of the measurements.* So accuracy and precision are different concepts. *Accuracy is a measure of how close the measured value is to the true value. Precision is a measure of the reproducibility (or dispersion) of the measurements.* The average of several measurements might be accurate without being very precise, or a set of measurements might have good reproducibility, and therefore good precision, but the average (or mean) value might not be very accurate. Good-quality measurements will be both precise and accurate. Figure 1-2 shows graphically what is meant by accuracy and precision.

We've said that we determine accuracy by comparison with standards. How do we measure

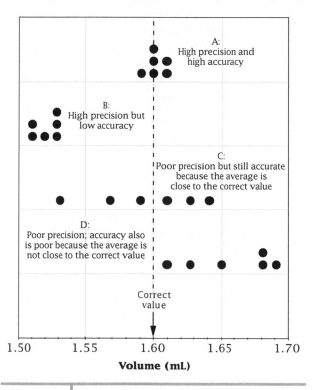

FIGURE 1-2 **The results of an acid–base titration exercise are shown for four individuals. Each person titrated six nominal 10-mL samples of the same 0.0400 M acid solution with 0.250 M NaOH (sodium hydroxide) dispensed from a 2-mL microburet. The horizontal position of the black dots shows the measured titration volumes. The vertical columns of two or three black dots mean that those measured volumes were the same. The results of individual A were both accurate and precise, while those of B were precise but not accurate. The results of C were accurate but not precise. The results of D were neither accurate nor precise. (There is a systematic error combined with substantial random errors.)**

precision? By making several measurements and determining the spread of the measurements. How many measurements? Somewhere between 3 and 30. Three is a minimum for making a rough estimate of the precision. Thirty measurements would give you a decent estimate. A single individual seldom has time in a general chemistry lab to make 30 replicate measurements of the same quantity, but we might aggregate the results of all the students in a lab section or in the whole class in order to get a better estimate of the precision. In today's world, computer-controlled instruments can make hundreds or thousands of repetitive measurements in a short time.

Human Factors It's been observed that students who are well-trained in a measurement technique and highly motivated will produce results with better precision (smaller spread in the measurements) than students who lack training and motivation. Likewise, a single skilled and motivated student will attain better precision than will a group of students in which there is a range of skills and motivation. Making measurements in the laboratory is not unlike learning a sport. Training, discipline, and practice produce superior performance.

The Mean (Average) Value and the Deviation Unless we have accurate standards against which we can test our measurement, we often do not know the true value of a measured quantity. If we do not, we can obtain only the mean, or average, value of a number of measurements and measure the spread or dispersion of the measurements. The mean value is obtained by adding all of the individual measurements and dividing by the total number of measurements:

$$\bar{x} = \frac{\Sigma x_i}{N} \qquad (1)$$

where Σ represents the operation of summation.[1]

A measure of the spread of individual values from the mean value is the *deviation*, δ—defined as the difference between the measured value, x_i, and the arithmetic mean, $\bar{x}$, of a number, N, of measurements (a bar over a quantity means the average value of the quantity):

$$\delta_i = x_i - \bar{x} \qquad (2)$$

Systematic Errors Errors are of two general types, systematic (determinate) and random (indeterminate). A *systematic error* causes an error in the same direction in each measurement and diminishes accuracy, although the precision of the measurement may

remain good. A miscalibrated scale on a ruler, for example, would cause a systematic error in the measurement of length.

Similarly, in Figure 1-2, the B group of measurements shows evidence of a systematic error, perhaps due to a misunderstanding of how to read the position of the meniscus in a graduated cylinder or pipet. If the volume is read at the top rather than the bottom of the curved meniscus, the measured volume is systematically too small.

Random Errors and Standard Deviation If a measurement is made a large number of times, you will obtain a range of values (like the distributions shown in Figures 1-2 and 1-3) caused by the *random errors* inherent in any measurement process (or deviations in a manufacturing process). For random errors, small errors are more probable than large errors and negative deviations are as likely as positive ones.

Figure 1-3 shows the results of weighing 100 different pennies (minted in 1983 or later). The resulting distribution of the measured masses of the pennies is displayed as a bar graph called a *histogram*.[2] (The root of the word *histogram* comes from the greek *histos*, which has several meanings, one being "mast," as in the mast of a sailing ship.) In the penny mass histogram of Figure 1-3, the main cause of the dispersion (or spread) is not the uncertainty in measuring the mass of the pennies, which is accurate to ±0.001 g, but rather the random deviations in the process for manufacturing the pennies.

Superimposed on the histogram is a bell-shaped curve called a *Gaussian error distribution*, the distribution that is approached when the number of measurements becomes very large (or infinite). The mean value of the set of measurements is the most probable value, corresponding to the center of the

---

[1]Therefore, Σx_i means to calculate the sum $x_1 + x_2 + x_3 + x_4 + \cdots$.

[2]A histogram is a bar graph that shows the frequency distribution of the measurements of a variable (*x*) in which the widths of contiguous vertical bars are proportional to an interval Δx of the variable and the heights of the bars are proportional to the number of *x*-values that fall in the interval Δx.

The process of constructing a histogram may be thought of as drawing a horizontal line representing the *x*-axis variable, dividing the line into intervals of width Δx, which we will call bins, sorting the *N* measurements of the variable *x* into the bins, and counting the number of values that fall in each bin. The number of *x*-values that fall in a bin is called the *frequency*. We then plot the frequencies as contiguous bars rising from the horizontal *x*-axis. If we sum the frequencies (bar heights) of all the bars, they will add up to *N*, the total number of *x*-values in the sample.

In a *normalized* histogram, each frequency is divided by *N*, the total number of measurements, to give the *fraction* of the measurements that fall into that bin.

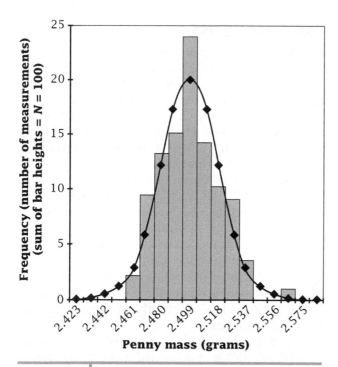

Penny mass (grams)

FIGURE 1-3 The measured mass distribution of 100 different pennies minted in the year 1983 or later is plotted as a bar graph (called a histogram). The mean (average) mass of the sample of pennies is 2.499 g with a standard deviation of 0.019 g. Think of the process of constructing the bar graph as dividing the horizontal mass axis into bins of arbitrary width and sorting pennies into the bins according to their masses and counting the number of pennies in each bin. The width of each bin, represented by the bar width, is arbitrarily made equal to one-half standard deviation (0.0095 g). On the horizontal mass axis, the mass value labels are the values in the center of each bin. For each bin, the bar height, called the frequency, is equal to the number of pennies having masses that fall within the bin boundaries.

The smooth bell-shaped curve is the Gaussian distribution for a large number of pennies whose masses are randomly distributed about a mean value of 2.499 g with a standard deviation of 0.019 g. As the total number of pennies in the sample increases, the bar graph approaches a Gaussian distribution.

Gaussian distribution curve. The spread, or dispersion, of the results is expressed by the *standard deviation, s:*

$$s = \left(\frac{\Sigma(x_i - \overline{x})^2}{N - 1} \right)^{1/2} \qquad (3)$$

This formula says: Sum the squares of the deviations, divide by $N - 1$, and take the square root of the result. So the standard deviation could also be called the *root-mean-square deviation.* This formula actually

gives only an estimate of the standard deviation unless the number of measurements is large (>50). We must recognize that when we repeat a measurement only two or three times,[3] we are not obtaining a very large sample of measurements, and the confidence that we can place in the mean value of a small number of measurements is correspondingly reduced.

Although the formula may look forbiddingly complex, the steps are very simple.

- First calculate the arithmetic mean or average value, $\overline{x}$, of the measurements.

- Then subtract the mean value, $\overline{x}$, from each one of the individual values, x_i, to obtain the deviation. There will be both positive and negative deviations.

- Square each deviation, and add all of the squares.

- Divide the total by $N - 1$, where N is the total number of measurements.

- Finally, take the square root of the result to obtain the estimate of the standard deviation.

The procedure for the calculation of the standard deviation is illustrated in Table 1-1.

We said that the maximum of the smooth bell-shaped curve (Gaussian error curve) shown in Figure 1-3 corresponds to the average (or mean) penny mass (2.499 g). The spread (or dispersion) of the curve is determined by the standard deviation (0.019 g). The standard deviation can be related to the *confidence interval,* or the range about the mean value in which one of a group of measurements may be expected to fall. If we recognize that the Gaussian error distribution is a distribution of the probabilities of obtaining a particular measurement, we see that the probability of measurements occurring close to the mean will be greater than the probability of a measurement occurring far away from the mean. In fact, there is a 0.68, or 68%, probability that a given measurement will fall within plus or minus *one* standard deviation of the mean value. There is a 0.95 (95%) probability that a measurement will fall within plus or minus *two* standard deviations of the mean value. This means that, for the curve of Figure 1-3, we can expect that 95% of the measurements will fall between 2.461 and 2.537 g. If we were to measure the total area under the curve, we would also see that

---

[3]If only two or three measurements are made, the standard deviation may be approximated by the average deviation, which is the mean value of the absolute values of the deviations, δ_i: $\quad \overline{\delta} = \dfrac{\Sigma|x_i - \overline{x}|}{N}$

TABLE 1-1 The Procedure for Calculating the Standard Deviation

Measured Value[a] (x_i)	Deviation $(x_i - \bar{x})$	Square of Deviation $(x_i - \bar{x})^2$
4.28	−0.01	0.0001
4.21	−0.08	0.0064
4.30	0.01	0.0001
4.36	0.07	0.0049
4.26	−0.03	0.0009
4.33	0.04	0.0016
$\sum = 25.74$	$\sum = 0.00$	$\sum = 0.0140$

The mean, $\bar{x} = \dfrac{\sum x_i}{N} = \dfrac{25.74}{6} = 4.29$ g

The standard deviation, $s = \left(\dfrac{0.0140}{6-1}\right)^{1/2} = 0.053$ g

The best value of the measurement is written as 4.29 ± 0.05 g.

[a]The measured values are those obtained from a series of six replicate measurements of the weight of a sample on a triple-beam balance.

this interval of $4s$ ($\pm 2s$ about the mean value) corresponds to 95% of the area under the Gaussian error curve. Note that we cannot make a definite prediction about any single measurement. We can only say that if we make a measurement a large number of times, we can expect that 95% of the values obtained will fall within plus or minus two standard deviations of the mean.

Personal Errors To the types of errors already described—systematic and random—we might add a third category, the personal error, or blunder. Such errors are all too common in student work. Thus, if the numbers on a scale are misread, recorded incorrectly, or if part of a solution is spilled in a titration, the result will contain an error. Careful work will not contain any blunders, and any work suspected of containing one should be repeated.

Precision of laboratory operations

The precision associated with various pieces of equipment you may use in the laboratory is summarized in Table 1-2. These uncertainties express limitations in the reading of the instruments and do not reflect systematic errors.

Significant Figures

The limited precision of any measurement must be considered whenever that measurement is to be used, whether that use is to report the value of the quantity, to constitute one of a series of measurements, or to be combined in a calculation with other measured quantities. The limited precision creates uncertainty, which limits the conclusions that can be drawn from a measurement.

Keeping track of this uncertainty would be rather cumbersome if the uncertainty needed to be reported each time the measurement itself were reported or used in a calculation. Introducing the concept of *significant figures* allows us to *imply* the precision of a measurement without having to state explicitly the uncertainty. Significant figures also allow us to easily estimate the precision of a value that is calculated from a combination of different measured quantities.

For example, when the mass of an object is reported as 56.78 g, it is implied that the object was weighed in such a way that there is no question that the mass is at least 56.7 g and that the value of the hundredths place is thought to be 8, but that some judgment was required in deciding this last digit. We should infer that there is an uncertainty of at least 0.01 g in the value 56.78 g.

The number 56.78 has four significant figures (that is, digits); three are completely certain (56.7), and one has some uncertainty in it (the 8), but all four digits contain useful information. When the significant figure convention is used, the last digit written is assumed to have an uncertainty of at least 1.

Note that although the use of significant figures to imply the uncertainty is practical, the convention

TABLE 1-2 Typical Instrument Uncertainties

Instrument	Typical Uncertainty
Platform balance	±0.50 g
Triple-beam (centigram) balance	±0.01 g
Top-loading semimicro balance	±0.001 g
Analytical balance	±0.0001 g
100-mL graduated cylinder	±0.2 mL
10-mL graduated cylinder	±0.1 mL
50-mL buret	±0.02 mL
25-mL pipet	±0.02 mL
10-mL pipet	±0.01 mL
Thermometer (10 °C to 110 °C, graduated to 1 °C)	±0.2 °C
Barometer (mercury)	±0.5 torr

has limitations. The values 35.4 ± 0.1 and 35.4 ± 0.4, for instance, would both be expressed as 35.4 to the correct number of significant figures. Nevertheless, the convention provides a convenient way of stating the approximate uncertainty in a number, and significant figures are especially helpful in estimating the uncertainty in a value calculated from a series of uncertain quantities.

Other examples of the number of significant figures in quantities are given in Table 1-3.

Zeros require special attention in order to determine whether or not they are significant figures in a number. The general rule is that zeros used to hold a decimal place are not significant. The implications of this statement are as follows:

1. Zeros in front of a number are not significant. In Table 1-3, the examples 0.0862 (with three significant figures) and 0.001740 (with four, the last zero being significant) illustrate this.

2. Terminal zeros are significant if they are to the right of the decimal point. Thus, the terminal zeros in 0.001740, 1.6300, 14.00, and 250.00 are significant digits. This also means that if a measurement was not made precisely, the terminal zeros must not be written. For instance, 14.00 implies a measurement made to ±0.01. If the measurement is only precise to ±1, only "14" should be reported.

3. Zeros as digits between other significant digits are significant (note 12.0004 with six significant figures).

4. By convention, terminal zeros are not significant if there is no decimal point, but a decimal point implies that all digits to the left of it are significant. This is the origin of the implied four significant figures in the quantity 1500., but only two in the 62,000

quantity shown in Table 1-3. To specify 62,000 as having three significant figures, it would be necessary to write it in scientific notation as 6.20×10^4.

Propagation of Error in Calculations

The limitations of each measurement must be recognized whenever that measurement is to be used or reported. When the value of a measurement is used to calculate another quantity, the uncertainty in the measurement limits the certainty of the calculated value. This uncertainty must be recognized, but the relationship between the uncertainty of the measurement and that of the calculated quantity is not always easy to predict. Significant figures are useful because they enable us to estimate the error in calculated quantities. Two simple rules allow us to produce this estimate for the basic mathematical operations.

1. For multiplication and division, the answer is rounded to the same *number* of significant figures as that of the *least* significant factor in the calculation. Examples include:

$$1.48 \times 3.2887 = 4.87$$

$$\frac{2.62}{8.1473} = 0.322$$

$$0.023 \times 1.482 \times 13.25 = 0.45 \qquad (4)$$

$$\frac{3.457}{0.00015} = 23,000 \quad \text{or} \quad 2.3 \times 10^4$$

$$\frac{1.918 \times 0.47523 \times 81.96}{53} = 1.4$$

In each case, the least significant factor determines the significance of the answer, no matter how significant the other factors are. If you perform the foregoing calculations on a calculator, you will notice that calculators are unaware of the significant-figure rules and are more than happy to create an abundance of nonsignificant digits. You must decide which of the digits are to be retained as significant and which are to be discarded.

2. For addition or subtraction, the last significant digit in the answer is the right-most common digit of all the terms in the sum. This rule is more easily seen and rationalized by considering an example, such as adding these two numbers:

$$
\begin{array}{ll}
12.34 & \text{addend} \\
\underline{+5.6} & \text{addend} \qquad (5) \\
17.94 & \text{sum} \\
\uparrow & \\
\text{not significant} &
\end{array}
$$

TABLE 1-3 **Number of Significant Figures**

Quantity	Number of Significant Figures
857.29	5
435	3
6.02×10^{23}	3
0.0862	3
0.001740	4
12.0004	6
1.6300	5
62,000	2
14.00	4
1500.	4
250.00	5

The digit "4" in the sum is not significant because the addend 5.6 does not have a significant digit in the hundredths place. This rule can be understood by realizing that the hundredths digit of 5.6 is unknown and could have any value. When this unknown value is added to the "4" of 12.34, the result is also an unknown (nonsignificant) digit.

Other examples of significant figure determinations in addition and subtraction are

$$
\begin{array}{r}
73.213 \\
+\ 14.84 \\
\hline
88.05
\end{array}
\qquad
\begin{array}{r}
1.00257 \\
+\ 0.0013 \\
\hline
1.0039
\end{array}
\qquad
\begin{array}{r}
8.5672 \\
+\ 153 \\
\hline
162
\end{array}
$$

(6)

$$
\begin{array}{r}
6.02 \times 10^{23} = 6.02 \times 10^{23} \\
+5.2\ \ \times 10^{22} = \underline{.52 \times 10^{23}} \\
6.54 \times 10^{23}
\end{array}
\qquad
\begin{array}{r}
1.76541 \\
-\ 1.75893 \\
\hline
0.00648
\end{array}
$$

Graphing

In many instances, the goal of making measurements is to discover or study the relationship that exists between two variables. The pressure and the volume of a gas, the volume and the temperature of a substance, or the color of a solution and the intensity of that color are examples of sets of variables that are related. As one variable changes, so does the other.

We often use graphs to visualize the relationship between two variables. If there are two variables, the graph will be a two-dimensional plot of the points that represent pairs of values of the two variables. An example of a well-drawn graph is shown in Figure 1-4. Notice that the graph has several features that help clarify the meaning of the graph.

Features

1. A Title The title on a graph should be a brief but clear description of the relationship under study. Titles like "Lab Number 1" or "Volume and Temperature" are not good practice because their meaning is clear only to those familiar with the experiment and because the meaning will be lost as memory fades with the passage of time.

2. Labeled Axes Each axis of the graph should be clearly labeled to show the quantity it represents and the units that have been used to measure the quantity. You should recognize the distinction between the quantity measured (pressure, volume, temperature, time, etc.) and the units that have been used to measure that quantity (atmospheres, liters, degrees Celsius, seconds, etc.).

It is convenient to label each axis with the name of the measured quantity followed by the unit (usually abbreviated) in parentheses—for example, Volume (L).

Then only numbers need to appear along each axis, and the axes are not cluttered with the units of each variable.

3. Scales The scale on each axis should be chosen carefully so that the entire range of values can be plotted on the graph. For practical reasons, 2, 4, 5, or 10 divisions on the graph paper should represent a decimal unit in the variable. This equivalence will make it easy to estimate values that lie between the scale divisions. For greatest accuracy and pleasing proportions, the scales selected should be chosen so that the graph nearly fills the page. Be sure, however, that no plotted points fall outside the borders of the graph.

Note that the lower left corner of the graph does not *have* to represent zero on either axis. If the range of measured values extends to zero, the latter may be a good choice; but if not, there will be much wasted space on the graph.

4. Data Points It is good practice to mark the location of each data point with a very small dot and then draw a small circle around the dot to make the point more visible.

5. The Curve A smooth curve should be drawn through the points. The curve should pass as close as possible to each of the points but should not be connected point-to-point with short line segments. If the relationship appears to be linear, the smooth curve should be a straight line. If the line is extended past the range of the measured values, this extension should be indicated by a dashed rather than a solid line.

Linear relationships

Although many variables in chemical systems may be related in a complex nonlinear way, some of the

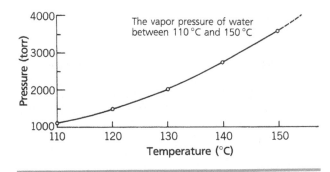

FIGURE 1-4 | **This is an example of a well-made graph. Notice the clear title, the properly labeled axes with variable names and units, the clearly marked data points, and the smooth curve showing the observed trend and the extrapolated trend.**

relationships turn out to be direct proportions; that is, the value of one variable is a constant factor times the value of the other variable plus or minus a second constant. When this proportional relationship exists, the graph of the two variables will be a straight line. An example of a linear graph is shown in Figure 1-5, which illustrates the volume of a sample of gas versus the temperature of the gas.

The *slope* of the line is the constant factor relating the two variables. It can be determined by using any two points on the line, as shown in Figure 1-5. Note that the slope is calculated in the units of the two variables.

In Figure 1-5, the point on the line where the value of the volume is zero (which is where the line crosses the temperature axis) is called the temperature *intercept.*

If a relationship exists between two variables, it is possible to estimate what the value of one variable would be for any value of the other variable by using the graph. If the point of interest lies within the range of measured values, this estimation is called *interpolation.* For example, from Figure 1-4 we can determine by interpolation that at 125 °C the vapor pressure of water is about 1800 torr.

If the estimation is made beyond the measured range, the process is called *extrapolation.* From Figure 1-4 we can see that the vapor pressure of water at 155 °C is estimated by extrapolation to be about 4000 torr.

Interpolation and extrapolation are useful techniques, but both are estimates and assume that the graph is accurate or that it extends beyond the measured values. For extrapolation, especially, this assumption may lead to incorrect conclusions.

Exercises in Measurement

Materials and Supplies: Sets of four pennies dated 1981 and earlier and sets of four pennies dated 1983 and later (plus some extra ones if you plan to do experiments on the pennies as described in Consider This); meter sticks; 30-cm metric rulers; cardboard milk cartons (quart or half-gallon size); beakers of five or six sizes; 60-cm lengths of string; fine-tipped marking pens.

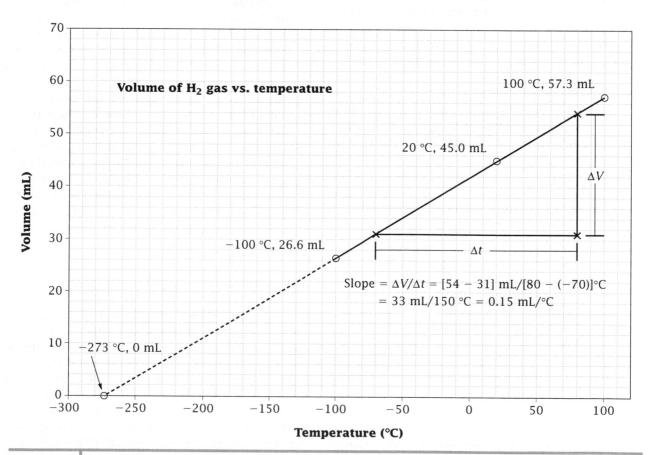

FIGURE 1-5 | **The relation between the temperature and the corresponding volume of a sample of hydrogen gas. A direct proportion, as represented by Charles's law, $V = kT = k(t\,°C + 273)$, plots as a straight line. (T is the Kelvin temperature.)**

1. Mass Measurement Pennies minted in the United States in 1981 and earlier years are significantly different from pennies minted in 1983 or later. Is it possible to tell the difference between these two kinds of pennies by weighing them?

Work in pairs. One student will receive from the instructor four pennies dated 1981 and earlier, and the other student will receive four pennies dated 1983 and later. Carefully check the zero on the analytical balance. Weigh and record the mass of each of the four pennies to the nearest milligram as well as the year in which each penny was minted. Then put all four pennies on the pan and weigh and record the total mass of the four pennies. Finally, record the masses and dates of the four pennies weighed by your partner.

In your report, calculate the mean (average) value and standard deviation for the masses of the pennies that you and your partner weighed. Can you tell from these data if there is a significant difference in the average mass of the two groups of pennies? If there is a significant difference, what hypothesis could you offer to explain the difference?

In part 3, your instructor may also ask you to plot the data (penny mass versus year in which the penny was minted) for the class data. This can be conveniently done using a spreadsheet program on a computer, if one is available, or it can be done by hand.

2. Volume Measurement

(a) Using a metric ruler or meter stick, measure the length of each of the three edges of a quart or half gallon cardboard milk carton to the nearest 0.5 mm. (Ignore the irregular portion of the carton at the top.) Record these values. In your report, calculate the volume of the milk carton in units of cubic centimeters, liters, and quarts.

(b) Read the volumes of partially filled 10-mL and 50-mL graduated cylinders provided by the instructor. Take care to avoid parallax error and note the position of the bottom of each meniscus as described in the Introduction. Read the value of the smaller cylinder to the nearest 0.02 mL, and read the larger one to the nearest 0.2 mL. Record these values. Report the two values to the instructor. If they are not correct, the instructor will show you how to read the meniscus correctly.

3. Graphing

(a) *Circumference versus diameter of different size beakers.* Measure the diameter of each of four different sizes of beakers. An easy way to do this is to place the edges of two books or blocks of wood against the beaker on opposite sides. Then carefully remove the beaker and measure the distance across the gap with a metric ruler or meter stick. Record each diameter measurement.

Now measure the circumferences using a length of string. Draw the string snugly around the beaker and mark the overlapped ends of the string with a fine-tipped marking pen. Measure the distance between the marks on the string with a meter stick to get the circumference of each of the four beakers.

In your report, make a graph of the circumferences of the beakers versus their diameters. Draw a line through the plotted points. Pick two points on the line (located near the ends of the line) and determine the slope of the line, where

$$\text{slope} = \frac{\text{change in circumference}}{\text{change in diameter}} \qquad (7)$$

Is the numerical value of the slope, the ratio of the circumference to the diameter of a circle, what you expected it to be?

(b) *Penny mass versus year of minting.* If your instructor so directs, use the accumulated data for the class to make a graph of the average mass of your pennies versus the year in which they were minted. The class data can be entered and plotted by hand or by using a computer and spreadsheet program. If you have several pennies minted in the same year, you can enter a data point for each penny or plot the average mass for that year. From looking at your graph, is it possible to see evidence of any change in the average mass of pennies over time? What year(s) does it appear that this change took place? What hypothesis would you suggest to explain your findings? Was there a change in size of the pennies? If not, how could they be the same size, but have different masses?

CONSIDER THIS

To get to the bottom of the question of how pennies could be different without looking much different to casual observation, some further experimentation is called for. Here are some experiments you can try. Take a penny from each group (1981 or earlier and 1983 or later) and use a triangular file to file four equally spaced notches on the edge of the penny. First, carefully observe the notches you have filed. Are there any obvious differences? Then, working in a fume hood, put two notched pennies in a small beaker and cover them with 10 mL of 6 M HCl. Observe the pennies for a few minutes and after standing for an hour or two. Do you see any difference in the way the pennies behave in HCl?

Or try this. First, file notches on the edges of the pennies. Then, using a pair of steel tongs to hold the pennies, try heating the pennies in the flame of a Bunsen burner. Be careful to tilt the burner so that if the penny melts, the molten metal won't fall into the burner and clog it, or worse, fall on your hand. Is there a difference in the way the pennies behave in the flame of a Bunsen burner? Now try to put these observations together to explain any differences you have found in the pennies.

Bibliography

The National Institute of Standards and Technology (NIST) Internet site (http://physics.nist.gov) provides a full discussion of the physical constants, including historical information and the best values of the constants and their uncertainties.

Mauldin, R. R. "Introducing Scientific Reasoning with the Penny Lab," *J. Chem. Educ.* **1997**, *74*, 952–955.

Miller, J. A. "Analysis of 1982 Pennies," *J. Chem. Educ.* **1983**, *60*, 142.

Ricci, R. W.; Ditzler, M. A. "Discovery Chemistry: A Laboratory-Centered Approach to Teaching General Chemistry," *J. Chem. Educ.* **1991**, *68*, 229. The authors describe a measurements experiment they call the Pennies Lab.

Sardella, D. J. "An Experiment in Thinking Scientifically: Using Pennies and Good Sense," *J. Chem. Educ.* **1992**, *69*, 933.

Stolzberg, R. J. "Do New Pennies Lose Their Shells? Hypothesis Testing in the Sophomore Analytical Chemistry Laboratory," *J. Chem. Educ.* **1998**, *75*, 1453–1455.

Name _____   Date _____

Data and Calculations

1. Mass measurement

(a) Your data for the weighing of four pennies:

Date of the Penny	Mass of the Penny (x_i)	Deviation $(x_i - \bar{x})$	Square of Deviation $(x_i - \bar{x})^2$
_____	_____ g	_____ g	_____
_____	_____ g	_____ g	_____
_____	_____ g	_____ g	_____
_____	_____ g	_____ g	_____

Calculate the sum: $\sum$ = _____ g $\sum$ = _____ g[a] $\sum$ = _____

Mass of your four
 pennies weighed together _____ g

Calculate the average (mean)
 mass of the four pennies.

Average mass: $\bar{x}$ = _____ g

Now calculate the deviations, the deviations squared, and their sums at the top of the next column.

Calculate the standard deviation for the four pennies following the example shown in Table 1-1.

Standard deviation s = _____ g

[a]If there are no errors in your calculation, the sum of the deviations should be 0.

(b) Your partner's data for the weighing of four pennies:

Date of the Penny	Mass of the Penny (x_i)	Deviation $(x_i - \bar{x})$	Square of Deviation $(x_i - \bar{x})^2$
_____	_____ g	_____ g	_____
_____	_____ g	_____ g	_____
_____	_____ g	_____ g	_____
_____	_____ g	_____ g	_____

Calculate the sum: $\sum$ = _____ g $\sum$ = _____ g[a] $\sum$ = _____

Mass of your four
 pennies weighed together _____ g

Calculate the average (mean)
 mass of the four pennies.

Average mass: $\bar{x}$ = _____ g

Now calculate the deviations, the deviations squared, and their sums at the top of the next column.

Calculate the standard deviation for the four pennies following the example shown in Table 1-1.

Standard deviation s = _____ g

[a]If there are no errors in your calculation, the sum of the deviations should be 0.

(c) Summarize the results of the calculations for your data and your partner's data.

	Your Data	Your Partner's Data
Average mass, $\bar{x}$	g	g
Standard deviation, s	g	g
Sum of individual masses	g	g
Mass of the four pennies weighed together	g	g

Do the data indicate that there is a significant difference between the masses of pennies minted in 1981 and earlier and those minted in 1983 and later years? Explain, using the averages and standard deviations.

Suggest a hypothesis for the difference in masses.

2. Volume measurement

(a) Milk carton dimensions:

Height: _____ cm   Width: _____ cm   Depth: _____ cm

Calculate the volume of the milk carton in units of cubic centimeters: $V =$ _____ cm^3

Convert this volume to units of liters: $V =$ _____ L

Convert this volume to quarts: $V =$ _____ qt

Does the calculated volume agree with the value printed on the milk carton?

(b) Volume of liquid in 10-mL cylinder: _____ mL

Volume of liquid in 50-mL cylinder: _____ mL

(Show your values to your instructor.)

Name _____   Date _____

3. Graphing

(a) Circumference vs. diameter for different size beakers

Beaker Size (mL)	Diameter (cm)	Circumference (cm)

On the graph paper provided, plot the circumference along the vertical axis and the diameter along the horizontal axis. Ensure that your graph has all of the features discussed on page 1-8.

Calculate the slope of your graph, following the example shown in Figure 1-5, and discuss the significance of the numerical value of the slope.

$$\text{slope} = \frac{\Delta \text{ circumference}}{\Delta \text{ diameter}} = \underline{\hspace{2in}}$$

(b) Penny mass vs. year of minting

Make a plot of the average mass of the pennies vs. the year in which they were minted for each year for which you have data. Expand the mass scale (a range of 2.3 to 3.3 g is suggested) so that you can easily see any trends. The horizontal axis should range over the years from your oldest to newest penny (there may be gaps in your plot if you have no data for a particular year).

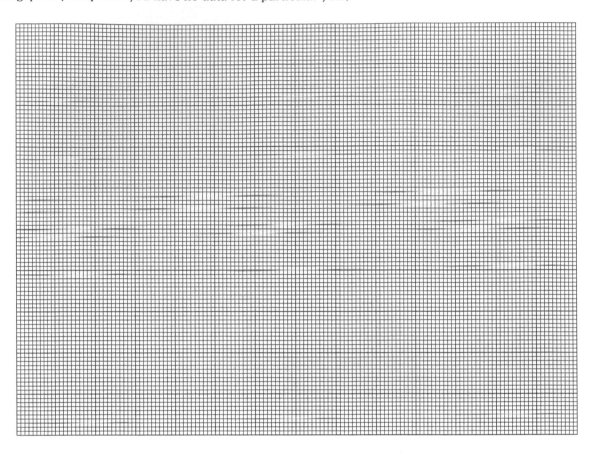

From looking at your graph, is it possible to see any evidence of a change in the average mass of pennies over time?

In what year(s) does it appear that this change took place?

Was there a change in the size of the pennies? If not, how could they be the same size but have different masses?

CONSIDER THIS

If you tried any of the experiments, briefly summarize your observations and what you concluded from them.

Mass and Volume Relationships

Purpose

• Become familiar with metric units of mass, length, and volume.

• Measure the densities of liquids and solids.

• Investigate the relationship between solution density and solution concentration.

Pre-Lab Preparation

Units of measurement

Chemistry is distinctly an *experimental* science. The establishment of the fundamental laws and theories of the nature and behavior of matter depends on careful measurements. In the experiments in this manual, we will measure various quantities, including mass, volume, length, temperature, time, and electrical magnitudes.

The *metric system* of units and its modern successor, the International System of Units (see Table 1 of the Appendix), is especially convenient because it is a decimal system based on powers of ten, like our numbering system. The standard metric units were originally intended to be related to certain quantities in nature. For example, the meter was intended to be one ten-millionth of the distance from the North Pole to the equator along the meridian line through Paris. The first international standard meter was defined as the distance between two fine lines engraved on a bar of platinum-iridium alloy that is kept in a vault at the International Bureau of Weights and Measures near Paris. In 1960, a new standard was adopted by the Eleventh General Conference on Weights and Measures, and the standard meter was defined to be 1,650,763.73 wavelengths of a particular orange-red line of the gaseous element krypton-86. This standard has the advantage that it permits comparisons that are ten times more accurate than is possible with the meter bar and can be precisely reproduced, so it is not subject to accidental loss or destruction, as is the meter bar.

A cornerstone of relativity theory is the notion that the speed of light is a universal constant that is the same for all observers, regardless of the motion of the observer or of the source of light. This makes it reasonable to treat the speed of light as a constant of nature and use it in conjunction with the unit of time to provide a unit of length. Because time can be measured accurately with atomic clocks and because the speed of light is now known accurately, the 1983 General Conference on Weights and Measures was able to redefine the meter in terms of the velocity of light as "the length of the path traveled by light in a vacuum during a time interval of 1/299,792,458 of a second."[1]

Multiples and decimal fractions of the meter (represented as m), as well as their symbols, are listed below.

1 km = 1 kilometer = 1000 meters

1 dm = 1 decimeter = 0.1 meter

1 cm = 1 centimeter = 0.01 meter

1 mm = 1 millimeter = 0.001 meter

1 μm = 1 micrometer = 10^{-6} meter

1 nm = 1 nanometer = 10^{-9} meter

1 pm = 1 picometer = 10^{-12} meter

A related unit of length sometimes encountered in the older chemical literature is the *angstrom* (1 Å= 10^{-8} cm = 10^{-10} m). The size of a hydrogen atom is of the order of 1 angstrom = 100 picometers.

The kilogram was intended to be the mass of a cubic decimeter (1000 cm^3) of water. The present standard of mass is a platinum-iridium cylinder kept at the International Bureau of Weights and Measures and assigned, by international agreement, a mass of 1 kilogram. Each nation has its own metric standards, but all are based on and have been carefully compared with the originals kept in France. The U.S. copy of the international standard of mass, known as Prototype Kilogram No. 20, is housed in a vault at the National Institute of Science and Technology (NIST). Practically all scientists working on research and development today use metric units.

[1]Pipkin, F. M., and Ritter, R. C. "Precision Measurements and Fundamental Constants," *Science,* **1983,** *219:* 913.

The common submultiples of the *kilogram* (kg) are the *gram* (g) and the *milligram* (mg). Chemists often use the terms *mass* and *weight* interchangeably, although they represent different concepts. Mass is defined in terms of the standard kilogram. Weight is a *force*, actually a mass times the acceleration due to the gravitational attraction of the earth. (On the moon, an object weighs less than it does on earth because the moon's mass and gravitational attraction are smaller than the earth's.) In chemistry we are concerned primarily with mass, but because virtually all weighings in the laboratory involve the comparison of an unknown mass with a standard mass, the operation we call *weighing* is actually a measurement of mass. This dual terminology seldom causes problems, and in this manual when we say "weigh an object," we mean "determine the mass of the object."

The practical laboratory unit of volume is the *liter* or cubic decimeter. This is a derived unit, having the units of length cubed, m^3. It is the volume enclosed by a 1-decimeter cube. The *milliliter* (mL) is the most commonly used volume unit in the chemical laboratory. The prefix *milli* tells us that it is one thousandth the volume of the liter. Recently, the milliliter was redefined slightly so that 1 mL $\equiv$ 1 cm^3. (The triple equal sign means "is *exactly* equal to.")

A condensed table of the most commonly used metric units of length, mass, and volume, and their English equivalents, is included in the Appendix as Table 1. Consult the Internet site listed in the Bibliography for more detailed information.

Precision of measurement

See Experiment 1 for a more complete discussion of the treatment of experimental errors, the concept of precision, and related topics.

When measuring physical quantities, it makes sense that the measuring devices should have a precision consistent with the overall precision desired. So, for a precision of 1% in the weighing of a 10-g sample, a balance that is accurate to 0.1 g (0.01 × 10 g = 0.1 g) is sufficient. For the same precision with a 1-g sample, a balance accurate to 0.01 g is necessary. In this experiment, balances that read to 0.01 g and pipets that can deliver with an accuracy of about 0.05 mL should not cause uncertainty greater than about 1% in the calculated density. In a determination that combines several measurements, there is no point in using a much greater precision for one measurement than for the other. Choose your measuring devices according to the precision desired.

Density Ordinary matter, the stuff of which our familiar world is made, has two important proper-

ties. We feel its weight when we hold it in our hands, and it takes up space. In the language of science, we say that a particular sample of matter has a definite *mass* and *volume*. The ratio of these quantities, or the mass per unit volume, is called the *density*, written $d = m/V$. In the metric system, this ratio is usually expressed as grams per cubic centimeter (g/cm^3) or the equivalent grams per milliliter (g/mL). Density is a fundamental physical property of matter.

The measurement of density is necessary for a variety of important procedures in the science of chemistry, such as the calculation of Avogadro's number from unit-cell dimensions of crystals, the determination of the molecular weight of a substance from its gas density, the conversion of the height of a column of liquid to hydrostatic pressure units, the conversion from mass to volume, and the determination of the concentration of a solute from density measurements.

An understanding of density is also useful in the everyday world. For example, the charge of an automotive battery can be determined by measuring the density (and hence the concentration) of the sulfuric acid solution in the battery. Also, a wine maker (even the amateur making wine at home) measures the density of the grape juice to determine whether the sugar content is sufficient for fermentation.

This experiment focuses on the density of liquids (aqueous solutions) and solids. To give you some idea about the volume of space occupied by 0.2-g samples of different substances, Figure 2-1 shows their relative volumes. The smaller the volume the greater the density of the substance.

Note in Figure 2-1 the difference between the density of gases and that of liquids or solids. In liquids and solids the atoms or molecules are about as close together as they can get. In gases there is a lot of empty space between the molecules. Liquids and solids are not very compressible. For example, doubling the pressure on any volume of water from one to two atmospheres decreases the volume by only 0.005%. Doubling the pressure on a gas will reduce its volume by about a factor of 2.

Most liquids and solids expand slightly on heating, and the volume of a sample of water increases about 4% on heating from 4 °C to 100 °C. Water is a curious substance. Its density *increases* slightly as it is heated from 0 °C to 4 °C. At 4 °C, water has its maximum density of 1.000 g/cm^3. As heating is continued above 4 °C, the density of water *decreases*, following the normal behavior of liquids. Because density changes with temperature (and pressure), it is necessary to specify the temperature when reporting the measured density of a liquid or solid. (The pressure is ordinarily assumed to be one atmosphere unless otherwise specified.)

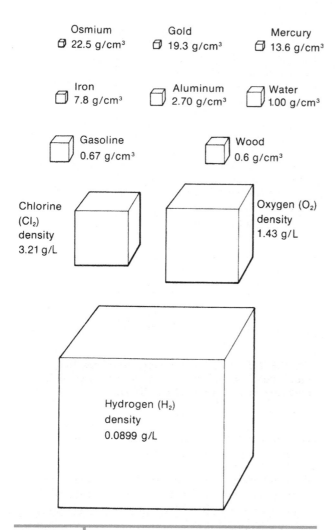

Osmium
$\square$ 22.5 g/cm³

Gold
$\square$ 19.3 g/cm³

Mercury
$\square$ 13.6 g/cm³

Iron
$\square$ 7.8 g/cm³

Aluminum
$\square$ 2.70 g/cm³

Water
$\square$ 1.00 g/cm³

Gasoline
$\square$ 0.67 g/cm³

Wood
$\square$ 0.6 g/cm³

Chlorine (Cl₂) density 3.21 g/L

Oxygen (O₂) density 1.43 g/L

Hydrogen (H₂) density 0.0899 g/L

FIGURE 2-1 | The relationship of density to volume. The cubes of different sizes represent the relative volumes of equal masses (about 0.2 g) of the various materials at 0 °C. Note the difference in density units of solids (g/cm³) and gases (g/L).

Specific Gravity Specific gravity is a property closely related to the concept of density. It is the mass of a substance compared to the mass of an equal volume of water. This is equivalent to the dimensionless ratio of the density of a substance measured at some temperature divided by the density of water at 4 °C or some other specified temperature. (The two temperatures do not need to be the same but must be specified.) For example, the specific gravity of a substance, symbolized by D_4^{20}, is equal to the density of the substance measured at 20 °C divided by the density of water at 4 °C.

$$\text{Specific gravity} (D_4^{20}) = \frac{d_{\text{substance}} \, (20\,°C)}{d_{\text{water}} \, (4\,°C)}$$

Because water has its maximum density of 1.0000 g/cm³ at 4 °C, the specific gravity, D_4^{20}, of the substance at 20 °C would be numerically equal to the density of the substance at 20 °C.

If the mass of the substance is compared to the mass of an equal volume of water at some temperature other than 4 °C, the specific gravity and density are no longer equal because the density of water is generally less than 1.0000 g/cm³ at temperatures other than 4 °C. For example, if the specific gravity of a solution, D_{20}^{20}, is specified as equal to 1.0400, its density at 20 °C would be equal to

$$1.0400 \times d_{\text{water}} \text{ at } 20\,°C \, (0.9982 \text{ g/cm}^3)$$
$$= 1.0381 \text{ g/cm}^3 \text{ at } 20\,°C$$

The Buoyancy and Density of Liquids and Solids
Buoyancy is the tendency of a substance to remain afloat or to rise in a fluid. A cork floats on water. An ice cube also floats on water, but it does not ride as high out of the water as a cork. Each substance that floats on water finds a level where the mass of water displaced is equal to the mass of the substance. Based on these observations, which is denser, ice or cork?

A hydrometer, shown in Figure 2-2, operates on this principle. The greater the density of the liquid in which it floats, the higher the stem of the hydrometer rises above the liquid. The density of the liquid is read from a scale in the stem of the hydrometer. In wine making, the density of the grape juice, which can be measured by the hydrometer, depends on the amount of sugar in the juice. In the process of fermentation, the density of the fermenting juice decreases as sugar is converted to alcohol.

Cork and ice don't mix with water to form a *homogeneous* solution. (A solution is said to be *homogeneous* if every part of the solution has the same composition.) What do you think would happen if we added two homogeneous solutions having different densities to one another? If the solutions are completely miscible, would you expect them to immediately mix? (Solutions are said to be *miscible* if they can be mixed together in all proportions to form a homogeneous solution.) We will be doing some experiments to discover answers to these questions.

Measurement techniques

Your instructor will demonstrate the correct techniques to use in reading the meniscus in a graduated cylinder and in careful weighing with the balance. See also Figures 2-3 and 2-4. For every weighing, observe the following rules and precautions:

1. *Keep the balance pans clean and dry.* Clean up *immediately* any chemical that is spilled.

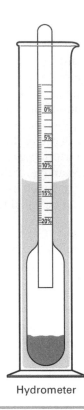

Hydrometer

FIGURE 2-2 The hydrometer is used by wine makers to monitor the progress of fermentation of the grape sugars to alcohol. The dual scale reads both density and mass% sugar (grams of sugar per 100 g of solution).

2. *Make certain the balance is level.* Sensitive balances usually have a bubble level. The balance is level when the bubble is centered in the black circle on the bubble level. Error will be introduced into the reading if the balance is not level.

3. *Check the rest point (or tare) of the empty balance.* Electronic balances have no swinging beam but usually have a tare button that allows you to zero the reading.

    If you have a platform or triple-beam balance, first be sure all movable beam weights are at their zero position. Then release the beam release (if the balance has one), give the pan a little push to cause it to swing gently, and note the central position on the scale about which the pointer oscillates. Use this point as the reference-zero rest point in your weighings.

    Never take readings with the beam and pointer at rest. When the balance beam is stationary, it is possible for the balance beam suspension to stick in a position that is not the true equilibrium position. If the pointer reading differs by more than two to three scale divisions from the marked zero point, have

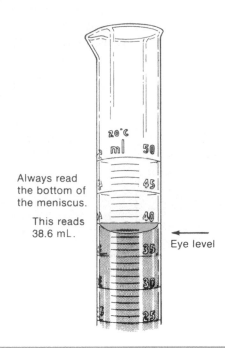

Always read the bottom of the meniscus.
This reads 38.6 mL.
Eye level

FIGURE 2-3 The proper method of reading a meniscus (curved surface of a liquid) in order to measure the volume of the liquid.

your instructor adjust the balance. *Do not change the balance adjustments yourself.*

4. *Never weigh an object while it is warm.* The convection currents of warm air will affect the mass reading of an electronic balance or the rest point of a balance beam.

5. *After weighing an object, return the beam weights to the zero position, and restore the beam release to its rest position.*

Experimental Procedure

Special Supplies: One six-pack each of diet and regular (sugared) cola drinks in 12-ounce aluminum cans; one or more buckets of 2–3 gallons capacity; fine-tipped (50 drops/mL) polyethylene transfer pipets; 10×75 mm and 13×100 mm test tubes; pipetting devices; red and yellow food coloring; aluminum metal pellets or lengths of aluminum rod ($\frac{3}{4}$ in. diameter $\times$ $2\frac{3}{4}$ in. long); other metal samples of regular shapes (cylinders or rectangular blocks); metric rulers and/or metric vernier calipers.

Reduced scale procedure: 10-mL polystyrene serological pipets.
Macroscale procedure: 50-mL polystyrene serological pipets.

Chemicals: Liquid samples: 8% and 16% by weight sucrose solutions, each given a different color by adding a small amount of red or yellow food coloring.

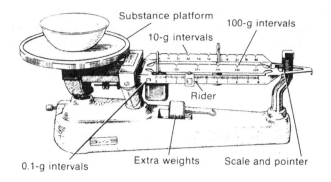

A

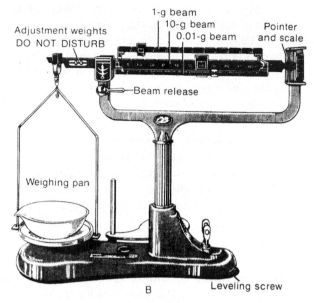

B

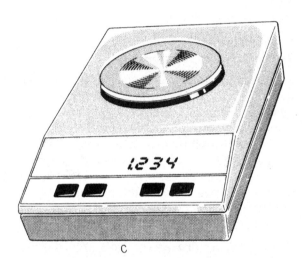

C

FIGURE 2-4 | **Laboratory balances. (A) A platform balance for weighing to the nearest 0.1 g. (B) A triple-beam balance for weighing quantities to the** **nearest 0.01 g. (C) An electronic top-loading balance for weighing to the nearest 0.001 g (the necessary draft shield is not shown).**

1. The Buoyancy and Density of Diet and Regular Coke (A Group Exercise) For this part of the experiment, you will need a six-pack of carbonated diet drinks and a six-pack of regular (sugared) drinks in 12-oz (355-mL) aluminum cans. (Either Diet Coke and Coca-Cola Classic or Pepsi will work fine.) This experiment will be done as a group exercise and need not be carried out by each individual. Weigh and record the mass (to the nearest 0.1 g) of each of the cans of sugared cola. (Make sure that the outsides of the cans are dry.) Then fill a 2–3-gal bucket with tap water and place the previously weighed cans in the bucket. Do they sink or float? Repeat the experiment with six previously weighed cans of diet cola. Do you observe a difference in the buoyancy of the cans of regular and diet drinks?

(a) Calculate the average mass for each sample of six cans of cola. Is there a significant difference in the average mass of a can of regular and diet cola?

(b) Next, determine the approximate density of diet and regular cola. To do this you need to know the average mass and volume of the beverage contained in the can. You can get the net mass of beverage in the can by subtracting the average mass of an empty aluminum can from the average mass of a can of the beverage. If you are supplied with some clean, dry, empty aluminum cola cans, you can determine this yourself. (We found the average mass of an empty 12-oz aluminum can to be about 13.7 g.)

Assuming that the average volume of the drink is that specified by the manufacturer on the can's label (355 mL), calculate the density of regular and diet cola. Compare this density to that of water at the same temperature. You can find the density of water in a handbook, such as the *CRC Handbook of Chemistry and Physics* or *Lange's Handbook of Chemistry*.

(c) Also calculate the mass percentage of sugar in the regular (sugared) cola, using the manufacturer's specified sugar content and volume (355 mL), and the density you calculated in part 1(b). The grams of cola = 355 mL × density (g cola/mL).

$$\text{mass\% sugar} = \frac{\text{g sugar}}{\text{g cola}} \times 100\%$$

2. The Buoyancy of Sugar (Sucrose) Solutions
We will see if two miscible solutions having different densities can be floated on one another. The two solutions will at first be identified only by their different colors, produced by adding a tiny amount of food coloring to the solution. One solution is red, and the other solution is yellow. One of the solutions contains 8.0% sucrose by mass (or weight), the other 16% sucrose by mass. (Sucrose is common table sugar.)

Your job is to try to find out which solution is the most dense by trying to float one solution on the other and later to make an accurate density measurement of the two solutions. At this point, your instructor will tell you the composition of the two solutions; you will then be able to see whether there is a correlation between the density of a solution and its composition, expressed as mass% sucrose. (An aqueous solution that is 8% sucrose by mass contains 8.0 g of sucrose for every 100 g of solution, meaning that 100 g of solution contains 8.0 g of sucrose and 92 g of water.)

For the experiment, you will need two fine-tipped (50 drops/mL) polyethylene transfer pipets, two 13 × 100 mm test tubes, and about six 10 × 75 mm test tubes. Obtain about 5 mL each of the red and yellow sucrose solutions in two clean 13 × 100 mm test tubes. Put a transfer pipet in each colored sucrose solution and draw up some solution into the pipets. As you fill the pipets, leave the tip below the surface of the solution and take your fingers completely off the pipet bulb, so that you do not draw air into the stem of the pipet. Put about a 1-cm deep layer of red solution in a clean, dry 10 × 75 mm test tube. Put the same amount of yellow solution in a second and the same amount of pure water in a third 10 × 75 mm test tube.

Now observe and record what happens when you add solution of the opposite color (red to yellow and yellow to red). Begin by adding red to the yellow in the following way. Put the tip of the pipet with red solution about 1 mm below the surface of the yellow solution and *gently* squeeze the bulb to add the red solution one drop at a time. You will be able to see what's happening more clearly if you view the tube against a white piece of paper. Do the red drops sink or rise as they are added? Keep adding solution drop by drop until you have added a volume of the red solution that is about equal to the volume of the yellow solution already in the tube. Repeat the process, adding yellow solution to a tube of red solution. After finishing, view both tubes against a white background. In which tube is one layer most clearly floating on top of the other? What conclusions do you draw from these observations? Which solution has the higher density, the red or yellow sucrose solution?

3. Quantitative Measurement of the Density of Sugar (Sucrose) Solutions

Macroscale Procedure If your balance has a sensitivity of ±0.1 g, the following procedure employing a 50-mL graduated cylinder can be used. Weigh the empty, clean, and dry graduated cylinder to the nearest 0.1 g. Put slightly less than 50 mL of solution in the graduated cylinder and reweigh the cylinder with solution to the nearest 0.1 g. Also read and record the exact volume of the solution, estimating the volume to the nearest 0.2 mL. The ratio of the mass of liquid to its volume gives the density.

Reduced Scale Procedure This procedure is preferred if your balance has a sensitivity of ±0.01 or 0.001 g. You will need a 10-mL polystyrene serological pipet and six 25- or 50-mL Erlenmeyer flasks to do parts (a), (b), and (c). (A polystyrene pipet is preferred because it is not wet by aqueous solutions, so that when the pipet drains, little or no solution will be left behind.)

(a) *Density determination of the red sucrose solution.* Number each flask; then weigh and record the masses of the clean, dry Erlenmeyer flasks. Use a pipeting device to draw up a little more than 10 mL of the red sucrose solution into the 10-mL serological pipet. Adjust it exactly to the full mark; then deliver the contents into a previously weighed Erlenmeyer flask. Blow out any liquid in the tip of the serological pipet into the flask. (The double ring at the top of the pipet means that it is designed to deliver 10 mL when all of the liquid is blown out.) Reweigh the Erlenmeyer flask to obtain the mass of sucrose solution. Calculate the density of the sucrose solution as mass/volume (g/mL). [A more accurate value for the volume of the pipet can be determined by calibration with pure water, as described later in part 3(c).]

Also measure and record the temperature of the solution.

No matter whether you are using the reduced scale or macroscale procedure, repeat all of these measurements with a fresh sample of the same liquid, including temperature. The average of duplicate determinations will be more reliable and will provide you with a check on gross errors in measuring mass and reading volumes correctly.

(b) *Density determination of the yellow sucrose solution.* Now repeat the entire process described in part 3(a) in duplicate using the yellow sucrose solution. (Be sure that you have rinsed out all of the previous liquid before measuring the volume and mass of the next liquid.)

When you have finished the measurements on both sucrose solutions, your instructor will tell you which is the 8.0 mass% and which is the 16.0 mass% sucrose solution.

(c) *Density determination of water and pipet volume calibration.* As a final check on your technique, and to provide a way to determine the volume of your pipet, repeat the entire process in duplicate using pure water, if your instructor so directs. Record the temperature of the water and compare your calculated density for water with the value (at the same temperature) found in the *CRC Handbook of Chemistry and Physics* or some other reference source. If the density you measure is significantly different (± 0.01 g/mL) from the literature value, it may indicate that the volume of the polystyrene pipet was significantly different from the nominal value, since manufacturing tolerances on the pipet volumes are on the order of 0.5–1%. Use the average mass of water delivered divided by the known (literature) value for the density of water at the same temperature to calculate the corrected volume of the pipet. You can use the corrected volume of the pipet in all of your density calculations.

$$\text{corrected pipet volume (mL)} = \frac{\text{mass of water delivered (g)}}{\text{density of water (g/mL)}}$$

4. Making a Graph of the Data Is there a simple relationship between density and mass% sucrose? To help visualize it, construct a graph of density (g/mL) versus mass% sucrose. Include a data point for pure water (0% sucrose), so that you have a total of three data points. If you did not determine the density of pure water as described in part 3(c), use the literature value for the density of water at the same temperature as your sucrose solutions. If your instructor so directs, plot the average values of density using the combined data of all of the members of

your laboratory section. (The average of several values is statistically a more reliable measure of the true value of a quantity than any single measurement.) Draw a straight line that looks like a reasonable fit to the points. (Don't connect the points together by straight line segments.) Better yet, use a spreadsheet or graphing program to plot the points and to add the best-fit line to the data points.

The graph you make can be used to estimate the sugar content in regular or diet cola or any common beverage such as apple juice or other fruit juice drinks. (In these drinks, besides water, the main ingredient is sugar.) We will use it to estimate the sugar content in the regular (sugared) drink we used in part 1.

On the graph of density versus mass% sucrose, draw a horizontal line corresponding to the density of regular cola calculated in part 1. Where this line intersects the best-fit straight line (drawn through your three data points), drop a perpendicular line to the mass% sucrose axis and read off the corresponding mass% sucrose. Compare this value with the mass% sucrose you calculated in part 1 [(g sugar/g cola) × 100]. Do the two values agree, considering possible sources of uncertainty in the measurement of density and mass% sucrose?

5. The Density of Solid Aluminum

(a) Use the following technique to determine the density of an irregularly shaped solid. You will be supplied with aluminum metal pellets or some other sample that is selected by your instructor. (Lengths of aluminum rod 3/4 in. diameter × 2-3/4 in. long are also suitable.) Use about 50 g of the aluminum pellets (avoiding fine or powdered material), or an amount that will give a volume increase of around 20 mL.

Weigh a small beaker to a precision of 0.1 g. Add the sample of metal pellets and weigh again. Place about 25 mL of water in a 50- or 100-mL graduated cylinder and read the volume to the nearest 0.1 mL. Tilt the cylinder and slide the weighed sample pieces into it carefully, to avoid loss of water by splashing, then tap the sides to dislodge any adhering air bubbles. Again read the volume. The increase is the volume of the sample (see Figure 2-5). Calculate the density of the sample. If time permits, make a duplicate determination with another sample, to increase the accuracy of your measurements.

Save the aluminum pellets or other material you used, since they can be dried and reused.

(b) *Optional.* If you wish to determine the density of a regular-shaped solid such as a rectangular block or a cylinder of aluminum (or some other material), the volume can be calculated from

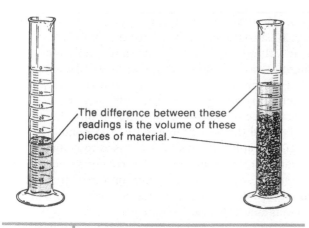

The difference between these readings is the volume of these pieces of material.

FIGURE 2-5 | A method of measuring the volume of an irregularly shaped solid.

measurements of the dimensions. The mass may be obtained directly on the balance, and the density can then be calculated. In making these measurements, note how the quality and precision of the rulers and balances you use affect the number of significant figures you retain, and hence determine the degree of precision you can expect to obtain in your measurement.

CONSIDER THIS

If you add ice at 0 °C to a container of water maintained at 0 °C, the ice will float on the water with the water finding a certain level in the container. As the ice gradually melts, forming liquid water at 0 °C, would you expect the water level in the container to go up, down, or remain the same? Explain the reasoning by which you arrived at your conclusion.

The density of diet cola is about the same as pure water. The cola was contained in an aluminum can, and if you did part 5 of the experiment, you know that the density of aluminum is considerably greater than that of water. Considering all this information and your own experience drinking carbonated beverages, explain how it is that cans of diet cola can float in water while cans of regular (sugared) cola sink. Wouldn't you expect both kinds of cola cans to sink since both diet cola and regular cola have a density equal to or greater than water and since they are

contained in an aluminum can that has a density greater than water?

The mass of the earth is known to be about 5.98×10^{24} kg. Assume that it is approximately spherical, with an average diameter of 1.276×10^7 m. From these data, calculate the average density of the earth (g/cm^3). Could the earth be made mostly of water (density = 1.00 g/cm^3), abundant elements like calcium silicate (Ca$_2$SiO$_4$, density = 3.27 g/cm^3), or aluminum oxide (Al$_2$O$_3$, density = 4.0 g/cm^3)? What hypothesis would explain why the earth has a considerably higher density than any of these materials, which are abundant at the surface of the earth?

Bibliography

The National Institute of Standards and Technology (NIST) Internet site (http://physics.nist.gov) provides a full discussion of the physical constants, including historical information and the best values of the constants and their uncertainties.

Henderson, S. K.; Fenn, C. A.; Domijan, J. D. "Determination of Sugar Content in Commercial Beverages by Density: A Novel Experiment for General Chemistry Courses," *J. Chem. Educ.* **1998,** *75*(9), 1122–1123.

Stein, M.; Miller, D. "Density Explorations," *The Science Teacher* February **1998,** 45–47.

Summerlin, L. R.; Borgford, C. L.; Ealy, J. B. "66. Sugar in a Can of Soft Drink: A Density Exercise," *Chemical Demonstrations: A Sourcebook for Teachers,* Volume 2, 2nd ed., American Chemical Society, Washington, D.C., 1988, pp. 126–127.

Wikipedia on-line encyclopedia has interesting articles on the history of the kilogram mass standard and the International System of Units (known by the French-language initials as "SI" units):

http://en.wikipedia.org/wiki/Kilogram

http://en.wikipedia.org/wiki/SI_units

Mass and Volume Relationships

Name _____

Date _____ Section _____

Locker _____ Instructor_____

Data, Calculations, and Observations

1. The buoyancy and density of diet and regular cola

Brand name(s) of beverages _____

Volume of cola listed on the label _____ mL

Sugar content of regular cola listed on the label _____ g

Describe the buoyancy behavior of cans of regular and diet cola. Which float and which sink?

(a) Determination of the average mass of cans of diet and regular cola and empty cans. (Optionally, assume that the mass of an empty 12-oz aluminum can is about 13.7 g.)

Mass of cans of regular cola	Mass of cans of diet cola	Mass of empty cans
g	g	g
g	g	g
g	g	g
g	g	g
g	g	g
g	g	g
Average: g	Average: g	Average: g

(b) After subtracting the average mass of an empty can from the average mass of cans of regular and diet colas, calculate the average density of regular and diet cola as the net mass of cola divided by the volume of the cola.

Calculated density of regular cola:

Calculated density of diet cola:

(c) Calculated mass% of sugar in cola: (g sugar/g cola) × 100%

2. The buoyancy of sugar (sucrose) solutions

Record what you observed when you layered the red sucrose solution on top of the yellow sucrose solution and vice versa. What are your conclusions about which solution is denser, the red or yellow sucrose solution?

3. Quantitative measurement of the density of sugar (sucrose) solutions

In calculating the density, use the nominal volume of the solution (10.00 mL) or preferably the corrected volume determined from the pipet calibration in part 3(c): V(mL) = mass of water (g)/density of water (literature value, g/mL).

Volume calculated from data of part 3(c): _____ mL

	(a) Red sucrose solution		(b) Yellow sucrose solution		(c) Water	
Temperature, °C						
	Trial 1	Trial 2	Trial 1	Trial 2	Trial 1	Trial 2
Mass of empty flask	g	g	g	g	g	g
Mass of flask + solution	g	g	g	g	g	g
Net mass of solution	g	g	g	g	g	g
Average mass of solution		g		g		g
Volume of solution		mL		mL		mL
Calculated density		g/mL		g/mL		g/mL

Name _____   Date _____

4. Making a graph of the data

Make a graph of density (g/mL) vs. mass% sucrose. (A density scale running from 0.99 to 1.08 g/mL and a mass% sucrose scale running from 0 to 20% will be suitable.)

There should be three data points, one for pure water (0 mass% sucrose) and data points for the red and yellow sucrose solutions. If possible, make your graph using the average of the data for your entire laboratory section. Draw the best straight line through the points.

Using a spreadsheet program, such as Microsoft Excel, makes the drawing of graphs and the addition of the best-fit line (called a *trendline* in Excel) easier, once you have learned how to use the program.

5. The density of solids

(a) Density of aluminum metal

	Trial 1	Trial 2
Initial volume of water in the graduated cylinder	mL	mL
Final volume of water after addition of the solid metal sample	mL	mL
Net volume of the solid metal sample	mL	mL
Mass of the metal sample	g	g
Calculated density of the metal sample	g/mL	g/mL

(b) *Optional:* Record the measurements and mass of any regular-shaped solid that you measure, and calculate its density.

CONSIDER THIS

Discuss in the space below the three questions posed at the end of the experimental procedure in the *Consider This* paragraphs.

Forensic Chemistry

Crime Scene Investigation

Purpose

• Experience some of the challenges inherent in the field of forensic science, including careful observation and thorough record-keeping.

• Explore the laboratory techniques involved in chromatography to illustrate the utility of multi-component analysis and identification.

• Demonstrate the utility of chemical spot tests, a form of qualitative analysis, for revealing the invisible and identifying the presence of previously unknown substances.

Introduction

Forensic science (science applied to legal settings) is a popular topic in our day, as shown by the weekly presence of television series like *CSI: Crime Scene Investigation, Law and Order, Forensic Files,* and *Bones*—all on the list of the most-watched shows. But in previous generations, the topic was just as popular, evidenced by Sir Arthur Conan Doyle's works featuring Sherlock Holmes in the late nineteenth century and Agatha Christie's famous investigators, Hercule Poirot and Miss Jane Marple, solving crimes of the twentieth century. One of the earliest stories in the history of science, Archimedes deducing the essential concept of buoyancy, was occasioned by a possible crime: whether the king's crown had been adulterated with metals less valuable (and less dense) than gold.

In these earlier eras as well as ours, the fictional investigators seem to require much less time and effort to solve their crimes than real-life forensic scientists do. Reading about or watching forensic scientists at work may be more enjoyable than performing the careful collection and analysis procedures that are required. But the reward of setting an innocent person free or helping to convict a guilty party can be quite motivating, as is the satisfaction of helping to assemble the pieces of a puzzle—these are all aspects of forensic science.

Forensic science spans most of the natural sciences, including chemistry, biochemistry, biology, physics, geology, and medicine. The field extends into engineering, anthropology, and psychology as well. By far, the most common techniques are those of chemists: chromatography and other techniques for analysis of inks, toxins, adulterants, fibers, drugs, explosives, gunshot residues, and flammable materials. Biochemical analyses have recently become important forensic tools, including the identification of blood, semen, and DNA. Fascinating examples of biological applications to forensics include the entomology and microbiology of decay processes. Fingerprint analysis, the study of patterns of blood spatter or glass fragmentation, and bullet trajectories are examples in which the physics and physical properties of materials are studied. The investigation of computer crimes, forgeries, sound recordings, and serial crimes require techniques from a variety of other disciplines. It's no wonder, then, that forensic scientists typically specialize in only one aspect of this broad field and are usually trained in an apprentice setting after obtaining degrees in chemistry, biochemistry, or biology.

The training of all scientists includes emphasis on the need for accurate records, but the demands of the courtroom and the high stakes of legal proceedings place extra demands on forensic scientists for systematic record-keeping and sample storage. These urgent duties also dictate rigorous attention to instrument calibration, blank and standard samples, and sources of possible contaminants or interferences. The "chain of custody" requirement not only demands that the laboratory personnel keep thorough records but that they document the entire history of the sample from collection through trial (and preserve the samples in long-term storage in case further investigation is necessary). Custody issues also demand that a sample be protected from contamination or degradation, a special challenge for biological samples.

Pre-Lab Preparation

Observations and record-keeping

Making observations is the starting point of any scientific enterprise, even if it isn't always the first step chronologically. This is certainly true in the case of forensic science, where the preservation of a crime scene is second in priority only to the preservation of human life. Careful recording of observations is essential in any lab setting, but especially so in the context of forensic science where a person's life or freedom may be at stake. Accurate and understandable records are also necessary in criminal or civil proceedings because so many different participants need to access and process the information: police officers, detectives, crime scene investigators, lawyers for all parties involved, and judges.

Many of you may use the Report form pages of this lab book to record your observations, make your calculations, and state your conclusions. In real laboratory settings, especially in a forensic laboratory, this would not suffice as a lab notebook. A permanently bound (not spiral bound or glued) notebook would be required; data would be written in ink, on dated pages that include all original measurements, notes, calculations, comments, and conclusions. Electronic equivalents of this type of record are now accepted, but not in a simple document or spreadsheet form, as these are easily modified.

In this experiment, we will explore your powers of observation and record-keeping in the laboratory as a pair of necessary, but not sufficient, skills to perform forensic analysis.

Chromatography

Separating two substances that have quite different physical properties is usually a simple and straightforward task. A mixture of sand and salt can easily be separated by adding water to dissolve the salt, filtering the mixture to remove the sand, and recovering the salt by evaporating the water. But what if the substances to be separated have similar physical and chemical properties? Separation of a mixture of several closely related substances presents a more difficult problem, but the problem has been solved by the development, over the past 100 years, of a family of powerful separation techniques that collectively are called *chromatography.*

The word *chromatography*—formed from the Greek words *khroma,* meaning "color," and *graphein,* meaning "to draw or to write"—was coined by the Russian botanist M. S. Tswett around 1906 to describe his process of separating mixtures of plant pigments. He allowed a solution of plant pigments to percolate down a glass column of absorbent powder, where they separated into colored bands—literally, a "color drawing" on the powder. Even though most chromatographic methods are now used on solutes that do not produce bands of color visible to the human eye, the term *chromatography* is applied to any separation procedure employing the same principle as the method described by Tswett. Chromatography is widely used for qualitative or quantitative analysis or for isolation and recovery of the components of a mixture.

All forms of chromatography employ the same general principle: A mixture of solutes in a *moving phase* passes over a selectively absorbing medium, the *stationary phase.* Separation occurs because the solutes have different affinities for the stationary and moving phases. Solutes that have a greater affinity for the moving phase will spend more time in the moving phase and therefore will move along faster than solutes that spend more time in the stationary phase.

There are several ways to achieve a flow around or through a stationary phase. In *column chromatography,* small particles of the stationary phase are packed in a tube; the moving phase (a liquid or gas) flows around the particles confined in the tube. In *thin-layer chromatography (TLC),* the stationary phase is a thin layer of fine particles spread on a glass or plastic plate. In *paper chromatography,* the paper itself forms the stationary phase, along with any solvent adsorbed to the paper. Figure 3-1 illustrates the separation of a sample mixture as a moving liquid phase flows through a stationary phase (such as paper).

In this experiment, we will employ paper chromatography. The cellulose structure of paper (shown in Figure 3-2) contains a large number of hydroxyl groups, —OH, that can form weak bonds (called *hydrogen bonds*) to water molecules. So the stationary phase can be regarded as a layer of water, hydrogen-bonded to cellulose. If the solvent is water, the moving phase is also aqueous; but if a mixture of an organic solvent is used with water, the moving phase is apt to contain a high proportion of the organic solvent.

The solvents employed in chromatography must wet the stationary phase so that the mobile phase will move through the paper fibers by capillary attraction. A solute in the mobile phase moves along

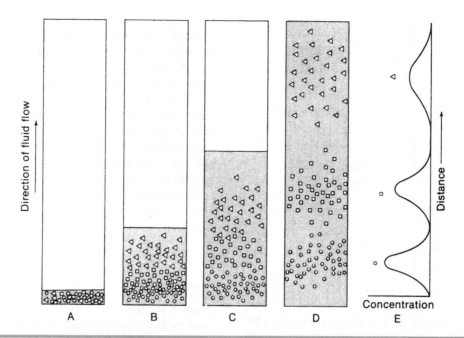

FIGURE 3-1 Schematic illustration of the separation of a three-component mixture by differential migration on a strip of paper. Parts (A) through (D) represent a time sequence showing how the components move along at different rates. (A) At the beginning of a separation, the three components are clustered together at the point of sample application. (B–D) As the sample migrates along the paper strip, the three sample components are gradually separated, the fastest-moving component spending the most time in the moving liquid phase. (D) The final distribution of substances. (E) A smoothed plot of the concentration of each substance as a function of the distance from the origin, for the distribution shown in (D).

with the solvent during the chromatographic *development* (the term used to describe the process that occurs as the solute moves along and is partitioned between two phases). As the solute moves, it undergoes many successive distributions between the mobile and stationary phases. The fraction of time it spends in the mobile phase determines how fast it moves along. If the solute spends all of its time in the moving phase, it will move along with the solvent front (the leading edge of the solvent moving through the stationary phase). If it spends nearly all of its time in the stationary phase, it will stay near the point of application.

After the chromatogram has been developed and the solutes on the paper have been located, the movement of the solute on the paper is expressed by the R_f value (called the *retention factor*), where

$$R_f = \frac{\text{distance traveled by the solute}}{\text{distance traveled by the solvent front}}$$

The distances used in calculating R_f values are measured as shown in Figure 3-3. The distance

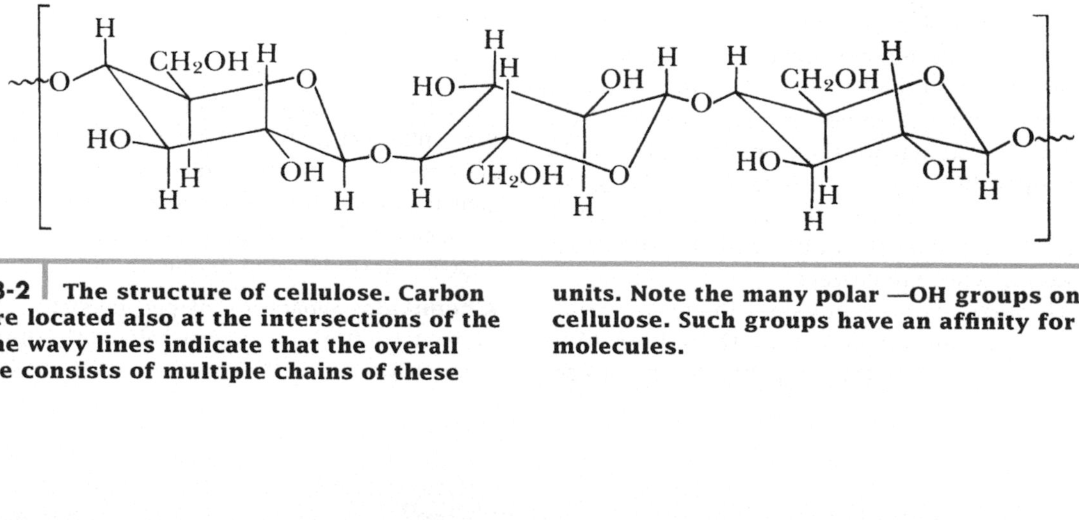

FIGURE 3-2 The structure of cellulose. Carbon atoms are located also at the intersections of the lines. The wavy lines indicate that the overall structure consists of multiple chains of these units. Note the many polar —OH groups on the cellulose. Such groups have an affinity for water molecules.

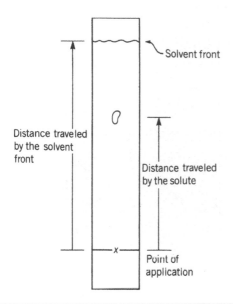

FIGURE 3-3 | **The movement of the solute on the paper chromatogram is expressed by the R_f value, the distance traveled by the solute divided by the distance traveled by the solvent front.**

traveled by the solute is measured from the point of application to the center or densest part of the spot; the distance traveled by the solvent front is measured from the point of application to the limit of movement of the solvent front (which must be marked immediately after the paper is removed from the developing chamber because it may be invisible after the solvent evaporates).

If all conditions could be maintained constant, R_f values would be constant. However, variations in temperature or in the composition of the solvent phase or changes in the paper can alter the R_f value. The R_f value is useful mainly for expressing the relative mobility of two or more solutes in a particular chromatographic system. The absolute R_f values may change from day to day, but their values in relation to each other remain nearly constant.

Paper chromatography can be performed on strips or sheets of paper, with the solvent being allowed to flow either upward or downward. Horizontal development may also be used, in which case the solvent is allowed to flow radially outward from the center. Two-dimensional chromatography can be used to separate complex mixtures by applying two different solvent systems. The paper is first developed in one direction, then turned by 90° and developed again in a second solvent.

Development is accomplished in the following way. First, the solution containing the solutes to be separated is spotted on the dry paper. If ink pens are used, they are perfectly designed for placing small amounts of ink on the paper. Then the spots are allowed to dry, and the filter paper is placed in a tightly closed receptacle containing the developing solvent, whose vapor saturates the atmosphere of the container. The paper is supported so that the developing solvent moves through the spots that have been applied. The point of application must not be immersed in the developing solvent; otherwise, the solute spots will diffuse into the solvent and be greatly diluted.

The solvent flow is allowed to continue for a fixed time or until the solvent front reaches a specified point on the paper. Then the paper is removed, the solvent front is marked with a pencil, and the paper is dried; the result is that the separated solute spots are fixed in their position at the end of development.

If the solutes are colored, they can be readily seen on the paper. Colorless spots are located by various means. Often the paper is exposed to the vapor of a reagent (such as iodine, which makes organic compounds visible), or it can be dipped in or sprayed with a reagent that reacts with the colorless substances to form a visible spot. When placed under an ultraviolet light, many substances can be located by their fluorescent glow.

After the spots are located and their R_f values measured, they are compared with standard known materials run under the same conditions. If possible, known and unknown substances are run on the same paper. If the known and unknown compounds have the same R_f values in different solvent systems and give the same reactions with various chromogenic (color-forming) reagents, they are probably identical.

Experimental Procedure

Special Supplies and Equipment: *Part 1*: Pennies or other coins (at least one per student); 5–10× magnifiers (plastic magnifiers, loupes, and low-power microscopes are available from Edmund Scientifics, http://scientificsonline.com, or Educational Innovations, Inc., http://www.teachersource.com). Optional procedure: 1.5 × 1.5 in. × 2 mil ziplock poly bags, http://discountplasticbags.com; or coin envelopes.

Part 2: Assorted blue or black ink pens; Whatman No. 1 filter paper, 46 × 57 cm sheets (pre-cut each sheet into twelve 11.5 × 19 cm pieces); 30-cm ruler; soft lead pencils; plastic food wrap; Size 33 rubber bands; scissors; stapler. Optional procedure: 9-in. Pasteur capillary pipets; ultraviolet lamp (UV light).

Chemicals: Separation of pen inks: The developing solvent solution is "rubbing alcohol" (70% isopropyl alcohol/water, v/v). Optional procedure: methanol.

NOTES TO INSTRUCTOR

The degree of difficulty of part 1 can be increased by supplying extra coins to the "evidence" pool or limiting the selection of coins to a single type, a small number of mint marks or dates, or by withholding from students information about the range of the coin characteristics, e.g., year and mint. The degree of forensic realism can be increased by instructing students in appropriate sample-handling procedures, such as wearing gloves and using plastic tweezers (to avoid fingerprints and other contamination) and sample storage (a small ziplock bag or coin envelope.)

1. Observations and Record-Keeping Choose a coin of your own or one supplied by your instructor. Handle the coin carefully from this point forward so that you do not contaminate the "evidence." If your instructor directs, obtain an evidence container (paper envelope or plastic bag) and mark it with appropriate identifiers.

Make as many observations of your coin as you can and record them on the Data Sheet of your laboratory notebook. At a minimum, you will record the date and mint mark of your coin, but your goal is to note as many unique aspects of your coin as possible so that you can distinguish it from all other coins in your laboratory. You may wish to use a magnifying glass or loupe or other available laboratory equipment.

When you have finished with your observations and recording tasks, return the coin to your instructor. Your instructor will return a coin to you. Observe this coin and compare it to the description in your lab notebook. Determine whether you believe this is the same coin you originally studied or a different coin.

Form groups of 4 or 5 students. One at a time, attempt to convince your group that this is or is not your original coin. The only evidence you may cite is that in your laboratory records. The members of your group may challenge you to show them your recorded observations.

At the end of your presentation, pass a slip of paper to each of the group members and ask them to render their verdict; they should write "is" or "not" on the ballot, indicating whether they are convinced "beyond a reasonable doubt" that your coin is or is not your original coin. You may interview the members of your jury group to help you determine which of your observations were critical to the case and which were not.

2. Chromatography of Pen Inks If pre-cut sheets are not already prepared for you, cut an 11.5 × 19 cm piece of Whatman No. 1 filter paper. Using a soft lead *pencil* (not an ink pen), draw a line parallel to the long dimension about 2 cm from the edge. Still using a pencil, put five small X's on the line, beginning 5 cm from the edge of the paper and spacing the marks about 2 cm apart. (The X's locate the points where you will later spot the inks.) Put 70 mL of the developing solvent (70% isopropyl alcohol in water) in the bottom of a clean, dry 800- or 1000-mL beaker and cover the beaker with plastic wrap held in place by a rubber band—that way, the enclosed air will become saturated with the vapors of isopropyl alcohol and water.

Using a scrap of filter paper, practice applying small (2–3 mm diameter) ink spots on the paper. For ballpoint pens, this will involve marking a small circle of ink, going over it several times without enlarging it beyond 3 mm; for felt-tip pens, you should hold the pen to the paper until the ink has spread to form a 2–3 mm circle. Allow the paper to dry for a minute, then repeat this process 5–6 times. When you are confident you have mastered the technique of applying the spots, lay the 11.5 × 19 cm piece of paper down on a clean paper towel or sheet of clean paper and carefully apply a small (2–3 mm) spot of ink on the first X. Continue spotting the different inks, using a different pen on each X and writing in pencil near each X a letter symbol to identify the pen spotted there. Record the identity of each spot. Allow the spots to dry by gently waving the paper in the air.

After the spots have dried, gently curve the paper into a cylinder shape, as shown in Figure 3-4, and staple the edges together, leaving a gap so that the edges do not quite meet. Put the paper cylinder in the beaker containing the developing solvent, placing the spots at the bottom and taking care that the paper does not touch the wall of the beaker. *The spots of ink must be above the surface of the solvent in the beaker.* Immediately seal the beaker with plastic wrap secured with a rubber band and allow the solvent to rise to within 1 cm of the top of the paper. (This usually takes 30–40 minutes.) Note the position of the *solvent front* (the wet-looking leading edge of the solvent) every few minutes so that you do not let the solvent front rise all the way to the top of the paper. Also note how the spots begin to separate into bands of color as they move along.

When development is finished, remove the paper from the beaker and immediately mark the solvent front with a pencil line before the solvent evaporates. Then dry the paper thoroughly, preferably in the hood.

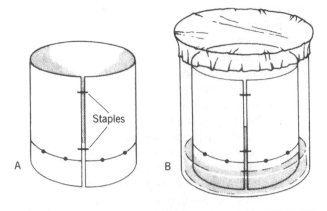

FIGURE 3-4 Gently curve the chromatographic paper into a cylinder shape and staple as shown in (A). Put the paper cylinder carefully into the beaker containing the solvent, then cover with plastic film held in place by a rubber band, as shown in (B).

After the paper has dried, locate the densest part of the band for each color and draw a pencil line through it. Measure the distance (in millimeters) from the pencil line at the point of application to the densest part of the band of color, as well as the distance from the point of application to the solvent front. Record the values in your report. Calculate and record the R_f values for every colored spot on the paper. For each original spot of ink, describe the number of components you find and their colors. Can you identify each of the inks by its chromatogram? If your neighbor used the same set of pens, can you match them by comparing your tables of R_f values without comparing the chromatograms? (Note that this is how chromatographic analyses are typically compared.)

Pencil your name and the date on the chromatogram and turn it in with your lab report if your instructor so directs.

Matching the Ink: If you wish to test your forensic skills, select four pens with the same ink color; have your instructor select one of the pens and mark one spot on a new piece of chromatograph paper (set up as previously, with the pencil line and five X's). Next, make a spot with each of the four pens on the chromatograph paper. Develop the chromatogram as before. Determine the identity of the instructor's pen.

When you are finished developing all of your chromatograms, add water to the beaker until it is nearly full. This diluted solution can be poured down the drain unless your lab instructor directs otherwise.

Optional: For an even more realistic test of your forensic skills, have your instructor sign a piece of ordinary paper; you can think of it as an important letter or a will that might have been forged. (Or your instructor could sign two documents with the same pen or two different pens.) Carefully cut the signature from the document and then cut the signature into small pieces, small enough to fit into a 20 × 150 mm test tube. (If this were a real-life situation, a forensic laboratory technician would use a syringe needle to punch small samples of ink-laden paper to be extracted. The original document needs to be preserved for courtroom display and further analyses if the judicial process so requires.)

Add 2–3 mL of methanol to the test tube, enough to cover the pieces. Heat the test tube in a water bath if needed to increase and hasten the dissolution of the ink, but do not get the tube too hot or else the methanol will evaporate. You can also use a Pasteur pipet to squirt the hot methanol over the pieces of paper. When most of the ink has been extracted from the paper, use a transfer pipet to transfer the methanol solution to another small test tube. Heat this test tube until most, but not all, of the methanol has evaporated. Then use a transfer pipet to spot this sample onto your chromatograph paper. Add spots from the suspect pens to the chromatograph and develop as before.

⚠ SAFETY PRECAUTIONS:
The isopropyl alcohol and methanol used in part 2 are flammable. No open flames (e.g., Bunsen burners) should be used in the laboratory during this experiment. Dilute solutions of the alcohols are not flammable and readily biodegrade in the wastewater disposal process of the community.

☠ WASTE COLLECTION:
The alcohol solvents used in this experiment are ordinary household chemicals and can be disposed of in the diluted form described at the end of the experimental procedure.

CONSIDER THIS

Are any of the pigments used in the pen inks fluorescent? If an ultraviolet light is available in the laboratory, test for this property. Are the fluorescent pigments in addition to the ones you saw in the chromatogram or are some of the pigments both colored and fluorescent? Why might a manufacturer add fluorescent pigments to its inks?

Can you distinguish between an original signature and a copy made on a color photocopier or printer? Can you distinguish an inkjet-printed copy from a conventional photocopy?

Experiment with other developing solvents, such as 5% acetic acid (household vinegar). Does an ink developed with rubbing alcohol give the same number of spots as the same ink developed with vinegar? Do they give the same R_f values? Why or why not?

Bibliography

Bergslien, E. "Teaching to Avoid the 'CSI Effect': Keeping the Science in Forensic Science," *J. Chem. Educ.* **2006,** *83*(5), 690–691.

Berry, K.; Ishii, G. G. "Forensic Chemistry: An Introduction to the Profession," *J. Chem. Educ.* **1985,** *62*(12), 1043.

Breedlove, C. H. "The Analysis of Ball-Point Inks for Forensic Purposes," *J. Chem. Educ.* **1989,** *66*(2), 170–171.

Quigley, M. N.; Qi, H. "A Chemistry Whodunit," *J. Chem. Educ.* **1991,** *68*(7), 596–597.

Strain, H. H.; Sherma, J. "Michael Tswett's Contributions to Sixty Years of Chromatography," *J. Chem. Educ.* **1967,** *44*, 235. This article is immediately followed on page 238 by a translation of Michael Tswett's article, "Adsorption Analysis and Chromatographic Methods," published in 1906 in *Berichte der deutschen botanischen Gesellschaft.*

Forensic Chemistry

Crime Scene Investigation

Name _____

Date _____ Section _____

Locker _____ Instructor_____

1. Observations and record-keeping

Sample Identification: _____

(a) Describe the key identifying marks of your sample evidence.

(b) Do you believe that the coin returned by your instructor was the same as the coin that you initially studied? What were the key identifiers?

(c) Did you persuade the jury? (Remember, in a criminal trial you must obtain a unanimous verdict; in a civil trial, only $\frac{3}{4}$ of the jury usually needs to agree.)

(d) Which of your observations proved crucial to your case? Which were not needed?

2. Chromatography of pen inks

Identity of Pens:

1. _____ 2. _____

3. _____ 4. _____ 5. _____

	Pen:	1	2	3	4	5
Component 1						
Color of spot		____	____	____	____	____
Distance traveled by spot		____ mm	____ mm	____ mm	____ mm	____ mm
Distance traveled by solvent front		____ mm	____ mm	____ mm	____ mm	____ mm
R_f value		____	____	____	____	____
Component 2						
Color of spot		____	____	____	____	____
Distance traveled by spot		____ mm	____ mm	____ mm	____ mm	____ mm
Distance traveled by solvent front		____ mm	____ mm	____ mm	____ mm	____ mm
R_f value		____	____	____	____	____
Component 3						
Color of spot		____	____	____	____	____
Distance traveled by spot		____ mm	____ mm	____ mm	____ mm	____ mm
Distance traveled by solvent front		____ mm	____ mm	____ mm	____ mm	____ mm
R_f value		____	____	____	____	____
Component 4						
Color of spot		____	____	____	____	____
Distance traveled by spot		____ mm	____ mm	____ mm	____ mm	____ mm
Distance traveled by solvent front		____ mm	____ mm	____ mm	____ mm	____ mm
R_f value		____	____	____	____	____

Question

1. Does your neighbor's chromatogram have an ink that shares the same set of R_f values as one or more of your inks?

Matching the ink

Identity of Pens: 1. ___Instructor's___ 2. _____

3. _____ 4. _____ 5. _____

	Pen:	1	2	3	4	5
Component 1						
Color of spot		_____	_____	_____	_____	_____
Distance traveled by spot		___ mm	___ mm	___ mm	___ mm	___ mm
Distance traveled by solvent front		___ mm	___ mm	___ mm	___ mm	___ mm
R_f value		_____	_____	_____	_____	_____
Component 2						
Color of spot		_____	_____	_____	_____	_____
Distance traveled by spot		___ mm	___ mm	___ mm	___ mm	___ mm
Distance traveled by solvent front		___ mm	___ mm	___ mm	___ mm	___ mm
R_f value		_____	_____	_____	_____	_____
Component 3						
Color of spot		_____	_____	_____	_____	_____
Distance traveled by spot		___ mm	___ mm	___ mm	___ mm	___ mm
Distance traveled by solvent front		___ mm	___ mm	___ mm	___ mm	___ mm
R_f value		_____	_____	_____	_____	_____
Component 4						
Color of spot		_____	_____	_____	_____	_____
Distance traveled by spot		___ mm	___ mm	___ mm	___ mm	___ mm
Distance traveled by solvent front		___ mm	___ mm	___ mm	___ mm	___ mm
R_f value		_____	_____	_____	_____	_____

Identity of the Instructor's Pen: _____

Are any of the pigments used in the pen inks fluorescent? If an ultraviolet light is available in the laboratory, test for this property. Are the fluorescent pigments in addition to the ones you saw in the chromatogram or are some of the pigments both colored and fluorescent? Why might a manufacturer add fluorescent pigments to its inks?

Can you distinguish between an original signature and a copy made on a color photocopier or printer? Can you distinguish an inkjet-printed copy from a conventional photocopy?

Experiment with other developing solvents, such as 5% acetic acid (household vinegar). Does an ink developed with rubbing alcohol give the same number of spots as the same ink developed with vinegar? Do they give the same R_f values? Why or why not?

Observing the Reactions of Household Chemicals

Deductive Chemical Reasoning

Purpose

• Sharpen your powers of observation.

• Practice describing reactions of ordinary household chemicals with several reagents.

• Deduce from observed reaction patterns the identity of a household chemical presented as an unknown.

Pre-Lab Preparation

One of the important goals of the first course in general chemistry is to make connections or build bridges between three worlds of human experience:

• **The world of everyday life experience.** This includes the natural world around us and increasingly the products of human technology, including a vast array of chemical products not previously found in the natural world even as recently as 60 years ago.

• **The visible world of the laboratory.** The laboratory is where we perform experiments under conditions that we control, then observe what happens. Our observations are often assisted by instruments that work as extensions of our five senses (sight, hearing, smell, touch, and taste). Instruments allow us to measure things that we ordinarily can't sense at all, such as weak electric or magnetic fields, or the flux of particles from a radioactive source.

In the general chemistry lab, touching and tasting chemicals are almost never done, because of unknown hazards. Smelling chemicals is sometimes cautiously done, with supervision of a knowledgeable instructor and when there is every reason to believe that there will be no harmful consequences.

• **The invisible microscopic world of atoms.** This is largely a world that we construct in our imaginations to provide an atomic or molecular interpretation of the visible world.

Experimentation is a cycle of planning, observation, and interpretation. The planning of this experiment already has been done. You are coming in on the observation and interpretation steps of the cycle. We call it a cycle because the planning, observation, and interpretation of one experiment often naturally lead into the next experiment.

When you see something happen in a chemical reaction, try to begin thinking about how it can be understood or interpreted at the atomic level. In simple language, ask yourself the question, "What are the atoms doing, and why are they doing it?" To answer these questions, you need to find out what chemical substances are present and to begin constructing mental ideas and pictures of the possible ways that they can interact. An important part of this process is learning to write chemical equations and chemical formulas and structures to explain your laboratory observations.

We are not born with any knowledge of chemical formulas and reactions. One step at a time, we must patiently build our knowledge about what atoms can do and why they do it. To make connections between the observations and our atomic and molecular interpretation of them, we must actively think about, or conceptualize, what the atoms and molecules are doing. The quicker and better we learn to do this, the greater the probability that we will be successful students of science and ultimately practicing scientists, if that is our goal. Conceptualizing chemical processes is the hardest thing to do at the beginning, when we have the least information stored in our heads and the least experience in interpreting things at the atomic and molecular level. The rewards are well worth the effort.

In this first introductory experiment, we will focus primarily upon learning to make careful observations by studying the reactions of various household chemicals with a series of reagents that test for different chemical properties. These household chemicals are available at most supermarkets.

Studying chemical reactions

Before we begin our experiments, let's first think about the best way to carry them out. Usually we don't try to mix together two solid substances. Chemical reaction requires intimate contact between the atoms of the reacting substances. When the reactants are solids, we can achieve this only by grinding them together. This is less convenient and often is unsafe because pure solid reactants are in a very concentrated form. A reaction that evolves a lot of heat could lead to fires or even explosions. If the reactants are in a more dilute form, dissolved in water, the water can absorb any heat evolved and also can provide a clear, transparent medium for observing any reaction that takes place. Also, reactants in solution are easily mixed together. So, usually one or both of the substances that we are going to mix together will be a liquid or will be dissolved in water to form an aqueous solution.

Observing chemical reactions

Relying mainly on sight, what things should we look for when we mix two chemical substances? The most common visible changes that accompany chemical reactions are the following:

• **The formation of an insoluble solid when two solutions are mixed.** Besides the fact that it does form, the insoluble solid, or *precipitate,* may appear to have particular characteristics. Formation may occur immediately, or it may take some time. The precipitate may be a finely divided suspension, or it may coagulate to form larger particles that rapidly settle out. It may be *flocculent* (forming cottony looking tufts called *flocs*) or *gelatinous* (looking like a thick starch suspension). So you may need to include a number of specific adjectives to give a more complete and accurate description, for example, "a brown gelatinous precipitate formed."

The word *precipitate* comes from the Latin word meaning to "throw down suddenly," describing the tendency of most precipitates to be formed as soon as the substances are mixed and for the solid material to eventually settle to the bottom. Note: Be careful not to mispronounce and misspell "precipitate" as "percipitate."

• **A visible color change.** The color change may take place in solution or may be associated with the formation of a precipitate.

• **The evolution of a gas.** Generally, you will see bubbles of the gas forming in solution or on the surface of a solid. If the evolution of the gas is sufficiently vigorous, you may see the foam as it forms and hear a fizzing sound. We must be careful about incorrect assumptions, however. If you heat a solution to speed up the reaction, do not be fooled into thinking a gaseous reaction product is being formed, when in actuality you may only be observing the boiling of water to form bubbles of water vapor (or steam).

• **Evolution or absorption of heat.** If the reaction is carried out in a test tube, the tube may get warm or cold, depending on whether the reaction evolves or absorbs energy (called *heat* in this context). Temperature changes involving liberation or absorption of small amounts of heat may be detectable by using a thermometer, an example of using a simple instrument as an extension of the senses.

• **You may detect no changes.** This result is ambiguous. It may mean that no chemical reaction is taking place, or it could mean that although a chemical reaction is taking place, it is invisible to the eye. Many chemical substances, when dissolved in water, do not absorb visible light and therefore appear colorless.

Pattern recognition

The human eye and brain are very good at recognizing subtle features and discriminating between different patterns. For instance, babies very soon learn to recognize the characteristic features of their parents' faces and to recognize the differences between them and the faces of strangers. We can make use of this inborn human skill by thinking of our collected observations in terms of patterns. By comparing chemical reaction patterns of various substances with several test reagents, we can learn to distinguish one substance from another. Or, if your instructor presents to you as an unknown one of the substances whose reaction pattern you have studied, you can identify the substance from its characteristic reactions. When applied to chemical reactions, this form of pattern recognition is called *qualitative analysis.*

What you will do

In this experiment, you will be provided with a set of household chemicals purchased at the supermarket. Each is composed predominantly of one active ingredient that is specified on the label. You will mix each of these substances with a series of reagents whose

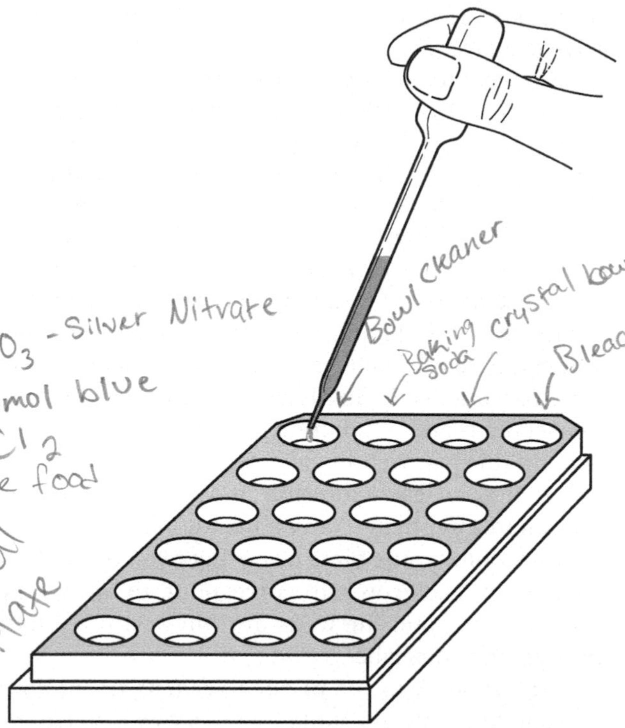

Observations + record keeping
Qualitative data
Drawing conclusions
make a commitment

disposable Pipet

AgNO₃ - Silver Nitrate
Thymol blue
BaCl 2
Blue food

Bowl cleaner
Baking Soda
crystal bowl
Bleach

well Plate

FIGURE 4-1 | **Four household chemicals and five reagents are mixed in a 24-well reaction plate.**

Household Chemicals: Clorox or other brand of liquid bleach (5% sodium hypochlorite); Lysol Toilet Bowl Cleaner (9.5% hydrochloric acid); Arm & Hammer Baking Soda (100% sodium bicarbonate); Vanish Crystal Bowl Cleaner (62% $NaHSO_4$); S & F Iris or other brand of blue food color (FD & C Blue #1, listed in the *Merck Index* as Brilliant Blue FCF).

Chemicals: 0.1 M $AgNO_3$; 1.0 M $CaCl_2$; 6 M HCl; 0.1% thymol blue indicator (1.0 g/L thymol blue, sodium salt).

NOTE TO INSTRUCTOR

It's helpful to provide students with preprinted labels for the four household chemicals and five reagent solutions. Students can tape the labels to 13 × 100 mm test tubes that contain the household chemicals and reagents. Fine-tipped polyethylene transfer pipets can be stored in the test tubes when they are not being used. Alternatively, the polyethylene transfer pipets can be labeled by taping a label to the bulb with a circle of cellophane tape. If time is short, the pipets can be pre-filled with the household chemicals and reagents by the stockroom assistant.

SAFETY PRECAUTIONS:
Several of the household chemicals are concentrated corrosive solutions, as is the 6 M HCl reagent. WEAR EYE PROTECTION AT ALL TIMES. If you spill any of the household chemicals or reagents on your skin, immediately rinse the affected area with water under the tap. Neutralize any acid or base spills with sodium bicarbonate, and sponge them up immediately.

WASTE COLLECTION AND CLEAN-UP: **Your instructor will tell you if you are going to be given two unknowns to identify. If so, you need to save the solutions and reagents that you will use in part 1.**
   The contents of the reaction plate wells can be disposed of by emptying them into a large beaker, neutralizing the solutions with about 2 g of sodium bicarbonate (baking soda), adding enough water to dissolve the sodium bicarbonate,

composition is specified. (If the household chemical is a solid or concentrated liquid, you will first make an approximately 5% solution of the substance in water.) You will carefully observe and record what happens when you add each reagent to a fresh sample of the household chemical, working on a reduced scale in a 24-well reaction plate, shown in Figure 4-1. Using the observed reaction patterns, you will be asked to identify two unknowns. Each unknown will be one of the household chemicals whose reactions you observed. All of the unknowns will be aqueous solutions, so if the unknown is a solid, it will be presented to you as a 5% solution of the unknown in water. If you are keeping a separate lab notebook, read through the experimental procedure and prepare a table to record your observations.

Experimental Procedure

Special Supplies: 24-well reaction plates; ten polyethylene transfer pipets (fine-tipped, 50 drops/mL); Handi-Wrap plastic film; preprinted labels; cellophane tape (see *Note to Instructor*); ten 13 × 100 mm test tubes.

read!

then flushing the solution down the drain with plenty of water. Then rinse out the reaction plate with tap water, using a test-tube brush or paper towel and detergent to clean out the wells in which a precipitate was formed, and follow with a final rinse with deionized water. The reaction plate can be used for part 2, identifying two unknowns, if your instructor so directs.

At the end of the experiment, the unused household chemicals can be collected together in a beaker, along with the unused 6 M HCl, thymol blue indicator, blue food color, and 1 M $CaCl_2$. Neutralize the mixture with 2–3 g of sodium bicarbonate (about one level teaspoon), and flush it down the drain. The unused 0.1 M $AgNO_3$ (silver nitrate) should be collected in a waste container labeled "silver nitrate waste."

1. Observing Reaction Patterns for Household Chemicals Obtain a 24-well reaction plate, nine clean 13 × 100 mm test tubes in a test-tube rack, a 30-mL beaker, and ten fine-tipped polyethylene transfer pipets. Now obtain samples of the household chemicals you will study: 2–3 mL of Clorox Liquid Bleach, approximately 0.5 mL of Lysol Toilet Bowl Cleaner, and 0.5 g of baking soda. Also place about 0.5 g of Vanish Crystal Bowl Cleaner in a 30-mL beaker. (We're using a beaker for the Vanish rather than a test tube because it foams a lot when water is added to it.) The Clorox will be used as received (without dilution).

Read the labels of the four household chemicals, and record in your lab notebook the following: the full commercial name of the product, the intended purpose of the product, the active ingredient(s) listed on the label and their amounts or concentrations, and any special warnings about the proper use of the product.

Next, prepare solutions of the four household chemicals you will study. Add about 8 mL of deionized water to each of the two test tubes containing the Lysol Toilet Bowl Cleaner and baking soda and to the beaker containing the Vanish Crystal Bowl Cleaner. Cover the mouths of the test tubes with a small piece of plastic film (Handi-Wrap) and, putting your thumb over the mouth of the tube, shake them until the solution is thoroughly mixed and any solids have dissolved. Stir the Vanish Crystal Bowl Cleaner

in the beaker with a glass stirring rod until the solids have dissolved.

Next, obtain samples of each of the five reagents in five clean 13 × 100 mm test tubes. About 3 mL of each solution (a test tube approximately 1/3 full) will be sufficient. The five reagents are: blue food color (FD & C Blue No. 1 dye), 0.1 M $AgNO_3$ (silver nitrate), 0.1% thymol blue indicator, 1.0 M $CaCl_2$ (calcium chloride), and 6 M HCl (hydrochloric acid).

Now take the 24-well reaction plate. Think of it as having four columns of six wells each. Each column will correspond to one of the four household chemicals. To work as efficiently as possible, we will put a household chemical solution in each of five wells in a column (the last well in each column will be empty). Put enough of each solution in a well to cover the bottom to a depth of 2–3 mm (about 0.5 mL).

Now think of the plate as having five rows of four wells each. Each row will correspond to one of the five reagents, and the wells are filled with the four household chemicals.

Add about 3 drops of each reagent to the four household chemicals contained in a row. ***Caution!*** *A noxious substance (chlorine) is produced when 6 M HCl is added to Clorox Liquid Bleach. Do the addition of 6 M HCl in the fume hood.* It may help to view the reactions against a white sheet of paper, or to lift up the plate and view the wells through transmitted light reflected from the white sheet of paper. The formation of white precipitates is often best seen by viewing against a dark background. As you add the test reagents, note what happens when you first add the reagent, then watch for 30 seconds or so to see if any further reaction takes place; finally, come back and look at each well after 5 or 10 minutes have passed and note any further changes. When you finish, you will have observed 20 possible reactions.

If you were well-prepared, you would already have read this and prepared a data table in your lab notebook before you came to lab, with columns and rows to provide an organized space to record your observations. Leave plenty of space—a full page in your lab notebook—for recording the observations.

When you have finished recording all of your observations of the 20 possible reactions, look carefully at your data table. Is there a unique pattern of observations for each household chemical that would serve to distinguish that chemical from the others? For example, how is the pattern for baking soda different from the pattern for Lysol Toilet Bowl Cleaner? Could you tell the difference between them with just one test reagent? If so, which reagent would you use, and what differences would you look for?

Save your solutions of household chemicals and reagents if you are going to do part 2, identifying two unknown household chemicals.

2. Identifying an Unknown by Its Reaction Pattern Obtain a clean 24-well reaction plate, or clean and use the same plate used in part 1. Put samples of the first unknown in 5 of the cells in one column of the plate and samples of the second unknown in the next column, and react them with the same five reagents used in part 1. Compare the reaction patterns used for the unknowns with those previously determined for the four household chemicals in part 1. Record your observations, your deductions based on the reaction patterns, and your conclusions about the identity of the unknown(s) in your lab notebook. Also include them in your final laboratory report.

C O N S I D E R   T H I S

The label on Clorox Liquid Bleach has several warnings. One says, "Do not mix with other household chemicals, such as toilet bowl cleaners, rust removers, acids, or products containing ammonia. To do so will release hazardous gases."

Likewise, the label on Lysol Toilet Bowl Cleaner contains the following warning: "*Danger: Corrosive—* produces chemical burns. Contains hydrochloric acid. . . . Do not use with chlorine bleach or any other chemical products."

What did you observe when you added 6 M HCl to Clorox Liquid Bleach? Let's begin in a small way to try to answer the question, "What are the atoms doing in the reaction of Clorox with hydrochloric acid?"

First, let's provide a little background information. The label says Clorox contains sodium hypochlorite (NaOCl). Clorox can be produced by bubbling chlorine gas into an aqueous solution of sodium hydroxide. In the process, some NaCl is also produced. A chemical equation describing the process is

$$Cl_2(g) + 2NaOH(aq) \rightarrow NaCl(aq) +$$
$$NaOCl(aq) + H_2O \text{ (Clorox bleach)}$$

The sodium hypochlorite gives Clorox its bleaching action. Recall what happened when you added the blue food color and the thymol blue indicator to Clorox. Did these colored materials get bleached?

The process can be reversed, forming noxious chlorine gas, when hydrochloric acid (or any acid) is added to Clorox:

$$NaOCl(aq) + 2HCl(aq) \rightarrow Cl_2(g) + H_2O + NaCl(aq)$$

Now we have a clearer picture, at the atomic level, of what the atoms are doing, and why it is a bad idea to mix chlorine bleach with acid-containing products to clean toilet bowls.

Now you try one: What happens when you add a few drops of 0.1 M $AgNO_3$ (silver nitrate salt) to Lysol Toilet Bowl Cleaner, in which the active ingredient is HCl (hydrochloric acid)? What did you see happen?

What combinations of positive ions (Ag^+ and H^+) with negative ions (NO_3^- and Cl^-) could lead to formation of a precipitate of an insoluble salt? *Hint:* The possible combinations we need to consider are AgCl and HNO_3. We already know that the other two possible combinations, $AgNO_3$ and HCl, are soluble in water. (How do we know that?) So, a good guess might be that the precipitate is AgCl (silver chloride) or HNO_3 (nitric acid). Try looking up the properties of AgCl and HNO_3 in your chemistry text, or in the *CRC Handbook of Chemistry and Physics.* Which one is least soluble? Now try to write a plausible set of products for the reaction:

$$Ag^+ + NO_3^- + H^+ + Cl^- \rightarrow ?$$

This is an example of a very common type of reaction in aqueous solution: the combination of a positive ion (called a *cation*) with a negative ion (called an *anion*) to form an insoluble salt (or precipitate).

Bibliography

Bosma, W. B. "Using Chemistry and Color to Analyze Household Products: A 10–12 Hour Laboratory Project at the General Chemistry Level," *J. Chem. Educ.* **1998,** *75,* 214–215.

Myers, R. L. "Identifying Bottled Water: A Problem-Solving Exercise in Chemical Identification," *J. Chem., Educ.* **1998,** *75,* 1585–1587.

Ricketts, J. A. "A Laboratory Exercise Emphasizing Deductive Chemical Reasoning," *J. Chem. Educ.* **1960,** *37,* 311–312.

A Cycle of Copper Reactions

Purpose

• Observe a sequence of reactions of copper that form a cycle, along with the color and physical property changes that indicate those reactions.

• Gain skill in recording observations and interpreting them in terms of chemical equations.

• Use a simple classification scheme for grouping chemical reactions by reaction type.

• Practice quantitative laboratory techniques by determining the percentage of the initial sample of copper that is recovered.

Pre-Lab Preparation

To a beginning student of chemistry, one of the most fascinating aspects of the laboratory is the dazzling array of sights, sounds, odors, and textures that are encountered there. Among other things, we believe that this experiment will provide an interesting aesthetic experience. You will be asked to carry out a series of reactions involving the element copper and to carefully observe and record your observations. The sequence begins and ends with copper metal, so it is called a *cycle of copper reactions*. Because no copper is added or removed between the initial and final steps, and because each reaction goes to completion, you should be able to quantitatively recover all of the copper you started with if you are careful and skillful. This diagram shows in an abbreviated form the reactions of the cycle of copper:

$$Cu \xrightarrow[(1)]{HNO_3} Cu(NO_3)_2 \xrightarrow[(2)]{NaOH} Cu(OH)_2$$
$$\uparrow \text{ } \xleftarrow[(5)]{Zn, HCl} CuSO_4 \xleftarrow[(4)]{H_2SO_4} CuO \xleftarrow[(3)]{heat}$$

Like any good chemist, you will probably be curious to know the identity of each reaction product and the stoichiometry of the chemical reactions for each step of the cycle. Here they are, numbered to correspond to the steps shown in the chemical cycle.[1]

$$8\ HNO_3(aq) + 3\ Cu(s) + O_2(g) \rightarrow$$
$$3\ Cu(NO_3)_2(aq) + 4\ H_2O(l) + 2\ NO_2(g) \quad (1)$$

$$Cu(NO_3)_2(aq) + 2\ NaOH(aq) \rightarrow$$
$$Cu(OH)_2(s) + 2\ NaNO_3(aq) \quad (2)$$

$$Cu(OH)_2(s) \rightarrow CuO(s) + H_2O(l) \quad (3)$$

$$CuO(s) + H_2SO_4(aq) \rightarrow CuSO_4(aq) + H_2O(l) \quad (4)$$

$$CuSO_4(aq) + Zn(s) \rightarrow ZnSO_4(aq) + Cu(s) \quad (5)$$

These equations summarize the results of a large number of experiments, but it's easy to lose sight of this if you just look at equations written on paper. You can easily be overwhelmed by the vast amount of information found in this lab manual and in chemistry textbooks. It is in fact a formidable task to attempt to learn or memorize isolated bits of information that are not reinforced by your personal experience. This is one reason why it is important to have a laboratory experience. Chemistry is preeminently an experimental science. As you perform this—and any other—experiment, watch closely and record what you see. Each observation should be a little hook in your mind on which you can hang a more abstract bit of information, such as the chemical formula for the compound you are observing.

It is also easier to remember information that is organized by some conceptual framework. Observations and facts that have not been assimilated into some coherent scheme of interpretation are relatively useless. It would be like memorizing the daily weather reports when you have no knowledge of or interest in meteorology.

Chemists look for relationships, trends, or patterns of regularity when they organize their observations of chemical reactions. The periodic table, which groups the elements into chemical families, is a product of this kind of thinking. Each element bears a strong resemblance to other members of the same chemical family but also has its own unique identity and chemistry.

---

[1]Labels specify the states of the reactants and products: (s) means a solid, (l) means a liquid, (g) means a gas, and (aq) means an aqueous (water) solution.

In a similar fashion, it is useful to classify reactions into different types. Because no one scheme is able to accommodate all known reactions, several different kinds of classification schemes exist. A simple classification scheme we will use at the beginning is based on ideas of precipitation (ion combination), acid–base (proton transfer), and redox (electron transfer). Here we present an outline and some examples of this kind of classification.

A simple scheme for classifying chemical reactions

1. Precipitation Reactions (the Combination of Positively Charged Ions with Negatively Charged Ions to Form an Insoluble Neutral Compound That Precipitates from Solution)

If we add a solution of sodium chloride (NaCl) to a solution containing silver nitrate ($AgNO_3$), an insoluble white solid forms. The solid, called a *precipitate*, is silver chloride (AgCl), and we may write the following chemical equation to describe the reaction, using symbols to represent the substances in solution.

$$NaCl(aq) + AgNO_3(aq) \rightarrow AgCl(s) + NaNO_3(aq)$$

When ionic compounds dissolve, they dissociate into their component ions, and we can write the equation in its ionic form:

$$Na^+(aq) + Cl^-(aq) + Ag^+(aq) + NO_3^-(aq) \rightarrow$$
$$AgCl(s) + Na^+(aq) + NO_3^-(aq)$$

Next we eliminate the ions called *spectator ions*, which appear on both sides of the equation but do not participate in the precipitation reaction. What is left is the *net ionic equation*:

$$Ag^+(aq) + Cl^-(aq) \rightarrow AgCl(s)$$

The net ionic equation concisely summarizes the net result of mixing the two solutions: the formation of an insoluble precipitate when a positively charged silver ion is combined with a negatively charged chloride ion. The solid AgCl precipitate is easily separated from the solution containing the soluble sodium nitrate salt. If we desired, we could also recover the sodium nitrate by evaporating the water from the solution.

2. Acid–Base Reactions (Proton Transfer Reactions)

An acid is a substance that reacts with water to form hydronium ions (H_3O^+) by transferring a proton to a water molecule.

(a strong acid) $HCl(g) + H_2O \rightarrow H_3O^+(aq) + Cl^-(aq)$

(a weak acid) $CH_3COOH(aq) + H_2O \leftrightarrow$
$$H_3O^+(aq) + CH_3COO^-(aq)$$

HCl is a strong acid, completely dissociated in aqueous solution; acetic acid (CH_3COOH) is a weak acid

that only partially dissociates into hydronium ion and acetate ion.

A base is a substance that forms hydroxide ions when dissolved in water:

(a strong base) $NaOH(s) \rightarrow Na^+(aq) + OH^-(aq)$

(a weak base) $NH_3(g) + H_2O \leftrightarrow NH_4^+(aq) + OH^-(aq)$

NaOH is a strong base, completely dissociated in aqueous solution; NH_3 is a weak base, only partially dissociating into ammonium ion and hydroxide ion.

Acids react with bases to form salts and (usually) water. Both are neutral compounds, being neither strongly acidic nor strongly basic. So acid–base reactions are also called *neutralization reactions*. Two examples are

(a strong acid + a strong base)

$$H_3O^+(aq) + Cl^-(aq) + Na^+(aq) + OH^-(aq) \rightarrow$$
$$2 H_2O + Na^+(aq) + Cl^-(aq)$$

(the net ionic equation)

$$H_3O^+(aq) + OH^-(aq) \rightarrow 2 H_2O$$

(a weak acid + a weak base)

$$CH_3COOH(aq) + NH_3(aq) \rightarrow NH_4^+(aq) + CH_3COO^-(aq)$$

Note that in these two examples of acid–base reactions a proton is transferred from the acid to the base.

3. Redox Reactions (Electron Transfer Reactions)

Oxidation–reduction reactions, called *redox reactions*, are reactions that involve the shift or transfer of electrons from one kind of atom to another. In some reactions, the transfer is obvious, as in the reaction

$$Mg(s) + 2 H_3O^+(aq) + 2 Cl^-(aq) \rightarrow$$
$$H_2(g) + Mg^{2+}(aq) + 2 Cl^-(aq) + 2H_2O$$

Here, each magnesium atom is giving up two electrons to two hydrogen ions, forming magnesium ion and hydrogen gas.

Sometimes the transfer of electrons between atoms is less obvious, as in the reaction

$$2 SO_2(g) + O_2(g) \rightarrow 2 SO_3(g)$$

Here the reactants and products are all gases, and no ions are formed. In classifying this reaction as an oxidation–reduction reaction, we use the concept of assigning oxidation numbers (also called oxidation states) to each atom in the compound. A simple set of rules defines the procedure for assigning the oxidation number. For a simple binary compound (a compound composed of two different elements), we imagine that all of the electrons in the chemical bonds are assigned to the atoms with the greatest affinity for electrons. The ability of an atom to attract electrons to itself is called *electronegativity*. The most electronegative elements—fluorine, oxygen, and chlorine—are found in the upper right-hand corner of the periodic table.

In sulfur dioxide, SO_2, and sulfur trioxide, SO_3, we imagine that all of the electrons in the S—O bonds are assigned to the O atoms, giving each oxygen atom a full valence shell; this formally gives each oxygen atom a net charge of -2. So if oxygen is assigned oxidation number -2, the sulfur in SO_2 must have oxidation number $+4$ and the oxidation number of S in SO_3 would be $+6$, since the sum of the oxidation numbers on all the atoms must add up to the net charge on the molecule (zero, in this case). The oxidation number of atoms in their elemental form is always assigned zero. This seems reasonable for O_2 because the oxygen atoms are equivalent so that there would be no tendency for one oxygen atom to transfer electrons to its partner in the O_2 molecule.

Once we have assigned oxidation numbers to each element in the chemical reaction, we will see that in this particular reaction the oxidation number of the sulfur atoms increases from $+4$ to $+6$ while the oxidation number of the oxygen atoms decreases from zero (in O_2) to -2 (in SO_3). From this viewpoint, the change in oxidation number is formally equivalent to transferring electrons from sulfur to oxygen. We say that the sulfur atoms have been *oxidized* (because their oxidation number increases), while the oxygen atoms in O_2 have been *reduced* (because the oxidation number of oxygen atoms decreases from zero in O_2 to -2 in SO_3).

We must be careful to note, however, that the oxidation numbers we assign do not necessarily represent the real distribution of electronic charge in the molecule. By assigning the oxidation number according to fixed rules, we have artificially assigned integer changes in oxidation number to particular atoms (sulfur and oxygen, in this case), but the changes in the electron density on the sulfur and oxygen atoms may not be as large as implied by the assigned oxidation numbers. Nevertheless, it is reasonable to suppose that the sulfur atom in SO_3 is more positive than the sulfur atom in SO_2 because the added oxygen atom would tend to draw electrons away from the sulfur atom.

4. Decomposition Reactions (a Substance Breaking Down into Simpler Substances Under the Influence of Heat)

Although many chemical reactions can be classified into one of the three reaction types we described earlier, it is possible to find examples of chemical reactions that do not neatly fit into this scheme. For example, when calcium carbonate is heated, it breaks down into simpler substances:

$$CaCO_3(s) \rightarrow CaO(s) + CO_2(g)$$

This reaction might be called a decomposition or dissociation reaction. Any compound heated to a sufficiently high temperature will begin to decompose into simpler substances, so this kind of reaction is common.

Does a Comprehensive Classification Scheme Exist? If we searched we would find examples of other reactions that do not fit into these four categories. Indeed, it is probably fair to say that there is no completely comprehensive classification scheme that would accommodate all known chemical reactions. However, many, if not most, of the chemical reactions described in your general chemistry text will fit into this simple scheme. As you carry out each step of the cycle of copper reactions, think about what is happening in each reaction, and try to fit it into one of the four categories we described.

Experimental Procedure

Special Supplies: Infrared lamps or steam baths; porcelain evaporating dish.

Chemicals: 18- to 20-gauge copper wire; concentrated (16 M) HNO_3; 3 M NaOH; 6 M H_2SO_4; 30-mesh zinc metal; 6 M HCl; methanol.

> **SAFETY PRECAUTIONS:**
> **Concentrated nitric acid, HNO_3, is hazardous. It produces severe burns on the skin, and the vapor is a lung irritant. When you handle it, you should use a fume hood while wearing safety glasses (as always) and rubber or polyvinyl chloride gloves. A polyethylene squeeze pipet can be useful for transferring the HNO_3 from a small beaker to your 10-mL graduated cylinder. Rinse your hands with tap water after handling HNO_3. The dissolution of the copper wire with concentrated HNO_3 should be carried out in a fume hood. If no hood is available, construct the apparatus shown in Figure 5-1 to substitute for the fume hood. The brown NO_2 gas that evolves is toxic and must be avoided.**

NaOH solutions are corrosive to the skin and especially dangerous if splashed into the eyes—*wear your safety glasses.*
Methanol and acetone are flammable and their vapors are toxic. Use them in the hood to avoid breathing the vapor, and *keep them away from all open flames.*

⚠️ **WASTE COLLECTION:**
The supernatant solution that is decanted in step 5 contains zinc sulfate and zinc chloride and should be collected.
A waste container should also be provided for the methanol used to dry the product in step 5.

1. *Cu to Cu(NO₃)₂* Cut a length of pure copper wire that weighs about 0.5 g (about a 10-cm length of 20-gauge copper wire). If it is not bright and shiny, clean it with steel wool, rinse it with water, and dry it with a tissue. Weigh it to the nearest milligram, recording the weight in your laboratory book. Coil the wire into a flat spiral, place it in the bottom of a 250-mL beaker, and—in the fume hood—add 4.0 mL of concentrated (16 M) nitric acid, HNO_3. (If a fume hood is not available, use the apparatus shown in Figure 5-1.)

Record in your notebook a description of what you see. Swirl the solution around in the beaker until the copper has completely dissolved. What is in the solution when the reaction is complete? After the copper has dissolved, add deionized water until the beaker is about half full. Steps 2 through 4 can be conducted at your lab bench.

2. *Cu(NO₃)₂ to Cu(OH)₂* While stirring the solution with a glass rod, add 30 mL of 3.0 M NaOH to precipitate $Cu(OH)_2$. What is formed in the solution besides $Cu(OH)_2$? Record your observations in your lab book.

3. *Cu(OH)₂ to CuO* Stirring gently with a glass rod to prevent "bumping" (a phenomenon caused by the formation of a large steam bubble in a locally overheated area), heat the solution just barely to the boiling point on a hot plate, using the apparatus shown in Figure 5-2. If the solution bumps, you may lose some CuO, so don't neglect the stirring. Record your observations. When the transformation is complete, remove the beaker from the hot plate, but do not place the hot beaker directly on the lab bench. Continue stirring for a minute or so, then allow the CuO

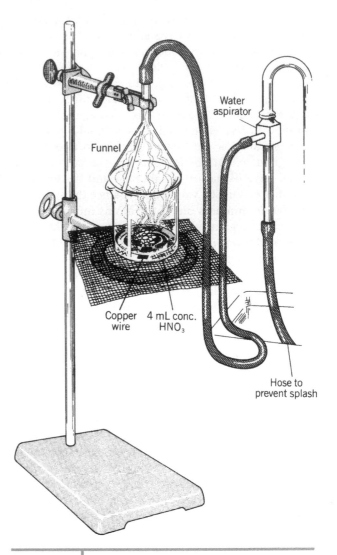

FIGURE 5-1 | **If a fume hood is not available, substitute this apparatus.**

to settle. Next, decant (pour off) the supernatant liquid, being careful not to lose any CuO. Add about 200 mL of hot deionized water, allow to settle again, and decant once more. What is removed by this washing and decantation process?

4. *CuO to CuSO₄* Add 15 mL of 6.0 M H_2SO_4, while stirring. Record your observations. What is in solution now? Now transfer operations back to the fume hood.

5. *CuSO₄ to Cu* In the fume hood, add all at once 2.0 g of 30-mesh zinc metal, stirring until the supernatant liquid is colorless. What happens? What is the gas produced? When the evolution of gas has become very slow, decant the supernatant liquid and pour it into the waste container provided. If you can see any silvery grains of unreacted zinc, add 10 mL of 6 M

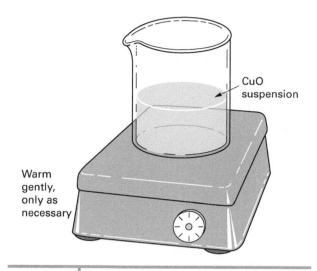

Warm
gently,
only as
necessary

CuO
suspension

FIGURE 5-2 Setup for heating Cu(OH)₂ to convert it to CuO.

HCl and warm, but do not boil, the solution. When no hydrogen evolution can be detected by eye, decant the supernatant liquid and transfer the copper to a porcelain dish. A spatula or rubber policeman is helpful for making the transfer. Wash the product with about 5 mL of deionized water, allow it to settle, and decant the wash water. Repeat the washing and decantation at least two more times. *Move to the hood, away from all flames.* Wash with about 5 mL of methanol, allow to settle, and decant. *Dispose of the methanol in the proper receptacle.* Place the porcelain dish under an infrared lamp or on a steam bath or hot plate, and dry the copper metal. What color is it? Using a spatula, transfer the dried copper metal to a preweighed 100-mL beaker and weigh to the nearest milligram. Calculate the mass of copper you recovered by subtracting the weight of the empty beaker from the weight of the beaker plus the copper metal.

Calculation of Percentage Recovery Express the percentage of copper recovered as

$$\text{percentage recovery} = \frac{\text{mass of recovered copper}}{\text{initial mass of copper wire}} \times 100\%$$

If you are careful at every step, you will recover nearly 100% of the copper you started with.

CONSIDER THIS

If you used a penny as the source of your original copper in this experiment, would it matter if you used a pre-1982 penny (essentially pure copper) or a post-1982 penny (copper cladding over a zinc core)? Describe what would happen in each step if you used a post-1982 penny. Test your predictions.

Why would it be hard to perform a cycle of oxygen or a cycle of hydrogen experiment similar to this cycle of copper exercise? Can you design an experimental apparatus for a cycle of oxygen lab?

Bibliography

Bailar, Jr., J. C. "A Further Improvement on the Copper Cycle Experiment," *J. Chem. Educ.* **1983**, *60*, 583.

Condike, G. F. "Near 100% Yields With the 'Cycle of Copper Reactions' Experiment," *J. Chem. Educ.* **1975**, *52*, 615.

Todd, D.; Hobey, W. D. "An Improvement in the Classical Copper Cycle Experiment," *J. Chem. Educ.* **1985**, *62*, 177.

Umans, T.; de Vos, W. "An Improved Copper Cycle Experiment," *J. Chem. Educ.* **1982**, *59*, 52.

Synthesis of a Chemical Compound: Making Alum from Aluminum

Purpose

• Apply the principles of synthetic chemistry, the process by which one pure substance is transformed into another.

Pre-Lab Preparation

Anyone who lives in a modern city surrounded by the necessities that make urban life possible and the luxuries that make it enjoyable should ponder this question: Where does all this stuff come from? The answer, of course, is that it comes from the elements found on earth. The same elements that make up the mountains and valleys, the seas and the air, and all living and nonliving matter can be transformed into new substances that are used in a variety of ways in our technological society.

The process of making new substances from other, usually simpler, substances is called *chemical synthesis.* The word *synthesis* implies the notion of putting things together. The opposite of synthesis is *analysis,* which implies taking things apart, that is, determining the chemical composition of substances.

In this experiment, we will transform the aluminum from a beverage can into a substance known as *alum,* more formally designated "aluminum potassium sulfate dodecahydrate," which has the chemical composition $AlK(SO_4)_2 \cdot 12\ H_2O$. Alum is a versatile substance used in dyeing fabrics, tanning leather, and clarifying water in municipal water treatment plants. Medicinally, it has applications as an astringent and a styptic; for example, if you nicked yourself while shaving, you might use a stick of alum to coagulate the blood and stop the bleeding. Derivatives of alum are also widely used in formulating antiperspirants, thus exploiting the astringent properties of Al^{3+} ions, which bind water strongly.

Since aluminum cans have some intrinsic value, empty cans are normally recycled, that is, collected, melted down, and used to make more cans or other aluminum metal products. So this experiment represents another way of recycling scrap aluminum into something potentially useful. (To be truthful, we would have to say that if you wanted to make tons of alum, it would be cheaper to start with natural ore already containing Al^{3+} ions. This is because a lot of chemical reagents and electrical energy go into making aluminum metal from natural ores. On a large scale, making alum from aluminum metal would be uneconomical.)

Now let's look at the chemistry involved in transforming an aluminum can into alum. Aluminum cans are made of fairly pure aluminum, except for the decorative paint on the outside and the thin coating of plastic on the inside, which prevents acidic chemicals in the beverage from reacting with the aluminum can. The first step in this experiment is to remove these coatings before the chemical reactions with aluminum can begin.

The next step is to react the clean aluminum with potassium hydroxide solution:

$$2\ Al(s) + 2\ K^+(aq) + 2\ OH^-(aq) + 6\ H_2O(l) \rightarrow$$
$$2\ K^+(aq) + 2\ Al(OH)_4^-(aq) + 3\ H_2(g)\quad (1)$$

This reaction can be classified as an oxidation–reduction reaction, in which aluminum is oxidized and water is reduced, producing hydrogen gas and hydroxide ion by transfer of electrons from aluminum to water. The KOH helps to dissolve the thin oxide coating on the aluminum that normally protects it from further oxidation. The hydroxide ion produced when aluminum replaces hydrogen forms a complex ion, the tetrahydroxyaluminate ion, $Al(OH)_4^-$.

Equation (1) also explains why alkaline products like liquid detergents and bleaches are never stored in aluminum containers. The aluminum would slowly dissolve, producing hydrogen gas. In a sealed container, a high pressure of hydrogen gas might build up, possibly causing the container to rupture or explode if the hydrogen ignited.

Addition of a slight excess of sulfuric acid to the products produced in Equation (1) neutralizes the excess KOH from the previous step and neutralizes the four OH^- groups on the tetrahydroxyaluminate complex ion:

$$K^+(aq) + Al(OH)_4^-(aq) + 2\,H_2SO_4(aq) \rightarrow$$
$$K^+(aq) + Al^{3+}(aq) + 2\,SO_4^{2-}(aq) + 4\,H_2O(l) \quad (2)$$

This is an acid–base (or neutralization) reaction in which the H^+ ions in H_2SO_4 react with the OH^- ions in $Al(OH)_4^-$ to form water, leaving behind $Al_2(SO_4)_3$. The reaction is strongly driven to the right by the formation of the very stable H_2O molecule.

The resulting solution contains K^+, Al^{3+}, and SO_4^{2-} ions in the proportions in which they are found in alum: $AlK(SO_4)_2$. After sufficient cooling, crystals of hydrated potassium aluminum sulfate, or alum, will form.

There are 12 molecules of water associated with each formula unit of the salt, so the formula of the crystalline alum that forms is $AlK(SO_4)_2 \cdot 12\,H_2O$. This is typically true of compounds that contain metal ions, but in a beginning chemistry course we often ignore these water molecules as we write chemical formulas—unless they are critical to the results of an experiment, as they are here. For example, what you usually see written as $FeCl_3$ is actually $FeCl_3 \cdot 6\,H_2O$ (or $[Fe(H_2O)_6]Cl_3$) to show that the waters surround the metal). X-ray crystallographic studies show that each K^+ ion and each Al^{3+} ion is surrounded by six water molecules. In solution, each is also surrounded by a cluster of approximately six water molecules. As you will study later in greater detail, water is a polar molecule, with the oxygen atom having a slightly negative charge while the hydrogen atoms are slightly positively charged. So the negative oxygen atoms in the water molecules are attracted to the positively charged K^+ and Al^{3+} ions, and there is just enough room for about six water molecules to crowd around each positive ion.

The solution containing the dissolved alum is cooled with ice because alum is quite soluble in water, but its solubility decreases as the temperature is lowered. So cooling the alum solution increases the amount of solid alum that can be recovered.

To get pure alum, it is necessary only to filter off the crystals and wash them to remove the excess H_2SO_4 used in the second step. The washing is done with methanol, which does not dissolve the alum or remove any of the water of crystallization. Methanol also has the advantage of evaporating quickly.

Experimental Procedure

Special Supplies: Aluminum beverage can; sturdy scissors or snips; coarse steel wool (or metal pot scrubber); smooth board or other material on which the aluminum metal can be scrubbed; crushed ice; glass filter funnel; Whatman No. 4 filter paper; filter flask; Büchner filter funnel; filter paper to fit Büchner funnel; aspirator or house vacuum; hot plate or steam bath; rubber policeman; container for the alum produced.

Chemicals: 1.4 M KOH; 9 M H_2SO_4; methanol; $NaHCO_3$.

SAFETY PRECAUTIONS:
WEAR EYE PROTECTION AT ALL TIMES. Potassium hydroxide, KOH, and sulfuric acid, H_2SO_4, are very corrosive. Both can cause burns if left in contact with the skin. If either chemical gets on your skin, immediately rinse with plenty of cool water for several minutes. Your instructor will discuss emergency measures to follow if your eyes are affected. If spills occur, notify the instructor, who will know how to neutralize the spill with sodium bicarbonate and can advise you on proper cleanup procedures. Wear gloves if there is any chance of contact with strong acids or bases.

Methanol vapors are toxic, and liquid methanol is toxic if swallowed. Use methanol in the hood, and avoid getting it on your skin. If it does contact your skin, rinse it off immediately with water. Because methanol is flammable, no open flames should be allowed in the laboratory.

The synthesis of alum

To begin the synthesis of alum, cut a piece of aluminum about 5 × 5 cm from an aluminum beverage can. Be careful with the sharp-edged piece of metal. Place it on a smooth board surface, and scour off the coatings on both sides with coarse steel wool or a metal pot scrubber. When the piece of aluminum is clean and bright, rinse and wipe it with a paper towel. Using scissors, cut the clean aluminum into small squares about 0.5 × 0.5 cm.

Tare,[1] or weigh, an empty 150-mL beaker to the nearest milligram, and record the mass. Add between 0.5 and 0.6 g of the small pieces of aluminum, and record the mass to the nearest milligram.

Obtain 25 mL of 1.4 M KOH solution.

CAUTION: KOH will dissolve aluminum and your skin! Make certain you are wearing your eye protection, and immediately rinse off with water any KOH that gets on your skin.

Working in the hood, carefully add the 25 mL of 1.4 M KOH to the beaker containing the small pieces of aluminum. Use a stirring rod to guide the solution stream as you pour it.

Place the beaker on a hot plate set at low heat or on a steam bath. Warm the solution, but do not allow it to boil. With occasional stirring, all the aluminum should react in about 20 min. While the aluminum is dissolving, set up a glass funnel using a ring stand and iron ring or funnel support. See Figure 6-1 for an example of the equipment setup. Place the stem of the funnel in a 150-mL beaker. Fold a piece of coarse filter paper (Whatman No. 4 or equivalent) into quarters, and place it in the funnel. Pour a little water into the filter paper so that it will stay in place. Discard any water that runs through the filter into the beaker.

When there are no visible pieces of aluminum or bubbles of hydrogen gas, you will know that all of the aluminum has reacted. Remove the beaker from the hot plate temporarily. Obtain 10 mL of 9 M H_2SO_4, and carefully but rapidly pour all of the 9 M H_2SO_4 into the beaker, stirring steadily. If the H_2SO_4 is added too slowly, you may see the transient formation of insoluble $Al(OH)_3$ that will precipitate when just enough acid has been added to react with one of the OH^- groups of the $Al(OH)_4^-$ complex ion. Because you are adding a strong acid to a solution of a strong base, there is an exothermic neutralization reaction, making the solution hotter.

Stir the solution thoroughly. If white crystals are visible, return the beaker to the hot plate or steam bath, and stir until all such material dissolves. Any tiny black specks may be disregarded.

---

[1]To "tare" a container is to weigh it on a balance while adjusting the balance to read "zero" while the container is on the balance. You can then place an object in the container, and the weight of the object can be read directly from the balance display. If you are to use this feature on your balance, your instructor will need to instruct you in the taring process. The alternative is to weigh the container, then weigh the container with the object in it, and then subtract one weight from the other to determine the weight of the object. This method is referred to as "weighing by difference." Under no circumstances should reagents be weighed by placing them directly on the balance pan.

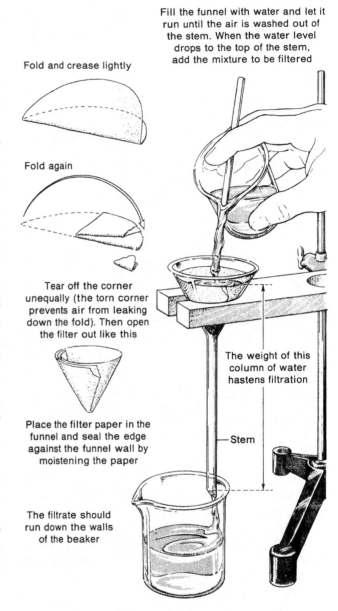

Fold and crease lightly

Fold again

Tear off the corner unequally (the torn corner prevents air from leaking down the fold). Then open the filter out like this

Place the filter paper in the funnel and seal the edge against the funnel wall by moistening the paper

The filtrate should run down the walls of the beaker

Fill the funnel with water and let it run until the air is washed out of the stem. When the water level drops to the top of the stem, add the mixture to be filtered

The weight of this column of water hastens filtration

Stem

FIGURE 6-1 | **The process of filtration.**

When the beaker can be handled safely, place your stirring rod against the lip of the beaker to guide the hot solution into the previously prepared glass filter funnel. Do not touch the filter paper with the stirring rod or you may puncture the paper. Don't overfill the filter funnel. The level should not be closer than 1 cm to the top of the filter paper. The filtrate should be clear and colorless.

Allow the 150-mL beaker containing the filtrate to cool. Fill a 600-mL beaker about half full of crushed ice. Place the 150-mL beaker in the ice bath for about 20 min, stirring frequently as crystals of alum form.

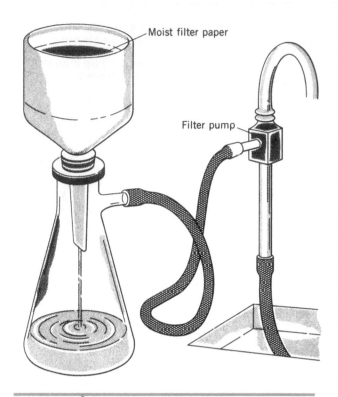

Moist filter paper

Filter pump

FIGURE 6-2 Vacuum filtration technique using a Büchner funnel. To prevent aspirator water from entering the flask, disconnect the rubber tube before turning off the aspirator.

While the solution is cooling, set up a filter flask and Büchner filter funnel attached to an aspirator or house vacuum (see Figure 6-2).

When the crystallization is complete, stir *carefully* with a thermometer, and record the temperature when the reading stabilizes. The solution temperature should be 6 °C or less.

Place a circle of filter paper (preferably Whatman No. 4, although No. 1 will do) of the appropriate size in the Büchner funnel. Moisten the filter paper to hold it in place. Turn on the aspirator or house vacuum. Remove the 150-mL beaker from the ice bath, and stir the contents of the beaker with a stirring rod to loosen all of the alum crystals. Transfer all of the suspended crystals to the Büchner funnel. A rubber policeman on the end of your stirring rod is helpful in scraping out as many crystals as practical.

Add two 10-mL portions of methanol in succession to wash the excess H_2SO_4 out of the crystals. Continue to draw air through the alum for another 10 to 15 min to dry the crystals.

Weigh a clean, dry 150-mL beaker to the nearest milligram. Record the mass. Shut off the source of vacuum, and remove the Büchner funnel. Carefully transfer the dried alum crystals and filter paper circle to the weighed 150-mL beaker. Use a stirring rod or metal spatula to scrape off any crystals adhering to the filter paper and remove the filter paper. Reweigh the beaker containing the alum crystals, and record this mass.

Follow the instructor's directions for disposing of the sample of alum that you prepared.

Weigh 10 g of sodium bicarbonate, $NaHCO_3$. Add about 80% of the $NaHCO_3$ to the solution left in the filter flask. Swirl the flask until the bubbling ceases and the $NaHCO_3$ has dissolved. Now add the rest of the $NaHCO_3$, and note whether any additional bubbles of CO_2 are produced. If additional bubbles form, add 2 g more of $NaHCO_3$. When there is no more bubbling upon further addition of a small amount of $NaHCO_3$, rinse the solution in the filter flask down the drain.

CONSIDER THIS

What does the *CRC Handbook of Chemistry and Physics*, the *Merck Index*, or another reference tell you about what it takes to remove the 12 water molecules from alum? Why do we often "forget" to include the "•12 H_2O" when we write the formula of alum or similar hydrated salts? Would it be better to write the formula as $Al(H_2O)_6K(H_2O)_6(SO_4)_2$? Comment on other ways the formula could be written, such as $AlH_{24}KO_{20}S_2$ or $AlK(SO_4)_2(H_2O)_{12}$.

When we combine salt (NaCl) and water, we don't describe the process as a synthesis or the product, "salt water," as a compound. What is the difference between the synthesis of a compound such as alum and the production of a mixture or solution such as salt water? As you write your description, suppose you are an early chemist and cannot use the concepts of an element or a formula in your explanation.

Clean, dry, and weigh an aluminum can. Find the price of aluminum in a newspaper. (It's usually found on the stocks page in the business section.) What is the value of the aluminum in a beverage can? Then determine how much a local recycling center is paying for aluminum cans. How does the value of aluminum compare to its "worth" in your community? Why are the two prices so different?

Synthesis of a Chemical Compound: Making Alum from Aluminum

Name _____

Date _____ Section _____

Locker _____ Instructor_____

Data

1. Weighing the aluminum

(a) Mass of empty 150-mL beaker	g
(b) Mass of 150-mL beaker + aluminum	g
(c) Temperature of cooled alum crystals	°C

2. Weighing the alum

(a) Mass of empty 150-mL beaker	g
(b) Mass of 150-mL beaker + alum	g

Calculations

Mass of aluminum metal used	g
Mass of alum obtained	g

Calculate the number of moles of aluminum that you used. Show your calculations:

_____ moles Al

Calculate (or look up in a reference such as the *CRC Handbook of Chemistry and Physics*) the formula mass (molar mass) of alum, $KAl(SO_4)_2 \cdot 12\ H_2O$, including its water of crystallization. (You might find it under the name potassium aluminum sulfate.)

_____ g/mole

Noting that one aluminum atom is in each formula unit of alum, we know that each mole of aluminum will give rise to one mole of alum. Based on the number of moles of aluminum you used, calculate the maximum number of grams of alum that you could produce, assuming that aluminum is the limiting reagent:

_____ g alum

6-5

Using the number of grams of alum that you actually obtained, calculate the percentage yield of alum for your synthesis:

$$\% \text{ yield} = \frac{\text{g alum from experiment}}{\text{maximum g alum from theory}} \times 100\% = \underline{\hspace{2cm}} \%$$

Show your calculation.

Questions

1. Experiments show that the solubility of alum in 25 mL of 1.4 M KOH plus 10 mL of 9 M H_2SO_4 is about 1.0 g at 1.0 °C and 1.7 g at 6.0 °C. Using your measured solution temperature, estimate the amount of alum left in your chilled solution.

Alum left in solution: \underline{\hspace{3cm}} g

2. Does the amount of alum left in solution account for most of the difference between the maximum grams of alum and your experimental grams of alum? Explain.

CONSIDER THIS

What does the *CRC Handbook of Chemistry and Physics,* the *Merck Index,* or another reference tell you about what it takes to remove the 12 water molecules from alum? Why do we often "forget" to include the "•12H$_2$O" when we write the formula of alum or similar hydrated salts? Would it be better to write the formula as $Al(H_2O)_6K(H_2O)_6(SO_4)_2$? Comment on other ways the formula could be written, such as $AlH_{24}KO_{20}S_2$ or $AlK(SO_4)_2(H_2O)_{12}$.

When we combine salt (NaCl) and water, we don't describe the process as a synthesis or the product, "salt water," as a compound. What is the difference between the synthesis of a compound such as alum and the production of a mixture or solution such as salt water? As you write your description, suppose you are an early chemist and cannot use the concepts of an element or a formula in your explanation.

Clean, dry, and weigh an aluminum can. Find the price of aluminum in a newspaper. (It's usually found on the stocks page in the business section.) What is the value of the aluminum in a beverage can? Then determine how much a local recycling center is paying for aluminum cans. How does the value of aluminum compare to its "worth" in your community? Why are the two prices so different?

The Chemistry of Oxygen: Basic and Acidic Oxides and the Periodic Table

Purpose

• Learn a method for the laboratory preparation of dioxygen, O_2.

• Observe the reactions of O_2 with several metallic and nonmetallic elements to form oxides.

• Determine whether aqueous solutions of these oxides are acidic or basic.

• Discover the relationship between acidic or basic properties of the oxides and the position of the element in the periodic table.

Pre-Lab Preparation

Oxygen and life

Oxygen is one of the most abundant elements on earth. Large amounts of it are found in the molten mantle, in the crust that forms the great land masses, in the water of the vast oceans that cover most of the earth's surface, and in the gaseous atmosphere that surrounds the earth. Oxygen is found in elemental form, as molecular O_2, mainly in the atmosphere, with smaller amounts of dissolved oxygen found in the oceans, lakes, rivers, and streams, where it is vital to all water-dwelling creatures. Elsewhere, oxygen is found in combination with other elements.

Dioxygen, O_2, reacts so avidly with both metals and nonmetals that the presence of a large amount of free O_2 in the atmosphere raises the question, where did all of that O_2 come from? Studies over the past 200 years have provided the general outline of the answer, although some details are still being actively investigated. Most of the dioxygen on earth has been produced by plants, from the smallest single-celled algae to the majestic redwoods. Plants use water, carbon dioxide (CO_2), and sunlight to form carbohydrates and oxygen in a complex process called *photosynthesis.* Animals reverse the process. They react carbohydrates with

oxygen inside their cells in a process called *respiration,* forming CO_2 and water. So plants and animals exist together in a grand symbiotic cycle, each supplying the other's needs. The energy involved in this biological cycle is about 30 times the amount of energy expended each year by all of the machines on earth.

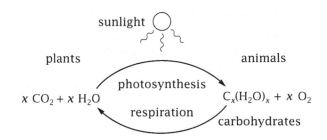

The chemical properties of oxygen

To observe and understand some of the chemistry of oxygen, we will first prepare some pure oxygen by catalytic decomposition of hydrogen peroxide, then react the O_2 with a variety of elements. The oxides formed will be dissolved in water, and the solutions tested to see if they are acidic or basic.

Ask yourself these questions while doing the experiment:

1. Is there a difference in the vigor of reaction (rate, light and heat evolved, etc.) of the different elements with oxygen? If so, do those differences relate to the positions of these elements in the periodic table?

2. Is there a relation between the acidity or basicity of the water solution of these oxides and the positions of the elements in the periodic table?

The elements whose oxides (or hydroxides) you will study are circled in the abbreviated periodic table shown in Table 7-1. The heavy zigzag line approximately divides the metals from the nonmetals, so you will study samples from each group.

TABLE 7-1 **Main Groups of the Periodic Table**[a]

1	2			13	14	15	16	17	18
I	II			III	IV	V	VI	VII	VIII
H									He
		Metals ←→ Nonmetals							
Li	Be		(Transition metals)	B	C	N	O	F	Ne
Na	Mg			Al	Si	P	S	Cl	Ar
K	Ca	Sc ... Fe ... Zn		Ga	Ge	As	Se	Br	Kr
Rb	Sr	Y ... Ru ... Cd		In	Sn	Sb	Te	I	Xe

[a]The heavy zigzag line divides the metals from the nonmetals. The oxides (or hydroxides) of the elements that are encircled will be studied in the experiment.

Acids and bases from oxides: a conceptual framework

Let's begin with a microscopic view of what the atoms are doing and why they are doing it. Take a look at Figure 7-1. It's a picture of how we might imagine a solid element E reacting with gaseous dioxygen molecules, O_2.

For reaction to occur, the O_2 molecules have to collide with the atoms of E in the surface of a piece of the solid element. What happens next depends on the nature of the atoms of E and of oxygen. We're going to construct two scenarios that lead to two different outcomes, depending on a property of atoms called *electronegativity*. This is defined qualitatively as *the ability of atoms to attract electrons to themselves.*

Elements on the left side of the periodic table (the metals) have less ability to attract electrons to themselves and therefore exhibit smaller electronegativity than elements on the right-hand side of the periodic table. Why is this? The increase in electronegativity, moving from left to right in a period (or row) of the periodic table, is largely the result of an increasing nuclear charge. The greater the nuclear charge, the more attraction there will be to draw in valence electrons.

The polarity of the bonding between an element and oxygen correlates with the difference in electronegativity between the element and oxygen. If the difference is large, the E—O bond will have a more ionic character. If the electronegativity difference is relatively small, the E—O bond will be less polar, so there will be more electron sharing, which we call *covalent bonding.* Thus, reaction products involve two kinds of bonding, ionic and covalent, as shown in Figure 7-1.

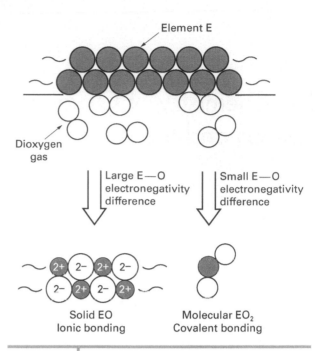

FIGURE 7-1 | **A piece of solid element E reacts on its surface with gaseous dioxygen, O_2. If the electronegativity of O is much larger than that of E, the reaction product will be an ionic solid. If the electronegativity difference is smaller, E and O_2 react to form covalently bonded molecules—$EO_2(g)$ in this example. E atoms shrink when they lose electrons to form positive ions, whereas O atoms expand when they add electrons to form negative ions.**

In the first scenario, suppose we have an element oxide, EO, in which there is mostly covalent character in the E—O bond. Such compounds tend to be molecular solids rather than ionic solids. Oxide ions, as such, don't exist in the covalent oxide (EO); when the covalently bonded oxides are dissolved in water, they too form a covalently bonded hydrated form of the oxide, $E(OH)_2$:

$$EO + H_2O \rightarrow E(OH)_2$$

This reaction could be viewed as a near simultaneous transfer of a proton from a water molecule to the oxygen atom and the formation of a second E—OH bond to give $E(OH)_2$.

A covalent bond implies more electron density concentrated in the bond between the E and O atoms, drawing electron density from the O—H bond and tending to make the H atom more positive. This makes it easier for the $E(OH)_2$ molecule to give up (or donate) a proton to a second water molecule:

$$E(OH)_2 + H_2O \rightarrow H_3O^+ + E(O)OH^-$$

and by further dissociation

$$E(O)OH^- + H_2O \rightarrow H_3O^+ + EO_2^{2-}$$

Therefore, because it forms hydronium ions (H_3O^+), $E(OH)_2$ is an *acid*.[1] Acids are called *proton donors* in the Brønsted–Lowry definition of acids and bases.

In the second scenario, let's imagine that the element E transfers all of its valence electrons to oxygen atoms, forming an ionic oxide, EO, which we schematically represent by formulating the oxide as $E^{2+}O^{2-}$. What might we expect to happen if we put this oxide in water? A plausible possibility is described by the reaction

$$EO + H_2O \rightarrow E^{2+} + 2\ OH^-$$

This reaction can be viewed as a water molecule donating a proton (H^+) to the oxide ion, forming two OH^- ions:

$$O^{2-} + \overset{+}{H}\text{---}\overset{-}{O}H \rightarrow 2\ OH^-$$

where we have deliberately exaggerated the polarity of the H—O bond in water.

The net result is that the compound EO reacts with water to form hydroxide ions. Compounds that form hydroxide ions in water are called *bases*. Bases are called *proton acceptors* in the Brønsted–Lowry definition of acids and bases.

To summarize, the striking difference in the outcome of the two scenarios, one producing an acid and the other a base, is determined by the nature of the element E and the polarity of the E—O bond. That polarity is in turn related to the electronegativity difference between elements E and O. The polarity of the E—O bond, ranging over a continuum from ionic to covalent, correlates well with the position of E in the periodic table.

Acid and base anhydrides

If the oxide of an element forms an acid in water, it is termed an *acidic oxide* or *acid anhydride.* If the oxide in water forms a base, we speak of a *basic oxide* or *base anhydride.* To determine the formula of the anhydride of an oxyacid or base, simply subtract water to eliminate all hydrogen atoms. For example,

$$2\ NaOH(s)\ \text{minus}\ H_2O \rightarrow Na_2O(s)$$

$$Mg(OH)_2(s)\ \text{minus}\ H_2O \rightarrow MgO(s)$$

$$2\ B(OH)_3(s)\ \text{minus}\ 3\ H_2O \rightarrow B_2O_3(s)$$

There are some borderline elements whose oxides are not very soluble in water but that can display both weakly acidic and basic properties by dissolving in both strong bases and acids. Examples of these *amphoteric oxides* include SnO_2, ZnO, Al_2O_3, and PbO.

Predicting chemical formulas for oxides of the elements

When an element combines with oxygen, there is a tendency for electrons to be transferred from atoms of the element E to atoms of oxygen. A molecule of dioxygen has the capacity to accept a maximum of four electrons to form two oxide ions:

$$O_2(g) + 4\ e^- \rightarrow 2\ O^{2-}$$

An oxide ion, O^{2-}, corresponds to an oxygen atom with a filled valence shell. Assuming that the element E gives up all its valence electrons to oxygen, the formula of the resulting compound will depend on the number of electrons in the valence shell of the element E.

When a metal atom forms an ion by transferring its valence electrons to a nonmetal such as oxygen, the charge of the metal cation can be determined by its position in the periodic table for main-group elements. It is simply equal to the group number: +1 for Group I, +2 for Group II, and so on. The charge corresponds to the loss of the valence electrons from the metal atom.

A second important principle is that anything stable enough to be collected and put in a bottle must be electrically neutral, which means that the formula for the compound must show zero net charge. (This is sometimes called the *electroneutrality principle.*) In practice, this means that we must determine the least common multiple of the cation and anion charges to find the formula of the compound.

An Example. Cesium oxide contains cesium ions (cesium cations). Cesium is in Group I, so it forms Cs^+ cations; the oxide ion is O^{2-}, and the requirement that every stable substance be electrically neutral requires that two Cs^+ ions combine with one O^{2-} ion to form neutral cesium oxide, Cs_2O. Aluminum is from Group III, so the neutral product between Al^{3+} and O^{2-} ions would be Al_2O_3.

In contrast, the bonding in nonmetal oxides is best described by the *covalent bonding model,* which involves the sharing of electron pairs rather than the complete transfer of electrons implied by the ionic bonding model. The covalent model is compatible with the ability to form more than one oxide of an element. For example, carbon can form the oxides CO and CO_2; sulfur can form the oxides SO_2 and SO_3. This obviously creates some difficulties when trying to predict the formulas of the oxides of the elements. How do we know which oxide will form? In truth, for a particular set of reaction conditions, we have to do the experiment to find the answer to this question.

---

[1] A number of important acids are not derived from oxides, such as the binary hydrides of the nonmetals: HF, HCl, HBr, and H_2S.

Some of the nonmetal oxides appear to follow the rules that we used to determine the formulas of the metal oxides. From their formulas, the molecules CO_2, SO_3, and N_2O_5 all look as if they could be ionic (but they are not). They correspond to the oxides we would predict if C, S, and N each gave up all of their valence electrons to oxygen.

However, the molecules CO, SO_2, NO, and NO_2 also exist. If they were ionic compounds (they are not), we would be tempted to say that these compounds resulted from incomplete transfer of electrons from the element to oxygen (CO resulting from a C^{2+} ion combining with an O^{2-} ion, SO_2 resulting from S^{4+} combining with two O^{2-} ions, etc.). It's as if there were formally only a partial transfer of the valence electrons from the element to oxygen. In a sense, this is what we mean by covalent sharing of electrons. Covalent bonding implies a tendency for electron density to partially shift from C, S, and N atoms to O atoms, without complete transfer of electrons to form ions. This allows more than one way for C, S, and N to share their valence electrons with oxygen atoms.

For purposes of this exercise, you will be given the formulas of the nonmetal oxides that are already prepared: B_2O_3, CO_2, P_2O_5, SO_2, and Cl_2O_7. In this group, only SO_2 does not follow the pattern of the other members of the group. If we pretend that the oxygen atoms in SO_2 have full valence shells, formally equivalent to oxide ions, O^{2-}, then by the electroneutrality principle we would have to assign a charge of +4 to the sulfur atom. Later on, we will see that this is exactly what we do when assigning oxidation numbers to each element in a binary compound (a binary compound is one composed of two different elements).

If a sulfur atom gave up all of its six valence electrons to oxygen atoms, the predicted formula for the oxide would be SO_3. This is a known oxide of sulfur, but it is a molecular and not an ionic compound. *A good general rule is that the nonmetal oxides are better described by a covalent bonding model than by an ionic bonding model.*

Hydration of the oxides—summary concepts

We've now established a conceptual framework for understanding what happens when oxides come in contact with water. The outcome depends on the polarity of the E—O bond. If it is strongly ionic, the oxide ions react with water to form a base. If the bonding is largely covalent, the oxide will form a molecular hydrate with a stronger E—O bond, leaving the H atom more positive, so that the hydrate acts as an acid by donating a proton to a water molecule. Now we illustrate these concepts with examples of the reaction of representative elements.

Ionic Oxides. For an ionic metal oxide, a plausible prediction is that one water molecule is added for each oxide ion present in the original oxide compound. Thus calcium oxide, CaO, would react with a water molecule to form calcium hydroxide:

$$CaO + H_2O \rightarrow Ca(OH)_2$$

Lithium oxide, Li_2O, reacts to give lithium hydroxide, LiOH; a plausible equation describing the reaction would be

$$Li_2O + H_2O \rightarrow 2\,LiOH$$

We do not write $Li_2O_2H_2$ because we always prefer the simplest empirical formula for ionic compounds.

For Fe_2O_3 the reaction would be

$$Fe_2O_3 + 3\,H_2O \rightarrow 2\,Fe(OH)_3$$

Covalent Oxides. Because the bonding in the nonmetal oxides is much less polar than the bonding in ionic compounds, the behavior of the hydrated oxides is different. Not all of the nonmetal oxides react with one water molecule per oxygen atom, as the metal oxides do; the actual structure of the final compound must be determined by experiment. Therefore, for this part of the exercise you will be given the formulas of the hydrated forms of the nonmetal oxides.

For example, when the acidic oxide N_2O_5 is added to water, we might expect the following reaction to occur:

$$N_2O_5 + 5\,H_2O \rightarrow 2\,N(OH)_5$$

But we find that molecules with a lot of —OH groups bonded to the same atom are often unstable and tend to dehydrate. So the partially hydrated oxyacid $HONO_2$ proves to be the most stable configuration, which you may recognize as nitric acid when written as the more familiar formula HNO_3.

Although acidic oxides seldom form fully hydrated oxyacids, it is still easy to determine the formula of the anhydride by subtracting water so as to leave no hydrogen atoms. For example, the formula of the anhydride of perchloric acid is obtained by subtracting one mole of water from two moles of perchloric acid.

$$2\,HClO_4 \text{ minus } H_2O \rightarrow Cl_2O_7$$

A warning about the structures of the oxyacids

Don't let the conventional way of writing the formulas for nitric acid and perchloric acid as HNO_3 and $HClO_4$ mislead you into thinking that the hydrogen atoms in the acids are directly bonded to the nitrogen or chlorine atoms. The structural formula for nitric acid might more accurately be written as $HONO_2$ or

and that for perchloric acid as $HOClO_3$ or

It is a good general rule for oxyacids that the hydrogen atoms that are acidic (or dissociable as protons) are bonded to oxygen atoms.

Conventions for writing the formulas of acids and bases

If a hydroxide compound is basic in water, it is because it releases hydroxide ions into the water. For example,

$$LiOH \rightarrow Li^+ + OH^-$$

and

$$Ca(OH)_2 \rightarrow Ca^{2+} + 2\ OH^-$$

If, instead, the hydrated oxide were acidic, it would react with water by donating a proton to a water molecule to form hydronium ions, H_3O^+:

$$LiOH + H_2O \rightarrow H_3O^+ + LiO^-$$

or

$$Ca(OH)_2 + H_2O \rightarrow H_3O^+ + CaO_2H^-$$

Which set of reactions is most plausible? Do LiOH and $Ca(OH)_2$ behave as bases or as acids? You will discover the answer by making the oxides of lithium and calcium and testing solutions of them in water.

If the compounds are acids, there is a convention that we write the acidic hydrogen atoms first in the formula. So if LiOH were an acid, it would be written HLiO, and $Ca(OH)_2$ would be written as H_2CaO_2. Here are some examples:

Base formula	Conventional formula for an acid
KOH	HKO
$Sr(OH)_2$	H_2SrO_2
$B(OH)_3$	H_3BO_3
$SO(OH)_2$	H_2SO_3
$ClO_2(OH)$	$HClO_3$

Reminder: This convention of writing the formulas for acids can be misleading. The hydrogen atoms written first are not bonded to the element that immediately follows. As we have already pointed out, you need to remember that, *in oxyacids, the hydrogen atoms dissociable as protons are bonded to the oxygen atoms.*

Experimental Procedure

Special Supplies: Deflagrating spoon; porcelain crucible and clay triangle; crucible tongs; six glass squares; plastic basin; six wide-mouthed 250-mL (8-oz) bottles; 250-mL Erlenmeyer flask with 1-hole rubber stopper; pH meter and/or universal indicator; cylinder of oxygen (if students do not generate O_2 by decomposition of hydrogen peroxide).

Chemicals: 3% (0.9 M) H_2O_2; 1 M $FeCl_3$; $MnO_2(s)$ (powder); active dry yeast (Fleischmann's or other brands used in baking are satisfactory); Ca metal (turnings); Li metal (shot, oil coated); Mg metal (ribbon); phosphorus (red); sulfur (powder); carbon (small pieces of charcoal); universal indicator;[2] $H_3BO_3(s)$ (labeled "H_3BO_3 or $B(OH)_3$"); 0.1 M $HClO_4$ (labeled "0.1 M $HClO_4$ or $ClO_3(OH)$").

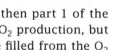

NOTES TO INSTRUCTOR

If $O_2(g)$ is available in cylinders, then part 1 of the experiment can be done to study O_2 production, but the bottles of O_2 for part 2 can be filled from the O_2 cylinder.

⚠ SAFETY PRECAUTIONS:
Although we expect you to ALWAYS WEAR EYE PROTECTION in the laboratory, it is especially important while doing this experiment.
Magnesium ribbon burns so brightly that the light can damage your eyes. *Do not* stare directly at the burning ribbon. Instead, glance out of the corner of your eye to note

[2]Universal indicator can be purchased from a laboratory supply house (Fisher). To prepare 200 mL of Yamada's universal indicator, dissolve 0.005 g thymol blue, 0.012 g methyl red, 0.060 g bromthymol blue, and 0.100 g phenolphthalein in 100 mL of ethanol. Neutralize the solution to green with 0.01 M NaOH and dilute to 200 mL with distilled water. From L. S. Foster and I. J. Gruntfest, *J. Chem. Educ.* **1937,** *14,* 274.

the flame.

Do not use white (or yellow) phosphorus in part 2(e) of this experiment. It spontaneously ignites in air, is highly irritating to the skin, and is acutely toxic. The red form of phosphorus used in this experiment is not spontaneously flammable but can be ignited by friction. Do not handle it with your bare hands because it is irritating to the skin.

The combustion of red phosphorus produces a choking, irritating smoke of P_2O_5. (The actual molecular species produced is P_4O_{10}.) Use a fume hood for the combustion. If a fume hood is not available, the laboratory instructor will perform the experiment as a demonstration.

The combustion of sulfur in part 2(f) produces SO_2, a choking, irritating gas. Use a fume hood for the combustion. If a fume hood is not available, the laboratory instructor will

⚠ **WASTE COLLECTION:**
The solutions containing $FeCl_3$ and MnO_2 in part 1 should be collected in a bottle marked "$FeCl_3$ and MnO_2 waste." All solutions produced in part 2 can be collected in a beaker, neutralized with sodium bicarbonate, $NaHCO_3$, and then flushed down the drain.

1. Preparation of Oxygen To observe the effect of *catalysts* (what is a catalyst?) on the rate of decomposition of hydrogen peroxide,

$$H_2O_2(aq) \rightarrow H_2O(l) + \tfrac{1}{2} O_2(g)$$

place four 13 × 100 mm test tubes in a beaker or rack and put 3 mL of 3% H_2O_2 in each test tube. To the first one, add 0.5 mL (10 drops) of 1 M $FeCl_3$. To the second, add a pinch of solid MnO_2 powder. To the third, add 10–20 grains of active dry yeast. Add nothing to the fourth tube; it will be your control. Shake the tubes to mix the contents, and observe the rate of evolution of oxygen. Are there any visible changes in the colors of the solutions? Does the MnO_2 dissolve or appear to react in any way? Record your observations.

Now assemble the apparatus shown in Figure 7-2.[3] Place about 0.2 g of active dry yeast in the 250-mL Erlenmeyer flask. Fill six 250-mL wide-mouthed bottles with deionized water, and invert these in the basin of water as needed. Add 200 mL of 3% H_2O_2 to the flask and replace the one-hole stopper. After about 15 s, when air has been expelled from the generator, fill the six bottles with oxygen by displacement of water. As soon as each bottle is full, leaving a few mL of water in the neck, cover it with a glass square and place it right side up on the bench top.

2. Preparation of Oxides Prepare oxides of the following elements (except calcium) by burning them in the bottles containing oxygen gas, making sure to keep the bottles covered as much as possible (see Figure 7-3). Number or label each bottle to avoid confusion. Just before each combustion, slide the glass square far enough aside to quickly add 30 to 50 mL of deionized water. Then replace the glass square, perform the combustion, shake the bottle to dissolve the oxide formed, and set it aside for later use.

(a) Lithium. Take three or four pieces of the lithium metal shot and place them in the deflagrating spoon. Heat the lithium in the spoon over a Bunsen burner. You may note an initial wisp of smoke as the mineral oil coating evaporates. When the lithium starts to burn, as indicated by a violet glow, stick the spoon into the bottle until the reaction stops. Leave the spoon and its contents in the bottle to cool, and then use water from your wash bottle to rinse the contents of the spoon into the solution at the bottom of the bottle.

(b) Magnesium. Grip a 10-cm length of magnesium ribbon with crucible tongs, ignite it using a match or Bunsen burner, and at once thrust it into a bottle of oxygen. (*Do not look directly at the brilliant light,* because it can injure your eyes.)

(c) Calcium. Calcium metal is difficult to ignite but burns brilliantly. Place a shaving of calcium metal in a crucible, heat it, cautiously at first and then very strongly, in the air over a Bunsen burner for 15 min. After the crucible has cooled, wash out the product with water into a beaker.

(d) Carbon. Ignite a small piece of charcoal in a Bunsen burner, holding it with tongs or in a clean

---

[3]Alternatively, the bottles of oxygen can be filled from an oxygen cylinder if one is available.

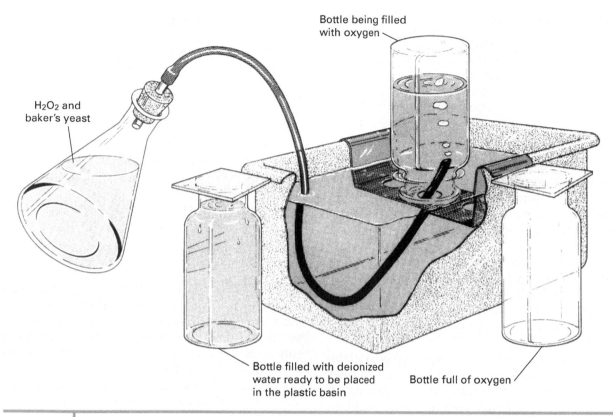

Bottle being filled with oxygen

H₂O₂ and baker's yeast

Bottle filled with deionized water ready to be placed in the plastic basin

Bottle full of oxygen

FIGURE 7-2 | **Collection of O₂ prepared by catalytic decomposition of hydrogen peroxide.**

deflagrating spoon, and thrust the charcoal into a bottle of oxygen.

(e) Phosphorus. Caution: *Carry out the ignition in a fume hood!* If a fume hood is not available, your instructor will demonstrate the combustion. Put a bit of red phosphorus (the yellow form is too toxic) no larger in volume than half a pea in a clean deflagrating spoon. Ignite it over the burner, then thrust it into a bottle of oxygen. After the combustion dies down, reheat the deflagrating spoon (in the fume hood) to burn out all the remaining phosphorus. *Avoid breathing the residual combustion products.*

(f) Sulfur. Caution: *Carry out the ignition in a fume hood!* Prepare the oxide of sulfur by the same procedure that you used for phosphorus and take the same precautions. Your instructor will demonstrate this combustion if a fume hood is not available.

3. Acids and Bases from Oxides For each bottle in which the oxide of an element has been formed by the foregoing procedures, and to which water has been added to form a solution, either use your pH meter to determine the acidity of the solution or add

FIGURE 7-3 | **The use of a deflagrating spoon to burn a substance in oxygen gas.**

a few drops of universal indicator solution to the bottle. The results of adding the universal indicator are interpreted as follows:

Indicator color	pH	pH value
Red	Acidic	≤ 4
Orange	Acidic	5
Yellow	Slightly acidic	6
Green	Neutral	7
Blue	Slightly basic	8
Indigo	Basic	9
Violet	Basic	≥ 10

Is there a relationship between the chemical character of the reaction product and the position of the element in the periodic table? Is there any difference between the behavior of the metal and the nonmetal oxides?

The preceding tests include representative elements from Groups I, II, IV, V, and VI in the periodic table. To complete the series, let us examine an oxide or hydroxide from each of the other principal groups, III and VII. (Why not include an element from Group VIII?)

The *oxide of boron*, B_2O_3, is not readily available and in some forms is very insoluble. Dissolve a small amount (0.1 g) of "boron hydroxide," labeled "$B(OH)_3$ or H_3BO_3," in 15 mL of hot deionized water. Cool the solution and test with universal indicator or a pH meter. Is this substance an acid or a base? How is the substance usually named? How is its formula usually written, $B(OH)_3$ or H_3BO_3?

The *oxides of chlorine*—$Cl_2O(g)$, $ClO_2(g)$, and $Cl_2O_7(g)$—are quite unstable. They react with water. A solution of the hydrated form of Cl_2O_7, labeled 0.1 M "$ClO_3(OH)$ or $HClO_4$," will be available to you. Test some of this solution with universal indicator or a pH meter, and in your report write the formula as an acid, $HClO_4$, or base, $ClO_3(OH)$, according to your observations of its properties.

CONSIDER THIS

Let's try out the concepts we have been learning in this experiment on a real molecule, sulfur trioxide (SO_3). First, locate sulfur in the periodic table. Is it among the metals or the nonmetals? Would you expect sulfur trioxide to form an acidic or a basic solution when it reacts with water?

Based on your answer to the previous question, what would be the preferred conventional way of writing the formula of a solution of sulfur trioxide in water: $SO_2(OH)_2$ or H_2SO_4?

If you reacted SO_3 with one molecule of water, how would you write the formula for the resulting hydroxide?

$$SO_3 + H_2O \rightarrow ?$$

If we think of the reaction of a Group I metal with oxygen as involving transfer of electrons from the metal atom to an oxygen molecule, what would happen if the addition of electrons to the oxygen molecule occurred in a step-by-step fashion? Each of the ions named below is known to exist. What would be the formulas for the resulting compounds if these ions were combined with sodium ions (Na^+)?

$$O_2(g) + e^- \rightarrow O_2^- \text{ (superoxide ion)}$$
$$O_2(g) + 2e^- \rightarrow O_2^{2-} \text{ (peroxide ion)}$$
$$O_2(g) + 4e^- \rightarrow 2\,O^{2-} \text{ (two oxide ions)}$$

Bibliography

Emsley, J. *The Elements,* 3rd edition, Oxford University Press, 1998. A most useful handbook summarizing properties of all known elements.

Mingos, D. M. P. *Essential Trends in Inorganic Chemistry,* Oxford University Press, 1998.

The Chemistry of Oxygen: Basic and Acidic Oxides and the Periodic Table

Name _____

Date _____ Section _____

Locker _____ Instructor_____

Observations and Data

1. Preparation of oxygen

(a) Write the equation for the reaction by which you prepared oxygen.

(b) Hydrogen peroxide slowly decomposes, even when no $FeCl_3$ is present. Compare the decomposition of H_2O_2 in the presence and absence of $FeCl_3$, $MnO_2(s)$, or yeast.

(i) Does the amount of oxygen that may be obtained when reaction is complete depend on whether $FeCl_3$ is added?

(ii) What is the purpose of adding $FeCl_3$, $MnO_2(s)$, or yeast to the H_2O_2 solution?

(iii) What is the name applied to a substance used as $FeCl_3$ in this reaction?

2. Preparation of the oxides

Describe the changes that occur during the reaction with oxygen of each of the following elements, and note any distinctive characteristics of the products formed. Write a chemical equation for each reaction. (For carbon, phosphorus, and sulfur, the formula of the oxide cannot be determined from charge considerations. The formulas of these oxides are CO_2, P_2O_5, and SO_2.)

(a) Lithium _____

Equation _____

(b) Magnesium _____

Equation _____

(c) Calcium _____

Equation _____

(d) Carbon _____

Equation _____

(e) Phosphorus _____

Equation _____

(f) Sulfur _____

Equation _____

3. Acids and bases from oxides

(a) Prediction of the formulas of the hydrated oxides

Write balanced chemical equations to describe the formation of the hydrated oxides (hydroxides) from the respective oxide, for example, oxide + water → hydroxide compound. For elements that are classified as nonmetals whose oxide and hydroxide reaction product formulas are more difficult to predict, two formulas are given, the first written as a base formula, and the second written as the conventional acid formula. You must balance these equations.

(a) Lithium	_____	_____	_____	_____
(b) Magnesium	_____	_____	_____	_____
(c) Calcium	_____	_____	_____	_____
(d) Carbon	$CO_2(g)$ +	$H_2O \rightarrow$	$CO(OH)_2$	or H_2CO_3
(e) Phosphorus	$P_2O_5(s)$ +	$H_2O \rightarrow$	$PO(OH)_3$	or H_3PO_4
(f) Sulfur	$SO_2(g)$ +	$H_2O \rightarrow$	$SO(OH)_2$	or H_2SO_3
(g) Boron	$B_2O_3(s)$ +	$H_2O \rightarrow$	$B(OH)_3$	or H_3BO_3
(h) Chlorine	$Cl_2O_7(l)$ +	$H_2O \rightarrow$	$ClO_3(OH)$	or $HClO_4$

(b) Summary of experimental results

(i) Write in the first column the periodic table group to which the element belongs.

(ii) In the next two columns write the formulas of the oxides and hydrated oxides [see part 3(a) above for the formulas]. If the hydrated oxide is acidic, write its formula in such a way as to indicate this fact, using the conventional acid formula.

(iii) From your pH measurements or the indicator color, describe the acidic or basic properties of the solutions formed from each oxide compound (or hydroxide, if the oxide was not available and you did not make it yourself).

(iv) Write a chemical equation showing how the hydrated oxide forms H_3O^+ or OH^- in the solution.

Element symbol	Periodic table group	Formula of oxide	Formula of hydrated oxide	Acidic or basic properties	Equation
Li					
Mg					
Ca					
C					
P					
S					
B					
Cl					

Name _____ Date _____

Application of Concepts

1. Comment on the acidic or basic character of the oxide of phosphorus, P_2O_5 (which actually exists as molecular P_4O_{10}), as predicted by its position in the periodic table and its classification as a metal or nonmetal.

2. Considering the positions of the elements in the periodic table, write the formulas of three other acidic oxides and three other basic oxides. On the longer line write the equation for the reaction of the oxide with water.

Acidic oxides: _____   _____

_____   _____

_____   _____

Basic oxides: _____   _____

_____   _____

_____   _____

3. What does the term *anhydride* mean?

4. Deduce and write the formulas of the anhydrides of the following.

H_2SO_3: _____   HClO: _____

H_2SO_4: _____ $HClO_3$: _____

$La(OH)_3$: _____ HNO_2: _____

$Sn(OH)_4$: _____ HNO_3: _____

RbOH: _____ H_2CO_3: _____

$Ba(OH)_2$: _____ H_4SiO_4: _____

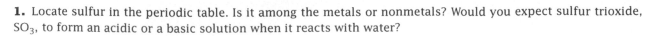

1. Locate sulfur in the periodic table. Is it among the metals or nonmetals? Would you expect sulfur trioxide, SO_3, to form an acidic or a basic solution when it reacts with water?

Based on your answer to the previous question, what would be the preferred conventional way of writing the formula of a solution of sulfur trioxide in water: $SO_2(OH)_2$ or H_2SO_4?

$$SO_3 + H_2O \rightarrow \underline{\hspace{3cm}}?$$

2. If we think of the reaction of a Group I metal with oxygen as involving transfer of electrons from the metal atom to an oxygen molecule, what would happen if the addition of electrons to the oxygen molecule occurred in a step-by-step fashion? Each of the ions named below is known to exist in a compound with sodium ion. What would be the formulas for the resulting compounds if these ions were combined with sodium ions (Na^+)?

Formula of sodium salt

$O_2(g) + e^- \rightarrow O_2^-$ (superoxide ion) _____

$O_2(g) + 2\,e^- \rightarrow O_2^{2-}$ (peroxide ion) _____

$O_2(g) + 4\,e^- \rightarrow 2\,O^{2-}$ (two oxide ions) _____

Determination of a Chemical Formula

The Reaction of Iodine with Zinc

Purpose

• Determine the iodine/zinc mass ratio for zinc iodide, which is prepared by reacting zinc with iodine in solution.

• Use the mass ratio and the iodine/zinc atomic mass ratio to determine the simplest chemical formula for zinc iodide.

• Write a balanced chemical equation for the reaction of zinc with iodine.

• Show that mass is conserved in a chemical reaction.

Pre-Lab Preparation

When you react two elements to form a binary compound, it is necessary to keep track of the masses of the reactants and products in order to write a balanced chemical equation. This atom bookkeeping is called chemical *stoichiometry*, a word derived from the Greek words *stoikheion* (element) and *metron* (to measure).

In this experiment, you will make zinc iodide by heating granular zinc metal with iodine that is dissolved in methanol. We want to know two things about this compound: (1) the mass of iodine that reacts with a given mass of zinc, that is, the *mass ratio*, g I/g Zn, and (2) the *atom ratio*, atom I/atom Zn, which defines the simplest chemical formula for the compound. The *mole ratio*, mol I/mol Zn, is exactly the same as the atom ratio, atom I/atom Zn, because a mole is just an Avogadro's number of the particles defined by the simplest chemical formula. Taking an Avogadro's number of particles doesn't change the ratio of iodine atoms to zinc atoms. We will represent the simplest chemical formula as Zn_aI_b or ZnI_x, where

$$x = b/a = \text{atom I/atom Zn} = \text{mol I/mol Zn}.$$

There are at least two approaches we can use to determine the stoichiometry of the reaction. We can react zinc and iodine, then analyze a sample of pure zinc iodide to determine the masses of zinc and iodine that are present. In the second approach (the one used in this experiment), we can synthesize zinc iodide by reacting a known mass of iodine with an excess of zinc. Because zinc is a granular solid and the reaction product is soluble in the solvent (methanol), the reaction product can be easily separated from the unreacted zinc. If we know the mass of iodine reacted and the mass of zinc before and after the reaction, we can determine the amount of each that reacted. By evaporating the volatile solvent, we can recover the zinc iodide as a solid and weigh it, too. This gives us a way to see whether the mass of the zinc iodide formed equals the mass of zinc and iodine that reacted.

From our data we will be able to determine the mass of zinc that reacts with a known mass of iodine and to calculate the mass ratio, g I/g Zn, for the reaction. We will also be able to answer this question: Is mass conserved in a chemical reaction?

The second ratio we want to determine is the atom ratio, atom I/atom Zn. As we pointed out, this ratio is equal to the mole ratio, mol I/mol Zn. Determining the atom or mole ratio is more difficult than determining the mass ratio because it requires knowledge about the relative masses of Zn and I atoms.

Pretend for the moment that you don't know the relative mass of a zinc atom and an iodine atom. (Put away your periodic table or table of atomic masses.) This, of course, was exactly the situation that existed during the time of John Dalton (1766–1844) and his contemporaries: In the early 1800s no one knew the relative masses of atoms of different elements.

Dalton found it useful to think in terms of a model, and we will do the same. Imagine that atoms are small indivisible spheres that can link together in various combinations to form compounds with a definite atom ratio that is equal to the ratio of small integers. To be specific, let's assume that zinc and iodine form a compound with the formula Zn_2I, so that $x = b/a = \frac{1}{2} = 0.50$. The formula $ZnI_{0.50}$ would be an equally correct way of writing the chemical formula, but this format doesn't lend itself to drawing a simple picture as Zn_2I does. We will picture the simplest particle of Zn_2I in the following way, using circles to represent the atoms:

Atoms are too small to see and count, so we can't directly determine the atom ratio by counting, but we can perform the experiment we described to determine the mass ratio, g I/g Zn. How are the two ratios related? To see the connection, let's write an equation, making use of the dimensions of each quantity to verify that the units come out correctly:

$$\text{mol ratio} \times \text{atomic mass ratio} = \text{mass ratio}$$

$$\frac{\text{mol I}}{\text{mol Zn}} \times \frac{\text{g I/mol I}}{\text{g Zn/mol Zn}} = \frac{\text{g I}}{\text{g Zn}} \qquad (1)$$

If we multiply the mole (or atom) ratio by the atomic mass ratio (*mass of a mole of I*)/(*mass of a mole of Zn*), we get the mass ratio for the compound, g I/g Zn. The mass ratio can be directly determined by a chemical experiment.

Dalton and his contemporaries could measure the mass ratio, g I/g Zn, as you can, but they didn't know either the mole ratio (1 mol I/2 mol Zn in our hypothetical example) or the atomic mass ratio [(g I/mol I)/ (g Zn/mol Zn)]. But if two of the three ratios shown in Equation (1) were unknown, how could they be determined? To be perfectly truthful, there is no simple way to do this, and in the early days scientists sometimes made mistakes. For example, Dalton knew of only one compound formed between hydrogen and oxygen, namely, water. When he thought that only one compound existed, he usually made the simplest possible assumption: One atom of one element reacted with one atom of the other. So he assumed that water had the formula HO. He knew that the mass ratio, g O/g H, in water was about 8, so in his first table of atomic weights he assigned 1 to the lightest element, hydrogen, and 8 to oxygen. It was only after many compounds had been synthesized and analyzed that inconsistencies began to show up; these inconsistencies could be resolved only by assigning the formula H_2O to water. As a result, the proper mass ratio for the mass of an oxygen atom relative to the mass of a hydrogen atom was found to be 16.

Now look at your periodic table or table of atomic masses, and find the values for the relative masses of zinc and iodine. Calculate for Zn_2I the ratio g I/g Zn from Equation (1). You should get the value 0.97 g

I/g Zn. Now do a second calculation, in reverse: Suppose there is an iodide of zinc that has the composition 1.94 g I/g Zn. By rearranging Equation (1), you can calculate the mole ratio, $x = b/a$, for this compound. You should get a value of 1.0, so we could write the formula for this compound as ZnI. What is the true formula for zinc iodide? Let's find out by doing the experiment.

Experimental Procedure

Special Supplies: Variable-temperature electric hot plate; aluminum foil; glazed weighing paper; desiccator (optional).

Chemicals: 20-mesh granular zinc metal; $I_2(s)$; methanol.

SAFETY PRECAUTIONS:
Handle solid iodine with care. The solid and its concentrated solutions can cause skin burns. Wash thoroughly with soap and water if iodine is spilled on your skin. Iodine is also corrosive to stainless steel balance pans, so clean up spills quickly. The action of iodine on skin and clothing is similar to chlorine bleach, so wash it off with water if it spills. Methanol is flammable and toxic. Avoid breathing methanol vapor or spilling the liquid on your hands. There must be no open flames in the laboratory. Working in a fume hood, conduct the reaction of zinc with iodine dissolved in methanol and the subsequent evaporation of the methanol to recover the zinc iodide.

1. The Reaction of Iodine with Zinc Weigh a clean, dry 125-mL Erlenmeyer flask as precisely as possible on an analytical balance (at least milligram sensitivity). Record the mass of the empty flask. Preweigh 2.0 ± 0.1 g of 20-mesh granular zinc metal on a glazed weighing paper using a lab balance (0.1-g sensitivity). Add the zinc metal grains to the Erlenmeyer flask and reweigh the flask, using the more accurate analytical balance. Record the mass of the flask plus added zinc. Using a lab balance, preweigh 2.0 ± 0.1 g of iodine crystals on a glazed weighing paper, taking care to avoid spilling any iodine crystals. If any iodine does spill, clean it up immediately as directed by your instructor. Iodine is quite corrosive to balance pans if left in prolonged contact. Add the iodine crystals to the Erlenmeyer flask containing the zinc metal and reweigh as accurately as possible on an analytical balance. Record the mass of the flask plus added zinc and iodine.

Working in the fume hood, obtain 25 mL of methanol and add it to the Erlenmeyer flask. (**Caution:** Methanol vapors are flammable and toxic.) Put an aluminum foil cap on the flask, crimping it around the edges to hold it in place. Place the flask on a hot plate in the fume hood. Adjust the temperature setting of the hot plate to a low level, sufficient to heat the methanol to boiling. Swirl the flask contents occasionally to dissolve the iodine. After 15 to 20 min, the reaction between the zinc and iodine will be complete, as shown by the disappearance of the dark brown color of the dissolved iodine. When the reaction is complete, the methanol solution should be pale yellow or nearly colorless, and the unreacted zinc granules should be easily visible.

Accurately weigh a clean, dry 250-mL beaker on the analytical balance while waiting for the zinc–iodine reaction to go to completion. Record the mass of the empty beaker. (The weighed beaker will be used in part 2 of the experiment.)

When the reaction is complete, remove the Erlenmeyer flask from the hot plate. (A folded strip of paper towel wrapped around the neck of the flask will protect your fingers from the heat.) Remove the aluminum foil cap. Working in a hood, and using a glass stirring rod to guide the solution as shown in Figure 8-1, pour the warm liquid contents of the flask into the previously weighed 250-mL beaker. Be careful not to let any of the solid zinc particles escape from the Erlenmeyer flask.

Add 5 mL of methanol to the Erlenmeyer flask and swirl for 15 s. Allow the solid zinc to settle; then pour the methanol into the beaker. Repeat with two additional methanol rinsings of 5 mL each. The three rinses ensure that all of the zinc iodide reaction product gets transferred to the beaker.

Place the Erlenmeyer flask containing the unreacted zinc back on the hot plate to dry. Leave the foil cap off. Adjust the hot plate to a medium setting.

Also, put the beaker containing the zinc iodide in methanol on the hot plate, following the procedure described in part 2.

Swirl the Erlenmeyer flask containing the zinc granules occasionally to redistribute them and to speed up the drying of the zinc. When all the methanol has evaporated from the flask and the zinc particles look dry and move freely on the bottom of the flask, remove the flask from the hot plate. Allow it to cool to room temperature (10 to 15 min). Does the dry product look like zinc granules, or does it contain a white residue of zinc iodide? If white residue is present, a few more rinses are in order. Be

FIGURE 8-1 Use a glass stirring rod to guide the methanol solution into the beaker so that the solution does not run down the side of the Erlenmeyer flask.

sure to transfer these rinses to the collection beaker as well, then redry the zinc granules. Reweigh the flask containing the unreacted zinc on the analytical balance. Record the mass.

Calculate the exact masses of iodine and zinc that reacted. From these masses, calculate the ratio g I/g Zn in the zinc iodide product. Using the atomic masses of zinc and iodine, calculate the ratio mol I/mol Zn and the simplest formula for zinc iodide.

2. Isolation of the Zinc–Iodine Reaction Product

Carry out the following procedure to isolate and weigh the zinc iodide produced in part 1 when zinc and iodine reacted. Place on the hot plate the 250-mL beaker (whose mass you have previously recorded) containing the methanol solution of zinc iodide and washings. Adjust the hot plate to medium heat, and allow the methanol to evaporate. (**Caution:** This must be done in the fume hood.) If it shows a tendency to boil irregularly or spatter, stir it occasionally with a glass stirring rod until the methanol has nearly all evaporated. It normally takes about 25 min to evaporate all of the methanol. When the last traces of methanol have completely disappeared, allow the beaker to heat 2 min more; then remove it from the hot plate. Allow the beaker to cool to room temperature for 15 min. Reweigh the beaker containing the dry zinc iodide (but not the stirring rod!) on the analytical balance. Record the mass.

Subtract the mass of the empty beaker to get the mass of the zinc iodide reaction product. Compare this mass with the sum of the masses of zinc and iodine that reacted as determined from your data for part 1. Do your results confirm the conservation of mass principle? Record the appearance of the zinc iodide product. Is the appearance distinctly different from that of the zinc and iodine reactants?

NOTE Zinc iodide is hygroscopic (that is, it readily absorbs moisture from the air), which means its mass may increase by 1 to 2 mg/min, depending on the relative humidity. If a desiccator is available, put the beaker in the desiccator while it is cooling. Otherwise, cover the beaker with a watch glass and then weigh the sample after it has cooled for 15 min.

CONSIDER THIS

Develop a procedure to decompose the zinc iodide product back into zinc metal and solid iodine. (A 9-V battery can serve as the power supply to electrolyze a solution of zinc iodide.) Can you recapture the zinc and iodine? Are the masses consistent with the laws of constant composition and conservation of mass? (See the DeMeo article listed in the Bibliography for hints and a possible procedure.)

Other binary compounds that can be prepared from the elements include the sulfides of copper, iron, nickel, and lead, as well as the oxides of magnesium, tin, and copper. Antimony iodide can also be prepared by a procedure similar to that used in this experiment. Prepare some of these other compounds using procedures from the sources discussed in the Bibliography or other lab books, and compare your results to those obtained for zinc iodide.

Bibliography

DeMeo, S. "Synthesis and Decomposition of Zinc Iodide: Model Reactions for Investigating Chemical Change in the Introductory Laboratory," *J. Chem. Educ.* **1995,** *72,* 836–839.

Walker, N. "Tested Demonstrations, Synthesis and Decomposition of ZnI_2," *J. Chem. Educ.* **1980,** *57,* 738.

Wells, N.; Boschmann, E. "A Different Experiment on Chemical Composition," *J. Chem. Educ.* **1977,** *54,* 586.

Determination of a Chemical Formula	Name Jessica Mueller
The Reaction of Iodine with Zinc	Date 9/22/15 Section Lab 3
	Locker _____ Instructor Drolet

Reaction

Zn(s) + I₂(s) $\xrightarrow{\text{methanol } \Delta}$ Product (dissolved in methanol) + Zn(s)] in erlenmeyer

Separation - Product (in methanol) + Zn(s) $\xrightarrow{\text{decant}}$ Product + methanol into beaker + Zn in erlenmeyer

wash Zn

Data and Calculations

Isolation washings + Product in beaker $\xrightarrow{\Delta}$ Product

↳ white - brown

1. The reaction of iodine with zinc

Before Reaction

(a) Mass of empty Erlenmeyer flask	93.89	g
(b) Mass of flask + zinc	95.89	g
(c) Mass of flask + zinc + iodine	97.91	g

After Reaction

(d) Mass of flask + unreacted zinc	95.37	g

(e) Describe the appearance of the solution after the reaction is complete.

Calculations

Calculate the mass of iodine that was put into the flask. Assume that all of the iodine reacted.

(f) Mass of iodine that reacted	2.02	g
(g) Moles of iodine that reacted	0.00796	mol

Subtract the mass of the flask + unreacted zinc (after reaction) from the mass of the flask + zinc (before reaction) to get the mass of zinc that reacted.

(h) Mass of zinc that reacted	0.52	g
(i) Moles of zinc that reacted	0.00795	mol

(j) Calculate the **mass** ratio, g I reacted/g Zn reacted [see parts (f) and (h) above].

(k) Calculate the **mole** ratio, mol I reacted/mol Zn reacted [see parts (g) and (i) above].

(l) Write the simplest formula for zinc iodide.

(m) Write a balanced chemical reaction for the reaction of zinc with iodine, assuming that zinc and iodine each exist in the solid as atoms, i.e., iodine is not I_2 but I.

_____ Zn + _____ I → Zn ____ I ____

2. Isolation of the zinc–iodine reaction product

Data

(a) Mass of empty beaker	g
(b) Mass of beaker + dry reaction product	g

(c) Describe the appearance of the zinc–iodine reaction product.

Mass of the Reaction Product

(d) Mass of zinc iodide reaction product	g

Mass of Reactants

(e) Mass of iodine reacted [from part 1(f)]	g
(f) Mass of zinc reacted [from part 1(h)]	g
(g) Total mass of reactants (zinc + iodine)	g

Calculate the percentage difference between the mass of the product and the mass of the reactants.

$$\text{percentage difference} = \frac{(\text{mass of product}) - (\text{mass of reactants})}{(\text{mass of reactants})} \times 100\% = \underline{\hspace{2cm}} \%$$

Name Jessica Mueller   Date _____

Questions

1. Iodine is known to exist in the solid state as a diatomic molecule, I_2. Using the simplest formula you found for zinc iodide, write a balanced chemical equation for the reaction of Zn with I_2.

2. What observations support the assumption that all of the iodine reacted with the zinc? Explain.

3. Comment on the statement, "The properties of a chemical compound are the averages of the elements that make up the compound." What evidence do you have from this experiment about the truth (or falsity) of this statement?

4. Do your observations support the principle that mass is conserved in a chemical reaction, that is, that the mass of the products equals the mass of the reactants? Explain.

5. If at the end of the reaction some of the unreacted zinc escaped from the Erlenmeyer flask, the mass of the flask plus the unreacted zinc would be lower than it should be. Would this cause the calculated mass ratio, g I/g Zn, to be higher or lower than the value predicted?

CONSIDER THIS

Develop a procedure to decompose the zinc iodide product back into zinc metal and solid iodine. (A 9-V battery can serve as the power supply to electrolyze a solution of zinc iodide.) Can you recapture the zinc and iodine? Are the masses consistent with the laws of constant composition and conservation of mass? (See the DeMeo article listed in the Bibliography for hints and a possible procedure.)

Other binary compounds that can be prepared from the elements include the sulfides of copper, iron, nickel, and lead, as well as the oxides of magnesium, tin, and copper. Antimony iodide can also be prepared by a procedure similar to that used in this experiment. Prepare some of these other compounds using procedures from the sources discussed in the Bibliography or other lab books, and compare your results to those obtained for zinc iodide.

Determination of a Chemical Formula by Titration

The Reaction of Calcium with Water

Purpose

• Learn what happens when active Group I or Group II metals react with water.

• Use an acid–base titration to determine the chemical formula of the product formed when calcium reacts with water.

Pre-Lab Preparation

The reaction of an active metal with an acid produces hydrogen gas. For example, zinc reacts with hydrochloric acid to produce hydrogen gas and zinc chloride:

$$Zn(s) + 2\, HCl(aq) \rightarrow H_2(g) + ZnCl_2(aq) \qquad (1)$$

Hydrogen is a lighter-than-air flammable gas. At one time it was used in rigid, hydrogen-filled luxury airships that made regular Atlantic crossings. One of them, the German dirigible *Hindenburg,* caught fire and burned as it was landing in New Jersey in 1937. World War II ended the era of the rigid airship.

In the chemical reaction shown in Equation (1), the element hydrogen has been replaced by zinc, in the sense that hydrogen chloride has been converted to zinc chloride.

What if we put zinc in water instead of in an HCl solution? We would see that over the course of a few minutes or even an hour, very little reaction would take place. But if we put a more active metal, like sodium, in water, we would observe a violent reaction that produces hydrogen gas and a strong base, sodium hydroxide:

$$2\, Na(s) + 2\, H_2O(l) \rightarrow H_2(g) + 2\, NaOH(aq) \qquad (2)$$

By boiling the solution to evaporate the water, we could recover the sodium hydroxide as a white, translucent solid that is used in the manufacture of paper and soaps and in petroleum refining. It is extremely corrosive to the skin and other tissues and is sometimes called *caustic soda* or *lye.*

The heavier Group II metals, such as calcium, strontium, and barium, also react readily with water at room temperature to form hydrogen and the metal hydroxide. (Magnesium does not react rapidly with water at room temperature but will react with high-temperature steam.)

The reaction of calcium with water is classified as an oxidation–reduction reaction (sometimes also called a *single replacement reaction*) in which calcium replaces the hydrogen in water by transferring its two valence electrons to water molecules:

Calcium + water →
        hydrogen gas + calcium hydroxide   (3)

We did not write the chemical formulas for this reaction because the point of this experiment is to determine the chemical formula for the calcium hydroxide produced when calcium reacts with water. We don't want to give away the end of the story, *but your first clue has already been revealed in Equation (2).*

Like sodium hydroxide, calcium hydroxide is corrosive to the skin. When calcium hydroxide is heated to drive off water, calcium oxide (often called *lime*) is produced. Lime is an important ingredient in the mortar used in plastering walls.

Let's look at the reactions of active metals with water in greater detail. The reactions are classified as oxidation–reduction reactions because they involve changes in the charges on the metal atom and on the oxygen (and hydrogen) atoms in water.

In the reaction of sodium with water, a sodium atom loses an electron to a water molecule. The sodium becomes a positively charged sodium ion with an electron configuration like that of neon, a noble gas.

Having received the electron from the sodium atom, the water molecule becomes unstable. It splits into a neutral hydrogen atom and a hydroxy group with a negative charge, called a *hydroxide ion.* That means the product of the reaction is ionic, and solid

NaOH is best characterized as containing sodium ions, Na^+, and hydroxide ions, OH^-.

The neutral hydrogen atoms, each with one electron, wander around until they find each other. (This random wandering around is called *diffusion*.) When two hydrogen atoms bump into each other, they combine to form molecular hydrogen gas, H_2. (The "wander and combine" process occurs in a tiny fraction of a second.) The two hydrogen atoms each contribute an electron to form a shared electron pair bond, called a *covalent* bond. The net result is that when an active Group I or Group II metal reacts with water, an ionic metal hydroxide is formed along with molecular hydrogen.

Here is your second clue: If a calcium atom tends to give up electrons until it has the same number of electrons as the noble gas nearest to it in the periodic table, what would be the charge on the calcium ion that is left after the electrons have been removed?

Calcium ion charge: _____

Now for the third and last clue: If every substance that is stable enough to be collected and stored in a bottle must be electrically neutral, what would you predict for the formula of calcium hydroxide, keeping in mind that each hydroxide ion has a −1 charge?

Expected formula: _____

Now let's put your predictions to the test. If we react a known amount of metal with water, we will obtain the metal hydroxide, a base. If we then react the base with a known amount of acid, we can determine how many moles of the base were present. Knowing the moles of base produced and the mass of the metal and its atomic mass, we can easily calculate how many moles of base were produced for every mole of metal. This allows us to write the correct formula for the metal hydroxide produced in the reaction.

To determine the amount of hydroxide ion produced from a given amount of metal, we will use a procedure called an *acid–base titration*. The base is contained in a reaction flask. To the base, we add a known concentration of acid dispensed from a buret. Knowing the volume and concentration of the acid, we can calculate the moles of acid added, which will be equal to the moles of hydroxide because they react one-to-one, as illustrated by the reaction of sodium hydroxide with hydrochloric acid:

$$NaOH(aq) + HCl(aq) \rightarrow NaCl(aq) + H_2O(l) \quad (4)$$

This is an acid–base reaction, driven to completion by the formation of water, a very stable compound.

The reactants and products shown in Equation (4) are all colorless, so we cannot see anything happening as the reaction proceeds. We need something else to indicate when we have added exactly enough hydrochloric acid to react with the hydroxide. This is called the *end point* of the titration. To detect the end point, we use a substance called an *acid–base indicator*. It has the property of changing color when the solution goes from slightly basic to slightly acidic (or vice versa).

Experimental Procedure

Special Supplies: Glazed weighing paper; forceps for handling calcium turnings[1]; sturdy side-cutting pliers to cut calcium turnings into smaller pieces.

Microscale: 1-mL buret
Macroscale: 25–50-mL buret

Chemicals: Mossy zinc (see footnote 1); fresh calcium turnings; 6 M HCl; 0.02 M HCl; 1% phenolphthalein indicator; 0.1% thymol blue indicator; $NaHCO_3(s)$.

Microscale: Standardized 6.00 M HCl for titration
Macroscale: Standardized 0.300 M HCl for titration

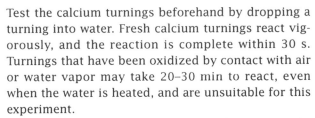

NOTES TO INSTRUCTOR

Test the calcium turnings beforehand by dropping a turning into water. Fresh calcium turnings react vigorously, and the reaction is complete within 30 s. Turnings that have been oxidized by contact with air or water vapor may take 20–30 min to react, even when the water is heated, and are unsuitable for this experiment.

If you prefer to use conventional 25- or 50-mL burets, use standardized 0.300 M HCl titrant in place of the 1-mL buret and standardized 6.00 M HCl titrant.

 SAFETY PRECAUTIONS:
HCl solutions are corrosive and irritating to the eyes and skin.
WEAR EYE PROTECTION, and wash your skin immediately with cool water if contact with HCl solution should occur. Spills should be neutralized with bicarbonate solution and cleaned up according to your instructor's directions (wear protective gloves).

Calcium metal is corrosive and irritating to the skin. Use forceps or tongs to handle the calcium turnings.

---

[1]The term "mossy zinc" is descriptive of the pieces of zinc that are formed by dropping molten zinc metal into water. Calcium "turnings" are the curly chips that form when the metal is turned on a lathe.

1. The Reaction of Zinc and Calcium with HCl and Water Place 5 mL of 6 M HCl in a 15 × 125 mm test tube. Fold a piece of glazed weighing paper to make a crease in it. Weigh out one or two pieces of mossy zinc (approximately 0.5 g) on the creased weighing paper; then slide the zinc into the test tube containing the 6 M HCl. Observe what happens and record your observations.

While the reaction is proceeding briskly, invert over the reaction tube a second 15 × 125 mm test tube, holding the two tubes so they are mouth-to-mouth with no gap between them. After about 30 s, ask a neighbor to light a match for you. Quickly move the mouth of the inverted test tube close to the flame of the match, as shown in Figure 9-1. Be prepared to hear a "woof" or sharp "bark" if the hydrogen ignites. Describe what happens. If you neither see nor hear anything, try the procedure again.

Next, put 5 mL of 6 M HCl into another 15 × 125 mm test tube. Using forceps to handle the calcium, put one or two calcium metal turnings (approximately 0.1–0.2 g) on a creased sheet of glazed weighing paper. Add the calcium turnings to the 6 M HCl in the test tube. Observe what happens and record your observations. Repeat the procedure for collecting the hydrogen in an inverted test tube and igniting it. Again, record your observations.

Next, on sheets of creased weighing paper weigh out fresh samples of zinc and calcium of the same size as before. Record the weight, then add each metal by itself to a separate 15 × 125 mm test tube half filled with deionized water. Observe what happens and record your observations. Did both the zinc and calcium react with water? Do your observations confirm the notion that active metals react more rapidly with acid solutions than with water? What is the

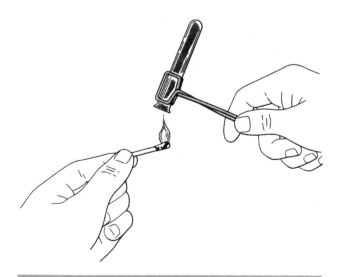

FIGURE 9-1 **Testing for hydrogen gas.**

insoluble white compound that forms when calcium reacts with water?

Add two drops of phenolphthalein indicator to the water in the test tube containing the calcium metal. Describe the color of the indicator. Phenolphthalein is colorless in acidic solutions and red or pink in basic solutions. Does the color indicate that the solution containing calcium and water is acidic or basic?

Transfer the solution containing the calcium–water reaction products and any unreacted calcium to a 250-mL beaker. Add 6 M HCl one drop at a time while stirring until all of the calcium has dissolved and the solution has just become colorless. At this point, the solution has been neutralized and may now be discarded into the sink drain.

2. Titration of the Calcium–Water Reaction Product (Read the instructions for the use of a micro- or macroburet found in the Introduction of this manual.)

Fold a crease in a piece of glazed weighing paper and weigh it on a balance to the nearest milligram. Record the mass of the paper. (If you have an electronic balance with a taring feature, it is convenient to tare the paper to zero before putting the calcium on the paper. Then the balance reading gives the mass of the calcium directly.) Using forceps, get a piece of calcium metal weighing between 0.08 and 0.10 g and put it on the paper, again weighing to the nearest milligram. Record the mass of the paper + calcium metal. If your instructor has directed you to carry out two titrations, weigh a second sample of calcium metal, recording the masses as before.

In the procedure that follows, we will assume that you are making a single determination. If you are making two determinations, it will save time if you weigh out both your samples and carry out the reactions at the same time, as described in the following paragraph.

Add the previously weighed calcium metal sample to a clean 250-mL Erlenmeyer flask. Mark the flask with your initials. Add about 150 mL of deionized water to the flask. Swirling the flask gently every few minutes may speed the reaction. If the calcium turnings are fresh, the calcium should react vigorously and completely with the water. If they are not fresh, the reaction may be slow, and it may be necessary to heat the flask containing the calcium and water until the calcium has completely reacted.

Microscale procedure: Obtain about 5 mL of standardized 6.00 M HCl in a clean, dry 13 × 100 mm test tube. Fill a 1-mL buret with the solution, following the directions given to you by the instructor.

Read the initial buret reading, estimating to the nearest 0.01 mL. Record this volume.

Macroscale procedure: Obtain about 75 mL of standardized 0.300 M HCl in a clean flask. Fill a 25- or 50-mL buret with the solution, rinsing the buret with two or three 5-mL portions before finally filling it. Read the initial buret reading, estimating to the nearest 0.02 mL. Record this volume.

When the calcium metal previously added to the Erlenmeyer flask has completely reacted, add 10 drops of 0.1% thymol blue indicator. With the tip of the buret slightly below the mouth of the flask, begin adding the 6.00 M HCl one drop at a time, swirling the flask as you add the HCl solution. (If you are using 0.300 M HCl in a 25- or 50-mL buret, begin by adding about 1 mL at a time.)

As you near the end point of the titration, you will notice that the blue color of the indicator turns yellow at the point where the acid enters the solution. If you approach the end point carefully, the solution will change within 1 to 2 drops from a blue color to a yellow color that remains for at least a full minute when the solution is swirled (swirling is very important here). Read and record the final volume. Also record the exact molarity of the HCl titrant used, which may be different from the nominal 6.00 M (or 0.300 M) value.

If you add the HCl solution too rapidly near the end point, you may overrun it and spoil the measurement. As you near the end point, take care to add the titrant drop by drop because, if you add a few drops in excess, the solution will have the same yellow color as if the correct volume had been added. A pink or red color indicates that you have grossly overrun the end point. If you overrun the end point by more than 2 to 3 drops, you should repeat the entire procedure with a fresh sample of calcium metal.

If you are carrying out two titrations, repeat the procedure with the second sample.

When you have finished all of your titrations, drain the HCl from the buret into a 250-mL beaker. Add to the beaker the unused HCl titrant contained in the test tube or flask. Neutralize the HCl by adding solid sodium bicarbonate, $NaHCO_3$, until the solution stops foaming (about 2 g is sufficient). The neutralized solution may now be discarded down the drain. The contents of the Erlenmeyer flask(s) that you have titrated may also be discarded down the drain. Rinse the buret with water. Use deionized water for the final rinse.

From your titration data, you will calculate the moles of HCl used to titrate the hydroxide ion produced in the reaction and the moles of hydroxide per mole of calcium for each sample.

▬▬ CONSIDER THIS ▬▬

Look up the first and second ionization energies of zinc and calcium in your text or in the *CRC Handbook of Chemistry and Physics.* If we make the hypothesis that the most active metals will have the smallest ionization energies, which metal might we expect to be most active, that is, which metal atoms give up their electrons more easily, zinc or calcium?

Write a plausible balanced chemical equation for the reaction of calcium with water. In doing this, think about what happens when a calcium atom gives up its two electrons to two water molecules. Which atoms in a water molecule are the more positive and therefore the most likely to accept the electrons donated by a calcium atom? What could be formed when you add an electron to a water molecule? What form of the calcium atom is left when it gives up two electrons?

Then write a second equation in which the products of the titration reaction (the calcium–water reaction, that is, calcium hydroxide) react with the HCl titrant. The two reactions may be combined (added together) to get the net chemical reaction that shows the initial reactants (calcium and water) and the final reaction products after the titration is completed (calcium chloride, $CaCl_2$). What is the net chemical reaction?

**Determination of a Chemical Formula
by Titration**

Name _____

Date _____ Section _____

The Reaction of Calcium with Water

Locker _____ Instructor_____

Observations and Data

1. The reaction of zinc and calcium with HCl and water

(a) Describe what happened when you added zinc to 6 M HCl in the test tube.

(b) What happened when you brought the hydrogen gas in the inverted test tube near a flame?

(c) What happened when you added calcium metal to 6 M HCl in the test tube?

(d) Compare the reaction of the gas collected from the calcium–HCl reaction with the reaction of hydrogen collected from the zinc–HCl reaction when each is brought near a flame.

(e) Describe what happened when you added the zinc and calcium metal to test tubes containing only water. Which metal is more reactive with water?

(f) What gas is produced by the reaction of Group I and Group II metals with water? (Consult your text or ask your instructor if you are uncertain.)

(g) What color results when phenolphthalein indicator is added to the solution in which calcium is reacting?

(h) The base produced in the reaction of Group I and Group II metals with water is called hydroxide ion. Write the formula for hydroxide ion. (Consult your text or ask your instructor if you are uncertain.)

2. Titration of the calcium–water reaction product

Weighing data	Trial 1	Trial 2
Mass of calcium + glazed weighing paper:	_____ g	_____ g
Mass of glazed weighing paper:	_____ g	_____ g
Mass of calcium metal:	_____ g	_____ g

Titration data

	Trial 1	Trial 2
Final buret reading:	_____ mL	_____ mL
Initial buret reading:	_____ mL	_____ mL
Net volume of HCl titrant:	_____ mL	_____ mL

Record the exact molarity of the HCl titrant used
(the nominal value is 6.00 M or 0.300 M): _____ mol/L

Calculations

Look up the molar mass and calculate the moles of calcium in each sample. Show your calculations below:

_____ mol Ca   _____ mol Ca

Calculate the moles of HCl that were required to titrate each sample. Show your calculations below:

_____ mol HCl   _____ mol HCl

Calculate the ratio mol HCl/mol Ca for each sample. Show your calculations below:

_____   _____

Round the ratio mol HCl/mol Ca to the nearest integer: _____

If one mole of HCl reacts with one mole of hydroxide [recall Equation (2)], how many moles of hydroxide must have been produced for every mole of calcium that reacted?

From your experimental result, what must be the formula for the calcium hydroxide produced in the reaction of calcium with water?

Name _____   Date _____

 **CONSIDER THIS**

1. Look up in your text or the *CRC Handbook of Chemistry and Physics* the first two ionization energies for the elements calcium and zinc. Record them below. Based on the values of the ionization energies, which metal would you expect to be the most active, calcium or zinc?

	Ca	Zn
IE(1)	_____	_____
IE(2)	_____	_____

2. Complete and balance the following sequence of chemical reactions that takes place when the calcium reacts with water and when the calcium reaction products are titrated with HCl solution:

$$Ca(s) + \quad H_2O \rightarrow$$

$$Ca(OH)_2 + \quad HCl \rightarrow$$

The two reactions may be combined (added together) to get the net chemical reaction that shows the initial reactants and the final reaction products after the titration is completed. Write the net chemical reaction:

3. If some of the calcium metal remains unreacted when the titration is begun, the following reaction would occur as HCl is added:

$$Ca(s) + 2HCl(aq) \rightarrow CaCl_2(aq) + H_2(g)$$

Would this cause an error in the titration? If so, would more or less HCl be required?

The Language of Chemistry: Chemical Nomenclature

Purpose

• Learn the basic design of nomenclature systems for simple covalent, ionic, and acidic compounds.

• Understand how the periodic nature of chemical properties creates useful patterns that are used in chemical names.

Introduction

This study assignment provides a basic introduction to inorganic chemical nomenclature and explains how to write simple chemical formulas, using the periodic table to correlate the ionic charges or oxidation states of the elements. Rules for naming compounds are essential to avoid the massive confusion that would result if each person invented his or her own names and symbols for the elements and their compounds. The importance of this task is underscored by noting that in January 1965, *Chemical Abstracts* (a periodical that provides short summaries of all articles appearing in chemistry journals) started a compound registry index. By 2007, this index contained about 33 million different chemical substances. Since then, the list has been growing at the rate of 4000 new substances each day.

It is estimated that about 35 million chemical substances are known, the great majority of which are organic compounds (those containing carbon). Mastering the details necessary to name all of these substances would require a great deal of study. Fortunately, most of the time we work with a limited number of chemical substances, so you will not need to learn more than a few hundred names. This task is made even easier because (1) most of the names are established according to simple rules and (2) the elements are grouped in chemical families of the periodic table, and members of each family bear strong resemblances to one another. In addition, you do not have to learn all of the names at once. Just as you are able to learn the names of new friends a few at a time, you will find it easy to learn the names of new chemical substances a few at a time, as you encounter them.

However, one problem cannot be avoided. The grand traditions of chemistry go back several centuries, and some compounds are commonly referred to by names coined long ago, as well as by their more systematic names. Consequently, for some substances it is necessary to learn both the *common,* or *trivial,* name and the systematic name. Fortunately, the use of older common names is diminishing: For example, in industrial commerce it is common to refer to sodium carbonate (Na_2CO_3) as soda ash, but in today's chemistry lab you are not likely to find a bottle of sodium carbonate labeled as soda ash.

Chemical Nomenclature

In many respects, the best way to represent a chemical substance is to make a three-dimensional model of it, showing the arrangement of all the atoms in space and their relative sizes. With the advent of molecular modeling programs and fast computers, this has now become a possibility. Most of the time, however, this process is too cumbersome, requiring us to have a computer always at hand. So we usually represent atoms and chemical substances in a more compact way, using symbols (such as Na) and formulas (such as Na_2CO_3) and using names (such as sodium and sodium carbonate) for these symbols and formulas. We call the system of naming substances *chemical nomenclature.* In rational nomenclature, the name of a compound would indicate (1) the elements of which it was composed and (2) the relative proportions of each element. Recent changes in chemical nomenclature have moved in this direction but still fall considerably short of this goal. Nomenclature, like language, is not static. It continually changes as needs and current usage dictate.

The standards of nomenclature are established by the International Union of Pure and Applied Chemistry (IUPAC). Since this international body must coordinate different languages and conflicting views, it is not surprising that the rules reflect a compromise. As a result, American chemists occasionally employ usages not officially sanctioned by the IUPAC rules. In this study assignment, we have followed the IUPAC rules[1] except for those conflicting with current American usage. Most of the conflicts are minor, and the beginning student of chemistry is not likely to encounter any serious conflicts in applying the rules listed here.

Chemical Symbols and Formulas

You may have already used chemical symbols and the formulas of substances and may have written equations for a number of chemical reactions. Now we must pause to emphasize the exact meaning and correct usage of these and other terms that constitute the unique language of chemistry.

Each element is represented by a *chemical symbol.* The symbol consists of either one or two letters, such as C for carbon and Ba for barium. Several of the elements have symbols derived from their ancient Latin names: Cu for copper from *cuprum,* Fe for iron from *ferrum,* Au for gold from *aurum,* Pb for lead from *plumbum,* and Ag for silver from *argentum.* Several elements discovered after 1780 have names derived from Latin or Germanic stems: Na for sodium from *natrium,* K for potassium from *kalium,* and W for tungsten from *wolfram.* We recommend that you train yourself to recognize an element name when you see its symbol and vice versa, at least those up to radon, element 86.

A *chemical formula* represents the composition of a given *substance,* which may be either an element or a compound. Thus H and O are the symbols for the elements hydrogen and oxygen, and they can also represent the atomic state of the elements, whereas H_2 (hydrogen gas) and O_2 (oxygen gas) represent the more stable molecular forms of the elements hydrogen and oxygen. When we speak of the chemical properties of oxygen, it is usually the stable molecular form of oxygen, O_2, that is meant. In this manual we will always try to specify the chemical formula of a substance in order to avoid any confusion or misinterpretation.

A *chemical compound* is a substance formed from two or more elements, such as H_2O (water), H_2O_2 (hydrogen peroxide), or NaCl (sodium chloride).

---

[1]For the most recent recommendations of the International Union of Pure and Applied Chemistry (IUPAC), see N. G. Connelly, ed., *Nomenclature of Inorganic Chemistry: IUPAC Recommendations 2005,* http://old.iupac.org/publications/books/author/connelly. html (accessed July 2008).

Classifying and Naming Chemical Compounds

We will introduce three important classes of chemical compounds: *covalent compounds, ionic compounds,* and *acids.* This classification is based on generalizations regarding the physical and chemical properties of compounds. Each of these classes has its own naming system, so the first order of business is to train yourself to identify the compound type. This classification system requires that you know the distinctions between the metallic elements and the nonmetals. The bold stairstep-shaped line that you see on most periodic tables, including the one on the inside front cover of your lab book and in Figure A-1, serves as the divider between the elements that are metals and the nonmetals. The metallic elements are those found below and to the left of this divider; the nonmetals are above it and to the right.

The distinction between metals and nonmetals is based on shared characteristics. If you saw samples of the metals, you would quickly notice their typical shiny surfaces (even when they are colored, such as gold and copper). They also have the typical properties of metals: They conduct electricity and heat, and they are malleable and ductile. The nonmetals cannot be so simply described; some are gases when pure (for example, hydrogen, oxygen, nitrogen), others are solids (for example, sulfur, iodine, phosphorus), and bromine is a liquid at room temperature. In contrast to the metals, the nonmetals are electrical insulators (that is, nonconductors), none of them have shiny surfaces, and the solids are not readily formed into shapes.

The elements lying near the stairstep line separating metals from nonmetals have intermediate properties so they are often referred to as *semimetals* or *metalloids.* For purposes of our naming scheme, however, we will treat them as either metals or nonmetals based on which side of the stairstep line they fall. Hydrogen plays a unique role in chemistry, mimicking the behavior of metals in some contexts, but for naming purposes you should classify it as a nonmetal.

I. Covalent compounds

Covalent compounds are composed of nonmetals bonded to nonmetals. They are easily recognized from their names or formulas because they contain no metal elements.

More than 90% of the known compounds are covalent compounds, and their naming system gets quite complicated because of the various forms and structures of these compounds. We will limit this discussion to the *binary covalent compounds,* that is,

Periodic Table of the Elements

FIGURE A-1 The periodic table showing the categories of metals (in white) and nonmetals (gray).

those containing only two nonmetal elements. The system for naming these binary covalent compounds is simple; the formula is transliterated into a name. The numbers in the formula are converted to the Greek prefixes, such as

1 → mono	6 → hexa
2 → di	7 → hepta
3 → tri	8 → octa
4 → tetra	9 → nona
5 → penta	10 → deca

The elements are listed in the same order as in the formula (which is from the least to the greatest *electronegativity*, a concept that will be discussed as an important part of chemical bonding ideas), and the suffix of the last element is changed to *-ide*. Examples are

N_2O_5, dinitrogen pentaoxide

P_4O_{10}, tetraphosphorus decaoxide

ClO_2, monochlorine dioxide or chlorine dioxide

Na_2O, NOT A COVALENT COMPOUND!

NOTE The prefix *mono-* is usually *not* used (although it is not incorrect). The one exception to this rule is CO, called *carbon monoxide* (with the elision of the last vowel in "mono" when followed by a vowel). NO is called *nitrogen oxide.*

It is critical to identify that the compound is a covalent compound before this system is used. Na_2O is simply called *sodium oxide,* not disodium oxide, because it is an ionic compound composed of metal and nonmetal ions. The charge on each elemental ion correlates with the position of the element in the periodic table and dictates the atom ratio, so the number prefixes are not necessary.

Trivial Names Some common binary compounds are designated by trivial names that have been assigned historically. Examples are

H_2O, water

NH_3, ammonia

CH_4, methane

PH_3, phosphane (or phosphine)

AsH_3, arsane (or arsine)

II. Ionic compounds

Ionic compounds are those that form between a metal and one or more nonmetals. Ionic compounds contain a *metal cation* and a *nonmetal anion.* The name of the compound is simply the cation name followed by the anion name. Ionic compounds do *not* use Greek prefixes (except in rare cases where the ion itself contains one of these prefixes in its name, such as dichromate). The formula can be determined from the fact that all compounds are electrically neutral, so once the charge on the ions is known, the formula can be readily deduced.

Learning the names of ionic compounds then consists of learning the cation and anion names. To determine the formula from the name of a compound requires knowing the individual ion charges as well.

A. Cations (positive ions) Cations are metal atoms with electrons removed. Because it is easy to recognize whether a metal is in a compound or not (in a compound other elements are bonded to the metal and named with it), the metal name can be used for both the element and the ion. In a few cases, the name needs to be augmented to specify the charge on the metal cation.

1. The cations from periodic table Groups I, II, and III have charges of +1, +2, and +3, respectively. Because of this observation, the unmodified metal name is used for the ions. For example: Na^+, sodium; Ba^{2+}, barium; Sc^{3+}, scandium; and Al^{3+}, aluminum.[2]

2. All other metal cations are named by augmenting the metal name with a roman numeral in parentheses to indicate the charge on the ion. This is necessary because these metals have different ion charges in different compounds. For example: Fe^{2+}, iron(II); Fe^{3+}, iron(III); Pb^{2+}, lead(II); and Pb^{4+}, lead(IV).

NOTE In speaking, the roman numeral is pronounced as the number, such as "iron two," "iron three," "lead two," and "lead four" in these examples.

There is an older, nonsystematic naming process for cations that uses

[2]There are a few exceptions to the "rule" that Group III always has a +3 charge, such as thallium, which forms a few Tl^+ compounds.

suffixes to indicate the charge, such as ferr*ous* for Fe^{2+} and ferr*ic* for Fe^{3+}. However, this is a cumbersome system because the suffixes correspond *not* to a specific charge but rather to the possible range of charges; for example, stannous is Sn^{2+}, but stannic is Sn^{4+}, whereas cupric is Cu^{2+} and cuprous is Cu^+. You will not be responsible for knowing this naming system, but you should be aware that it exists because you will run across such names from time to time. If so, look them up! Here are a few of the most common examples.

-ous: Fe^{2+} Co^{2+} Cu^+ Au^+ Hg_2^{2+} Sn^{2+}
-ic: Fe^{3+} Co^{3+} Cu^{2+} Au^{3+} Hg^{2+} Sn^{4+}

3. The metal ions Ag^+, Zn^{2+}, and Cd^{2+} are the ions formed by the respective metals, so the roman numeral designations are not necessary. They are known as silver, zinc, and cadmium ions in their compounds. It is all right to use the roman numerals, but it is not necessary.

4. The mercury(I) ion, which you would expect to be Hg^+, is always found paired as Hg_2^{2+} in ionic compounds.

5. The polyatomic cation NH_4^+, known as the ammonium ion, mimics the behavior of the Group I metal ions and is thus treated as such in ionic compounds. This means it is possible to have ionic compounds that do not contain any metals, such as NH_4Cl or $(NH_4)_2S$, but this should not pose a problem because the distinctive formula of ammonium in a formula should immediately remind you of this intriguing ion.

B. Anions (negative ions) Anions can be grouped into two classes: those having a single atom (monoatomic) and those having more than one atom (polyatomic).

1. *Monoatomic anions.* Monoatomic anions are the simplest ions formed from the nonmetal elements when their atoms take up enough electrons to give them a noble gas electron configuration. They are all named by adding the ending *-ide* to the *stem* of the nonmetal element name. The stem is a shortened version of the element's name that is long enough to identify the element while at the same time forming a pronounceable name when a prefix or suffix is added to the stem.

The negative charge on a monoatomic nonmetal anion can be deduced from the position of the ele-

ment in the periodic table, by counting the groups (columns) required to get to the right-hand noble gas column—Group VIII (18). Here are some examples:

Group V (15)	Group VI (16)	Group VII (17)
		H^- hydride
N^{3-} nitride	O^{2-} oxide	F^- fluoride
P^{3-} phosphide	S^{2-} sulfide	Cl^- chloride
As^{3-} arsenide	Se^{2-} selenide	Br^- bromide
	Te^{2-} telluride	I^- iodide

The carbide (and silicide) ions, C^{4-} (and Si^{4-}), follow the noble gas pattern. Unfortunately, the C_2^{2-} ion is also commonly called *carbide*. For example, the compound CaC_2 is called *calcium carbide,* a nonsystematic use of the name *carbide.* [IUPAC rules recommend calling the C_2^{2-} ion *dicarbide(2−)* or *acetylide* since the reaction of CaC_2 with water produces acetylene $HC\equiv CH$.] We will use the name *carbide* to refer to the C^{4-} anion.

The hydrogen atom is an exception to the patterns of the other elements, as it reacts with almost all the elements to form compounds with a diverse range of properties. With Group I and II metals, it can form ionic (saltlike) hydrides, such as NaH and CaH_2, which contain discrete hydride ions, H^-. (The ionic character of these compounds is revealed by their high conductivities at or near their melting points.)

With chlorine, hydrogen forms acidic hydrogen chloride, HCl. Dissolved in water, HCl dissociates to form a +1 hydrogen cation, H^+, and chloride ion Cl^-. (Bare H^+ ions, or protons, do not exist as such in aqueous solution but always attach themselves to an electron pair on a water molecule to form H_3O^+.)

Thus, hydrogen acts as if it can give up an electron to form a hydrogen ion, H^+, or accept an electron to form hydride, H^-. For these reasons, hydrogen is sometimes placed in the periodic table either at the head of the Group I metals or at the head of the Group VII (17) nonmetal elements (halogens).

We see from these examples that the chemical character of hydrogen depends on whether hydrogen is bonded to a metal or a nonmetal. The bonding character is related to a property of atoms called *electronegativity,* defined in simple language as the ability of an atom to attract electrons to itself. Hydrogen has a greater electronegativity than sodium or calcium but a smaller electronegativity than chlorine.

2. *Polyatomic anions.* In addition to the above list of monoatomic ions, there are a few polyatomic anions that are named using the *-ide* ending.

OH^-	hydroxide	O_2^{2-}	peroxide	CN^-	cyanide
NH_2^-	amide	N_3^-	azide	I_3^-	triiodide

Group III (13)

BO_3^-	borate

Group IV (14)

CO_3^{2-}	carbonate
HCO_3^{2-}	hydrogen carbonate (or bicarbonate)
$CH_3CO_2^-$	acetate
OCN^-	cyanate
SCN^-	thiocyanate[3]

Group V (15)

NO_2^-	nitrite
NO_3^-	nitrate
PO_4^{3-}	phosphate
HPO_4^{2-}	hydrogen phosphate
$H_2PO_4^-$	dihydrogen phosphate
AsO_3^{3-}	arsenite
AsO_4^{3-}	arsenate

Group VI (16)

SO_3^{2-}	sulfite
SO_4^{2-}	sulfate
HSO_4^-	hydrogen sulfate (or bisulfate)
$S_2O_3^{2-}$	thiosulfate[3]

Group VII (17)

ClO^-	hypochlorite
ClO_2^-	chlorite
ClO_3^-	chlorate
ClO_4^-	perchlorate

Transition metal oxoanions (6, 7)

CrO_4^{2-}	chromate
$Cr_2O_7^{2-}$	dichromate
MnO_4^-	permanganate

NOTE The *per-* prefix of peroxide implies the addition of an oxygen atom to the O^{2-} oxide ion. (We discuss the *per-* prefix in greater detail later, in connection with the oxoanions of chlorine.) Peroxides containing the O_2^{2-} ion, such as H_2O_2 (hydrogen peroxide) or Na_2O_2 (sodium peroxide), should be distinguished from normal oxides that contain the O^{2-} oxide ion, such as MnO_2 [manganese(IV) oxide], TiO_2 [titanium(IV) oxide], and SiO_2 [silicon(IV) oxide]. These compounds are also known by their common names: manganese dioxide, titanium dioxide, and silicon dioxide.

Sodium ion, in Group I, always has a charge of +1, so we must not fall into the trap of calling Na_2O_2 an *oxide*, which would require sodium ion to have a +2 charge since an oxide ion has a −2 charge. (Sodium oxide is Na_2O.)

Polyatomic oxoanions. The most common polyatomic anions are oxoanions, in which a central nonmetal (or transition metal) atom is bonded to one or more oxygen atoms.

The oxoanions are named by adding the characteristic endings *-ate* and *-ite* to the stem of the central element's name. Ions with the *-ite* ending have one fewer oxygen atom than the ions with the *-ate* ending. We list them below, arranged according to the periodic table group to which the central atom belongs.

The family of oxoanions is larger for the Group VII (17) elements (called *halogens*) than for the other groups. To distinguish the four oxoanions known to exist for chlorine and the other halogens, the prefixes *hypo-* and *per-* are used to extend the *-ite, -ate* naming system.

hypo- The *hypo–ite* anions have one fewer oxygen atom than the corresponding *-ite* anion. (*Hypo* means "lower.")

per- The *per–ate* anions have one more oxygen atom than the corresponding *-ate* anion. The prefix *per-* is a truncation of (*hy*)*per*. (*Hyper* means "higher.")

Note the *-ite, -ate* anion pairs: nitrite, nitrate; sulfite, sulfate; chlorite, chlorate. The ions of each pair have the same net charge, but the ion with the *-ite* ending has one fewer oxygen atom than the ion with the *-ate* ending.

Also note how the *hypo-* and *per-* prefixes are used in naming the oxoanions of chlorine: *hypo-* goes with the *-ite* ending and *per-* goes with *-ate* ending.

The oxoanions of bromine and iodine are named the same way as the oxoanions of chlorine using the stems *brom-* and *iod-*. Fluoride ion, F^-, is the main anion of importance formed by fluorine. Oxoanions of fluorine are not known, although the molecule FOH is known to exist.

---

[3]The prefix *thio-* in chemical nomenclature means "substitute a sulfur atom for an oxygen atom." Thus, when we replace the O atom in cyanate ion, OCN^-, with an S atom to form the ion SCN^-, we change the name of the ion to *thiocyanate*. The same principle applies to naming sulfate and thiosulfate.

Hydrogen anions. Most oxoanions having a net charge more negative than −1 are weak bases that will accept one or more hydrogen ions (or protons), H^+, to become protonated anions. We name these anions by adding the word *hydrogen* ahead of the name of the parent anion. Here are some examples:

HS^-	hydrogen sulfide
$H_2PO_4^-$	dihydrogen phosphate[4]
HPO_4^{2-}	hydrogen phosphate
HCO_3^-	hydrogen carbonate
HSO_4^-	hydrogen sulfate

If enough hydrogen ions (protons) are added to neutralize all of the charge on the anion, the uncharged molecule will be an acid. The naming of acids is described in part III.

The anions in the four preceding lists include most of the common anions that you will encounter in general chemistry. Their names are an important part of the vocabulary of chemistry; mastering their principles of nomenclature will put you on the road toward success in general chemistry.

Hydrated Ionic Compounds Many ionic compounds can be found in two or more forms that incorporate water into their structures and formulas. These water molecules are not simply absorbed into pores or cracks in the solids or adsorbed onto the surface, but they are incorporated into specific positions in the solid lattice and are chemically bonded to the metal cation. The key feature of these *hydrates* is that they have a specific number of water molecules associated with each cation and thus are true compounds. Because a simple heating process can remove these water molecules and because we think of water as a neutral or inert substance in many contexts, we tend to think of these not as new compounds but as modified versions of the unhydrated compounds.

Given this context, a system for writing formulas and naming these compounds has arisen that treats them as slightly modified versions of the unhydrated compounds. Thus $CuSO_4$, which is found as a colorless (white) solid, turns distinctly blue when it is exposed to water vapor or when, by drying, it forms an aqueous solution of copper(II) sulfate. The blue compound has the formula $Cu(H_2O)_5SO_4$, which shows that the water molecules are bonded to the Cu^{2+} ion. However, most often you will see it written as $CuSO_4 \cdot 5H_2O$ and named as copper(II) sulfate pentahydrate. Upon heating to 150 °C, the blue pentahydrate returns to the colorless $CuSO_4$ form. Other examples of hydrated compounds include the alum of Experiment 6, which is aluminum potassium sulfate dodecahydrate, $AlK(SO_4)_2 \cdot 12H_2O$, and calcium chloride, $CaCl_2$, which is often used as a desiccant (drying agent) because it forms a dihydrate, $CaCl_2 \cdot 2H_2O$, and a hexahydrate, $CaCl_2 \cdot 6H_2O$.

III. The acids

The set of *covalent compounds* (that is, neutral molecules) formed nominally from H^+ cations in combination with any of the anions are known as *acids*. They are covalent compounds (they contain no metal cations and have covalent bonding properties), but they also have many properties related closely to ionic compounds. Because of this uniqueness, they are in a special class with a special naming system. The acids are easily identified by formula, because their formulas, by tradition, are written with the H first. As you will see, their names always contain the word *acid,* so they are simple to spot.[5]

Acid names are derivations of the anions from which they derive:

The *-ide* anions form acids with the naming structure *hydro–ic acid.*

The *-ate* anions (including the per–ates) form acids with the naming structure *-ic acid.*

The *-ite* anions (including the hypo–ites) form acids with the naming structure *-ous acid.*

Examples are

HCl (from Cl^-, chloride), hydrochloric acid

H_2S, hydrosulfuric acid

HNO_3, nitric acid

HNO_2, nitrous acid

---

[4]IUPAC rules recommend combining hydrogen and the name of the anion into one word, e.g., *dihydrogenphosphate.* In American usage, the single word is often separated into two words: dihydrogen phosphate.

[5]This tradition of writing the H first in the formula of an acid conflicts with another tradition of using the formula to convey information about structure. For example, in the case of the oxyacids (oxygen-containing acids) such as sulfuric or nitrous acids, the formulas H_2SO_4 and HNO_2 would lead you to believe that the hydrogen atoms are bonded to the sulfur or nitrogen atoms, whereas the hydrogens are invariably bonded to an oxygen atom. In this light, it would be better to write these formulas as $HOSO_3$ and HONO, although this arrangement is not usually seen.

H_3PO_4, phosphoric acid

$HClO_4$, perchloric acid

$HBrO_3$, bromic acid

HIO_2, iodous acid

$HClO$, hypochlorous acid

The property of *acidity* is really only displayed when a solvent like water is present. A few of the acids exist as gaseous molecules outside of solution and could be named as covalent molecules, such as hydrogen chloride, HCl, and dihydrogen sulfide, H_2S. In most cases, their acid names are used even in this context.

IV. Other classes of compounds

Two large arenas of chemistry do not fit under the naming systems described here. Because of carbon's ability to bond to itself in long chains, the compounds of carbon display a tremendous variety of formulas and structures. The naming system for these molecules is an elaborate system beyond the scope of general chemistry. Thus, we have not described any of the rules for naming these compounds, which are known as organic molecules.

Transition metal ions have the ability to form complex structures that themselves can be anions and/or cations in compounds. We have left the naming of these transition metal complexes to a later time as well.

The Language of Chemistry:
Chemical Nomenclature

Name _____

Date _____ Section _____

Locker _____ Instructor_____

Exercises on Formulas and Nomenclature

Note: These exercises will help you learn how to write correct formulas and name compounds. Check your answers, if necessary, with your instructor.

1. Name the following.

FeI_2 _____

$FeBr_3$ _____

I_2 _____

$AlCl_3 \cdot 6H_2O$ _____

$Fe_2(SO_4)_3$ _____

FeS _____

NCl_3 _____

H_2CO_3 _____

$CaCO_3$ _____

Be_2C _____

$Sn(NO_3)_2$ _____

$(NH_4)_2S$ _____

N_2O_4 _____

MgO_2 _____

2. Write the correct chemical formula.

Barium chloride	_____	Stannous chloride	_____
Ammonium sulfate	_____	Barium carbonate	_____
Stannic chloride	_____	Aluminum carbide	_____
Sodium carbonate	_____	Sodium hydrogen carbonate	_____
Magnesium phosphate	_____	Nitrogen dioxide	_____
Calcium hydrogen phosphate	_____	Disulfur dichloride	_____

3. Complete the following table. (Note that "salt" is a synonym for "ionic compound.")

Formula	Name as acid	Formula for sodium salt	Name of salt
HF			
HNO_2			
$HBrO_2$			
$HBrO_4$			
HNO_3			
HBrO			
$HBrO_3$			
H_3AsO_4		Na_3AsO_4	
H_3AsO_3		NaH_2AsO_3	

4. Name the following compounds.

NaIO _____

Mg_3N_2 _____

K_2HPO_4 _____

$Fe(NO_3)_3$ _____

P_4O_6 _____

NH_4ClO_4 _____

K_2CrO_4 _____

Na_2SO_3 _____

$KMnO_4$ _____

$BaSO_3$ _____

$Ca(ClO_2)_2$ _____

$FeSO_4$ _____

P_4O_{10} _____

$HClO_3$ _____

$K_2Cr_2O_7$ _____

Na_2SO_4 _____

Name _____ **Date** _____

5. The spaces below represent portions of some of the main groups and periods of the periodic table. In the proper squares, write the correct formulas for the chlorides, oxides, and sulfates of the elements of Groups I, II, and III. Likewise, write the formulas of the compounds of sodium, calcium, and aluminum with the elements of Groups VI and VII. Two of the squares have been completed as examples.

Period	Group I	Group II	Group III	Group VI	Group VII
2	LiCl Li_2O Li_2SO_4		(omit sulfate)		
3				Na_2S CaS Al_2S_3	
4					
5					

6. What is the systematic name of water? _____

What is the name of water when viewed as an acid (which it is)? _____

7. If sulfate has the formula SO_4^{2-}, then what is the formula of the thiosulfate ion? (See footnote 3 for the meaning of the prefix *thio-*.) _____

8. If glutamic acid has the formula $HC_5H_8NO_4$, then what is the formula of sodium glutamate, commonly known as MSG, or monosodium glutamate? _____

The Estimation of Avogadro's Number

Purpose

- Make an order-of-magnitude estimate of the size of a carbon atom based on simple assumptions about the spreading of a thin film of stearic acid on a water surface.

- Estimate the number of atoms in a mole of carbon (12.0 g C) using the estimated size of a carbon atom and the density of the diamond form of carbon.

Pre-Lab Preparation

John Dalton (1766–1844) was so taken with the notion of atomism that he never quite grasped the distinction between an atom and a molecule of an element. The most stable form of many elements, such as oxygen, hydrogen, and the halogens, is not a single atom but a diatomic molecule. Because Dalton believed that the smallest unit of an element must be an atom, he had a hard time accepting Gay-Lussac's data about the combining volumes of gases.

The reconciliation between Dalton's theory and Gay-Lussac's data was brought about by Amedeo Carlo Avogadro (1776–1856). He accomplished this by making the distinction between an atom and a molecule of an element and by making the hypothesis that equal volumes of gases contain equal numbers of molecules. For example, if you believe that hydrogen and chlorine are monoatomic and that equal volumes of gases (at the same temperature and pressure) contain equal numbers of atoms, then one volume of hydrogen should react with one volume of chlorine to form one volume of HCl:

$$H(g) + Cl(g) \rightarrow HCl(g)$$

But when you do the experiment, you will find that two volumes of HCl are produced. Avogadro interpreted this result by assuming that the smallest unit of hydrogen and chlorine is a diatomic molecule, not an atom, and by writing the equation as follows:

$$H_2(g) + Cl_2(g) \rightarrow 2\ HCl(g)$$

Many subsequent experiments have proved that Avogadro's explanation was correct.

The hypothesis, published in 1811, was perhaps ahead of its time and went virtually unnoticed. More than half a century passed before Stanislao Cannizzaro (1826–1910) demonstrated the general applicability of the hypothesis in an article published in 1858. When he distributed it in pamphlet form at the first international chemical congress at Karlsruhe, Germany, in 1860, the pamphlet so clearly and completely discussed atoms, molecules, atomic weights, and molecular weights that chemists were convinced of his views and quietly incorporated them into chemical thinking. Fifty years of pondering Dalton's atomic theory had created the right moment for the acceptance of Avogadro's hypothesis. Cannizzaro gave credit to Avogadro as well as to Ampère and Dumas, and no doubt it was Cannizzaro who saved Avogadro's hypothesis from oblivion. As a result, today we call the number of particles (atoms or molecules) in a mole *Avogadro's number.*

How many particles are there in a mole? Avogadro did not invent or use the concept of a mole and never knew the number of particles in a mole. In 1865, a few years after Avogadro's death, the German chemist Joseph Loschmidt (1821–1895) published an article estimating the size of a molecule and, although Loschmidt didn't present the calculation in his paper, it is possible from the calculations he did present to estimate the number of molecules in a cubic centimeter of gas. This is only one simple step away from determining the number of molecules in a mole. For this reason, German chemists often called the constant *Loschmidt's number* rather than *Avogadro's number,* and the International Union of Pure and Applied Chemistry (IUPAC) allows the use of either the symbol N_A or L for Avogadro's number, to honor the contributions of both men.

A French scientist, Jean Perrin (1870–1942), is often given credit for determining the first value of Avogadro's number in 1908. Perrin measured the vertical distribution in the earth's gravitational field of gamboge (a natural resin) particles suspended in water and obtained values in the range 5.4 to 6.0×10^{23}.

After the American physicist Robert Millikan determined the charge of an electron around 1915, a more accurate value was obtained by dividing the charge of a mole of electrons (the Faraday constant) by the charge of a single electron. More refined values have been obtained by accurate measurements of silicon crystals by X-ray diffraction: dividing the volume of one mole of silicon by the effective volume of a silicon atom yields Avogadro's number.

All of the refined measurements require sophisticated and expensive equipment and great care in experimental technique and treatment of data. The payoff is the most accurate value of Avogadro's number, N_A, that we have to date: $N_A = 6.02214 \times 10^{23}$ particles/mol. This is one of the numbers you will encounter in introductory chemistry that is worth remembering.

The mole is a fundamental unit in the International System (SI) of units. A mole of carbon is precisely 12.000000 g of carbon-12, and Avogadro's number is defined as the number of atoms in a mole of carbon-12.

In this experiment, you will make an approximate (order-of-magnitude) estimate of Avogadro's number by determining the amount of stearic acid that it takes to form a single layer (called a *monolayer*) on the surface of water. By making simple assumptions about the way the stearic acid molecules pack together to form the monolayer, we can determine its thickness; from that thickness we can estimate the size of a carbon atom. Knowing the size of a carbon atom, we can compute its volume; and if we know the volume occupied by a mole of carbon (in the form of diamond), we can divide the volume of a mole of carbon by the volume of an atom of carbon to get an estimate of Avogadro's number. The number won't be accurate to within 10% or even a factor of 2, but it will enable you to estimate to within about a power of 10 the number of particles in a mole. That's better than chemists could do only 100 years ago.

Concepts of the experiment

Matter can exist in three states: as a gas, liquid, or solid. The fact that a gas can be condensed to a liquid and a liquid frozen to a solid indicates that there are attractive forces between all molecules. We can schematically represent these forces at the surface of a liquid by the arrows in Figure 10-1. In the interior of the liquid, the forces exerted on a given molecule are uniform in all directions. At the surface, however, it is clear that there is a net force attracting each surface molecule inward. These molecules have higher energies than do interior molecules, thus giving rise to the force known as *surface tension*. It is because of this force that liquid droplets are spherical. A spherical shape presents the smallest surface area for a given volume.

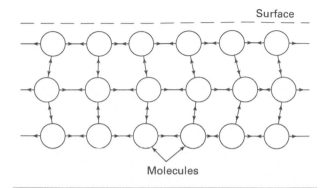

FIGURE 10-1 **The action of molecular forces at a liquid surface creates a surface tension that tends to minimize the surface area of the liquid.**

If the liquid is water, the surface tension is especially strong because particularly strong intermolecular forces, called *hydrogen bonds,* exist between the water molecules. (While strong for intermolecular bonds, these bonds are still considerably weaker than the intramolecular covalent bonds that bind atoms together in a molecule.) Hydrogen bonding arises whenever a hydrogen atom that is attached to a highly electronegative atom, such as oxygen, has access to an unbonded pair of electrons, such as those of another oxygen atom.

Another property displayed by water is polarity. Polar molecules possess a separation of charge. In ionic compounds, such as NaCl, there is a separation of a full unit charge, $Na^+ \cdots Cl^-$. Polar covalent molecules display a partial separation of charge, denoted by delta, δ. An arrow is used to denote this charge separation. This polarity, referred to as the *dipole moment,* is equal to the partial charge times the distance separating the charges; it is represented by an arrow pointing toward the negative end of the molecule. Figure 10-2 displays the polar nature of the water molecule.

Polar molecules attract each other. The negative end of the dipole of one molecule is attracted to the positive end of the dipole of another molecule. For this reason, water dissolves formic acid, H—COOH,

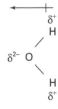

FIGURE 10-2 **The bond polarity and angular structure of a water molecule create a dipole moment.**

which has a dipole moment, but it does not dissolve butane, $CH_3CH_2CH_2CH_3$, which has a nearly uniform charge distribution. If a molecule possessing the properties of both of these molecules is brought up to the surface of water, the polar part of the molecule will be attracted to the surface and the nonpolar portion will be repelled. If the nonpolar part is much larger than the polar portion, the molecule will not dissolve in water but will simply stick to its surface. Consequently, it will lower the energy of the surface water molecules and of the adhering molecules.

The molecule we will use in this experiment, stearic acid, behaves in just this way. Stearic acid has a polar end consisting of a carboxyl group, —COOH, and a large nonpolar "tail" consisting of 16 methylene groups, —CH_2—, terminating in a methyl group, —CH_3. Figure 10-3 is a reasonably accurate representation of this molecule.

Adding a limited number of stearic acid molecules to a water surface results in the formation of a monolayer, as illustrated in Figure 10-4. However, after the surface is covered with a monolayer of

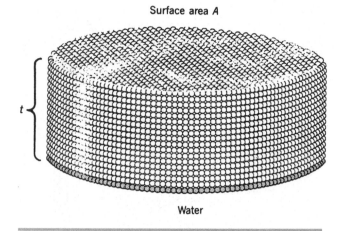

Surface area A

Water

FIGURE 10-4 Magnified view of a stearic acid film (not to scale). The stearic acid molecules orient themselves vertically because of the attraction of their polar ends (the —COOH groups) to the water molecules at the water surface and the weak attractive forces between their hydrocarbon tails. The volume of the stearic acid monolayer is the product of surface *area* times *thickness (A × t).*

stearic acid molecules, the addition of more molecules causes the stearic acid molecules to cluster in globular aggregates. The polar heads are attracted to the polar water molecules and form a lens-shaped surface, with the hydrocarbon tails pointing inward to form an "oily" interior.

The properties of the water surface and the stearic acid molecule permit us to perform what amounts to a titration of the water surface. We can add stearic acid molecules to the water surface until a monolayer covers the entire surface. Further addition will cause a convex lens to form on the liquid surface. At this point, the stearic acid molecules are stacked on the water surface like a layer of cordwood turned on end. If we know the area of the surface of water and have a way of measuring the volume of substance added to form the monolayer, we can calculate the thickness, *t,* of the layer. This thickness equals approximately the length of the stearic acid molecule.

As we saw in Figure 10-3, this molecule consists of 18 carbon atoms linked together. If we make the simple assumption that the atoms are like little cubes linked together (Figure 10-5), the edge length of a cube of carbon is given by $t/18$. If we cube the edge length, we have estimated the volume of a carbon atom. Now we have half the data needed to calculate Avogadro's number.

The other necessary data require no experimental work on your part. What we need is the volume of a mole of carbon. Diamond is pure carbon, and the density of diamond is known to be 3.51 g/cm^3. You

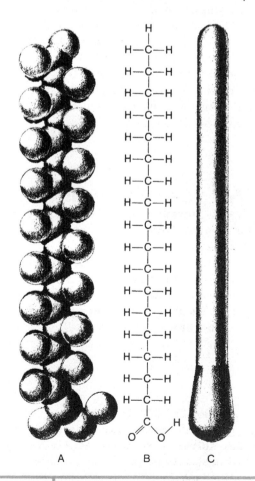

H—C—H
H—C—H
H—C—H
H—C—H
H—C—H
H—C—H
H—C—H
H—C—H
H—C—H
H—C—H
H—C—H
H—C—H
H—C—H
H—C—H
H—C—H
H—C—H

A   B   C

FIGURE 10-3 The stearic acid molecule.
(A) space-filling model; (B) structural formula;
(C) schematic "matchstick" representation.

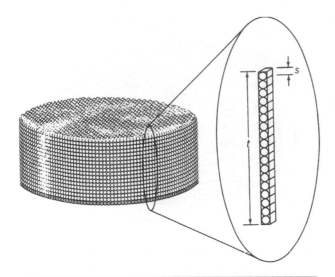

FIGURE 10-5 **For purposes of our estimate, each stearic acid molecule can be approximated as a vertical stack of 18 carbon atoms with a total height equal to the thickness, t, of the monolayer film. Each carbon atom then has a diameter, s, equal to $1/18$ of the film thickness. When the empty space between the atoms is included, the gross volume of each atom, $V_{atom} = s^3$.**

may recall that you can calculate the molar volume of an element by dividing the molar mass (g/mol) by its density (g/cm^3). Verify this idea by looking at the dimensions of this quotient.

If we assume that diamond consists of little cubes of carbon atoms stacked together, an Avogadro's number of them would equal the volume of one mole. Avogadro's number results from one final step:

$$N_A = \frac{\text{molar volume (cm}^3/\text{mole})}{\text{atomic volume (cm}^3/\text{atom})}$$

$$= \text{number of atoms/mol}$$

Experimental Procedure

Special Supplies: 14-cm watch glass; cm rule or meterstick; fine-tipped (50 drops/mL) polyethylene transfer pipets; 1-mL all-polypropylene syringes (Normject Luer Slip 1 × 0.01 mL syringes are available from Air-Tite Products Co., Inc., www.air-tite-shop.com, tel 800-231-7762); pure Type I water, free of surface active materials (preferably distilled or passed through a water purifier); disposable rubber gloves (if students will be cleaning their own watch glasses in 0.1 M NaOH in 50/50 methanol/water); 13 × 100 mm test tubes with corks to fit.

Chemicals: Hexane, 0.10 g/L stearic acid (purified grade) solution in hexane. The exact weight/volume

concentration of this solution will be provided by your instructor. 0.1 M NaOH in 50/50 methanol/water (volume/volume, or v/v), used for washing the watch glasses.

NOTES TO INSTRUCTOR

A purified (98%) grade of stearic acid gives the best results and is available commercially (Sigma-Aldrich).

Instead of using a dropper and counting drops, we prefer to use a 1-mL all-polypropylene syringe so that the volume of stearic acid solution in hexane can be read directly. Using the syringe also minimizes evaporation of the hexane solvent. The Luer tip of the syringe is fitted with a tip cut from a fine-tipped polyethylene transfer pipet, as shown in Figure I-14 of the Introduction. Another device for dispensing the stearic acid solution is described by Lane et al. (1984). (See the Bibliography.)

The watch glasses should be soaked in the methanolic NaOH solution in a polyethylene basin overnight or at least for an hour or two before the beginning of the laboratory period. The watch glasses should then be rinsed in distilled water before they are distributed to the students. Distilled water (or Type I water produced by a Millipore Milli-Q or Barnstead Nanopure purifier) is better for this experiment than deionized water, because deionized water often contains organic substances leached from the organic resins used to deionize the water.

SAFETY PRECAUTIONS:
Hexane is flammable! There must be no open flames in the laboratory while hexane is being used.
If you clean your own watch glasses by soaking them in methanolic NaOH, you must wear rubber gloves to protect your hands. As always, wear eye protection in the laboratory.

WASTE COLLECTION:
At the end of the experiment, unused hexane solvent and stearic acid in hexane solution should be placed in a waste container marked "Waste hexane/stearic acid solution in hexane."

Preparation of Equipment Stearic acid is a solid. It is conveniently measured and applied to the water surface by dropping a solution of stearic acid in hexane onto the water. The hexane is insoluble in water. Because it has a high vapor pressure, it rapidly evaporates, leaving a layer of stearic acid spread on the water surface one molecule thick, called a *monolayer.*

Because stearic acid is one of the fatty acids present in soaps, you must not clean your watch glass with soap. Any soap, grease, or dirt present can form a film on the water that prevents the stearic acid from spreading on the surface, leading to meaningless results because not even the first drop will spread.

Obtain a 14-cm watch glass. If it has not been soaked in the methanolic NaOH solution described in the Chemicals section, wash it thoroughly with detergent. Rinse the detergent off completely under a full stream of cold tap water for a minute, then rinse thoroughly with distilled water. Repeat the washing after each experiment, or obtain another watch glass that has been soaked in methanolic NaOH. Handle the watch glass only by the edges, being especially careful not to let anything touch the inside of it. Be particularly careful to keep your fingers (which are usually slightly greasy) off the glass.

1. Measuring the Volume of Stearic Acid Solution Required to Cover the Water Surface Obtain about 3–4 mL 0.10 g/L stearic acid solution in hexane in a clean, dry 13 × 100 mm test tube. Keep the tube corked when not in use. Fill the clean watch glass to the brim with distilled water. Carefully measure the diameter of the water surface with a centimeter rule or meterstick. Cut off the tip of a fine-tipped polyethylene transfer pipet about 25 mm from the tip end and push the cut-off tip onto the Luer fitting on the end of the syringe. Next, rinse and fill your 1-mL syringe with stearic acid solution, taking care to eliminate bubbles in the solution inside the syringe.

Read and record the initial volume reading of the syringe. Then add the stearic acid solution drop by drop to the water surface. Initially, the solution will spread across the entire surface, and it will continue to do so until a complete monolayer of stearic acid has been produced. As this point is approached, the spreading will become slower and slower, until finally a drop will not spread out but will instead sit on the surface of the water (looking like a little contact lens). If this "lens" persists for about 30 s, you can safely conclude that you have added 1 drop more than is required to form a complete monolayer. Now record the final volume reading of the syringe.

Thoroughly clean the watch glass (or get another one), and repeat the experiment. Repeat until the results agree to within 2 or 3 drops (0.04 mL).

When you have completed all of your measurements, rinse your syringe with pure hexane, and dispose of all the hexane-containing solutions in the waste collection bottle provided.

2. Calculating Avogadro's Number The calculation proceeds in several steps. First, we calculate the volume of stearic acid solution in hexane required to deliver enough stearic acid to form a monolayer. All of the hexane evaporates, leaving only the thin monolayer film of stearic acid, so we next calculate the actual mass of pure stearic acid in the monolayer.

Then we calculate the thickness of the stearic acid monolayer, using the known density of stearic acid and the area of the monolayer. Assuming the stearic acid molecules are stacked on end and are tightly packed, and knowing that there are 18 carbon atoms linked together in the stearic acid molecule, we calculate the diameter and volume of a carbon atom.

Finally, we calculate the volume of a mole of carbon atoms in diamond; dividing the molar volume of carbon (diamond) by the volume of a single carbon atom, we obtain an estimate of Avogadro's number.

C O N S I D E R   T H I S

The method we employed in this experiment is not especially accurate. There are other experimental ways to get values for Avogadro's number. For example, if you divide the electric charge on a mole of electrons (the Faraday constant) by the electric charge on a single electron, you will obtain the number of electrons in a mole of electrons, which is Avogadro's number. Try it, using values for the constants found in Table 2 of the Appendix.

Because atoms are so small, the magnitude of Avogadro's number is so large that it is almost impossible to comprehend it. Try this exercise to get a grip on it. First, imagine that you have 64 toy alphabet blocks arranged in a cube, where each block is itself a cube, 2 in. on an edge. The cube of 64 blocks would be 8 in. in three dimensions, with 4 blocks along each edge. Then flatten the cube into a square containing 64 blocks. How large would the square be, and how many blocks will be along its edge? Finally, arrange the 64 blocks in a line. How long would the line be?

Now that you have the idea, imagine an Avogadro's number of tiny cubes, the edge length of each tiny cube being about the size of a hydrogen atom (1.00×10^{-10} m). If you had an Avogadro's number of hydrogen atom cubes arranged into a large cube (like the alphabet blocks), what would the edge length of the cube be? If you then flattened the cube into a square, what would the edge length of the square be? Finally, if you arranged the Avogadro's number of tiny H-atom cubes into a line, how long would the line be? Compare it to the earth-to-sun distance of 93 million miles.

In the method we used, we made a number of simplifying assumptions. Some of these are discussed in the Consider This section of the Report form, and their effects are quantitatively assessed. The cumulative effect of the simplifying assumptions is to cause the final result for Avogadro's number to be a factor of about three times too large.

Bibliography

Boorse, H. A.; Motz, L. (eds.) "The Determination of Avogadro's Number," *The World of the Atom,* Basic Books, New York, 1966, Vol. 1, pp. 625–640. Describes Perrin's method.

Diemente, D. "Demonstrations of the Enormity of Avogadro's Number," *J. Chem. Educ.* **1998,** *75,* 1565–1566.

Feinstein, H. I.; Sisson, R. F., III. "The Estimation of Avogadro's Number Using Cetyl Alcohol as the Monolayer," *J. Chem. Educ.* **1982,** *59,* 751.

Hawthorne, R. M., Jr., "Avogadro's Number: Early Values by Loschmidt and Others," *J. Chem. Educ.* **1970,** *47,* 751–755.

Henry, P. S. "Evaluation of Avogadro's Number" (by the method of Perrin), *J. Chem. Educ.* **1966,** *43,* 251. Uses Perrin's method.

King, L. C.; Neilsen, E. K. "Estimation of Avogadro's Number," *J. Chem. Educ.* **1958,** *35,* 198.

Lane, C. A.; Burton, D. E.; Crabb, C. C. "Accurate Molecular Dimensions from Stearic Acid Monolayers," *J. Chem. Educ.* **1984,** *61,* 815.

Porterfield, W. W.; Kruse, W. "Loschmidt and the Discovery of the Small," *J. Chem. Educ.* **1995,** *72,* 870–875.

Robinson, A. L. "Metrology: A More Accurate Value for Avogadro's Number," *Science* **1974,** *185,* 1037.

Slabaugh, W. H. "Determination of Avogadro's Number by Perrin's Law," *J. Chem. Educ.* **1965,** 42, 471.

Slabaugh, W. H. "Avogadro's Number by Four Methods," *J. Chem. Educ.* **1969,** *46,* 40.

The Estimation of Avogadro's Number

Name _____

Date _____ Section _____

Locker _____ Instructor_____

Data and Calculations

Note: A number of calculations are required in this report. Show the calculations for only one of your trials, but give the results for both trials, those for trial 1 in the left blank and those for trial 2 in the right.

1. Measuring the volume of stearic acid solution required to cover the water surface

	Trial 1	Trial 2
(a) Record the diameter of the water surface:	_____ cm	_____ cm
(b) Record the volume of stearic acid solution required to cover the surface:	_____ mL	_____ mL
(c) Record the concentration of the stearic acid solution:	_____ g/L	_____ g/L

2. Calculating Avogadro's number

(a) *Calculating the thickness of a monolayer of stearic acid*

(i) Based on your part 1 data, what was the volume of stearic acid solution required to form a monolayer?

_____ mL   _____ mL

(ii) Calculate the mass of stearic acid contained in that volume of stearic acid solution (the concentration in grams per liter will be given to you):

_____ g   _____ g

(iii) Calculate the volume, V, of pure stearic acid in the monolayer on the water surface. The density of solid stearic acid is 0.85 g/mL (or g/cm^3):

_____ mL (cm^3)   _____ mL (cm^3)

(iv) Calculate the area of the monolayer ($A = \pi r^2$; r is the radius of the water surface):

_____ cm^2 _____ cm^2

(v) Calculate the thickness of the monolayer ($t =$ volume/area $= V/A$):

_____ cm   _____ cm

(b) *Estimating the size and volume of a carbon atom*

(i) A stearic acid molecule consists of 18 carbon atoms linked together. Assuming that the thickness, t, of a monolayer is equal to the length of the stearic acid molecule, calculate the size of a carbon atom, $s = t/18$:

_____ cm   _____ cm

(ii) Assuming that a carbon atom is a little cube, calculate the volume of a carbon atom, volume = s^3:

V_{atom}: _____ cm^3/atom   _____ cm^3/atom

(c) *Calculating the volume of a mole of carbon atoms*

(i) Calculate the molar volume of carbon (diamond) by using the density of diamond (3.51 g/cm^3) and the atomic mass of a mole of carbon:

V_{mol}: _____ cm^3/mol   _____ cm^3/mol

(ii) Is the volume of a mole of diamond the same as the actual volume of a mole of carbon atoms?

(d) *Calculating Avogadro's number: Divide the volume of a mole of carbon atoms (diamond) calculated in 2(c) by the volume of a single carbon atom calculated in 2(b).*

(i) Calculate Avogadro's number from the appropriate ratio of volumes:

_____ atoms/mol   _____ atoms/mol

(ii) Calculate the average value of N_A from your results:

Average: _____ atoms/mol

(e) *Express your results as a number $\times 10^{23}$. Are you within a power of 10 of the accepted value of 6.02×10^{23}?*

CONSIDER THIS

Calculate an accurate value of Avogadro's number as the ratio of the electric charge on a mole of electrons (Faraday's constant) divided by the charge on a single electron (see Table 2 of the Appendix).

Calculate the edge length of a cube containing an Avogadro's number (6.02×10^{23}) of hydrogen atoms stacked to form a cube. Assume a hydrogen atom is equivalent to a tiny cube with edge length 1×10^{-10} m.

Now imagine an Avogadro's number of hydrogen atoms formed into a square one atom-layer thick. What would be the edge length of the square?

Finally, if an Avogadro's number of hydrogen atoms were arranged in a line, how long would the line be? Compare the length of this line to the distance from the earth to the sun, about 93 million miles.

Name _____   Date _____

In the method of estimating Avogadro's number that we used, there are a number of implicit assumptions:

1. We assume that the stearic acid molecules form a close-packed monolayer on the water surface.

2. We assume that the packing of stearic acid molecules is the same in the monolayer as in solid stearic acid.

3. We assume that carbon atoms are like little cubes stacked up in a linear fashion. A more accurate picture is one in which the C atoms are more nearly spherical, having a zigzag linkage with tetrahedral 109.5° C—C—C bond angles, as shown in Figure 10-3(A). If we assume this structure instead of the linear one, show that the size (diameter, s) of a carbon atom is linked to the thickness, t, of the monolayer by the expression.

$$s = \frac{1}{\sin 54.75°} \times \frac{t}{18},$$ where $\sin 54.75° = 0.8166$, increasing the value of s by a factor of $\frac{1}{0.8166} = 1.225$

When this sine factor is cubed, the calculated volume of a carbon atom is increased by a factor of $(1.225)^3 = 1.84$.

4. We assumed that the volume of a carbon atom was equivalent to a cube of edge length s, giving a volume of s^3. Show that if the volume of a sphere is given by the expression $V = \frac{4}{3}\pi r^3$, where r is the radius of the sphere, the volume of a sphere will be equal to $\frac{\pi}{6} \times s^3 = 0.524\,s^3$, where s is the diameter of the sphere inscribed in a cube of edge length s.

When we apply the two correction factors calculated for assumptions 3 and 4 to the calculation of the volume of a carbon atom, we see that their effect is to approximately cancel one another, that is, $1.84 \times 0.524 = 0.96$.

5. Finally, we assume that diamond consists of close-packed carbon atom cubes, whereas it actually has a more open structure in which each carbon atom is surrounded by four other carbon atoms with tetrahedral bond angles. In this structure, only about 34% of the volume of the unit cell is occupied by approximately spherical carbon atoms. Thus, the actual volume of a mole of spherical carbon atoms is about 0.34 times the volume of a mole of diamond.

If we incorporate this factor in the calculation of Avogadro's number, show that the calculated value of Avogadro's number, based on the simple assumptions, would be reduced by about a factor of 3.

6. What do we mean by the size of an atom? There is good evidence that the electron clouds surrounding isolated atoms do not have sharp, well-defined boundaries. That is, there is good reason to believe that atoms aren't hard spheres like billiard balls. We can rather accurately measure the internuclear distance between two atoms in a chemical bond. If the two atoms are the same, such as two carbon atoms, then we could define the radius of a carbon atom as one-half the internuclear distance (the C—C distance), meaning that the diameter of an atom would be equal to the internuclear C—C distance. Defined this way, the size (diameter) of a carbon atom is 1.54×10^{-10} m. Compare this value to the value of s that you calculated.

The Molar Volume of Dioxygen and Other Gases

Purpose

• Determine the molar volume of a gas.

• Gain practical experience in making quantitative measurements of gas volumes, using Dalton's law of partial pressures and the ideal gas law to calculate the volume of a gas at the standard temperature and pressure.

Pre-Lab Preparation

Avogadro's principle and the molar volume

This experiment deals with one of the most useful and important quantitative relationships involving gases—a relationship that depends on the following two ideas:

1. Under identical conditions of temperature and pressure, equal volumes of all gases contain equal numbers of molecules. This is called *Avogadro's principle* or *Avogadro's hypothesis.*

2. If we define a mole as a fixed number of any particle (atoms, molecules, or any particle of defined composition), then one mole of any substance contains the same number of particles as one mole of any other substance.

So, if a mole contains a fixed number of particles, then according to Avogadro's principle, a mole of *any* gas will occupy the same fixed volume, called the *molar volume of a gas,* measured under identical conditions of temperature and pressure.

The kinetic theory of gases and its consequences

We say that matter can exist in three states—as a solid, liquid, or gas. Solids and liquids are often called *condensed states of matter*, and the fact that solids and liquids exist implies the presence of inter-molecular forces of attraction between atoms and molecules in the solid and liquid states. If there were no intermolecular forces, every particle (atom or molecule) would be free to wander about constrained only by the walls of its container or, in the case of the

molecules forming the earth's atmosphere, being bound only by the gravitational attraction of the earth.

So, if we grant that, in contrast to liquids and solids, gas molecules can occupy the full volume of their container, why should each volume element of the container contain the same number of particles, whether the gas molecules are large or small?

An explanation offered in the seventeenth century proposed that gas molecules were like same-sized balls of wool or cotton that could expand or be compressed depending on how tightly they were squeezed. An equivalent, but slightly more sophisticated, explanation proposed that gas molecules were held through opposing forces of attraction and repulsion at locations equal distances apart, about which they oscillated in a sort of stable equilibrium.

A better model, and one that has survived all tests to date, is called the *kinetic theory of gases.* It assumes that there are, on average, large distances between molecules, that the forces between molecules are negligible, and that the molecules move in straight lines without interacting with other molecules, except at moments of collision. When molecules collide with one another or a wall, they rebound or change direction of motion just as in the collision of elastic spheres—like the collision of two balls on a pool table. In fact, if you could imagine all the balls on a pool table in constant motion, the image would be a reasonable two-dimensional analogy for a gas.

A consequence of the frequent collisions that occur in the gas is that the molecules acquire an average kinetic energy of translation (linear motion from point to point), which is dependent only on temperature and which is the same for all molecules, large and small. If they have the same kinetic energy, smaller molecules must travel, on average, faster than larger molecules.

The pressure exerted by the gas is the cumulative result of the impacts on the walls of the container, smaller molecules making impacts of smaller magnitude but making more of them because they move faster. As a result, the total force exerted is the same for all molecules, whether they have a mass that is large or small.

In more quantitative terms, when a molecule collides with the wall, the force exerted on the wall is proportional to the molecule's momentum (mass × velocity = mv) multiplied by the number of impacts per unit time (proportional to the velocity, v), so the force is proportional to mv^2. But $\frac{1}{2}mv^2$ is just equal to the kinetic energy of translation. And we know from experiment that the pressure (force/unit area) is proportional to the kelvin temperature, so the average kinetic energy must be proportional to the (kelvin) temperature. The relation given by the kinetic molecular theory is

$$\text{average kinetic energy of a molecule} = \frac{1}{2}mv^2 = \frac{3}{2}kT$$

where k is called the *Boltzmann constant* and is equal to the gas constant, R, divided by Avogadro's number.

To summarize, the mass and average velocity of a gas molecule are linked in such a way that as the mass, m, decreases, the square of the velocity, v^2, increases just enough to keep the product mv^2 constant (at constant temperature). An important physical insight arising from the kinetic theory of gases is that, for a fixed pressure, *the average number of molecules per unit volume (or the average volume per molecule) is determined only by the average kinetic energy (or the temperature) of the gas and does not depend on the nature of the gas.*

For a pressure of one atmosphere at room temperature, the average distance between molecules of a gas is about ten times the size of a molecule, so a gas is mostly empty space.

The molar volume at standard temperature and pressure

In this experiment, we will measure the mass and volume of a sample of dioxygen gas, O_2, and from these data and the ideal gas law calculate the volume at standard temperature and pressure (STP) conditions (0 °C, 760 torr). Then we will calculate the volume of 32.0 g (1 mol) of dioxygen gas.

An *ideal gas* is one in which (1) the attractive forces between molecules are vanishingly small and (2) the molecules are equivalent to "point masses" occupying a negligible volume compared with that of the container. All real gases deviate more or less from the ideal gas behavior described by the general ideal gas law because their molecules have some slight attraction for one another, and they do occupy some small volume of the container in which they are confined.

From measurements on real gases, the molar volume of an ideal gas has been determined to be 22.414 L STP ($T = 273.15$ K; $P = 1$ atm). This was done by calculating the molar volume (STP) at a series of progressively lower pressures and extrapolating to zero pressure, where the attractive forces between molecules and the volume occupied by the molecules are negligible. Most common gases, unless they have a high molecular weight or are measured quite near their condensation point (boiling point of the liquid), have molar volumes that do not deviate more than about 1% from the ideal volume at pressures around 1 atm.

Calculation of gas volumes

The quantity of a gas sample can be measured more conveniently by volume than by weight. In measuring the volume of a gas, it is also necessary to measure its temperature and pressure. Why? The separate laws relating pressure to volume and relating either pressure or volume to absolute temperature are

$$PV = k_1 \quad \text{and} \quad P = k_2T \quad \text{or} \quad V = k_3T$$

For a given gas sample, these three laws may be combined into one equation that shows the way in which all three variables—pressure, volume, and absolute temperature—are interdependent:

$$PV = kT \quad \text{or} \quad \frac{PV}{T} = k \tag{1}$$

Since any two corresponding sets of PV/T measurements will be equal to k and to each other for a given amount of gas, we can write

$$\frac{P_1V_1}{T_1} = \frac{P_2V_2}{T_2} \tag{2}$$

This equation can be transposed to give

$$V_1 = V_2 \times \frac{T_1}{T_2} \times \frac{P_2}{P_1} \tag{3}$$

Note in Equations (2) and (3) that if the temperature is constant ($T_1 = T_2$), the inverse proportionality of pressure and volume (Boyle's law) is expressed. Likewise, for constant pressures ($P_1 = P_2$), the direct proportionality of volume and absolute temperature (Charles's law) is expressed. If any five of the quantities in Equation (2) are known, the sixth can be calculated by simple algebra.

The general ideal gas law equation

For a specific amount of any gas, Equation (1) can be restated in its most general form:

$$PV = nRT \tag{4}$$

Here, n is the number of moles of gas, and R is a proportionality constant, called the *gas constant,* that has the same value for all gases under all conditions—namely, 0.08206 L•atm/mol•K. In all calculations in which this constant is employed, pressure must be expressed in atmospheres, volume in liters, and temperature in kelvins (represented by the symbol K).

As an example of the application of the general gas law where both volume and weight of a gas sample are involved, consider the following. What weight of chlorine gas, Cl_2, would be contained in a 5.00-L flask at 21 °C and at 600 torr ($^{600}/_{760}$ atm) pressure? Substituting in Equation (4), transposed to give n, the number of moles, we have

$$PV = nRT \quad \text{or} \quad n = \frac{PV}{RT}$$

$$n = \frac{\frac{600}{760}\,\text{atm} \cdot 5.00\,\text{L}}{0.08206\,\frac{\text{L atm}}{\text{mol K}} \cdot 294\,\text{K}} = 0.164\,\text{mol}$$

and, in grams,

$$0.164\,\text{mol} \times 70.9\,\frac{\text{g}}{\text{mol}}Cl_2 = 11.6\,\text{g }Cl_2$$

Aqueous vapor pressure and Dalton's law of partial pressures

When any gas in a closed container is collected over liquid water or is exposed to it, the water evaporates until a saturated vapor results—that is, until opposing rates of evaporation and condensation of water molecules at the liquid surface reach a balance. These gaseous water molecules contribute to the total gas pressure against the walls of the container. Thus, of all the gas molecules, if 3% are water molecules and 97% are oxygen molecules, then 3% of the total pressure is due to water vapor and 97% of the total pressure is due to oxygen. *Each gas exerts its own pressure regardless of the presence of other gases.* This is Dalton's law of partial pressures. Stated as an equation,

$$P_{\text{total}} = P_{H_2O} + P_{O_2}$$

or, if transposed,

$$P_{O_2} = P_{\text{total}} - P_{H_2O}$$

To illustrate Dalton's law, Figure 11-1 shows a mixture of oxygen molecules and water vapor molecules. In flask B, the water molecules have been removed, but all the oxygen molecules are still present, in the same volume. The pressure has been reduced by an amount equal to the vapor pressure of the water.

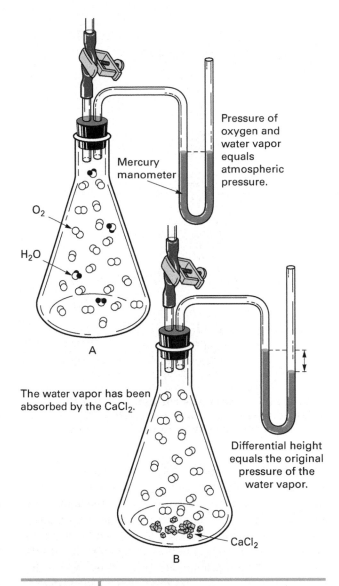

FIGURE 11-1 **When water vapor is removed from a gas mixture, the pressure is reduced. This is an illustration of Dalton's law of partial pressures. Note that in this figure the gas molecules are greatly exaggerated in size.**

Experimental Procedure

Special Supplies: 250-mL Erlenmeyer flask with a No. 5 one-hole rubber stopper; 500-mL Erlenmeyer flask with a No. 5 two-hole rubber stopper; 6- to 8-cm lengths of 6-mm-OD glass tubing; 30-cm length of 6-mm-OD glass tubing with a 90° bend 8 cm from one end; 10 × 75 mm test tube; 600-mL beaker; 3/16-in.-ID rubber tubing; pinch clamp; thermometer; 500-mL graduated cylinder; barometer; analytical balance with a 150-g capacity; pipet bulb.

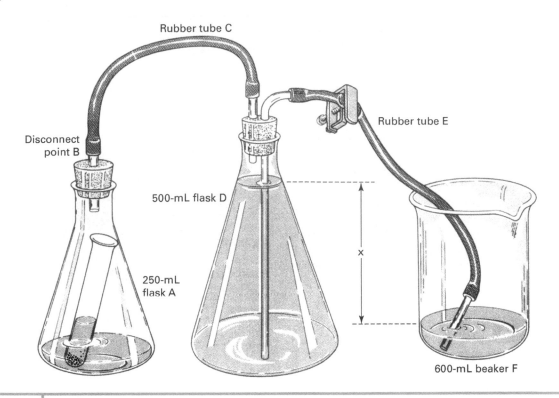

Rubber tube C

Disconnect
point B

Rubber tube E

500-mL flask D

250-mL
flask A

x

600-mL beaker F

FIGURE 11-2 **Apparatus for determining the molar volume of O₂. Flask A contains 3% aqueous H₂O₂. The inner test tube in flask A contains** **baker's yeast to catalyze the decomposition of H₂O₂ when the two are mixed.**

Chemicals: 3% (approx. 0.9 M) hydrogen peroxide; active dry yeast (Fleischmann's or other brands used in baking bread are suitable). Assemble the apparatus shown in Figure 11-2, taking care to lubricate each piece of glass tubing with a drop of glycerol and to wrap it in a towel while inserting it in the rubber stopper. (At this stage, the 250-mL Erlenmeyer flask—flask A—will be empty. The solution and test tube will be added later.)

The principle of the method is simple. We determine the mass of oxygen produced in a chemical reaction by weighing the reaction flask before and after the chemical reaction. The weight loss corresponds (to a good approximation) to the mass of oxygen produced. The oxygen that is produced flows into the middle flask (flask D of Figure 11-2), displacing an equal volume of water into the beaker. By measuring the volume of oxygen liberated at a known temperature and pressure, we can determine the molar volume of dioxygen. The oxygen is produced by the catalyzed decomposition of hydrogen peroxide (H₂O₂) solution, the same reaction we used to prepare oxygen in Experiment 7.

$$H_2O_2(aq) \xrightarrow[\text{catalyst}]{\text{yeast}} H_2O(l) + \tfrac{1}{2} O_2(g)$$

Before placing any solution in the 250-mL Erlenmeyer flask, slide an empty 10 × 75 mm test tube

into the flask and check that the test tube will stand at an angle inside the flask, without tipping over and lying flat. (Some older style 250-mL flasks will allow the test tube to lie flat; if this happens, it will be necessary to go to the next smaller size flask, 200 mL.) Now remove the test tube from the empty flask.

Measure 30 mL of 3% H₂O₂ solution in a graduated cylinder and pour it into the 250-mL Erlenmeyer flask (flask A of Figure 11-2).

Put 50 to 100 grains (0.050 g) of active dry yeast into a clean, dry 10 × 75 mm test tube, being careful not to get any on the outside of the tube. Holding the 250-mL Erlenmeyer flask at an angle of about 45°, slide the test tube into the flask or lower the test tube with a pair of forceps. Next, place the one-hole stopper, with a short length of glass tubing already inserted, in the flask, and seat the stopper firmly. During this step, take care not to allow any solution to run into the test tube—yeast must not contact the hydrogen peroxide solution before the stopper is inserted. If it does, the reaction will start prematurely, and some oxygen will be lost. If this happens, you must start over with a clean flask and test tube and a fresh portion of hydrogen peroxide solution.

Weigh the flask, with contents, stopper, and glass tube, as precisely as possible on the analytical balance (nearest 0.001 g). (The balance must have at least 150-g capacity.) Record the mass.

Fill the 500-mL Erlenmeyer flask D (in Figure 11-2) to the neck with tap water. Put in place the two-hole rubber stopper with the short glass inlet tube and the long glass siphon tube. Connect the rubber tubing C to the short glass inlet tube and the rubber tube E to the long siphon tube, as shown in Figure 11-2. (Do not yet connect the free end of tube C to the reaction flask containing the hydrogen peroxide.)

Fill beaker F about one-third full with tap water. Make sure the pinch clamp is open. Then, using a pipet bulb, blow into the open end of rubber tube C (whose free end is not yet connected at point B) to start water siphoning into the beaker. (A pipet bulb is preferred to blowing into tube C with your mouth; you want to avoid getting saliva into tube C because it would catalyze the decomposition of hydrogen peroxide.)

Raise and lower beaker F to rapidly siphon water back and forth through tube E to sweep out all air bubbles. Fill the 500-mL flask D with water until the water level is halfway up the narrow neck of flask D. Holding the beaker F so that its water level is the same as the water level in the neck of flask D, connect the free end of tube C to the reaction flask at point B.

Test the apparatus for leaks by raising beaker F as high as possible without removing the tubing from the beaker. The water level in flask D should move a little and then remain fixed. If the water level continues to change, a leak is present. Do not proceed until the leak is located and fixed.

Equalize the pressures inside and outside flask D by raising the beaker until both water levels are the same. While you continue to maintain the same levels in flask D and beaker F, have another student close the pinch clamp. Pour out all of the water in beaker F, but do not dry it. Replace the tube in the beaker, and *open the pinch clamp*. A little water will flow out, and this should be retained in the beaker. During the generation of O_2 that follows, see that the end of tube E is below the surface of the water in the beaker.

Check to make sure the pinch clamp is open. Then tip flask A carefully while tapping it so that the yeast grains fall into the solution, or solution runs into the test tube. (Avoid getting any solution on the rubber stopper.) Swirl the contents of the flask to mix the reactants. Note and record any changes you see as the reaction progresses. Frequently swirl or gently shake the flask to speed the release of oxygen from the solution. After a few minutes, put your hand on the bottom of the flask. Is heat released or absorbed in the reaction?

After a few minutes, the reaction will be nearing completion. Note very carefully the water level in the beaker (F). Continue frequent agitation of the reaction flask (A). When the water level remains unchanged for at least 5 min, you can assume that the decomposition of H_2O_2 is complete.

Adjust the water levels in flask D and beaker F until they are equal. Ask another student to close the pinch clamp on tube E.

Loosen the stopper and insert your thermometer into the gas in flask A. After a minute or two, note and record the temperature. Measure and record the temperature of the gas in flask D in the same way. (In the calculations that you will make later, you will take the average of these two temperatures as the temperature of the oxygen gas.)

Disconnect flask A at point B and reweigh it with its contents, rubber stopper, and glass tube. Record the mass.

Measure the volume of oxygen produced by carefully pouring the water in beaker F into a 500-mL graduated cylinder. (Alternatively, you can weigh the water by pouring it into a tared beaker on a 0.1-g balance. The volume of water in mL is equal to its mass in grams, within 0.3%, at room temperature.) Record the volume of water displaced by the oxygen gas.

Measure and record the barometer reading and the temperature of the barometer. [The corrected barometric pressure is obtained by subtracting the approximately 3 mm Hg (torr) barometer temperature correction found in the *CRC Handbook of Chemistry and Physics;* this will correct for the difference in the coefficients of expansion of the brass scale of the barometer and the column of mercury.] You will also need the vapor pressure of water, which can be found in Table 3 of the Appendix.

Repeat the experiment if time permits. From your data, calculate the volume at standard temperature and pressure of 32.0 g (1 mol) of dioxygen gas, O_2. All solutions may be safely disposed of down the drain.

CONSIDER THIS

DO OTHER GASES HAVE THE SAME MOLAR VOLUME AS DIOXYGEN? The method we employed to determine the molar volume of dioxygen can also be applied to other gases. For example, by placing 50 mL of 1 M HCl in the reaction flask and 1.0 ± 0.1 g of sodium bicarbonate ($NaHCO_3$) in the 10×75 mm test tube, you can generate $CO_2(g)$ by the reaction

$NaHCO_3(s) + HCl(aq) \rightarrow NaCl(aq) + H_2O(l) + CO_2(g)$

Does a mole of CO_2 gas measured at standard temperature and pressure occupy the same volume as a mole of oxygen (STP)? Try it and see. (The solutions may be disposed of down the drain after neutralization of all of the HCl with sodium bicarbonate.)

ESTIMATING MOLECULAR SIZE AND THE AVERAGE DISTANCE BETWEEN MOLECULES IN A GAS. The density of solid nitrogen is about 1.0 g/cm^3 at its freezing point. If we assume that in solid nitrogen the molecules are about as close together as they can get, we can estimate the approximate size of a molecule from the average volume occupied by one molecule. To do this, first determine the volume of a mole of solid nitrogen. Then divide this volume by Avogadro's number to get the volume occupied by one nitrogen molecule. If we assume this volume is a tiny cube containing one molecule of nitrogen, the edge length of the cube will be the approximate size of a nitrogen molecule. (If you've done the calculation correctly, you should get a value of about 0.36 nm.)

At room temperature and pressure, one mole of a gas occupies about 25 L (0.025 m^3). Knowing that there is an Avogadro's number of molecules in one mole, calculate the average volume per molecule. If we imagine that this volume is a little cube with, on average, one molecule located in the center of each cube, we see that the average distance between molecules will be equal to the edge length of the cube. Comparing the results of the calculation for liquid and gaseous nitrogen, show that the average distance between molecules in the gas is about ten times the size of a molecule.

Bibliography

Bedenburgh, J. H.; Bedenburgh, A. O.; Heard, T. S. "Oxygen from Hydrogen Peroxide: A Safe Molar Volume—Molar Mass Experiment," *J. Chem. Educ.* **1988,** *65,* 455–456.

The Molar Volume of Dioxygen

Name_____

Date _____ Section _____

Locker _____ Instructor_____

Data

Data	Trial 1	Trial 2
Mass of flask A, contents, stopper, and glass tube before reaction	g	g
Mass of flask A, contents, stopper, and glass tube after reaction	g	g
Temperature of the gas in flask A	°C	°C
Temperature of the gas in flask D	°C	°C
Volume of oxygen collected	mL	mL
Barometer reading (temperature of barometer _____ °C)	torr	torr
Aqueous vapor pressure at temperature of gas (average the two temperatures)	torr	torr

Observations

1. Describe the visible changes that take place during the reaction.

2. As the reaction proceeds, does flask A get warmer or cooler to the touch? Does the reaction absorb or evolve heat (energy)?

Calculations

Data	Calculations	Trial 1	Trial 2
Mass of oxygen		g	g
Temperature, absolute[a]		K	K
Corrected barometric pressure[b]		torr	torr
Pressure of oxygen alone, in flask A		torr	torr
Volume of oxygen at standard conditions		mL	mL
Moles of dioxygen, O_2		mol	mol
Molar volume of O_2 at standard conditions		L/mol at STP	L/mol at STP

[a]Average the two temperatures.
[b]Look in the *CRC Handbook of Chemistry and Physics* for the barometer temperature correction.

Sources of Error

There are two sources of error in this experiment, which fortuitously almost cancel one another. Circle the correct answer in the following statements.

(a) Initially the flask that you weigh is filled with air. At the end of the reaction it is filled with oxygen. Because oxygen is denser than air, at the end of the reaction the flask is (heavier, lighter) than it would be if filled with air.

(b) If the oxygen that is evolved is saturated with water vapor, some water will be carried over as well as evolved oxygen. At the end of the reaction, this effect would cause the flask to be (heavier, lighter) than if no water were carried over.

Calculate the molarity of a solution that is 3.0% hydrogen peroxide by weight (3 g H_2O_2/100 g solution), assuming that the density of the solution is 1.00 g/mL. For one of your trials, calculate the moles of hydrogen peroxide that you put in the flask, the moles of oxygen evolved, and the ratio mol O_2/mol H_2O_2. Is the value you calculate for the ratio consistent with the stoichiometry of the decomposition reaction?

$$H_2O_2(aq) \rightarrow H_2O(l) + \tfrac{1}{2}O_2(g)$$

Deviations from the Ideal Gas Law

All real gases deviate to some extent from the behavior of perfect gases. At standard conditions, the density of $O_2(g)$ is 0.0014290 g/mL, that of $H_2(g)$ is 0.00008988 g/mL, and that of $CO_2(g)$ is 0.0019769 g/mL.

(a) Using these values and the exact atomic weights, calculate the molar volume of each of these gases, in milliliters, to five significant figures.

O_2 _____

H_2 _____

CO_2 _____

(b) Correlate the molar mass, molecular size, and bond polarity of these three gases with the molar volume of a perfect gas.

Which gas deviates least? _____

Which gas deviates most? _____

(c) Suggest reasons to explain the deviations of O_2, N_2, and CO_2 from ideal gas behavior.

DO OTHER GASES HAVE THE SAME MOLAR VOLUME AS DIOXYGEN? Using the same apparatus and procedure as used for dioxygen, carry out measurements and calculations for determining the molar volume of carbon dioxide, as briefly described in the first paragraph of the Consider This section of the Experimental Procedure. Keep a record of your measurements and calculations in your laboratory notebook. Report the final result of your measurements here.

ESTIMATING THE SIZE OF A MOLECULE AND THE AVERAGE DISTANCE BETWEEN MOLECULES IN A GAS. Carry out the calculations as described in the Consider This section of the Experimental Procedure. Why are gases springy (easily compressible), while liquids like water are nearly incompressible?

The Molar Mass of a Gas

A Reduced Scale Method

Purpose

• Determine the molar mass of an unknown gas using two methods, one using the *ideal gas law* and the other using *Avogadro's law*.

• Explore the phenomenon of buoyancy in gases.

Pre-Lab Preparation

In a prior experiment, we estimated Avogadro's number and determined the molar volume of oxygen. To refresh your memory, let's restate Avogadro's law: *Equal volumes of gases contain equal numbers of molecules.* When Avogadro first published this statement in 1811, it was just a hypothesis. It took several decades for it to be accepted and only much later did his statement reach the status of a scientific law (a statement that summarizes a series of observations that has no known exceptions). This means that any time we take an Avogadro's number of molecules of a particular gas (which is the definition of one mole), we can expect that the volume of the gas will be the same as the volume of one mole of any other gas. (This expectation is, of course, subject to the usual restrictions that the gases must behave as ideal gases and that the volumes must be measured at the same pressure and temperature.) The ratio of the masses of the gas samples will match the ratio of the molecular masses of the gas molecules according to Avogadro's law.

We can also use the *ideal gas law* and measurements of the mass, pressure, volume, and temperature of a gas sample to calculate the molar mass.

In this experiment, we will first measure the mass and volume of a sample of gas contained in a plastic syringe at the laboratory temperature and pressure. This is fairly simple to accomplish using the syringe apparatus shown in Figure 12-1, but because our experiment takes place at the bottom of an atmosphere that creates a buoyant force on any container, we must think through the measurement process

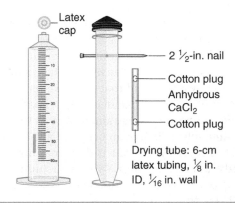

Latex cap

2 $\frac{1}{2}$-in. nail

Cotton plug

Anhydrous $CaCl_2$

Cotton plug

Drying tube: 6-cm latex tubing, $\frac{1}{8}$ in. ID, $\frac{1}{16}$ in. wall

FIGURE 12-1 | **The 60-mL syringe apparatus for gas molecular weight determination.**

carefully. From these measurements and the ideal gas law we can calculate the molar mass of our gas. We can also compare the mass of our gas sample to that of a known gas. This allows us to calculate the molar mass of our sample from Avogadro's law. We can use air for our "known gas" comparison and its average molar mass to calculate the molar masses of other gas samples.

When tanks of compressed gases are available, they are convenient, but by preparing your own sample of gas you will have an opportunity to learn some additional chemistry. Directions for the preparation of other gas samples can be found at the Web site of Bruce Mattson listed in the Bibliography.

Gas density and buoyancy

The density of a gas is equal to its mass divided by the volume it occupies at a particular temperature and pressure. The density is directly proportional to the molecular weight, as can be seen by deriving a modified form of the ideal gas law. Let's start with the ideal gas law in the form

$$PV = nRT \qquad (1)$$

where P is the pressure, V is the volume, n is the number of moles of gas, R is the gas constant, and T

is the Kelvin or absolute temperature (K). The molar mass of a gas (or any substance), M, is the mass of the sample, m, divided by the number of moles, n (take a moment to verify that the units are correct):

$$M = \frac{m}{n} \quad \text{or} \quad n = \frac{m}{M} \tag{2}$$

We can substitute m/M for n in Equation (1) and rearrange the resulting equation to obtain the density of a gas, m/V:

$$\text{density} = \frac{m}{V} = \frac{PM}{RT} \tag{3}$$

From Equation (3) we see that the density, m/V, is proportional to the molar mass of the gas, M.

If we fill a lightweight container, such as a balloon, with a gas that has a density much less than the density of air, we may find that when we try to weigh the balloon of gas on a balance it may float or even rise to the ceiling if not restrained. Why is this?

The atmosphere is like an ocean of air, and when an object is immersed in air it displaces air just as your body displaces water when you jump into a swimming pool. Just as you can float in water, the balloon will rise in the air if the combined mass of the balloon and the gas in it is less than the mass of the air that is displaced. Since the pressure (and density) of air decreases with altitude [note the effect on density of decreasing P in Equation (3)], the balloon will expand as it rises and will either explode or stop rising when it reaches an altitude where the mass of air displaced and the mass of the balloon plus the gas are equal. You can easily observe this buoyancy effect by using a lighter-than-air gas (such as natural gas, which is largely methane, CH_4) to blow bubbles in a soap solution, and then repeating the experiment with a heavier-than-air gas (such as carbon dioxide, CO_2).

Hot-air and lighter-than-air balloonists are able to enjoy their sport by filling their balloons with a gas that is less dense than air. Hot-air balloons use propane heaters at the base of the balloon to heat air. Hot air is less dense than cold air. [Note the effect on the density of a gas of increasing T in Equation (3).] Lighter-than-air balloonists use a gas with a low molecular weight, such as helium, to obtain the necessary lower density. [Note the effect of decreasing M in Equation (3).]

The effect of buoyancy must also be considered when very accurate weighings are to be made; since our gas samples weigh so little, all of our measurements need to be quite accurate. A sample placed on the balance pan displaces a certain amount of air. There will be a net effect of *buoyancy* (apparent mass is less than the true mass) whenever the density of the object being weighed is less than the

density of the standard masses used to calibrate the balance, because a sample with smaller density than the calibration masses displaces a larger volume of air. We can compensate for this error by weighing a syringe "filled" with a vacuum. The difference in masses between a syringe filled with a vacuum and a syringe filled with a gas gives the net mass of the gas. In most of the weighings you do in the laboratory, the buoyancy corrections will be negligible, but if you are weighing samples of gases, as in this experiment, the buoyancy effect is significant and can be larger than the sample masses.

If we substitute the definition of molar mass, Equation (2), into the *ideal gas law*, Equation (1), we obtain

$$PV = \left(\frac{m}{M}\right)RT \tag{4}$$

Rearranging gives an equation for molar mass in terms of measurable quantities:

$$M = \frac{mRT}{PV} \tag{5}$$

Avogadro's law states that equal volumes of two different gases (at the same pressure and temperature) contain the same number of molecules. If this is the case, then the mass ratio of two different gas samples should match the ratio of their molar masses, that is,

$$\frac{M_1}{M_2} = \frac{m_1}{m_2} \tag{6}$$

Air is 78.08% nitrogen, N_2; 20.95% oxygen, O_2; 0.93% argon, Ar; and 0.03% carbon dioxide, CO_2, and thus has an apparent molar mass of 28.96 g/mol. If we compare the mass of a gas sample to the mass of an air sample in the same-sized container at similar conditions, we should be able to determine the molar mass of the gas. This method is usually more accurate than the previous one because it involves fewer measurements and errors that tend to cancel out when Equation (6) is used.

Experimental Procedure

Special Supplies: 60-mL plastic Luer-lok syringes and latex syringe caps (available from Flinn Scientific); $2\frac{1}{2}$-in. finishing nails; 6-cm lengths of latex tubing ($\frac{1}{8}$-in. ID, $\frac{1}{16}$-in. wall thickness); cotton wool; the syringe is modified with the addition of a nail "stop" to ensure consistent volume determinations (see instructions below and Figure 12-1); drying tubes are made from the 6-cm lengths of latex tubing (see instructions below and Figure 12-1); natural gas supply; nonflammable, nontoxic compressed gases in cylinders or lecture bottles (suitable gases include He, Ar, N_2, O_2, N_2O, CO_2, SF_6, and freons); top-loading balance accurate to ± 0.001 g or analytical balance.

Chemicals: Silicone oil lubricant for the syringe plunger; anhydrous calcium chloride, $CaCl_2$, coarsely pulverized (for the drying tubes).

� ▬ NOTES TO INSTRUCTOR ▬ ▪

It is recommended that you prepare beforehand the 60-mL syringes (available from Flinn Scientific) so that the volume can be held reproducibly at a volume at or just below 60 mL even under "vacuum" conditions. This can be accomplished by adjusting the plunger to the 60-mL mark, marking the plunger even with the flange of the syringe barrel, and drilling a hole in the plunger at a 45° angle to the plunger ribs so that a $2\frac{1}{2}$-in. finishing nail will fit smoothly into the hole without much play. Alternatively, the nail can be held with pliers, heated, and then forced through the plastic plunger, but this method is not as safe or precise as drilling.

You may wish to predetermine the actual volume of each syringe when it is filled to the stop to avoid the need for your students to fill them with water for this procedure, with the necessity of drying the syringes thoroughly before gas weighings are made. (Or direct them to determine the volume by filling the syringe with water at the end of the experiment after they have completed all of the gas weighings.)

Lubricate the rubber plunger with a few drops of silicone oil to insure that the plunger remains airtight and moves easily in the syringe.

The drying tubes can be made beforehand by placing a small plug of cotton at the end of a 6-cm-long section of $\frac{1}{8}$-in. ID rubber tubing using a stirring rod. Then approximately 3 cm of anhydrous calcium chloride (coarsely pulverized) is added, with another cotton plug tamped in at the end, leaving enough space at both ends of the tube to connect to the syringe and the gas supply.

⚠ SAFETY PRECAUTIONS:
It is recommended that no toxic or flammable gases be used in this experiment.

All cylinders should be chained or securely fastened so that they cannot topple, and you should be instructed in the proper way to use the gas regulators or valves.

Carefully adjust the flow rates of compressed air or unknown gas *before* connecting the tubing to the drying tube to avoid accidents resulting from sudden pressure surges.

If you use Bruce Mattson's procedures for generating gas samples (http://mattson.creighton.edu/Microscale_Gas_Chemistry.html), be sure to follow the safety instructions included in his directions.

☠ WASTE COLLECTION:
If nontoxic gases are used as recommended, they can be vented into the laboratory; otherwise a hood will be necessary.

1. Explorations of Buoyancy Remove the latex cap from the syringe and pull the plunger all the way out. Lubricate the rubber seal of the syringe with a thin coating of silicone oil. (There should be no silicone oil residue remaining on the syringe, just a very thin coat on the rubber seal.) Reinsert the plunger fully into the barrel and replace the cap. Weigh the empty syringe, with its latex cap and nail stop, to an accuracy of at least ± 0.001 g and record the value. [Note that *all* weighings in this lab should include the entire apparatus (the syringe, latex cap, and nail stop) and should be made to at least ± 0.001-g precision.] You also must be careful not to lose the cap or nail and to use the same cap and nail stop throughout the entire experiment.

Next remove the cap, attach the drying tube, and then slowly draw the plunger out of the syringe until you reach the 60-mL mark; reinsert the nail stop, remove the drying tube, and replace the cap. Before you weigh the syringe apparatus, predict whether the syringe full of air will weigh more, less, or the same as the empty syringe and explain your prediction. Weigh the syringe full of dry air and record the mass. If your prediction and reasoning were incorrect, explain your observed results.

When weighing the syringe with plunger extended on a top-loading balance, you can weigh the syringe in a vertical position on the balance pan, as shown in Figure 12-2. If you are using an analytical balance, you may have to weigh the syringe in a horizontal position on the balance pan with the balance

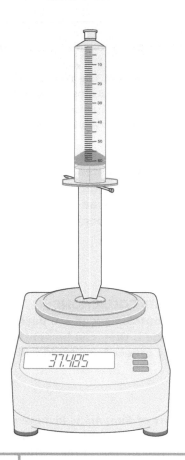

FIGURE 12-2 | **Weighing the syringe apparatus on a top-loading balance.**

chamber doors ajar, allowing the ends of the extended syringe to stick out. Using either method, you must be careful to avoid air currents that might disturb the mass measurement.

In the next step, you and your partner are going to pull back the plunger with the cap in place, forming a "vacuum" in the syringe, and then you will weigh it.[1] Before you perform this operation, have a conversation with your partner and predict whether you expect the apparatus to weigh more, less, or the same as the previous weighings (the empty syringe and the syringe containing 60 mL of air). Record your prediction and your reasoning.

Remove the cap and press the plunger fully into the syringe, then replace the cap. With your lab partner's help, pull the plunger out of the syringe with the cap in place until it reaches the 60-mL mark and you can reinsert the nail stop to hold the syringe in this configuration. Why is it so hard to pull the plunger out when the syringe is capped?

---

[1] It is a bit odd to talk about "forming a vacuum" or the syringe "containing a vacuum," since a vacuum consists of nothing, but it is useful to refer to this "nothing" as something.

Weigh the syringe (including the cap and nail) and record the value. Was your prediction correct? If not, rethink your hypothesis and explain the results of your three weighings.

With your partner's help, draw back the plunger far enough to remove the nail stop and let the plunger slowly move back into the syringe.

2. The Molar Mass of "Natural" Gas Connect your syringe to the lab's "natural" gas supply jet using the rubber drying tube. Open the gas valve and slowly fill the syringe at least three-fourths full of gas by slowly drawing back the plunger. Shut off the gas valve and disconnect the syringe from the drying tube momentarily to empty the syringe by pushing the plunger fully in. (This step should purge the syringe of most of the air that may have remained in it.) Then reconnect the drying tube to the syringe, reopen the gas valve, and slowly pull back the plunger of the syringe until it is a little beyond where you can reinsert the nail stop. Insert the nail, press the plunger in until the nail is tight up against the syringe body, and recap your syringe.

Weigh the syringe (plus cap and nail stop) and record the mass. Measure the room temperature and record this value. Determine the room's barometric pressure as directed by your instructor and record this value.

Did the syringe with natural gas in it weigh more, less, or the same as the syringe filled with air? How does it compare to the mass of the syringe containing a vacuum? Explain these results.

Using Equation (4), calculate the molar mass of natural gas, noting that the m in Equation (4) refers to the mass of the gas sample only, which would be the mass of the syringe containing the natural gas sample minus the mass of the syringe containing the vacuum (your third weighing). Are your results consistent with the statement that natural gas is almost pure methane, $CH_4(g)$?

Repeat the experimental steps and calculation for a second sample of natural gas.

3. The Molar Mass of an Unknown Gas Ask your instructor to identify the source of your second gas sample. If your supply is a cylinder containing the unknown gas, you will use the same procedure that you just used for natural gas to make your measurements. Because the gas pressure may be much greater than the natural gas supply, you will need to exercise caution when opening the valves on the cylinder to avoid blowing the plunger out of the syringe. If the gases are toxic, you may need to perform the fill and purge steps in the hood, but once the syringe is capped you can perform the measurements in the room.

You will also follow the measurement procedures in part 1 if you are going to generate samples of gas to weigh using directions from the Web or your instructor. In this case, high pressures should not be encountered, but you will need to work carefully to ensure that your samples are dry and are not contaminated with air.

As you weigh your gas sample, be sure to note whether the syringe containing your sample is heavier or lighter than the air-filled syringe and the syringe containing a vacuum.

4. Determining the Volume of the Gas Syringe (optional if determined previously)

Your instructor may give you a previously measured syringe volume to avoid the need to wet the syringes, or you may be directed to determine the volume after completion of all the gas weighings.

If you are to determine the syringe volume, remove the cap from the syringe and insert the plunger fully into the syringe. Replace the latex cap and weigh the empty syringe, cap, and nail stop to an accuracy of ±0.1 g. Fill a 250-mL beaker with water and place the tip of the syringe halfway into the beaker. Remove the cap and slowly pull the plunger back until the syringe is three-fourths full of water. Remove the syringe from the water and hold it with the tip up. Tap the syringe to dislodge any air bubbles from the sides and then gently push the plunger into the syringe until all air bubbles have been forced from the syringe. (You will probably want to perform this step over a sink or a container to catch the water.)

Reinsert the syringe, tip down, into the beaker and draw the plunger back slightly beyond the point that would allow you to insert the nail stop into its hole. Remove the syringe from the water and hold it with the tip up while you gently push the plunger until it stops with the nail stop against the barrel. Wipe the tip and exterior of the syringe with a paper towel and then recap the syringe. Weigh the syringe full of water, latex cap, and nail stop to the nearest ±0.1 g. The mass difference in the two weighings is the mass of water required to fill the syringe. Measure and record the temperature of the water used to fill the syringe. Calculate the volume of the syringe from the mass and known density of water at the temperature of the water.

5. Molar Mass Calculations Using Avogadro's Law

Since our syringe with its stop ensures a consistent volume in all of our samples, and the pressure and the temperature of the room do not change significantly between measurements, we should be able to calculate the ratio on the right-hand side of Equation (6) and use it to calculate the molar mass of an unknown gas if we know the molar mass of one of our gases.

You can measure the mass of a known gas in your apparatus and use it as the standard to calculate the molar mass of a second gas using Equation (6) or you can use the molar mass of air as your standard. Even though air is not a pure compound, the composition of dry air is essentially constant at 78.09% nitrogen, N_2; 20.95% oxygen, O_2; 0.93% argon, Ar; and 0.03% carbon dioxide, CO_2, giving an apparent molar mass of 28.96 g/mol.

CONSIDER THIS

1. Weigh an inflated basketball, volleyball, or soccer ball to at least the nearest 0.1 g. Deflate but do not collapse the ball and reweigh. Collapse the ball as much as possible and reweigh. Discuss your results.

2. You observed the effects of buoyancy as you weighed the natural-gas-filled syringe. Determine the balloon diameter needed to create a buoyant force sufficient to lift the mass of your body and the balloon mass if the balloon contains methane. (Assume that the balloon is spherical, the temperature is equal to 25 °C, and the combined mass of your body and the balloon is 120 kg.) Compare this diameter to the balloon diameters needed for helium-filled and hydrogen-filled balloons. Why is helium usually used? The syringe with the vacuum inside had the largest buoyant force; why not build a "vacuum balloon"?

Bibliography

Alyea, H. N. "Syringe Gas Generators," *J. Chem. Educ.* **1992,** *69,* 65.

Mattson, B.; Greimann, J.; Dedhia, R.; Saunders, E. "Microscale Gas Chemistry, Part 17. Molar Mass of Gas Determination," *Chem13 News,* **September 2001,** 4–6. *Chem13 News* is published nine times a year by the Department of Chemistry, University of Waterloo, Waterloo, Ontario, Canada N2L 3G1.

Mattson, B. *Microscale Gas Chemistry,* 2nd ed., Educational Innovations, Norwalk, CT, 2001. (A collection of articles that appear as a continuing series in *Chem13 News.*) The bound collection may be purchased from Educational Innovations, Flinn Scientific or Fisher Scientific. Professor Bruce Mattson of Creighton University maintains a Web site describing all of the gas experiments using syringes to prepare and manipulate gases: http://mattson.creighton.edu/Microscale_Gas_Chemistry.html.

Ramette, R. W. "Why Does a Helium-Filled Balloon 'Rise'?" *J. Chem. Educ.* **2003,** *80,* 1149–1150.

The Reactivity of Metals with Hydrochloric Acid

Purpose

• Illustrate the concept of equivalent mass by determining the mass of a metal that reacts with hydrochloric acid to form 0.5 mol of hydrogen gas, H_2.

• Gain experience with making measurements on gas samples and relating these values to the stoichiometry of the chemical reactions taking place in the system.

Pre-Lab Preparation

The concept of equivalent mass

Equivalent mass, as the adjective "equivalent" implies, designates the relative amounts of substances that are chemically equivalent—that is, the masses of those substances that react with or replace one another in chemical reactions. When a metal reacts with oxygen to form an oxide, the metal's equivalent mass is defined as the mass of metal that reacts with 0.25 mol (8.000 g) of oxygen gas (O_2). Here we will consider the equivalent mass of a metal to be the mass of metal that reacts with 1 mol of hydrogen ions (H^+) to produce 1 mol of hydrogen atoms, or 0.5 mol of hydrogen gas (H_2) (1.008 g). The reaction is a replacement reaction: The metal reacts with hydrochloric acid, HCl, to replace hydrogen atoms. It can also be classified as an *oxidation–reduction* reaction because the metal is *oxidized* to a positively charged metal ion (*increase* in oxidation state) and the H^+ is reduced to H_2 (*decrease* in oxidation state). We will measure the volume of hydrogen gas produced when a weighed sample of metal is added to an excess of hydrochloric acid. For example,

$$Ca(s) + 2\,HCl(aq) \rightarrow CaCl_2(aq) + H_2(g)$$

Note that for each mole of calcium that reacts to form Ca^{2+} ions (ionic charge +2), 1 mol of hydrogen gas is formed. Therefore, here the equivalent mass of calcium is half its atomic mass. Equivalent mass is always the atomic mass (for elements that form ions with ±1 charge) or a fraction of it, depending on the

charge of the ion produced. So, a general definition of equivalent mass is

$$\frac{\text{equivalent}}{\text{mass}} = \frac{\text{atomic mass}}{\text{change in charge on atom}} \quad (1)$$

Some elements, particularly transition metals like iron, can have more than one ionic charge in their compounds and therefore can have *more than one possible equivalent mass.* Consider the following possible reactions:

$$Fe(s) + 2\,HCl(aq) \rightarrow FeCl_2(aq) + H_2(g) \quad (2)$$

$$2\,Fe(s) + 3\,Cl_2(g) \rightarrow 2\,FeCl_3(aq) \quad (3)$$

$$2\,FeCl_2(aq) + Cl_2(g) \rightarrow 2\,FeCl_3(aq) \quad (4)$$

In Reaction (2), the change in charge is +2; in Reaction (3), it is +3; and in Reaction (4), it is +1. The corresponding equivalent masses of iron are, respectively, 27.92 (half the atomic mass), 18.62 (one-third of the atomic mass), and 55.85 (the atomic mass). *Thus, for elements that can have multiple ionic charges (or oxidation states), you must always specify the reaction before determining equivalent mass.* An alternative definition of equivalent mass is *that mass of substance that gives up or accepts 1 mol of electrons.* The defining equation then becomes

$$\frac{\text{equivalent}}{\text{mass}} = \frac{\text{atomic mass}}{\text{moles of electrons transferred}} \quad (5)$$

Thus, in Reaction (2), iron changes from an oxidation state of 0 to oxidation state +2. It loses 2 mol of electrons per mol of Fe in doing so. The equivalent mass of Fe *in this reaction* is therefore

$$\frac{55.85}{2} = 27.92$$

This is the same result obtained by means of Equation (1). Verify that the equivalent masses for Reactions (3) and (4) are 18.62 and 55.85, using Equation (5).

Equation (5) emphasizes that the replacement reactions studied in this experiment can also be classified as oxidation–reduction reactions, in which a transfer of electrons has caused changes in oxidation states. These concepts will be studied in greater detail in later experiments.

You should keep the three (related) definitions of equivalent mass in mind as you proceed:

• The mass of a metal that reacts with acid to produce ½ mole of $H_2(g)$

• The mass of a metal that reacts with acid to produce 11,200 mL of $H_2(g)$ at standard temperature (0 °C) and pressure (760 torr)

• The atomic mass of a metal divided by the change in charge that occurs in the reaction

Experimental Procedure

Special Supplies: Thermometer; pieces of metals (cut to size or issued as unknowns); 50-mL buret (for Method A only); fine copper wire; 25 × 250 mm test tube; two 500-mL Erlenmeyer flasks (for Method B).

Chemicals: Concentrated (12 M) hydrochloric acid, HCl; $Na_2CO_3(s)$.

 SAFETY PRECAUTIONS: Concentrated hydrochloric acid is a lung irritant and causes skin and eye burns. Handle with care. Dispense it in a well-ventilated fume hood. Clean up any spills immediately. Protect your hands with plastic gloves.

NOTE **The analysis samples for this experiment may be issued (1) as unknown metals, for you to calculate the equivalent mass; (2) as preweighed samples, for you to calculate and report the sample mass from the known equivalent mass; or (3) as Al–Zn alloys of different compositions, for you to calculate the percentage composition from the known equivalent masses. Prepare the units in your report accordingly, *before beginning the experiment.***

Your instructor will designate which of the preceding analyses and calculations you will do and which of the following alternate procedures you will follow. Method A is faster, but it limits you to smaller samples, to a fixed acid concentration, and to room temperature. Method B enables you to use larger samples and to control both temperature and acid concentration during the reaction.

NOTES TO INSTRUCTOR

Samples may be preweighed by a stock assistant on a rapid single-pan balance or, if they are in wire or ribbon form, cut to exact length to give a known mass and then be individually coded. The student may then report the corrected volume of hydrogen as a preliminary check on his or her work and finally report the calculated equivalent mass, or the mass of the sample from the known equivalent mass.

Obtain two samples of the metal to be used. Weigh these precisely on the analytical balance, at the same time taking care that the masses do not exceed the maximum permitted, so you do not generate more hydrogen than your apparatus can accommodate in either Method A or Method B. (For all laboratories except those at high elevation, maximum masses are as follows: for Method A, with a 50-mL buret, 0.12 g Zn, 0.032 g Al, 0.042 g Mg, 0.10 g Mn; for Method B, with a 500-mL flask, 1.10 g Zn, 0.40 g Mg, 0.30 g Al, 0.90 g Mn, 0.90 g Fe, 1.90 g Sn, 1.90 g Cd.)

Method A

Compress the weighed samples into compact bundles and wrap each sample in all directions with about 20 cm of fine copper wire, forming a small basket or cage, leaving 5 cm of the wire straight as a handle. This confines the particles as the metal dissolves and also speeds the reaction.[1]

Obtain and clean a 50-mL buret. Next measure the uncalibrated volume of the buret between the stopcock and the 50-mL graduation. Measure by filling the buret with water and draining the buret through the stopcock until the liquid level falls exactly to the 50-mL mark. Then use a 10-mL graduated cylinder to measure the volume delivered when the water level is lowered to the top of the stopcock. Using a funnel, pour into the buret the required amount of concentrated hydrochloric acid. Be careful not to allow the acid to touch your skin or clothing. Because of differences in the activity of the metals used, it is necessary to vary the amount of acid. For magnesium, use about 3 mL; for aluminum or zinc, use about 20 mL; for manganese, use about 7 mL.

---

[1]As the more active metal dissolves, it gives up electrons that move easily to the less reactive copper, where they react with hydrogen ions of the acid to form hydrogen gas. Note that the bubbles of gas form on the copper wire, thus keeping a larger surface of active metal exposed to the acid.

Fill the buret completely with water slowly and carefully, to avoid undue mixing of the acid. Insert the metal sample about 4 cm into the buret and clamp it there by the copper wire handle, using a one- or two-hole rubber stopper. Make certain no air is entrapped in the buret. Cover the stopper hole(s) with your finger[2] and invert the buret (Figure 13-1) in a 400-mL beaker partly filled with water. The acid, being more dense, quickly sinks, diffuses down the buret, and reacts with the metal. As the H_2 is generated, it collects at the top of the buret, expelling the HCl and water solution out the hole in the stopper at the bottom. (**Caution:** If the reaction is too rapid and the metal too close to the end of the buret, small bubbles of hydrogen may also escape from the buret as the acid solution is expelled. If so, repeat the experiment using less acid.)

After complete solution of the metal, let the apparatus cool to room temperature, since heat is generated by the reaction. Tap the apparatus to free any hydrogen bubbles adhering to the sides of the vessel or the copper wire. Measure the volume of gas liberated[3] and—without changing the position of the buret—measure the difference in height of the two water levels with a metric rule. Then calculate the equivalent pressure in millimeters of mercury (torr; mm Hg = mm solution × density of solution/density of mercury). Take the temperature of the gas by holding a thermometer in contact with the side of the buret. Raise the buret up out of the HCl solution in the beaker and allow the remainder of the solution to drain out of the buret into a large beaker. Slowly add solid sodium carbonate, Na_2CO_3, to the HCl solution until there is no further fizzing. Then flush the HCl solution down the drain with plenty of water and discard the copper wire in the wastebasket.

Obtain the barometer reading for the day. Repeat the determination with your second sample.

Method B

Set up the apparatus as sketched in Figure 13-2, utilizing a 25 × 250 mm test tube and a 500-mL flask to contain the evolved hydrogen. The exit tube C from the test tube, connected to the flask E, must not extend below either rubber stopper (so that gas will not be trapped). The longer glass tube in the test tube should extend nearly to the bottom and should be constricted to a small capillary and bent, as

---

[2]Use a plastic glove for this step to ensure that no acid comes in contact with your skin.

[3]If a 50-mL buret is used, the volume of gas liberated equals 50 minus the final buret reading plus the volume of the uncalibrated portion of the buret.

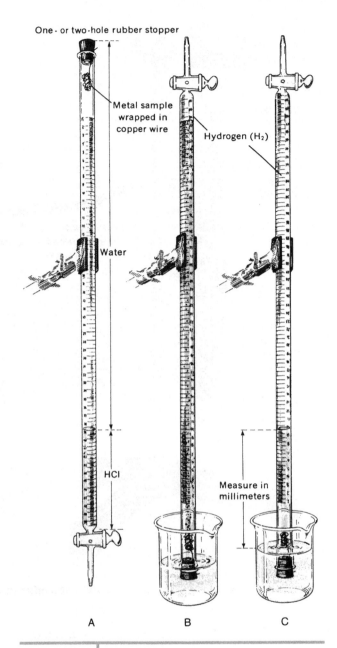

FIGURE 13-1 **The appropriate volume of concentrated (12 M) HCl is added to the buret, and water is then layered on top of it until the buret is completely filled, as in A. Inversion of the buret in a beaker partly filled with water begins the reaction (B), which continues until all of the metal is gone and the buret is nearly full of hydrogen gas, as shown in C.**

illustrated. Completely fill flask E with water and put a little water into flask F. Fill the siphon tube D, which extends to the bottoms of both flasks E and F, by blowing into the tube C.

Place the first carefully weighed sample of metal in the 25 × 250 mm test tube, as indicated. A fine copper wire, wrapped about the sample in all directions like a cage, as in Method A, may be of some

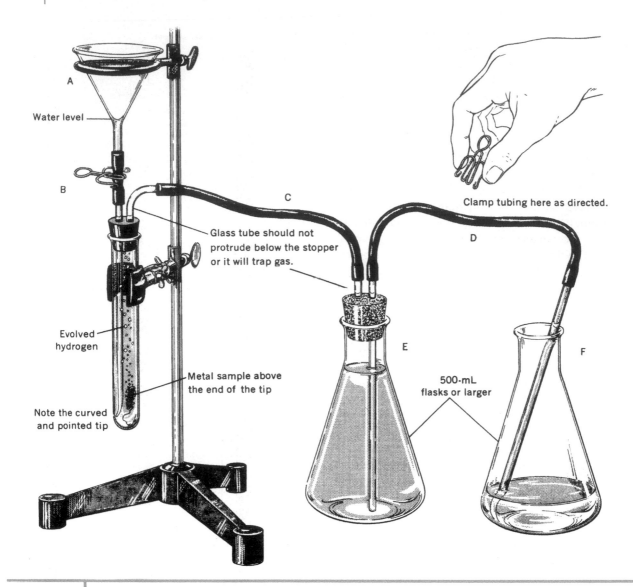

FIGURE 13-2 | **An alternate apparatus for determining the equivalent mass of a metal.**

help in increasing the rate of reaction and in confining small bits of metal as the sample dissolves, but this is not absolutely necessary. With the stopper in flask E loosened to permit air to escape, pour water into the funnel to completely fill the test tube and tubes B and C, then close clamp B. If flask E is not completely filled with water, raise flask F so that water siphons back into it, then push in the stopper in E tightly.

When all air bubbles have been thus removed from the apparatus, release clamps B and D just enough to permit the water level in the funnel to fall just to the stem top *but no farther;* then close clamp B. Now empty and drain flask F, but do not dry it, leaving tube D filled with water. Clamp D is open. When all is ready, pour exactly 25.0 mL of concentrated hydrochloric acid into the funnel. Release

clamp B momentarily to permit a little acid to flow into the test tube and react with the metal at a moderate rate.

A volume of water equal to the volume of hydrogen generated will siphon from E to F. You can control the rate of reaction by regulating the amount of acid you add and by immersing the tube (for less reactive metals) in a beaker of warm water as needed. When all the metal has dissolved and the reaction mixture has cooled to room temperature, release clamp B to permit the acid level to fall to the funnel stem, *but no farther.* Carefully measure 25.0 mL of water; add this to the funnel and again release the clamp to let the level fall exactly to the funnel stem. Then close the clamp tightly.

All gas should now be displaced from the test tube and connecting tubes into flask E. If it is not, repeat the addition of a measured amount of water. Adjust the levels in flasks E and F by raising or lowering one of them until they are even (avoid warming the gas in E with your hands on the flask); then close clamp D tightly. Obtain the temperature of the hydrogen by loosening the stopper enough to fit a thermometer in flask E. Measure the volume of water in flask F by pouring it into a 500-mL graduated cylinder. This volume, minus the volumes of acid and water added to the funnel, will be the volume of the hydrogen generated. Record the barometric pressure. Pour the HCl solution from the test tube and flask E down the drain with plenty of water. Repeat the determination with your second sample.

Calculations

Recall that the standard molar volume of any gas is 22.4 L. The volume of 0.5 mol of hydrogen gas (1.008 g) at standard conditions is thus 11.2 L, or 11,200 mL. *The equivalent mass of your metal sample is therefore that mass which will produce 11,200 mL of hydrogen gas at standard conditions from an acid.* Also note that you must consider the difference in liquid levels (Method A) and the vapor pressure of water in arriving at the correct hydrogen gas pressure.

Calculate and report your result according to the type of sample you have been given: (1) the equivalent mass of a metal (g metal/equiv), (2) the mass of a preweighed sample, or (3) the percentage composition of an Al–Zn alloy. The percentage of Al in the alloy is given by

$$\text{percentage of Al} = \frac{\text{g Al}}{\text{g sample}} \times 100\%$$

To do this, you must know *both* the mass of aluminum (g Al) and the mass of zinc (g Zn), and you will need two independent equations to solve for the two unknown masses. These two equations may be written as

$$\text{g sample} = \text{g Al} + \text{g Zn}$$

and

$$\text{vol H}_2 \text{ (STP)} =$$
$$11{,}200 \frac{\text{mL H}_2\text{(STP)}}{\text{equiv}} \times \left(\frac{\text{g Al}}{\text{g Al/equiv}} + \frac{\text{g Zn}}{\text{g Zn/equiv}} \right)$$

where g Al/equiv and g Zn/equiv are the equivalent masses of Al and Zn.

For your reported result, calculate the percentage of relative error given by

$$\text{percentage relative error} =$$
$$\frac{\text{your value} - \text{accepted value}}{\text{accepted value}} \times 100\%$$

CONSIDER THIS

Which metals would work in this experiment—in other words, which metals react with 12 M hydrochloric acid? Collect as many different samples of metals as you can and test. Do you note different speeds of reaction? What happens if you use acetic acid (vinegar) instead?

Does the concentration of the acid make much difference in the reaction? Make up 6 M, 3 M, 1.5 M, 0.75 M, and other concentrations of hydrochloric acid solutions and place a piece of zinc in each. Note any correlation. Is there a threshold concentration below which the reaction doesn't proceed?

Can you determine the amount of zinc in a post-1982 U.S. penny by measuring the volume of hydrogen gas produced in an experimental setup like the one used here? Does this technique work with a pre-1982 penny?

Bibliography

Masterton, W. L. "Analysis of an Aluminum–Zinc Alloy," *J. Chem. Educ.* **1961,** *35,* 558.

The Reactivity of Metals with Hydrochloric Acid

Name _____

Date _____ Section _____

Locker _____ Instructor _____

Data and Calculations

Data: Unknown Sample Identifier _____	Trial 1	Trial 2
Mass of sample (except for preweighed sample)	g	g
Volume of uncalibrated portion of buret (Method A)	mL	mL
Final buret reading (Method A)	mL	mL
Volume of hydrogen	mL	mL
Temperature of hydrogen	°C	°C
Barometric pressure	torr	torr
Difference in water levels inside and outside tube (Method A)	mm H_2O	mm H_2O
Aqueous vapor pressure at temperature of hydrogen	torr	torr

Data	Calculations	Trial 1	Trial 2
mm Hg (torr) equivalent of difference of water levels (Method A)		torr	torr
Pressure of H_2 after correction for difference in H_2O levels and for vapor pressure		torr	torr
Temperature, absolute		K	K
Volume of dry H_2 at standard conditions, i.e., corrected for vapor pressure, pressure, and temperature		mL	mL
Equivalent mass[a]		g/equiv	g/equiv
Percentage of relative error = $\dfrac{\text{your value} - \text{accepted value}}{\text{accepted value}} \times 100$		%	%

[a]NOTE: The units in the boxes provided for recording your calculated values should be changed as necessary, depending on whether you are reporting (1) the mass of a preweighed sample, or (2) the percentage composition of an Al–Zn alloy.

Problems

1. The atomic mass of tin is 118.69. What is its equivalent mass under the following conditions?

  (a) When it reacts with chlorine to form $SnCl_4$

  (b) When it reacts with hydrochloric acid to form $SnCl_2$

2. Write balanced chemical equations for the reactions of the following three metals with sulfuric acid, H_2SO_4. Underneath each reaction, write, in order, (i) the group number of the metal element in the periodic table; (ii) the oxidation number of the metal in the sulfate salt produced as a reaction product; and (iii) the volume (mL) of hydrogen gas, at standard conditions, that will be liberated by the reaction of sulfuric acid with the metal.

  (a) 0.0230 g Na: Equation: _____

  (b) 0.0243 g Mg: Equation: _____

  (c) 0.0270 g Al: Equation: _____

3. A 0.955-g sample of a certain metal reacts with acid to liberate 380 mL of hydrogen gas when collected over water at 20 °C and 720 torr total pressure. What is the equivalent mass of the metal? What metal could it be?

4. Calculate the maximum mass of calcium metal that should be used in an equivalent mass determination with a 100-mL gas measuring tube for collection. For the purposes of calculation, assume 80.0 mL of hydrogen gas produced at standard conditions.

▮▮▮▮▮ C O N S I D E R   T H I S ▮▮▮▮▮

Which metals would work in this experiment—in other words, which metals react with 12 M hydrochloric acid? Collect as many different samples of metals as you can and test. Do you note different speeds of reaction? What happens if you use acetic acid (vinegar) instead?

Does the concentration of the acid make much difference in the reaction? Make up 6 M, 3 M, 1.5 M, 0.75 M, and other concentrations of hydrochloric acid solutions; place a piece of zinc in each. Note any correlation. Is there a threshold concentration below which the reaction doesn't proceed?

Can you determine the amount of zinc in a post-1982 U.S. penny by measuring the volume of hydrogen gas produced in an experimental setup like the one used here? Does this technique work with a pre-1982 penny?

The Heat Capacity of Metals

Purpose

- Understand the terms *heat, energy, work,* and *heat capacity.*

- Gain an understanding of how heat is related to molecular motion.

- Use a simple Styrofoam-cup calorimeter to measure the heat capacity of a metal and confirm the law of Dulong and Petit.

Pre-Lab Preparation

The nature of heat, energy, and work

What is heat? Naming something is not the same thing as understanding it, and in science a familiar name often stands for a concept that, upon close examination, proves elusive and difficult to define in a precise way.

Heat is an effect produced by the transfer of energy. We might even call it "energy in transit." Energy naturally flows from warmer to cooler bodies. This is equivalent to saying that heat spontaneously passes from a hot to a cold body.

In describing heat, we used another familiar word—*energy*—that also stands for an elusive concept. What is energy? We will define it as the *capacity to produce useful work.* Work is defined as the product of *force × distance,* such as the work done in lifting a weight. The word *energy* was first employed in this sense by Rudolf Clausius (1822–1888), but its use has been traced back to the ancient Greeks (probably Aristotle), who coined the word by combining the Greek words *en* and *ergon,* meaning "at work."

Energy manifests itself in many forms. Part of the energy in coal and oxygen is released when the two combine chemically to form carbon dioxide. This is called *chemical energy,* and it produces heat that can be used to turn water into high-pressure steam. The energy stored in the steam is converted into kinetic energy by allowing the steam to expand through a turbine, causing it to rotate. The energy of motion in the rotating turbine is converted into *electrical energy* when the turbine is coupled to a generator.

The electrical energy distributed through wires can produce *mechanical work* (another form of energy), such as the shaft rotation of an electric motor, or it can be converted into heat by passing electric current through a resistor.

The rotation of the shaft of an electric motor can be used to turn a drum, winding up a cord to which a weight is attached. In raising the weight, work is done. The work done in raising the weight increases the weight's *potential energy.* If the motor is uncoupled from the drum and the weight allowed to descend, the potential energy stored in the raised weight is converted into kinetic energy, rotating the drum as the cord unwinds. If the rotating drum is now connected by gears to a paddle wheel churning a container of water, the temperature of the water increases. The potential energy stored in the weight is converted into work, which is in turn converted into heat.

James Prescott Joule (1818–1889) used a falling weight to churn water in exactly this way to determine that there is a fixed relationship between the amount of work done and the heat effect it produces. Joule called this quantity the "mechanical equivalent of heat." In modern units, 4.184 joules (J) = 1 calorie (cal) of thermal energy, where 1 cal is the quantity of energy required to increase the temperature of 1 g of water by 1 Celsius degree.

Modes of energy transfer

There are several ways to heat an object by energy transfer. Energy can be transferred through empty space (a vacuum) by *radiation.* You feel this effect when you turn your face to the sun on a warm spring day. Light is a form of electromagnetic radiation, and the carriers of light energy are called *photons.* Although they have no rest mass, when they are absorbed by the skin they set into motion the atoms and molecules in the skin, producing the sensation of warmth.

Heat can pass from one body to another by *convection* (movement of a fluid). If you hold the back of your hand near your mouth and exhale gently on it, you feel the sensation of warmth caused by the flow of warm air from your lungs. Finally, energy can be transferred by *conduction* through both fluids and solids. If you grasp the end of a copper rod and stick the other end in a flame, the end you are holding soon gets uncomfortably hot. The same thing happens if you spill hot coffee on yourself. The hot water heats your skin, perhaps enough to produce a painful burn.

Heat and molecular motion

The accumulated knowledge about heat's effects provides a rather detailed picture of what happens when you heat (or transfer energy to) a substance. On an atomic or molecular scale, heating increases the motion of the atoms. If the atoms are linked by chemical bonds to form a gaseous molecule, we can focus our attention not on individual atoms but on the molecule as a whole. If we heat a gas, there are three effects:

1. The individual molecules move about faster from place to place. Their kinetic energy of *translation* is increased.

2. Individual atoms in molecules are linked by chemical bonds that are "springy," constraining atoms that are bonded together to move like masses connected by springs. As the molecules are heated, the amplitudes of their *vibrations* increase.

3. Finally, molecules can rotate about their centers of mass; as a gas is heated, the rate of *rotation* of the molecules increases.

To summarize, gaseous molecules can store energy in *translational, vibrational,* and *rotational* motions.

What about solids such as metals? When you heat a solid metal, which consists of atoms closely packed together, the atoms, on average, are not going anywhere. An atom occupies a fixed position in the solid (unless the temperature gets too near the melting point); when heated, it just increases the amplitude of its vibration about this fixed position. Because atoms have nearly all of their mass located in a very small nucleus, they have a very small moment of inertia. Therefore, individual atoms have an insignificant amount of rotational energy.

Latent heat

When you heat a substance (thus putting energy into it), will a thermometer immersed in the substance always indicate a temperature increase? Not necessarily. In some circumstances, the temperature may stay constant. For example, if you drop a small piece of hot metal into a mixture of ice and water at 0 °C, some ice will melt, but the temperature will remain

constant until all of the ice is melted. Heat that produces a phase change such as the melting of ice (or the boiling of water) without producing a temperature change is called *latent* heat. A thermometer, therefore, can be used to measure *temperature* changes but cannot always be used to measure *energy* changes. This observation shows the important but sometimes subtle distinction between heat and temperature.

Heat capacity

When you heat a substance and its temperature increases, the input of energy divided by the temperature increase is called the *heat capacity.* It could just as well be called the energy capacity, since the quantity has the dimensions of *energy per unit temperature change.*

The heat capacity of metals and the law of Dulong and Petit

In 1819, as part of a systematic study of the heat capacity of metals, Pierre Dulong (1785–1838) and his collaborator Alexis Petit (1791–1820) published their discovery that the product of the molar mass (grams per mole) of a metal times its heat capacity (calories per gram per Celsius degree) is a constant. (This product could be called the *molar heat capacity,* and it is approximately equal to 6.0 cal • mol^{-1} • K^{-1}, or 25 J • mol^{-1} • K^{-1}.) At the time, there was no explanation of why this should be true, but Berzelius (and later Cannizzaro) used this empirical result to arrive at correct atomic masses and empirical formulas for compounds of metals.

It was not until 1907 that Albert Einstein published a theory of the heat capacity of metals. He treated atoms as equivalent to little oscillators and showed that the heat capacity decreases to zero as you approach zero on the absolute temperature scale, but it approaches a value of 3R if the temperature is sufficiently high. R is the universal gas constant. This constant has the value 8.314 J • mol^{-1} • K^{-1}. Therefore, it produces the predicted molar heat capacity of 24.9 • J • mol^{-1} • K^{-1}.[1] (In measuring heat capacities, you are considering only temperature changes, so it makes no difference whether you measure these changes on the Kelvin or Celsius scale; the degree intervals on both scales are the same size.)

[1]You've been using a value for R, the gas constant, of 0.08206 L• atm • mol^{-1} • K^{-1}. The value used here is the same constant but in different energy units, joules • mol^{-1} • K^{-1}, rather than L•atm, which also has the units of energy.

The law of Dulong and Petit became one of the first ways to estimate the molar mass of a metallic element. We will use it to try to determine the identity of our unknown. If the law gives us the **molar** heat capacity, $C_{P,M}$, as 25 J • mol^{-1} • K^{-1}, and we can measure the **gram** heat capacity (also known as the *specific heat*) in the laboratory with its units, J • g^{-1} • K^{-1}, then the molar mass is the ratio of the two. For example, the measured heat capacity of iron is 0.473 J • g^{-1} • K^{-1}; the law gives us $C_{P,M}$ as 25 J • mol^{-1} • K^{-1}. Thus,

$$\frac{25 \text{ J} \cdot \text{mol}^{-1} \cdot \text{K}^{-1}}{0.473 \text{ J} \cdot \text{g}^{-1} \cdot \text{K}^{-1}} = 53 \text{ g/mol} \approx 55.847$$

In its day, this measurement was as accurate as most laboratory determinations.

The heat capacity of metal salts and nonmetal compounds

When you measure the heat capacity of something like limestone ($CaCO_3$), an organic polymer such as Teflon ($-CF_2-)_x$, or any substance composed of light atoms, you find that the heat capacity per mole of atoms is typically about 60–80% that of a mole of heavy metal atoms. [If you multiply the specific heat of any substance (J/g • K) by the molar mass (g/mol), you will get the molar heat capacity (J/mol • K). Now if you divide this result by the number of atoms in the empirical formula you used for the compound, you will get the heat capacity per mole of atoms.] In general, atoms with smaller masses vibrate with higher frequencies in the solid and have lower heat capacities per mole of atoms, compared to heavy atoms. This result is in general agreement with Einstein's theory of the heat capacity of solids.

Heat and enthalpy

The amount of heat it takes to increase the temperature of a substance by a given amount depends on whether the heating is done under conditions of constant volume or constant pressure. If volume is constant, all of the energy goes into increasing the internal energy of the substance (increasing the motion of the atoms of the substance). If the volume is allowed to change and the pressure is constant (for example, equal to the atmospheric pressure), some work will be expended in pushing back the atmosphere; this work will require a greater amount of heat for a given temperature increase. If we represent the quantity of heat by the symbol q, we can in general say that q_P will be larger than q_V.

The heat added to a system at constant pressure is called the *enthalpy* increase (from the Greek *enthalpein*, "to heat in"). Changes in enthalpy are directly related to changes in temperature by the relation

$$\Delta H = q_P = m C_P \Delta T \qquad (1)$$

where m is the mass of the substance heated (in grams), C_P is the heat capacity at constant pressure (in joules per gram per kelvin), and ΔT is the temperature change. We will consistently define ΔT as the final temperature minus the initial temperature: $\Delta T = T_f - T_i$. (The temperature change ΔT will be negative if T_f is less than T_i. Also note that the temperature *change* in kelvins is equal to the *change* in Celsius degrees.)

Measuring the heat capacity of a metal with a Styrofoam-cup calorimeter

In this experiment, we will measure the heat capacity of an unknown metal and try to identify it from the measured heat capacity. We will first measure the effective heat capacity of a simple calorimeter containing cool (room temperature) water by adding a known mass of hot water to it and measuring the temperature increase. Then, in a second experiment, we will add a known mass of hot metal to the calorimeter under similar conditions.

Employing the principle of the conservation of energy, we will write an equation that states that the sum of the enthalpy change of the calorimeter plus the enthalpy change of the added hot water must be zero. (The change in enthalpy of the calorimeter is equal in magnitude to the change in enthalpy of the added hot water, but they have opposite signs.) The change in enthalpy of the calorimeter will be divided into two parts: the Styrofoam cups with the thermometer that makes up the calorimeter, and the cool water originally present in the calorimeter. Thus,

$$\underset{\substack{\text{Styrofoam cups} \\ \text{and thermometer}}}{\Delta H_{\text{cal}}} + \underset{\substack{\text{Cool water in} \\ \text{calorimeter}}}{\Delta H_{\text{CW}}} + \underset{\substack{\text{Added hot} \\ \text{water}}}{\Delta H_{\text{HW}}} = 0 \qquad (2)$$

In view of the definition of the enthalpy change given in Equation (1), we can rewrite Equation (2) as

$$\underset{\substack{\text{Styrofoam cups} \\ \text{and thermometer}}}{(mC_P) \Delta T} + \underset{\substack{\text{Cool water in} \\ \text{calorimeter}}}{mC_P \Delta T} + \underset{\substack{\text{Added hot} \\ \text{water}}}{mC_P \Delta T} = 0 \qquad (3)$$

It should be recognized that not all parts of the Styrofoam cups and thermometers come into contact with and are heated by the added hot water. Therefore, we can determine only the *effective* product (mC_P) for the Styrofoam cups and thermometer, which we will call the *calorimeter constant, B* (joules per kelvin). So we will rewrite Equation (3) as

$$B \Delta T_{CW} + m_{CW} C_P \Delta T_{CW} + m_{HW} C_P \Delta T_{HW} = 0 \quad (4)$$

where m_{CW} is the mass of cool water originally in the calorimeter, ΔT_{CW} is the temperature increase of the calorimeter and water, m_{HW} is the mass of added hot water, and ΔT_{HW} is the temperature decrease of the added hot water. These four quantities are all measured in the experiment. If we also know the heat capacity of water, C_P, we can solve Equation (4) to obtain the calorimeter constant, B.

A second experiment is then carried out, adding heated metal instead of hot water to the water in the calorimeter. We can write an equation similar to Equation (4) to describe the enthalpy changes in this experiment:

$$\underset{\substack{\text{Styrofoam} \\ \text{cups and} \\ \text{thermometer}}}{B \Delta T_{CW}} + \underset{\substack{\text{Cool water in} \\ \text{calorimeter}}}{m_{CW} C_P \Delta T_{CW}} + \underset{\substack{\text{Added hot} \\ \text{metal}}}{m_{HM} C_{PM} \Delta T_{HM}} = 0 \quad (5)$$

In this second experiment, we again measure the four quantities (m_{CW}, ΔT_{CW}, m_{HM}, and ΔT_{HM}). We substitute these values and the previously determined value of B into Equation (5) and solve for $C_{P,M}$, the heat capacity of the unknown metal.

If you perform the experiments carefully, the measured value of the heat capacity will enable you to identify the unknown metal from the list of metals in Table 14-1.

TABLE 14-1 **The Heat Capacity of Some Metals**

Metal	Heat Capacity ($J \cdot g^{-1} \cdot K^{-1}$)
Magnesium	1.04
Aluminum	0.904
Iron	0.473
Nickel	0.444
Copper	0.387
Zinc	0.386
Silver	0.236
Antimony	0.207
Gold	0.129
Lead	0.128

Experimental Procedure

Special Supplies: Hot plate; large iron ring or safety clamp and ring stand [or burner, iron ring, metal gauze, safety clamp (or large iron ring), and ring stand]; 1000-mL

beaker, two 0 to 110 °C thermometers (if available, use a 0 to 50 °C thermometer graduated in tenths of a degree in place of one of the 0 to 110 °C thermometers);[2] fine copper wire to suspend the thermometer; two 12- to 16-oz Styrofoam cups per student; 25 × 200 mm test tubes (two to contain water and one for each unknown-metal heat capacity determination); corks or rubber stoppers to fit test tubes; beaker tongs or pliers with padded jaws; 100-mL graduated cylinder.

Chemicals: Metals (in the form of shot or pellets so that they will pour easily from a test tube); for example, aluminum, magnesium, nickel, zinc, antimony, lead.

⚠ SAFETY PRECAUTIONS:
Boiling water can cause painful and serious burns. Even if you are using a hot plate, put a large iron ring or safety clamp around the beaker and securely fasten the ring or clamp to a ring stand, as shown in Figure 14-1(A). The safety clamp is even more important if you use a gas burner and ring stand to heat the water. Keep the hot plate or burner and ring stand well back from the edge of the laboratory bench.

A 25 × 200 mm test tube containing 80 g of metal is too heavy to handle with an ordinary test tube holder. Use beaker tongs or ordinary pliers whose jaws have been padded by having large diameter plastic tubing slipped over the jaws.

The thermometer used to measure the temperature in the Styrofoam cups should be suspended by copper wire from a ring stand support [see Figure 14-1(B)]. If not suspended, the thermometer may tip over the cups, spoiling the measurement and possibly breaking the thermometer. If the thermometer does not have a suspension ring at the top, wrap several turns of tape around the wire near the top of the thermometer.

---

[2]Moderately priced electronic digital thermometers (cost approximately $30) with a rugged stainless steel temperature sensing probe and 1-m cable, a range of −50 to +150 °C, and temperature resolution of 0.1 °C, are available from Hanna Instruments Inc. (Model Checktemp 1C, HI98509) or Flinn Scientific (Catalog No. AP8559).

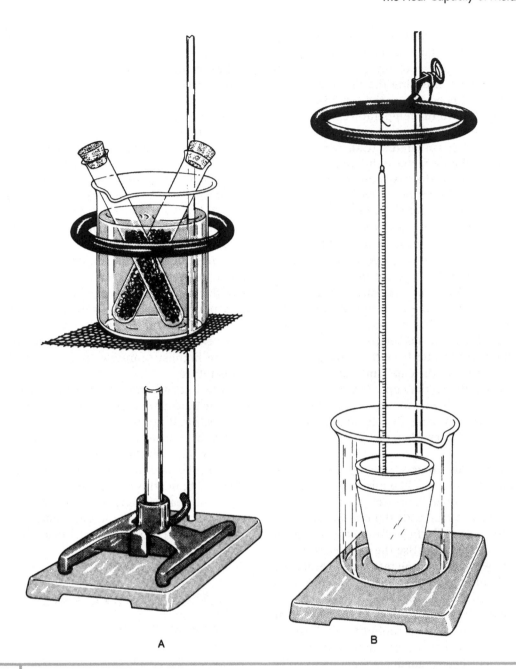

A

B

FIGURE 14-1 Apparatus for measuring the heat capacity of a metal. (A) Samples of metal pellets are heated in the test tubes. (B) The heated metal pellets are then poured into the Styrofoam cup calorimeter.

1. Determining the Calorimeter Constant Put about 700 mL of water in a 1000-mL beaker, clamp it securely in place on the hot plate or over the burner, and begin heating it to boiling [see Figure 14-1(A)]. Put 30 mL of water in each of two 25 × 200 mm test tubes and place them in the beaker so that they can be heated to the temperature of boiling water.

To save time, also prepare the weighed and stoppered tube(s) containing the unknown metal(s) as described in part 2, and put them in the beaker of boiling water.

Obtain two 12- or 16-oz Styrofoam cups, put one inside the other so that they are nested, and weigh the nested cups to the nearest 0.1 g on a platform balance. Record the mass of the empty cups. Add 70 mL of cool water to the inner cup, reweigh, and record the mass.

When the water in the 1000-mL beaker has reached the boiling point, wait 15 min longer, then stir the water in the nested cups with a thermometer (0 to 50 °C graduated in tenths of a degree, if available) until a constant temperature is reached. Leave the thermometer in the cup suspended by a copper wire attached to a ring stand support. (The wire suspension will keep the thermometer from tipping over the cups and possibly breaking the thermometer. If your thermometer has no ring at the top, use tape to fasten the wire.) Record the temperature.

Determine the temperature of the boiling water with a 0 to 110 °C thermometer. (**Caution:** Do not place a 0 to 50 °C thermometer in boiling water or it will break.) The bulb of the thermometer should be held 2 to 3 cm above the bottom of the beaker. Record the temperature of the boiling water. Be sure to record all temperature readings to the nearest 0.1 °C.

Using beaker tongs or pliers with padded jaws, grasp the test tube containing the 30 mL of water near the top and quickly but carefully pour the hot water into the Styrofoam cups. Be careful that no hot water on the outside of the test tube gets into the cups. Stir the water steadily, noting the highest temperature attained. Record this temperature.

Remove the thermometer from the cups. Reweigh, then record the mass of the cups and water.

Pour out the water in the cups. Dry the inside of the cups gently with a towel, and carry out a duplicate set of measurements by using the remaining test tube containing 30 mL of water. Then proceed to part 2.

2. Determining the Heat Capacity of an Unknown Metal

Weigh a 25 × 200 mm test tube, fitted with a cork or rubber stopper, to the nearest 0.1 g on a platform balance. (If the platform balance has a flat pan, use a cork ring with V-grooves cut in the top to support the tube. Use the same cork ring in all subsequent weighings of the test tube.) Record the mass. Place about 80 g of a *thoroughly dry* unknown metal in the test tube, stopper it, reweigh it, and record the mass. (If you are to determine the heat capacity of more than one sample of metal, weigh these metals in similar test tubes now so that all of the samples can be heated adequately before use. Put a label on each test tube so you can identify it.)

Loosely stopper the test tube and place it in the beaker of boiling water prepared as described in part 1. (The purpose of the cork or rubber stopper is to prevent condensation of water inside the tube while it is being heated. Any hot water in the metal would cause a serious error in the measurement of heat capacity.) Each test tube must be immersed in boiling water at least 15 min before the metal is transferred to the calorimeter. Determine the temperature of the boiling water with a 0 to 110 °C thermometer, as you did in part 1. Record the temperature.

Using a graduated cylinder, measure out exactly 100 ± 1 mL of cool water and pour it into the nested Styrofoam cups. Stir the water gently with a thermometer (0.1 °C if available) until a constant temperature is reached. Leave the thermometer in the cup, suspended by a wire or string as described in part 1. Record the temperature.

Using beaker tongs or pliers with padded jaws, grasp the test tube containing the heated metal near the top, remove the cork, and quickly but carefully pour the hot metal into the Styrofoam cups. Be careful that no hot water on the outside of the tube gets into the cups. Stir the water steadily, noting the highest temperature attained. Record this temperature.

Pour out the water, saving the metal as directed by your instructor. (The metal must be *thoroughly dried* before it can be reused.)

If your instructor so directs, repeat the determination with either the same metal or a different one.

From the data of part 1, calculate the calorimeter constant. Using this value and the data of part 2, calculate the heat capacity of the metal(s). Compare the heat capacity value(s) you determined with those in Table 14-1 to try to identify the unknown metal(s).

CONSIDER THIS

Search the Web for sites that have periodic tables. Find one that gives thermodynamic data for the elements. Use the data given for the specific heat (the heat capacity in units of $J \cdot g^{-1} \cdot K^{-1}$) to test the law of Dulong and Petit, that is, find the molar heat capacity in units of $J \cdot mol^{-1} \cdot K^{-1}$. Does the law hold up? Can you distinguish a metallic element from a nonmetal this way? What about the metalloids?

The Heat Capacity of Metals

Name _____

Date _____ Section _____

Locker _____ Instructor_____

Data and Calculations

1. Determining the calorimeter constant

Data	Trial 1	Trial 2
(a) Mass of empty Styrofoam cups	g	g
(b) Mass of cups + 70 mL water	g	g
(c) Mass of cups + 70 mL water + 30 mL hot water	g	g
(d) Initial temperature of water in the calorimeter (cups)	°C	°C
(e) Temperature of the boiling water bath	°C	°C
(f) Final temperature of calorimeter + added hot water	°C	°C

Calculations		
(g) Mass of cool water in cups, m_{CW}	g	g
(h) Mass of added hot water, m_{HW}	g	g
(i) Temperature change of cool water in the calorimeter ($\Delta T_{CW} = T_{lf} - T_{ld}$)	+ K	+ K
(j) Temperature change of added hot water ($\Delta T_{HW} = T_{lf} - T_{le}$)	− K	− K
(k) Calculate the heat *lost* by the hot water $\Delta H_{HW} = m_{HW}C_P \Delta T_{HW}$	− J	− J
(l) Calculate the heat *gained* by the cold water $\Delta H_{CW} = m_{CW}C_P \Delta T_{CW}$	+ J	+ J
(m) Calculate the heat *gained* by the calorimeter $\Delta H_{cal} = -\Delta H_{HW} - \Delta H_{CW}$	+ J	+ J
(n) Calculate the calorimeter constant, B $B = \Delta H_{cal}/\Delta T_{CW}$	J/K	J/K

2. Determining the heat capacity of an unknown metal

Data Unknown Identifier(s)	Trial 1	Trial 2
(a) Milliliters of water placed in cups	mL	mL
(b) Mass of empty stoppered test tube	g	g
(c) Mass of stoppered test tube + metal	g	g
(d) Initial temperature of water in the calorimeter (cups)	°C	°C
(e) Temperature of the boiling water bath	°C	°C
(f) Final temperature of water + added hot metal	°C	°C

Calculations		
(g) Mass of water in cups, m_{CW} (assume that the density of water is 1.00 g/mL)	g	g
(h) Mass of added hot metal, m_{HM}	g	g
(i) Temperature change of the water in the calorimeter ($\Delta T_{CW} = T_{2f} - T_{2d}$)	K	K
(j) Temperature change of the added hot metal ($\Delta T_{HM} = T_{2f} - T_{2e}$)	K	K
(k) Calculate the heat *gained* by the cool water $\Delta H_{CW} = m_{CW} \times C_P \times \Delta T_{CW}$	J	J
(l) Calculate the heat *gained* by the calorimeter $\Delta H_{cal} = B \times \Delta T_{CW}$	J	J
(m) Calculate the *total* heat gained $\Delta H_{gained} = \Delta H_{CW} + \Delta H_{cal}$	J	J
(n) Since $\Delta H_{gained} = -\Delta H_{lost}$ calculate the heat *lost* by the hot metal $\Delta H_{HM} = -\Delta H_{gained}$	J	J
(o) Calculate the heat capacity of the metal from $\Delta H_{HM} = m_{HM} \times C_{PM} \times \Delta T_{HM}$	$J \cdot g^{-1} \cdot K^{-1}$	$J \cdot g^{-1} \cdot K^{-1}$

The law of Dulong and Petit states that the heat capacity of a metal in joules per gram per kelvin multiplied by the molar mass of the metal in grams per mole is a constant: $C_P \times M = 25 \ J \cdot mol^{-1} \cdot K^{-1}$. In step (o), you determined the heat capacity in joules per gram per kelvin. Solve for the molar mass (grams per mole) for each of the metals studied.

_____ g/mol   _____ g/mol

Name _____   **Date** _____

Questions

1. If a student used a sample of metal that was wet, a significant percentage of the apparent mass of the metal would be water. Would the resulting measured value of the heat capacity be too high or too low? Explain.

2. A particular metallic element, M, has a heat capacity of $0.36 \; J \cdot g^{-1} \cdot K^{-1}$. It forms an oxide that contains 2.90 g of M per gram of oxygen.

    (a) Using the law of Dulong and Petit ($C_P \times M = 25 \; J \cdot mol^{-1} \cdot K^{-1}$), estimate the molar mass of the metal M.

_____ g/mol

    (b) From the composition of the oxide (2.90 g M/g O) and the molar mass of oxygen (16.0 g/mol), determine the mass of M that combines with each mole of oxygen.

_____ g M/mol O

    (c) Use your estimate of the atomic mass from (a) and the information from (b) to determine the empirical formula of the oxide (the moles of M that combine with each mole of O).

_____

    (d) Now that you know the composition of the oxide *and* its formula, what is the accurate value of the atomic mass of the metal M? What is the identity of M?

_____ g/mol

▌▌▌▌▌ CONSIDER THIS ▌▌▌▌▌

Search the Web for sites that have periodic tables. Find one that gives thermodynamic data for the elements. Use the data given for the specific heat (the heat capacity in units of $J \cdot g^{-1} \cdot K^{-1}$) to test the law of Dulong and Petit, i.e., find the molar heat capacity in units of $J \cdot mol^{-1} \cdot K^{-1}$. Does the law hold up? Can you distinguish a metallic element from a nonmetal this way? What about the metalloids?

Enthalpy Changes in Chemical Reactions

Hess's Law

Purpose

• Test Hess's law (which says that the heat of reaction is the same whether the reaction is carried out directly or in several steps) by measuring the enthalpy changes for the same net chemical reaction carried out by two different paths.

• Make an approximate estimate, by an indirect measurement, of the enthalpy change for the dissociation of water into hydronium (H_3O^+) and hydroxide (OH^-) ions.

Pre-Lab Preparation

What accounts for the energy change in a chemical reaction? The simplest explanation is that in a chemical reaction, the positively charged nuclei and the negatively charged electrons of the reacting atoms rearrange themselves. It is the energy change associated with the rearrangement (or reconfiguration) of nuclei and electrons that produces the energy change of a chemical reaction.

Thermochemistry deals with the thermal energy changes that accompany chemical reactions. These energy changes are usually called *heats of reaction.* When the reaction is carried out at constant pressure, the heat of reaction is called the *enthalpy* change, ΔH. (The word *enthalpy* comes from a Greek root word *thalpein* meaning "to heat"). A reaction that produces heat is said to be *exothermic*. In an exothermic reaction, the enthalpy of the products is less than the enthalpy of the reactants. If we define the change in enthalpy as $\Delta H = H_{products} - H_{reactants}$, we see that ΔH will be negative for an exothermic reaction. Conversely, if energy is absorbed from the surroundings when a reaction proceeds, the enthalpy change for the reaction will be positive, and we say the reaction is *endothermic*.

Although the enthalpy change for a chemical reaction tells us whether heat will be released or absorbed, this change cannot be used to determine whether a reaction can proceed spontaneously. There are examples of spontaneous reactions that have either negative or positive enthalpy changes. For example, both solid NaOH and KNO_3 dissolve spontaneously in water, but with NaOH the solution warms (ΔH is negative) and with KNO_3 the solution cools (ΔH is positive). As you study thermodynamics further, you will be introduced to *entropy* and *free energy* functions, which can be used to determine whether a particular reaction is possible.

A chemical reaction may also involve energy changes other than a flow of heat to or from the surroundings. For example, an electrochemical cell can produce *electrical energy* equivalent to the product of the cell voltage multiplied by the quantity of electric charge transferred to an external circuit. A chemical reaction that produces a gas must push back the atmosphere; this mechanical *work* is another form of energy. (If the reaction is carried out in containers open to the atmosphere, the work of pushing back the atmosphere is included in the enthalpy change, ΔH.)

Enthalpy changes may be classified into more specific categories: (1) The *heat of formation* is the amount of heat involved in the formation of one mole of the substance directly from its constituent elements; (2) the *heat of combustion* is the amount of heat produced when a mole of a combustible substance reacts with excess oxygen; (3) the *heat of solution* of a substance is the thermal energy change that accompanies the dissolving of a substance in a solvent; (4) the *heats of vaporization, fusion, and sublimation* are related to the thermal energy changes that accompany changes in state; (5) the *heat of neutralization* is the enthalpy change associated with the reaction of an acid and a base.

In this experiment, you will measure the enthalpy changes for the following three reactions:

$$H_2SO_4 (10M) + NaOH (1M) \rightarrow$$
$$NaHSO_4 (0.5M) + H_2O \quad (1)$$

$$H_2SO_4 (10M) + solvent \rightarrow H_2SO_4 (1M) \quad (2)$$

$$H_2SO_4 (1M) + NaOH (1M) \rightarrow$$
$$NaHSO_4 (0.5M) + H_2O \quad (3)$$

In these reactions, the numbers in parentheses specify the molar concentrations of the reactants and products. Water is both the solvent and a reaction product. In order to distinguish these two roles, we write H_2O when we mean a reaction product and write *solvent* when water is used to dissolve the reactants and serve as the reaction medium. You will also see how the measurement of the enthalpy change of Reaction (3) leads to an approximate estimate of the enthalpy of dissociation of water into hydronium ions and hydroxide ions.

Reactions (1) and (3) can be regarded as neutralization reactions. Reaction (2) could be called a solution (or dilution) reaction, and we will call the enthalpy change for this process the *heat of solution*. Note also that if we add Reactions (2) and (3), we get Reaction (1). Thus, Reaction (1) and the sum of Reactions (2) and (3) represent two different pathways by which we can get from the same initial state to the same final state.

Hess's law

Germain Henri Hess (1802–1850) discovered the principle we now call *Hess's law:* The enthalpy change of a reaction is the same whether the reaction is carried out directly or in a number of steps. This means that the enthalpy changes for chemical reactions are additive, just like the chemical reactions themselves. If

Reaction (1) = Reaction (2) + Reaction (3)

then

$$\Delta H_1 = \Delta H_2 + \Delta H_3 \qquad (4)$$

To state Hess's law another way, if you start in enthalpy state A and proceed to enthalpy state B by two different paths, the sum of the enthalpy changes along the first path will equal the sum of the enthalpy changes by the second path. Figure 15-1 shows these relationships in a diagram.

The data you obtain for the enthalpy changes for Reactions (1), (2), and (3) will be used to test the relation shown in Equation (4), which is a mathematical statement of Hess's law. [Reaction (2), the heat of solution of H_2SO_4, was described by Hess in his first thermochemical publication in 1840.]

Enthalpy changes are usually measured in a calorimeter, a simple version of which is shown in Figure 15-2. The purpose of the calorimeter is to isolate thermally the reaction under study. All of the heat liberated (or absorbed) by the chemical reaction goes into heating (or cooling) the calorimeter and its contents.

The amount of energy (heat) required to raise the temperature of 1 g of substance by 1 K (or 1 °C) is called the *heat capacity* of that material. (The term

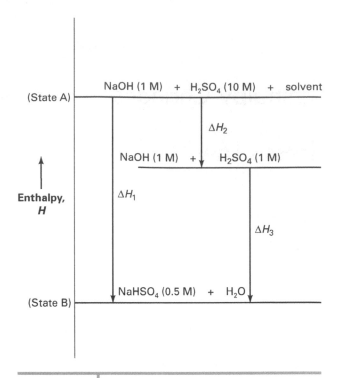

FIGURE 15-1 Hess's law says that if you go from State A to State B by any path (any sequence of chemical reactions), the net enthalpy change will be the same. For the two paths shown in the figure, this means that $\Delta H_1 = \Delta H_2 + \Delta H_3$.

specific heat is often used for this quantity, the gram *heat capacity*.) The heat capacity can be reported in units of joules per gram per kelvin, $J \cdot g^{-1} \cdot K^{-1}$, or joules per gram per Celsius degree, $J \cdot g^{-1} \cdot °C^{-1}$, since a Kelvin and a Celsius degree are the same size but are referenced to different points on the temperature scale. It requires 4.184 J to increase the temperature of 1 g of water by 1 °C. Heat capacities are generally measured experimentally. Knowing the volume and density (the product $V \cdot d$ is the mass) of a substance, its heat capacity, and the temperature change allows us to calculate the enthalpy change of the substance according to the equation

$$\Delta H = V \cdot d C_P \Delta T \qquad (5)$$

Dimensional analysis of Equation (5) shows that it gives the number of joules released or absorbed:

enthalpy change (ΔH, J) =

volume of solution (mL)

× density of solution (g/mL)

× heat capacity ($J \cdot g^{-1} \cdot K^{-1}$)

× temperature change (ΔT, K)

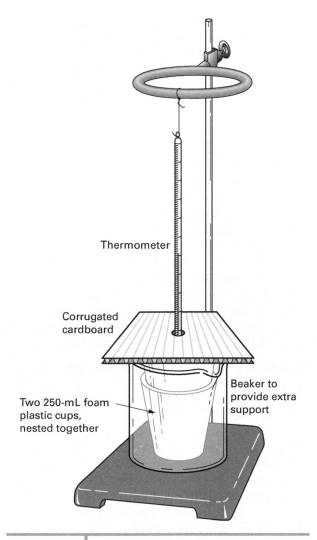

FIGURE 15-2 | **A simple calorimeter for measuring the heat of a reaction.**

Thermometer

Corrugated
cardboard

Two 250-mL foam
plastic cups,
nested together

Beaker to
provide extra
support

In principle, a correction should be made for the heat absorbed or evolved by the calorimeter itself and for the energy lost or gained from the surroundings. In the Styrofoam-cup calorimeter, portions of the cups and the thermometer are in contact with the solutions contained in the calorimeter. Through this contact, energy can leak into or out of the solution in the calorimeter, depending on whether the temperature of the solution is lower than or higher than the temperature of the surroundings. Because the energy losses or gains are small (typically about 3%) and are about the same order of magnitude as the uncertainty in the measurement of the temperature increase, we will omit these corrections so that you can focus your attention on the main event—the enthalpy changes of the reactant solutions that are mixed in the calorimeter. So, in effect, we will treat the Styrofoam cups and thermometer as if they were

perfect insulators, allowing no energy flow into or out of the solutions contained in the calorimeter.

Measuring Heats of Reaction with a Styrofoam-Cup Calorimeter We will carry out three experiments (each in duplicate) to measure the enthalpy changes for Reactions (1), (2), and (3). In these experiments, we mix aqueous solutions of the reactants, so water is the reaction medium, and any temperature changes result from the chemical reactions that take place in solution.

Employing the principle of the conservation of energy (which is a statement equivalent to the first law of thermodynamics), we will write an equation stating that the sum of the enthalpy changes of the solutions mixed in the calorimeter and the enthalpy change of the chemical reaction must be zero. Thus,

$$\Delta H_C \quad + \quad \Delta H_R \quad + \quad \Delta H_{\text{reaction}} = 0 \quad (6)$$

Enthalpy
change of
solution in
calorimeter

Enthalpy
change of
added
reactant
solution

Enthalpy
of
reaction

Substituting into Equation (6) the definition of the enthalpy change given in Equation (5), we can rewrite Equation (6) as

$$V_C d\, C_P \Delta T_C \quad + \quad V_R d\, C_P \Delta T_R \quad + \quad \Delta H_{\text{reaction}} = 0 \quad (7)$$

In these experiments, we measure V_C (the volume of reagent in the calorimeter), ΔT_C (the temperature change of the solution in the calorimeter), V_R (the volume of added reactant solution), and ΔT_R (the temperature change of the added reactant solution). We will take the product of the density and heat capacity of the solutions, $d\,C_P$, to be $4.10\ \text{J} \cdot \text{mL}^{-1} \cdot \text{K}^{-1}$, nearly the same as the product for pure water.[1] Substituting these values into Equation (7), we can solve for the enthalpy change for the chemical reaction. Knowing the number of moles of reactants mixed together, we can express the enthalpy change per mole of reaction, $\Delta H_{\text{reaction}}$, in kJ/mol.

Experimental Procedure

Special Supplies: Thermometer (1 °C divisions are suitable; if available, 0.1 °C divisions will give better accuracy; relatively inexpensive electronic thermometers with a temperature resolution of 0.1 °C are

---

[1]The density of the solutions is about 4 to 6% greater than that of pure water, but the heat capacity is about 4 to 8% less, so the product of density × heat capacity ($d\,C_P$) is nearly the same as the product for pure water (within 2%).

available);[2] four 7-oz Styrofoam cups per student for use as calorimeters (available from a supermarket); corrugated cardboard for calorimeter covers; polyethylene transfer pipets.

Chemicals: 10.0 M H_2SO_4; 1.00 M H_2SO_4; 1.00 M NaOH; $NaHCO_3$(s).

■ NOTES TO INSTRUCTOR ■

Prepare the solutions in advance and store them in the laboratory so they will be at the lab's ambient temperature.

⚠ SAFETY PRECAUTIONS:
Handle sulfuric acid (H_2SO_4) and sodium hydroxide (NaOH) with care. They are corrosive and can cause skin and eye burns. Wear safety goggles.

☠ WASTE COLLECTION:
The solutions produced by each of the reactions are first neutralized with 5–10 g of sodium bicarbonate (NaHCO3), then discarded down the drain.

Preparing the Calorimeters Prepare two calorimeters, each like that shown in Figure 15-2. (Two nested cups are used to provide greater insulation.) Using a pen, label the calorimeters as calorimeters 1 and 2.

In each experiment, you will be measuring initial temperatures of two solutions and the final temperature of the mixed solutions. You must be careful to use the same thermometer to measure the initial and final temperatures in an experiment because inexpensive lab thermometers often give readings that do not perfectly agree. If you are using a thermometer with 1 °C intervals, estimate the temperature reading to the nearest $\frac{1}{5}$ of a degree (±0.2 °C).

---

[2]Moderately priced electronic digital thermometers (cost approximately $30) with a rugged stainless steel temperature sensing probe and 1-m cable, a range of −50 to +150 °C, and temperature resolution of 0.1 °C, are available from Hanna Instruments Inc., (Model Checktemp 1C, HI98509) or Flinn Scientific (Catalog No. AP8559).

1. The Heat of Solution and Neutralization of 10 M H_2SO_4 and 1.00 M NaOH Place 50.0 mL of 1.00 M NaOH and 45.0 mL of deionized or distilled water in calorimeter 1. Measure out exactly 5.00 mL of 10.0 M H_2SO_4 in a 5- or 10-mL graduated cylinder. Measure the temperature of the 10.0 M H_2SO_4 and record the reading. Remove the thermometer and readjust the volume to exactly 5.00 mL, if necessary, using a polyethylene transfer pipet (or a medicine dropper). Then rinse and dry the thermometer and insert it into calorimeter 1, which contains the NaOH solution. Stir the solution gently; read and record the temperature. Then add the 5.00 mL of 10.0 M H_2SO_4. Stir for 30 s (or until the reading is steady or only slowly decreasing) and record the temperature (to the nearest 0.2 °C if you are using a thermometer with 1 °C intervals). Neutralize the solution with 5 g of sodium bicarbonate ($NaHCO_3$) and discard the solution down the drain. Rinse and dry the cups and carry out a duplicate set of measurements. Then rinse and dry the cups in preparation for part 2.

2. The Heat of Solution of 10 M H_2SO_4 Measure 90.0 mL of deionized or distilled water in a graduated cylinder and put it into calorimeter 1. Measure out 10.0 mL of 10.0 M H_2SO_4 in a 10-mL graduated cylinder. Measure and record the temperature of the 10.0 M H_2SO_4. Remove the thermometer and readjust the volume to exactly 10.0 mL, if necessary. Then rinse and dry the thermometer, place it in calorimeter 1 (which contains the water), and gently stir. Read and record the initial temperature of the water as soon as the reading is steady. Then pour the 10.0 M H_2SO_4 into calorimeter 1. Continue to stir for 30 s (or until the reading is steady), then read and record the final temperature of the solution. Neutralize the solution with 10 g of sodium bicarbonate ($NaHCO_3$) and discard the solution down the drain. Rinse and dry the calorimeter cups and carry out a set of duplicate measurements. Rinse and dry the cups to prepare for part 3.

3. The Heat of Neutralization of 1.00 M H_2SO_4 and 1.00 M NaOH Place 50.0 mL of 1.00 M NaOH in calorimeter 1 and 50.0 mL of 1.00 M H_2SO_4 in calorimeter 2. Put the thermometer in calorimeter 1 and stir gently until the temperature reading is steady; then read and record the initial temperature of the NaOH solution. Remove the thermometer, rinse and dry it, and put it in calorimeter 2. Stir the solution gently; when the temperature reading is steady, read and record the temperature of the sulfuric acid (H_2SO_4) solution. Then pour the H_2SO_4 solution in calorimeter 2 quickly and completely into calorimeter 1. Stir for 30 s (or until the reading is steady) and record the final temperature after stirring. Neutralize the solution with 5 g of sodium bicarbonate and discard the solution down the drain. Rinse and dry the cups. Carry out a duplicate set of measurements.

Calculations For each reaction in parts 1, 2, and 3, determine the temperature change for the solution in calorimeter 1 (ΔT_C) and the added reactant solution (ΔT_R) by subtracting the initial temperature from the final temperature ($\Delta T = T_{final} - T_{initial}$). Calculate the heats of reaction for parts 1, 2, and 3 by rearranging Equation (7) to solve for $\Delta H_{reaction}$. Assume that the product of the *density × heat capacity* ($d\,C_p$) for each solution is $4.10\ \mathrm{J \cdot mL^{-1} \cdot K^{-1}}$ (see footnote 1). Then calculate the enthalpy of reaction per mole for each reaction by dividing each $\Delta H_{reaction}$ by the number of moles of H_2SO_4 added in the reaction. The number of moles is calculated from the volume and molarity of the added H_2SO_4 [mol = V_R (liter) × M_R (mol/liter)]. Express the final calculated enthalpy changes in units of kJ/mol. Because most of the ΔT values are known only to two significant figures, the final average calculated enthalpy changes should be rounded to two significant figures.

CONSIDER THIS

The enthalpies of several common reactions or physical processes can be measured quite accurately with the Styrofoam-cup calorimeter. Try measuring:

(a) Mg(s), Zn(s), or Al(s) + HCl(aq)

(b) Dissolving of NaCl(s) in water

(c) Dissolving of KNO_3(s) in water

(d) Melting of ice

(e) Condensation of steam (as in making a latte)

(f) Vinegar + baking soda

Compare your results to those found from Hess's law and tabulated enthalpies of formation.

Does stirring the water in your calorimeter change the temperature? Try to duplicate Joule's original experiment to determine the mechanical equivalent of heat.

Enthalpy Changes in Chemical Reactions	Name _____
Hess's Law	Date _____ Section _____
	Locker _____ Instructor _____

Data and Calculations

1. The heat of solution and neutralization of 10 M H_2SO_4 and 1.00 M NaOH

Data	Trial 1	Trial 2
Initial temperature of 10 M H_2SO_4 (5.0 mL in a graduated cylinder)	°C	°C
Initial temperature of calorimeter 1 (50.0 mL of 1.00 M NaOH + 45.0 mL water)	°C	°C
Final temperature of the mixed solutions in calorimeter 1 (50 mL 1.00 M NaOH + 45.0 mL water + 5.0 mL 10 M H_2SO_4)	°C	°C

Calculations	Trial 1	Trial 2
(a) Temperature change of NaOH solution in calorimeter 1 ($\Delta T_C = T_{final} - T_{initial}$)	°C	°C
(b) Temperature change of added H_2SO_4 solution ($\Delta T_R = T_{final} - T_{initial}$)	°C	°C
(c) Use Equation (5) to calculate the heat gained by the NaOH solution. Assume that the product of *density* × *heat capacity* (dC_p) is 4.10 J·mL^{-1}·K^{-1}. $\Delta H_C = V_C dC_P \Delta T_C$	J	J
(d) Use Equation (5) to calculate the heat gained by the H_2SO_4 solution. Assume that the product of *density* × *heat capacity* (dC_p) is 4.10 J·mL^{-1}·K^{-1}. $\Delta H_R = V_R dC_P \Delta T_R$	J	J
(e) Calculate $\Delta H_{reaction}$ by using the principle of conservation of energy expressed by Equation (6), rearranged to solve for the enthalpy of reaction. $\Delta H_{reaction} = -\Delta H_C - \Delta H_R$	J	J
(f) Calculate the moles of added H_2SO_4 mol $H_2SO_4 = V_R(L) \times M$ (mol/L)	mol	mol
(g) Calculate $\Delta H_{reaction}$ per mole of reactants by dividing the results of part 1(e) by the number of moles of H_2SO_4 calculated in part 1(f). Express the final result in units of kJ/mol. Be sure to keep the proper sign of the calculated reaction enthalpy.	kJ/mol	kJ/mol
Average		kJ/mol

2. The heat of solution of 10 M H_2SO_4

Data	Trial 1	Trial 2
Initial temperature of 10 M H_2SO_4 (10 mL in a graduated cylinder)	°C	°C
Initial temperature of water in calorimeter 1 (90 mL deionized water)	°C	°C
Final temperature of calorimeter 1 (90 mL water + added 10 mL H_2SO_4)	°C	°C

Calculations	Trial 1	Trial 2
(a) Temperature change of water in calorimeter 1 ($\Delta T_C = T_{final} - T_{initial}$)	°C	°C
(b) Temperature change of added H_2SO_4 solution ($\Delta T_R = T_{final} - T_{initial}$)	°C	°C
(c) Use Equation (5) to calculate the heat gained by the NaOH solution. Use the same assumptions as in part 1. $\Delta H_C = V_C d C_p \Delta T_C$	J	J
(d) Calculate the heat gained by the H_2SO_4 solution. $\Delta H_R = V_R d C_p \Delta T_R$	J	J
(e) Calculate $\Delta H_{reaction}$ by using the principle of conservation of energy expressed by Equation (6), rearranged to solve for the enthalpy of reaction. $\Delta H_{reaction} = -\Delta H_C - \Delta H_R$	J	J
(f) Calculate the moles of added H_2SO_4. mol $H_2SO_4 = V_R(L) \times M$ (mol/L)	mol	mol
(g) Calculate $\Delta H_{reaction}$ per mole of reactants by dividing the results of part 2(e) by the number of moles of H_2SO_4 calculated in part 2(f). Express the results in units of kJ/mol. Be sure to keep the proper sign of the reaction enthalpy.	kJ/mol	kJ/mol
Average		kJ/mol

Name _____ Date _____

3. The heat of neutralization of 1.00 M H₂SO₄ and 1.00 M NaOH

Data	Trial 1	Trial 2
Initial temperature of NaOH solution in calorimeter 1 (50 mL of 1.00 M NaOH)	°C	°C
Initial temperature of H_2SO_4 solution in calorimeter 2 (50 mL of 1.00 M H_2SO_4)	°C	°C
Final temperature of mixture in calorimeter 1 (50 mL 1.00 M NaOH + 50 mL 1.00 M H_2SO_4)	°C	°C

Calculations	Trial 1	Trial 2
(a) Temperature change of NaOH solution in calorimeter 1 ($\Delta T_C = T_{final} - T_{initial}$)	°C	°C
(b) Temperature change of added H_2SO_4 solution ($\Delta T_R = T_{final} - T_{initial}$)	°C	°C
(c) Use Equation (5) to calculate the heat gained by the NaOH solution. Use the same assumptions as in part 2. $\Delta H_C = V_C d C_P \Delta T_C$	J	J
(d) Calculate the heat gained by the H_2SO_4 solution. $\Delta H_R = V_R d C_P \Delta T_R$	J	J
(e) Calculate $\Delta H_{reaction}$ by using the principle of conservation of energy expressed by Equation (5), rearranged to solve for the enthalpy of reaction. $\Delta H_{reaction} = -\Delta H_C - \Delta H_R$	J	J
(f) Calculate the moles of added H_2SO_4. mol $H_2SO_4 = V_R(L) \times M$ (mol/L)	mol	mol
(g) Calculate $\Delta H_{reaction}$ per mole of reactants by dividing the results of part 3(e) by the number of moles of H_2SO_4 calculated in part 3(f). Express the results in units of kJ/mol. (Be sure to keep the proper sign of the reaction enthalpy.)	kJ/mol	kJ/mol
Average		kJ/mol

Summary of Results

Write the chemical equations corresponding to the reactions of parts 1, 2, and 3, and summarize the calculated heats of reaction, $\Delta H_{\text{reaction}}$, in units of kJ/mol.

	Chemical equations	Average $\Delta H_{\text{reaction}}$(kJ/mol)
Part 1		
Part 2		
Part 3		

Add the chemical reactions for parts 2 and 3. Do they give the same net chemical reaction as the reaction for part 1? Compare the average $\Delta H_{\text{reaction}}$ for part 1 with the sum of the results for parts 2 and 3 (pay attention to the number of significant figures).

$\Delta H_{\text{reaction}}$ for part 1: _____ kJ/mol

Sum of the $\Delta H_{\text{reaction}}$ for parts 2 and 3: _____ kJ/mol

Difference: _____ kJ/mol

Calculate the percentage relative error for the difference between the two pathways:

$$\text{percentage relative error} = \frac{\text{difference}}{\Delta H_{\text{reaction}} \text{ for part 1}} \times 100\% = \text{_____} \%$$

Are your results in agreement with Hess's law? Explain.

(See Question 1 that follows for a discussion of the uncertainty in the measured values of the reaction enthalpies.)

Questions

1. *Measurement uncertainty.* To get some idea of the uncertainty in your measured reaction enthalpies, repeat the calculation of the ΔH_C (in joules) for part 1(c), Trial 1, assuming that ΔT_C is 0.4 °C larger than the original one you measured. (We are assuming that the worst case uncertainty in ΔT_C would be 0.4 °C, because ΔT is the result of subtracting two temperature measurements, each having an uncertainty of 0.2 °C when measured with a thermometer with 1 °C intervals.) We are using only the ΔH_C term because it is the largest and most important term. The uncertainty in this term is essentially the same as the overall uncertainty in $\Delta H_{\text{reaction}}$ calculated in part 1(g).

Name _____   Date _____

Calculated values of ΔH_C (joules):

Original value calculated for ΔT_C New value calculated for $\Delta T_C + 0.4\ °C$

_____ J   _____ J

Now compare the new value to the old (first) value by calculating the relative percentage uncertainty in ΔH_C:

$$\text{relative percentage uncertainty} = \frac{\Delta H_C(\text{new}) - \Delta H_C(\text{original})}{\Delta H_C(\text{original})} \times 100\% =$$

Typically, this relative percentage uncertainty is about 3%. Also keep in mind that our neglect of the heat losses to the calorimeter biases our results to values that are about 3% too low.

2. *The enthalpy of dissociation of water.* The H_2SO_4 and NaOH reactants and the $NaHSO_4$ product in part 3 actually exist as dissociated ions, so the equation

$$H_3O^+ + HSO_4^- + Na^+ + OH^- \rightarrow Na^+ + HSO_4^- + 2\ H_2O$$

better represents the reaction (the *total* ionic equation).

Write the equation that represents the net ionic equation by canceling out the spectator ions (those that appear on both sides of the total ionic equation).

Net ionic equation: _____

This net ionic equation represents the reaction of a strong acid and a strong base in water.

Based on this result, and the measured $\Delta H_{\text{reaction}}$/mol from part 3, what is the approximate enthalpy change for the dissociation of water into H_3O^+ and OH^- ions? (*Hint:* How is the net ionic reaction above related to the reaction below?)

$$2\ H_2O \rightarrow H_3O^+(\text{aq}) + OH^-(\text{aq}) \qquad \text{Estimated } \Delta H_{\text{reaction}}: \underline{\hspace{2cm}}\text{kJ/mol}$$

(*Note:* This is only an approximate estimate because HSO_4^- ion is nearly undissociated in 1 M H_2SO_4 but is about 14% dissociated in 0.5 M $NaHSO_4$ because the enthalpy change for this partial dissociation is included in our measurements. Compare your estimate for the dissociation of water with the literature value of about +57 kJ/mol.)

CONSIDER THIS

The enthalpies of several common reactions or physical processes can be measured quite accurately with the Styrofoam-cup calorimeter. Try measuring:

(a) Mg(s), Zn(s), or Al(s) + HCl(aq)

(b) Dissolving of NaCl(s) in water

(c) Dissolving of KNO_3(s) in water

(d) Melting of ice

(e) Condensation of steam (as in making a latte)

(f) Vinegar + baking soda

Compare your results to those found from Hess's law and tabulated enthalpies of formation.

Does stirring the water in your calorimeter change the temperature? Try to duplicate Joule's original experiment to determine the mechanical equivalent of heat.

The Enthalpy of Combustion of a Vegetable Oil

Purpose

- Explore the idea of food as fuel.

- Construct a simple calorimeter for measuring the energy produced by burning vegetable oil.

Pre-Lab Preparation

From a strictly material viewpoint, human beings are biochemical engines that require both a fuel and an oxidant. The oxidant is oxygen in the air. The fuel is the food we eat, which consists of carbohydrates (starches and sugars), fats, and proteins, plus traces of minerals, vitamins, and other substances we need for good nutrition. The big three (carbohydrates, fats, and proteins) all contain carbon and hydrogen, and most of the carbon and hydrogen we eat in our food is metabolized to carbon dioxide and water. This is equivalent to burning the food and suggests the possibility that we could measure the fuel equivalent of the food we eat by burning it in a calorimeter and measuring the energy produced as heat.

Energy needs of the body

What are the energy requirements of our bodies, and how do our bodies use the energy contained in the food we eat? Careful studies have shown that food energy requirements depend on factors such as age, weight, physical activity, and gender (males require about 10% more energy per kilogram of body weight than females).

A surprisingly large fraction of the food we eat is required to sustain basic cell functions necessary for the maintenance of life—such as the ongoing metabolic activities of each cell, the circulation of the blood by the beating of the heart, the rhythmic contraction of the lungs in breathing, and the maintenance of body temperature. These minimum energy needs, called the *basal metabolic rate* (BMR), must be met before any energy can be used for physical activity or for the digestion of food. As a rough rule of thumb, a college-age male requires about 1.0 kcal per kilogram of body weight per hour and a college-age female requires 0.9 kcal per kilogram per hour. (Problem: Verify that a young 140-lb woman would require 1375 kcal per day for her BMR. Assume that 1 kg = 2.2 lb. *Also note that the nutritional calorie used to express the caloric value of food portions is actually 1000 calories, or 1 kilocalorie.*)[1]

The second component of the total energy requirement is the energy required for physical activity that involves voluntary use of the skeletal muscles. The amount of energy required depends on how many muscles are involved, how much weight is being moved, and how long the activity lasts. Contrary to popular belief, mental effort requires very little extra energy, although it may make you feel tired. The energy requirements for a sedentary (mostly sitting) activity like that of a typist amount to about 20% of the BMR. Light activity (such as that of a teacher or student) requires about 30% of the BMR. Moderate to heavy activity, such as that of a nurse or athlete, may require energy amounting to 40 to 50% of the BMR.

The third component of the energy requirement has to do with digesting our food. When food is eaten, many cells that have been dormant begin to be active—for example, the cells that manufacture and secrete digestive juices and the muscle cells that move food through the intestinal tract. This requirement is about 10% of the total kilocalories used for BMR and physical activity.

Now let's go back and add up the daily energy requirements of our hypothetical 140-lb young woman. Her BMR requirement is about 1375 kcal/day. Assuming light activity, we will add 30% of the BMR for physical activity, or 412 kcal. Now add 10% of the sum (1375 + 412 kcal = 1787 kcal), or 179 kcal, for digestion of food to get a grand total of 1966 kcal/day as her approximate requirement. Of course, these figures are based on several estimates, so her needs might fall within ±100 kcal/day of the estimated value.

Note that 70% (1375 out of 1966 kcal) of her daily food requirement is required just to sustain her BMR. If she increases her physical activity but does not increase her food consumption, she will lose weight.

---

[1]One kilocalorie is the energy required to increase the temperature of one kilogram of water by 1 °C.

A deficit of 125 kcal/day brings about the loss of body fat at the safe rate of about a pound per month. (A pound of body fat is not pure fat but a mixture of fat, water, and protein, and yields about 3500 kcal when burned. A pound of pure fat would yield about 4040 kcal when burned.)

Combustion calorimetry

The caloric content of food is determined by burning it in a closed metal calorimeter pressurized with oxygen and completely surrounded by water. The calorimeter can be calibrated by burning a substance whose heat of combustion has been accurately determined. The method we will employ to measure the heat (enthalpy) of combustion of a vegetable oil is similar in principle, but the calorimeter we will use is a crude version of the combustion calorimeter used for accurate scientific work.

In this experiment, we will use a lamp that burns oil. The heat from the burning oil will be used to heat water, and the temperature increase of the water will be a measure of the energy produced in the combustion. The enthalpy change for a sample of water that is heated from temperature T_1 to temperature T_2 is given by

$$\Delta H = mC_P \Delta T \qquad (1)$$

where ΔH is the enthalpy change in calories (in this example, heat absorbed by water), m is the mass of water in grams, C_P is the heat capacity (at constant pressure) in $cal \cdot g^{-1} \cdot K^{-1}$, and ΔT is the temperature change ($\Delta T = T_2 - T_1$) in Kelvin (or Celsius) degrees.

When the oil burns, not all of the energy released finds its way into the water heated in the calorimeter, so it is necessary to determine the *efficiency* of the calorimeter (the fraction of energy that is captured) by burning a substance whose heat of combustion is known. We will use 1-decanol, $CH_3(CH_2)_9OH$, whose enthalpy of combustion has been determined to be 9.9 kcal/g, to calibrate the calorimeter. To do this, we will burn a known mass of 1-decanol under the same conditions used to burn a known mass of vegetable oil. The measured enthalpy change for the vegetable oil will then be divided by the efficiency to get a corrected value that will approximate the true value for the enthalpy of combustion of vegetable oil.

$$\Delta H_{combustion} = \frac{\Delta H_{measured}}{efficiency} \qquad (2)$$

Experimental Procedure

Special Supplies: 12-oz aluminum beverage can (cut the top out with a can opener); three-finger clamp with ring stand; alcohol lamps (filled with a vegetable oil); 0 to 110 °C thermometer; 100-mL graduated cylinder; steel wool; 18-in.-wide heavy duty aluminum foil; size 33 rubber bands; scissors.

Candles can be substituted for the 1-decanol lamps, cutting the number of lamps required in half (see Notes to Instructor). Votive candles or "tea" candles in metal cups (approximately 4 cm in diameter and 2–5 cm high) are the most suitable.

Chemicals: 1-decanol; vegetable oil (canola, cottonseed, safflower, sunflower, peanut, etc.)

NOTES TO INSTRUCTOR

Alcohol lamps with 4-oz (120-mL) capacity and larger-than-ordinary wicks are available from VWR Scientific (1-800-932-5000) as Catalog No. 17805-005. An analytical balance with 300-g capacity is required for weighing the lamps when they are filled.

Put a permanent label on each lamp, specifying the liquid that the lamp contains. If half of the lamps are filled with 1-decanol and the other half filled with vegetable oil, half of the class can do part 1 (burning 1-decanol) while the other half is doing part 2 (burning vegetable oil). Then students can exchange lamps to complete the experiment. If kept capped when not in use, lamps can be stored and reused without emptying and refilling.

The number of lamps can be reduced by half by using votive candles or "tea" candles in metal cups to measure calorimeter efficiency in place of lamps containing 1-decanol. If candles are used, take the enthalpy of combustion of the candles to be 10.0 kcal/g.

Increasing the energy capture efficiency. If the room is drafty, the flame may wander, allowing a significant amount of energy to escape. The efficiency of energy capture can be improved by doing two things: (1) Wrap a 5 × 9 in. (13 × 23 cm) sheet of heavy duty aluminum foil around the can to make a skirt that extends about 3 cm below the bottom of the can to capture a greater fraction of energy. The skirt can be held in place with a rubber band. (2) Make a draft shield by folding an 18 × 18 in. (45 × 45 cm) piece of heavy duty aluminum foil in half twice to make a 4-layer-thick piece of foil 4.5 × 18 in. Fold the foil into a 4.5-in.-high cylinder by wrapping it around a liter bottle. Overlap the ends, holding them in place with a paper clip. At the bottom of the cylinder, cut three slots about 2 cm wide and 2 cm high, folding the 2 × 2 cm flaps up to make apertures through which air can enter. Place the draft shield around the lamp, centering the aluminum can over the lamp. When the lamp is lit, the bottom edge of the skirt wrapped around the aluminum can should be positioned just above the tip of the flame. Taking these steps will typically increase the energy capture efficiency to >65% and improve the reproducibility of the measurements.

Using a can opener, cut the top out of a 12-oz aluminum beverage can. Mount the can vertically with a three-finger clamp held on a ring stand, shown in Figure 16-1. (To improve energy capture efficiency, make an aluminum foil skirt and draft shield as described in the Notes to Instructor.)

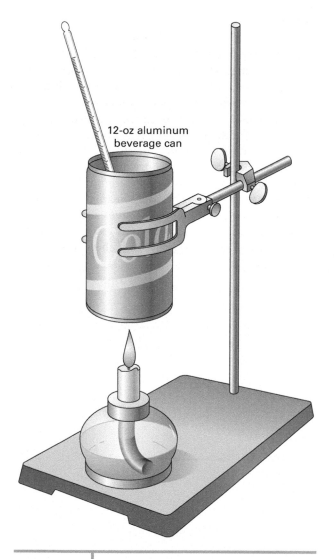

12-oz aluminum
beverage can

FIGURE 16-1 **A simple combustion calorimeter using an aluminum beverage can with the top cut out. An aluminum foil skirt on the can and a draft shield around the lamp (not shown) are recommended to improve the energy capture efficiency. (See Notes to the Instructor.)**

Parts 1 and 2 can be done in either order. Follow the directions of your instructor. Wipe off any oil on the metal cap surrounding the wick. Check that the wick is flush with the end of the metal tube holding it, and use scissors to trim off any stray threads. (If the wick protrudes from the metal shield, you will get a large, bushy, sooty flame, and combustion will be incomplete.) Light your lamp and see that it burns in a satisfactory manner. If the lamp is difficult to light, or burns with too sooty a flame, try pushing out the wick a bit, trimming it with scissors, then pushing it back in until it is flush with the metal tube.

If you put an aluminum foil skirt on the can, the bottom of the aluminum skirt should be positioned just above the tip of the flame. Blow out the flame, and allow the lamp to cool for a minute or two.

1. Calibration of the Calorimeter After the lamp has cooled, weigh the 1-decanol lamp (or a candle—see Notes to Instructor) on the analytical balance and record the mass to the nearest milligram.

Using your 100-mL graduated cylinder, measure out exactly 100 mL of water and pour it into the aluminum can, being very careful not to spill any water on the outside of the can. Stir the water thoroughly with a thermometer, then read the temperature to the nearest 0.2 °C, and record it.

Position the weighed 1-decanol lamp under the can, light the lamp with a match, and carefully center the flame under the bottom of the can. (A draft shield will keep the flame steadier (see Notes to Instructor). Lift up the draft shield to light the lamp, then quickly lower it into position.)

Stir the water continuously with a thermometer; when the temperature of the water in the calorimeter has risen about 18–20 °C above room temperature, remove the lamp and blow out the flame. Keep stirring the water in the calorimeter, recording the highest temperature attained to the nearest 0.2 °C.

When the lamp has cooled two or three minutes, reweigh it and record the mass. You now have the necessary data to calibrate the calorimeter, using the known heat of combustion of 1-decanol (9.9 kcal/g) and the known heat capacity of water.

At the end of the experiment, pour out the water and look at the bottom of the can. A light coating of soot is normal. If the coat is heavy, the wick should be trimmed and the measurement repeated. Take care that there is no water on the outside of the can when you start a measurement. Between each measurement, wipe off the excess soot with a paper towel or by scouring it with steel wool. If your instructor so directs, repeat the measurement. With care, the temperature change of the water divided by the mass change of the lamp should agree within 10% for the two trials.

2. Enthalpy of Combustion of a Vegetable Oil

Record in your report the type of vegetable oil used. Weigh to the nearest milligram and record the mass of the lamp containing vegetable oil. Using exactly the same procedure as in part 1, put exactly 100 mL of water in the calorimeter can, and carry out the temperature measurements, heating, and weighing as before. Record the initial and final temperatures of the water; also record the mass of the vegetable oil lamp after it has cooled.

By assuming that the fraction of the heat captured by the calorimeter is the same for the vegetable oil lamp as for the 1-decanol lamp, you now have the necessary data to calculate the heat of combustion of vegetable oil. If your instructor so directs, repeat the measurements to obtain a duplicate set of values. Take care that there is no water on the outside of the can when you start each measurement, and remember to keep the wick trimmed and to wipe off excess soot after each measurement.

When all of the measurements are completed, cap the lamps. Calculate the efficiency of the calorimeter from the data obtained in part 1 and the enthalpy of combustion of the vegetable oil from the measured efficiency and the data obtained in part 2.

CONSIDER THIS

Peanuts, once ignited, burn rather briskly. Burning half a Planter's brand peanut, weighing about 0.5 g, provides enough energy to heat 100 mL of water about 15 °C. To make a suitable holder for the peanut, wind a conical spiral on the end of a 15-cm length of 20-gauge nichrome wire. (The conical end of a cork borer sharpener makes an ideal form for winding the spiral.)

To determine the heat of combustion of a peanut half, weigh it, place it in the wire holder, ignite it by holding the flame of a match under it until it begins burning, then hold it under the aluminum can containing 100 mL of water. (If you're using a draft shield, you will need to cut a narrow slot in the shield through which you can thrust the wire holder and peanut.) The peanut burns with a large sooty flame and typically leaves behind a blackened bit of ash in the peanut's original shape. (Subtract the mass of the ash to get the net mass of peanut burned.) You can compare the value you get with the nutritional information on the container label.

Suppose that you did not put salad dressing equivalent to one tablespoon of vegetable oil on your salad at dinner. How many fewer kilocalories would you have eaten as fat? Assume that one tablespoon of oil has a volume of 15 mL, a density of 0.92 g/mL, and supplies 8.8 kcal/g of oil when consumed. (Also recall that 1 nutritional calorie is equal to 1 kcal.)

Bicycle riders in the Tour de France (a 20-day, 2000-mile bicycle race) require about 5000 kcal/day to maintain their energy needs—about 2.5 times the energy requirement for a person involved in more normal activities. How many loaves of bread would this be equivalent to, assuming 17 slices per loaf of bread, each slice (40 g) providing about 100 kcal?

Emission Spectra and the Electronic Structure of Atoms

The Hydrogen Atom Spectrum

Purpose

• Use a simple spectroscope to determine the wavelengths of the spectral lines emitted by hydrogen atoms excited by an electrical discharge.

• Show that these wavelengths fit a pattern of energy states described by a simple formula containing a single constant and integer quantum numbers.

• Show that atomic spectra also can be used qualitatively to identify elements.

Pre-Lab Preparation

The absorption and emission spectra of the elements' gaseous atoms do not look like the continuous rainbow-like spectrum of the heated filament of an incandescent bulb. They show instead many sharp lines separated by regions where there is no emission or absorption. This distinction provides the most convincing evidence that the electrons in atoms do not have a continuous range of energies but only certain discrete values of energy. In reaching this conclusion, studies of the emission spectrum of the simplest atom, hydrogen, have played a crucial role.

The spectroscope, introduced by Bunsen and Kirchhoff circa 1859, made it possible to observe the atomic spectra of excited gaseous atoms; scientists soon discovered that each element had a characteristic spectrum that could be used to identify it (or to detect previously undiscovered elements). These spectral studies produced much information on the spectral lines of the elements, but no real progress was made in explaining the origin of spectra until 1885, when Johann Balmer pointed out that the wavelengths of the lines of the hydrogen atom spectrum were given by a simple empirical formula containing one constant and the squares of small integers. Once this regular pattern was discovered, it became a challenge to human ingenuity to explain these regularities.

In 1913, the Danish physicist Niels Bohr used the notion of the quantum of energy (introduced by Max Planck) and the model of the nuclear atom (published in 1911 by Rutherford) to develop a theory of the behavior of an electron moving around a small, positively charged nucleus. He used this theory to calculate the wavelengths of the spectral lines of the hydrogen atom with remarkable accuracy and showed how the regularities could be explained by assuming that energy was absorbed or emitted only when an electron passed from one energy state to another. Each energy state had an energy given by a constant divided by the square of a small integer, called a *quantum number*.

Although Bohr's theory was quite successful in interpreting the spectrum of the hydrogen atom, it was not as successful for interpreting the spectra of atoms with more than one electron. This led others, such as Erwin Schrödinger, Werner Heisenberg, and Paul Dirac, to formulate more complete theories of quantum mechanics—the first of which was published in 1926. Since 1926, the development of these theories has shown that all aspects of atomic and molecular spectra can be explained quantitatively in terms of energy transitions between different allowed quantum states. (A quantum of energy is the smallest amount of energy that can be transferred to or from an atom or molecule. Transferred only in packets of discrete size, called *photons* or *quanta,* the energy is said to be "quantized.")

In this experiment, you will use a simple spectroscope to measure the wavelengths of the spectral lines generated by the light energy emitted by several elements when the atoms of those elements have been excited by an electric discharge or a hot flame. A commercial handheld spectroscope (like that shown in Figure 17-6) is constructed with a pre-calibrated scale. Readings from this scale can be converted directly into wavelength readings. If the spectroscope is homemade, the scale of the spectroscope can be calibrated using the known wavelengths of the lines observed in a low-pressure mercury lamp.

Once you have a spectroscope with a calibrated scale, you will then measure the wavelengths of hydrogen and some gaseous nonmetal or metallic elements. You will use data for hydrogen to construct quantitatively part of the energy diagram for hydrogen. As you do this experiment, think about the fact that you are following in the footsteps of scientific pioneers whose work ushered in the era of quantum mechanics and earned them Nobel prizes.

Waves and diffraction

Light is electromagnetic radiation. This means that it possesses electric and magnetic properties that vary sinusoidally and in phase with each other, as shown in Figure 17-1.

Because only the electric part of the wave interacts with the electrons in atoms, Figure 17-1 is often simplified to the representation in Figure 17-2. The wavelength of this wave, λ, is defined as the distance between any two repeating portions of the wave. The wavelengths of light visible to our eyes are 4×10^{-7} m (blue) to 7.5×10^{-7} m (red). Because these wavelengths are quite small, the units of wavelengths are often given in angstroms (Å), where 1 Å $= 10^{-10}$ m, or in nanometers (nm), where 1 nm $= 10^{-9}$ m. Demonstrate for yourself that the limits of the visible spectrum in these units are 4000 Å to 7500 Å, or 400 nm to 750 nm.

Another important property of waves is frequency, ν. The frequency of a wave is the number of wavelengths that pass a given point in a unit of time. The unit of frequency in reciprocal seconds (s^{-1}) is called the Hertz (Hz).

For a given wave, wavelength and frequency are not independent of each other. The higher the frequency, the smaller the wavelength. The reason for this relationship is that all light waves travel at the same velocity. This velocity is the speed of light, c, and is equal to 3.00×10^8 m/s. Some reflection on these three quantities leads to the correct mathemat-

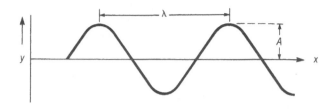

FIGURE 17-2 | **The amplitude (A) and wavelength (λ) of a light wave.**

ical expression (an analogy with a moving train in which λ represents the length of one car is helpful):

$$c\,(\text{m/s}) = \lambda\,(\text{m}) \times \nu\,(\text{s}^{-1}) \qquad (1)$$

One of the unusual properties of waves is diffraction. All waves display this phenomenon, whether they are ocean waves, light waves, or sound waves. Perhaps you have seen ocean waves striking a small opening or several openings in a breakwater and observed the interesting patterns formed by the waves after they have passed through these openings. One pattern that you might have observed would look like the schematic diagram in Figure 17-3.

These patterns are obtained only when both the openings and the spacing, d, between the openings are of the same magnitude as λ. Thus, in order for diffraction to occur with light waves, the light must pass between slits that are very narrow and very close together. The plastic replica holographic gratings used for the homemade spectroscopes are made from a laser-produced master grating that has 500 lines/mm. These plastic gratings are of the transmission type, which means that light does not

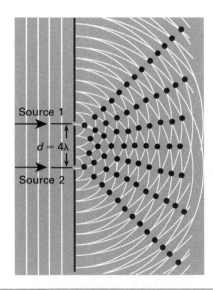

FIGURE 17-3 | **Diffraction of ocean waves by openings in a barrier.**

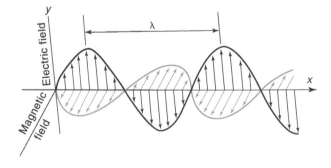

FIGURE 17-1 | **The electric- and magnetic-field components of a light wave.**

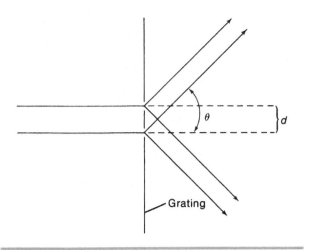

FIGURE 17-4 | **Diffraction of light by a transmission grating.**

pass through the grating where the lines have been drawn but rather through the spaces between the lines.

The quantitative description of the behavior of waves on passing through slits of the appropriate size is the famous Bragg equation:

$$m\lambda = d \sin \theta \qquad (2)$$

The quantity m is a small integer having values 0, 1, 2, 3, ... The value d is the distance between the lines of the grating. The meaning of the angle θ is displayed in Figure 17-4.

The usefulness of the Bragg equation, as applied to a diffraction grating, is that it shows that the longer the wavelength is, the greater the angle of deviation, θ, will be. In all measurements made in this experiment, you will be observing the first-order, or $m = 1$, diffraction patterns. This means that if light of differing wavelengths strikes the grating, it will be bent to varying angles and in fact separated into its component wavelengths. White light consists of light of all wavelengths. The diffraction pattern of an ordinary tungsten filament lamp therefore is just a lovely rainbow. One of the first observations you should make with your spectroscope is of the light from a fluorescent lamp. In addition to the rainbow pattern observed with the tungsten lamp, you may be able to discern a few discrete lines that are brighter than the general background of the rainbow spectrum. Some of these lines come from the small amount of mercury vapor that is in the fluorescent lamp.

Energy levels and line spectra of elements

Our present theory of atomic structure states that electrons in an atom can possess only discrete energy values. As we noted earlier, the word *quantized* is used to describe discrete values. Every element has a characteristic set of energy levels. The energy-level diagram for hydrogen is given in Figure 17-5. The energy levels are characterized by an integer n, which is called the *principal quantum number*. At the

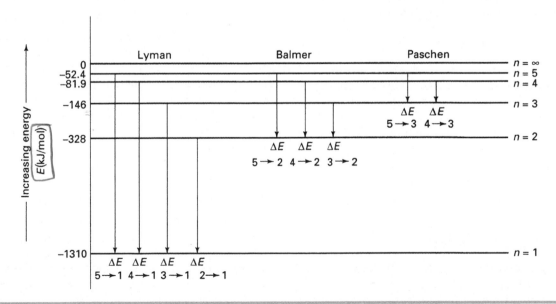

FIGURE 17-5 The energy levels of the hydrogen atom. The numbers on the left are the energy levels expressed in units of kJ/mol. The integers on the right are the principal quantum numbers. The vertical arrows correspond to the energy differences, ΔE, between levels described by the indicated quantum numbers. The energy differences represent observable lines in the hydrogen spectrum. Each observable line corresponds to a spectral transition, that is, the transition from one energy state to another state. Each of the three groups of spectral lines shown here is named for its discoverer.

temperature of the hydrogen discharge lamp, most of the hydrogen atoms have energy corresponding to the $n = 1$ level, which is called the *ground state.*

If energy is supplied to a hydrogen atom in the $n = 1$ state by an electrical discharge or by heat, a certain fraction of the atoms will absorb this energy and enter the $n = 2$, $n = 3$, or higher levels. These atoms having extra energy are called *excited* atoms. They can lose some of this extra energy in discrete amounts and drop back down to the $n = 3$ or $n = 2$ or $n = 1$ ground state, as indicated in Figure 17-5.

The energy that is lost in these transitions appears as a visible quantum of light called a *photon.* It is important to note that all light is quantized in these units, which can be described as both particles and waves, depending on the kind of experiment in which they are observed. Photons possess an energy proportional to the frequency of the light wave. The proportionality constant is Planck's constant, h, which has a value of 6.62×10^{-34} J • s. Thus the electron of a hydrogen atom in an excited energy state can lose energy by emitting a photon that has an energy corresponding to the transition from the n_1 to n_2 energy levels, where n_1 and n_2 represent the quantum numbers of the initial and final energy states. (Note that for the smallest energy transition in the Balmer series of Figure 17-5, $n_1 = 3$ and $n_2 = 2$.)

If a transition takes place from energy level n_1 to n_2, the absolute magnitude of the energy difference between the n_1 and n_2 levels is related to the wavelength of the photon by the equation

$$|E_{n2} - E_{n1}| = \Delta E = h\nu = \frac{hc}{\lambda} \qquad (3)$$

The last identity is derived from Equation (1), which expresses the relationship between the velocity of light, frequency, and wavelength. The energy ΔE can be expressed in joules/photon. This is a very tiny amount of energy. If we multiply this energy in joules/photon by Avogadro's number, we can express the energy in joules (or kJ)/mole of photons. The numbers shown on the energy scale in Figure 17-5 are expressed in kJ/mol.

Experimental Procedure

Special Supplies: Inexpensive spectroscopes like the one shown in Figure 17-6 may be purchased[1,2] or constructed (Method A); or a bare grating in a darkened

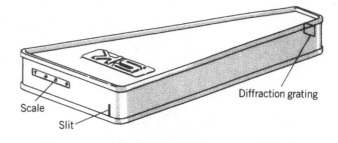

FIGURE 17-6 A handheld spectroscope for measuring wavelengths (Method A).

room may be used (Method B). The commercial spectroscopes have a calibrated scale that can be directly used to read wavelengths, and their higher-quality diffraction gratings give a brighter image. A helium-neon laser (or inexpensive laser pointer) is useful for demonstrating diffraction through a transmission-type plastic replica grating.

Method A: To construct a spectroscope like that shown in Figure 17-7, the following supplies will be needed: a cardboard box with lid (a No. 8 two-piece, set-up mailing box, $6 \times 4.5 \times 2.25$ in., is a convenient size); spray cans of flat black paint (have the laboratory assistant spray the inside of each box with flat black paint a day or two ahead of time; this saves much time and mess); single-edged razor blade; double-edged razor blade; small piece of translucent millimeter graph paper; metric ruler; transmission-type plastic replica grating mounted in a 2×2 in. cardboard slide mount;[2] masking tape.

Method B: The setup shown in Figure 17-8 requires a darkened room in order to see the spectral lines and works best with capillary-type gas discharge tubes. A Masonite grating holder and shield, constructed as shown in Figure 17-8, provide a convenient way to observe the spectra. A transmission-type plastic replica grating (see footnote 2) is taped to the grating holder. Black construction paper can be used to make a slit that is taped to the shield (which is painted flat black). A meterstick is used to measure the distance from the slit to the virtual image of the slit on the masonite shield.

Sources: The most effective spectral sources for calibration (mercury spectrum) and examination of the emission spectra of gaseous atoms are inexpensive spectrum tubes excited by a 5000-V, 7-mA current transformer-type power supply. These are available from a number of suppliers.[3]

---

[1]Science Kit and Boreal Laboratories, Inc., www.sciencekit.com. QA Spectroscope, Catalog No. WW16525M00.
[2]Edmund Scientific, www.scientificsonline.com. 3054251 Classroom Spectrometer with Scale. 3054512 Holographic Diffraction Grating, 2×2 in. Card Mounted Slides, 500 lines/mm.

[3]Edmund Scientific, www.scientificsonline.com 3071559 Spectrum Analysis Power Supply, 115 V. 3060906 Spectrum Tube, Hydrogen. Sargent-Welch, www.sargentwelch.com WL2393D Power Supply, Spectrum Tube. WLS68755-30G Spectrum Tube, Hydrogen.

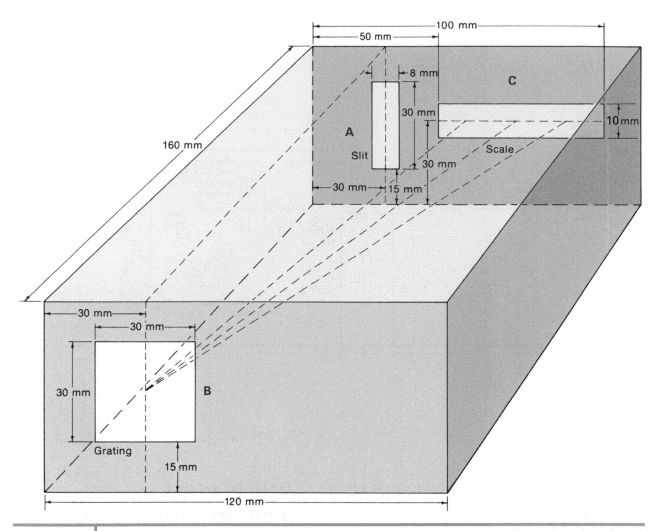

FIGURE 17-7 | **The construction of a handheld spectroscope for measuring wavelengths by Method A.**

In order to produce the spectra of different metals, solutions of metal salts can be aspirated into a nebulizer to form a fine mist that is directed into the base of a Bunsen or Fischer burner. Metal salt "pills" or pellets can also be heated on the grid of a Fischer or Meker burner. These can be formed with an inexpensive screw-type press of the kind used to make KBr pellets for taking infrared spectra. (Detailed procedures for constructing and using the flame sources are given in the references cited in the Bibliography.) The chloride salts of lithium, sodium, potassium, calcium, strontium, and copper are recommended. (Some metal vapors, e.g., barium and copper, are toxic, so it is preferable to place the gas burners used to heat the metal salts in a fume hood.)

Tungsten filament bulbs and fluorescent lamps should also be available so that students can compare their spectra (see part 3 of the Experimental Procedure).

SAFETY PRECAUTIONS:
The power supplies used for the spectrum tubes are sources of high voltage that represent a danger to the unwary user. If the spectrum tube power supply or holder (or both) has exposed electrical connections, a shield like that shown in Figure 17-8 should be placed between the source and the viewer. Cut a rectangular viewing port in the shield so that the source can be viewed through the port.

A low-pressure mercury lamp puts out a large fraction of its energy in the ultraviolet (UV) 254-nm line. Though not visible, this UV light can cause eye damage. It is absorbed strongly by the lens

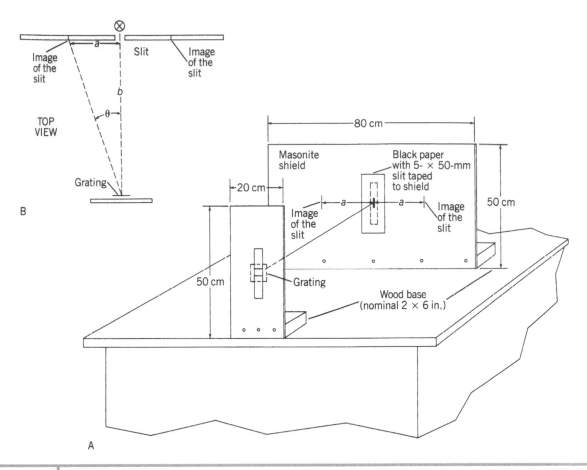

FIGURE 17-8 | **Plan (A) for the construction of a holder for the grating and slit for measuring wavelengths by Method B. The source and power supply are located behind the protective Masonite shield. The slit is a piece of black paper with a 5 × 50 mm aperture taped to the Masonite shield. (B) Top view of the setup.**

material in eyeglasses and safety goggles. Do not look at the lamp without wearing glasses, and limit the calibration time to 5 to 10 minutes. For greater safety, a piece of ordinary window glass can be taped over the viewing port on the back of the shield to act as a UV cutoff filter.

In order to prevent the breathing of any metal vapor (barium and copper are toxic), place the gas burners used to heat metal salts in a fume hood.

Your instructor will tell you if you are to use spectroscopes that are already prepared or if you are to construct your own (Method A, Figure 17-6 or 17-7), or if you are to use a bare grating in a darkened room (Method B, Figure 17-8).

Demonstrating the diffraction of light using a helium-neon laser. (Students should be cautioned not to look into the laser beam when using their spectroscopes.) If you have access to an inexpensive helium-neon laser (or laser pointer), the instructor can demonstrate how a monochromatic beam of light is diffracted by a plastic replica transmission grating. Direct the laser beam through the grating (with the laser beam perpendicular to the grating); allow the diffracted beam to fall on a sheet of white paper located about 30 cm away from the grating. You will see the bright undiffracted primary beam (zero order) in the center. Flanking it on either side will be less-bright spots of light that are the first-order diffraction; faint spots at twice the angle of the first-order diffraction are the second-order diffracted beams. The instructor may also take this opportunity to explain how the diffracted beam, entering the eye at an angle relative to the direction of the primary beam, appears as if it were coming from a point on the scale of the spectroscope even though no light is falling on the scale. It is an optical illusion. The image viewed is not real and is therefore called a *virtual image.*

1. Construction of a Spectroscope *Method A.* If you are to construct a spectroscope like that shown in Figure 17-7, obtain a small cardboard box (6 × 4.5 × 2.25 in.) that has been sprayed on the inside with flat black paint to reduce internal reflections and make the spectra easier to observe. Noting the dimensions given in Figure 17-7, mark the positions of the holes on the outside of the box and cut out the three holes for the *slit,* the *grating,* and the *scale.* Cut the holes through the box and lid simultaneously, using a single-edged razor blade.

Carefully (razor blades are quite sharp) break the double-edged razor blade lengthwise in two. At point A in Figure 17-7, tape the sharp edges facing each other to the outside of the box, with their edges 0.3 to 0.5 mm apart. (No diffraction is obtained from this slit because this opening is many times the wavelength of visible light.) Make sure that the edges of the slit are parallel to each other and perpendicular to the top and bottom of the box.

Observe the grating carefully. You will probably be able to observe some striations on the plastic surface. If you can see these reflections, mount the grating at point B in Figure 17-7, with the lines running perpendicular to the top and bottom of the box. (If you cannot observe these striations, close the box and hold it with the slit pointing toward a fluorescent light. Hold the grating next to the hole at point B and observe the diffraction pattern. The orientation of the grating that produces diffraction patterns on the right and left of the slit is the correct one.) Tape the grating to the inside of the box at point B, with the grating in the correct orientation.

Finally, tape a scale cut from a piece of millimeter graph paper at point C, Figure 17-7. If too much light comes in through the scale for you to see the lines clearly, mask off part of the scale with tape.

Method B. The spectroscope consists of the setup shown in Figure 17-8: a source (capillary spectrum tube), a Masonite shield with a slit taped on the front, a transmission grating taped to a Masonite holder, and a meterstick. The images you see to the left and right of the slit will be virtual, not real, images of the slit (or capillary source). The slit is a 5 × 50 mm rectangular hole cut in the center of a piece of black construction paper that is taped to the front of the Masonite shield (see Figure 17-8).

Tape the grating to a Masonite support like that shown in Figure 17-8. (The rectangular holes, 3 × 20 cm, cut in the Masonite grating support and shield are made long in order to allow vertical adjustment of the grating and slit to match the height of the source.) Position the grating about 50 cm from the slit. (The plane of the grating must be parallel to the capillary source or slit and form a

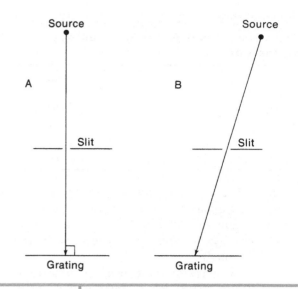

FIGURE 17-9 **The angle of the incident light should be 90° with respect to the grating: (A) correct; (B) incorrect.**

right angle with the incident light from the source, as shown in Figure 17-9.)

Make sure you are wearing glasses to protect yourself from UV radiation from the sources. Turn on the mercury source and darken the room. Position your eye directly behind the grating within eyelash distance so that you are looking through the grating toward the source. Look to the right and the left. You should see images of the slit symmetrically displaced on both sides of the slit, corresponding to $m = \pm 1$ in Equation (2). The images on one side may appear slightly brighter because of the construction of the grating. Use the brighter side for all further measurements. If you have a good grating, you may also be able to see the second-order images ($m = \pm 2$) further displaced on both sides of the slit.

For each line you see in the first-order spectrum, record in your report the distance *a* (distance between the slit and the image of the slit or capillary source on the shield), measured to the nearest millimeter (see Figure 17-8). It may be helpful to have a lab partner place a marker such as a pencil at the apparent position of the line. Also measure and record distance *b* (the distance between the grating and the slit).

2. Calibration of the Spectroscope Wear glasses at all times while observing lamps through the spectroscope to protect yourself from possible harmful UV radiation. If you are using a commercial spectroscope like that shown in Figure 17-6, you need not construct the calibration graph. Simply verify that the reading for the lines observed in the mercury lamp corresponds to the known wavelengths given in

TABLE 17-1 Wavelengths of the Mercury Spectrum Lines Used for Spectroscope Calibration

Color	Wavelength (nm)
Violet	404.7
Blue	435.8
Green	546.1
Yellow	579.0

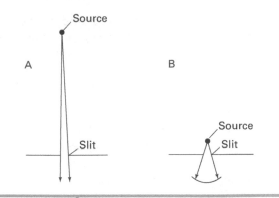

FIGURE 17-10 For the light to be well collimated when a handheld spectroscope is used (Method A), the slit should be at least 25 to 30 cm from the source. (A) Well-collimated light; (B) poorly collimated light.

Table 17-1. Each major division on the scale of the spectroscope marked by a numeral corresponds to an increment of 100 nm; for example, a reading of 4.2 on the scale corresponds to a wavelength of 420 nm (4200 Å).

If you are making your own spectroscope, you will use the emission lines of mercury as a "standard" to calibrate the spectroscope. Look at the spectrum of a mercury lamp. You should be able to see three or four lines. The colors and wavelengths of these lines are shown in Table 17-1.

Measure the positions of the four most prominent lines on the scale as accurately as possible (nearest millimeter). These four lines appear as two pairs of lines separated by an appreciable gap. The lines of each pair are separated by about 30 nm; the gap between the pairs of lines is about 110 nm. (Some may have difficulty seeing the less bright violet line of the violet and blue pair at 405 and 436 nm.) Record the average of several observations. Do you see any additional lines besides these four? If so, comment on them in the report form and measure their positions on the scale.

The following precautions and directions will enable you to measure reproducibly the positions of lines with handheld spectroscopes (Method A): (a) The lamp should be far enough away from the spectroscope (at least 25 to 30 cm) so that the direction of propagation of the light is well defined or collimated (Figure 17-10). If you do not observe this precaution, you will find that whenever you move your eye the spectral lines will appear to wander rather than remaining stationary on the scale. (b) On the other hand, the lamp should not be too far away (more than 40 cm), or the intensity of light falling on the slit will be too low for you to observe the spectrum clearly. (c) The light must be incident on the grating at 90° if the angle of diffraction is to satisfy the equation given previously (see Figure 17-9).

On the graph paper provided in your Report form, draw a calibration graph of wavelength versus scale reading. Do this during your laboratory period. In the event that a good straight line is not obtained, measure the positions of the lines again.

You will use your linear graph to determine the wavelength of unknown lines. First, measure the position of the unknown line on the scale of your spectroscope; then read from your calibration graph the wavelength that corresponds to the measured scale position.

3. Observations and Measurements of the Hydrogen Spectrum Measure the scale position of the observable lines of the hydrogen emission spectrum. Normally at least three lines can be seen: violet, green, and red. From your measurements and your calibration graph, determine the wavelengths of these lines.

Look up the accepted literature values of the wavelengths of these lines in the *CRC Handbook of Chemistry and Physics* (see the Bibliography). Calculate your relative error as a percentage:

$$\text{percentage relative error} =$$
$$\frac{\text{your value} - \text{accepted value}}{\text{accepted value}} \times 100\% \quad (4)$$

Convert your values of wavelengths for the hydrogen atom spectrum to units of kJ/mol. First calculate the energy of one photon in units of J from the relation $\Delta E = hc/\lambda$, where h is Planck's constant (6.626×10^{-34} J • s), c is the speed of light (3.00×10^8 m/s), and λ is the wavelength (m). Then multiply the result by Avogadro's number ($N_A = 6.023 \times 10^{23}$) to express the energy in J/mol of photons. Finally, divide by 1000 to express the result in units of kJ/mol. For each wavelength corresponding to a line in the hydrogen atom spectrum, the calculated energy difference in kJ/mol should correspond to a *difference* between two energy states shown in Figure 17-5. For example, the red line of hydrogen corresponds to the difference between −328 kJ/mol and −146 kJ/mol, which is 182 kJ/mol.

Draw that portion of the hydrogen atom energy level diagram that you have measured, using energy units of kJ/mol. Make your figure to scale on the graph paper provided.

Using a form of the Rydberg equation,

$$\Delta E = h\nu = \frac{hc}{\lambda} = R_\text{H}\left(\frac{1}{n_2^2} - \frac{1}{n_1^2}\right) \qquad (5)$$

calculate for each of the three lines you measured a value for the Rydberg constant, R_H, in units of kJ/mol of photons. This constant relates the energy change (kJ/mol) of a given transition to the principal quantum numbers for the transition. The values of the quantum numbers (n_1 and n_2) for each of the lines you observed can be obtained from Figure 17-5. Keep in mind that each line you observed corresponds to a transition between two energy states. Each state has an energy and a quantum number, shown in Figure 17-5 (n_1 is the quantum number of the initial state, while n_2 is the quantum number of the final state, so $n_1 > n_2$ for the $n_1 \rightarrow n_2$ transition).

The ionization energy of the hydrogen atom

The transition from $n = 1$ to $n = \infty$ describes the ionization of the hydrogen atom, the energy required to completely remove the electron from a hydrogen atom. The energy difference corresponding to this difference is equal to the Rydberg constant. Thus, the value of the Rydberg constant in units of kJ/mol can be read directly from Figure 17-5 as 1310 kJ/mol, the difference in energy between the $n = 1$ and $n = \infty$ levels.

CONSIDER THIS

Obtain other spectrum tubes containing various gaseous elements such as helium, neon, argon, or oxygen. Observe the emission spectra of the gases, and note and record the scale positions of the lines you see. [In the case of some gases such as neon, there may be too many lines to record. In this case, describe the wavelength region(s) where most of the lines fall, and give the corresponding predominant color of the lines.] Compare this information to that recorded in the *CRC Handbook*.

Obtain a variety of metal salts—for example, sodium, potassium, calcium, copper, lithium, or strontium chlorides. Ask your instructor for techniques for obtaining bright flames on a Bunsen or Fischer burner. (See the *Instructor's Manual*.) Determine whether you can detect the presence of one or more of these salts in an "unknown" solution.

Observe the spectrum of the light from a tungsten filament lamp and from a fluorescent lamp by using your spectroscope. How are the spectra similar? Different? How does the fluorescent lamp spectrum resemble that of the mercury lamp in part 2 of the experiment?

How are the ideas that are explored in this experiment used to identify elements in the sun and stars? How about planets that only reflect light? How does the relativistic Doppler effect (or redshift) come into play in this setting? How are these observations linked to the Big Bang?

Bibliography

Bernard, M. "Spectroscopic Cation Analysis Using Metal Salt Pills," *J. Chem. Educ.* **1980,** *57,* 153.

CRC Handbook of Chemistry and Physics. Published annually by CRC Press, Boca Raton, FL.

Edwards, R. K.; Brandt, W. W.; Companion, A. A. "A Simple and Inexpensive Student Spectroscope," *J. Chem. Educ.* **1962,** *39,* 147.

Harris, S. P. "Letter to the Editor," *J. Chem. Educ.* **1962,** *39,* 319.

Logan, K. R. "Some Experiments in Atomic Structures," *J. Chem. Educ.* **1974,** *51,* 411.

Smith, G. D.; Sanford, C. L.; Jones, B. T. "Continuous Liquid-Sample Introduction for Bunsen Burner Atomic Emission Spectrometry," *J. Chem. Educ.* **1995,** *72,* 438–440.

Ionic and Covalent Bonding

Conductivity of Solutions of Ionic and Covalent Compounds

Purpose

• Use electrical conductivity as a way of discriminating between substances with ionic and covalent bonding.

• Compare the electrical conductivity of substances in the pure solid (or liquid) state and in solution.

• Measure the effects of solvent properties on the dissociation of substances with polar covalent bonding, such as HCl.

• Measure changes in conductivity that result from the reaction of ionic substances to form products that are weakly ionized or insoluble.

Pre-Lab Preparation

Ionic, polar covalent, and covalent bonding

When an active metal reacts with an active nonmetallic element, the nonmetal atom becomes negatively charged when one or more electrons transfer from the metal, which in turn becomes positively charged. In general, when atoms act cooperatively in this fashion, each element winds up with a closed-shell electron configuration, like the electronic configuration of the noble gases of Group VIII (18). For the reaction of a sodium atom and a chlorine atom, we can represent the changes in the valence shells of the atoms by electron-dot formulas; for example,

$$\text{Na} \cdot + \cdot \ddot{\underset{\cdot\cdot}{\text{Cl}}} : \longrightarrow \text{Na}^+ \; : \ddot{\underset{\cdot\cdot}{\text{Cl}}} :^-$$

Such electrically charged atoms are called *ions,* and in the product Na^+Cl^-, the Na^+ ion has the same electron configuration as a neon atom, while the Cl^- ion has the electron configuration of argon. The bonding force between ions, due primarily to the attraction of unlike electric charges, is called an *ionic bond.*

However, when two elements of similar electronegative character react, they form electron pair bonds, mutually shared by both of the atomic nuclei. For example,

$$\text{H} \cdot + \cdot \text{H} \longrightarrow \text{H} : \text{H} \quad \text{or} \quad \text{H}_2$$

$$2\,\text{H} \cdot + \cdot \ddot{\underset{\cdot\cdot}{\text{O}}} : \longrightarrow : \ddot{\underset{\cdot\cdot}{\text{O}}} : \text{H} \quad \text{or} \quad \text{H}_2\text{O}$$
$$\phantom{2\,\text{H} \cdot + \cdot \ddot{\text{O}} : \longrightarrow : \ddot{\text{O}} :}\text{H}$$

Such a bond is called a *covalent bond.* In the second example, the water molecule, the oxygen atom is more electronegative than hydrogen, so the O—H bond is slightly polar, or partially ionic, in character. We could describe the water molecule as having *polar covalent* bonding. Note that the two types of bonds found in Na^+Cl^- and molecular H_2 are really two extremes of a continuum—the purely ionic bond at one extreme and the purely covalent bond at the other. In between are the polar covalent or partially ionic bonds.

Structure and bond type

Substances with covalently bonded atoms generally have discrete or separate molecules packed together in the crystal structure of the solid.[1] If soluble in water or other suitable solvent, these substances dissolve to form a solution of electrically neutral molecules; the resulting solutions do not conduct an electric current. Examples of such *nonelectrolytes* are common table sugar (sucrose, $\text{C}_{12}\text{H}_{22}\text{O}_{11}$) and acetone ($\text{CH}_3\text{COCH}_3$).

Salts, as well as *acids* and *bases,* have either ionic bonds or bonds that are quite polar. In solid salts, the ions are packed together so that ions of positive charge are surrounded by ions of negative charge and vice versa. As a result, ions form the structural units of the crystal. When melted by heat, vacancies

---

[1]Omitted from discussion here are substances called *network solids,* which form covalently bonded giant molecular networks such as silica (SiO_2) and carbon in the form of diamond, as well as metals and intermetallic compounds.

or holes are introduced into the rigid crystal structure, allowing the ions to move about so that the melted salt becomes fluid and a good electrical conductor. When these substances dissolve in water, the ions likewise separate as independently moving particles, and the solutions are electrical conductors. So, in equations throughout this manual, we write the formulas of such substances in solution, or in the molten state, as separate ions. Examples of such *electrolytes* are sodium hydroxide (NaOH), which dissociates into Na^+ and OH^- ions, and potassium sulfate (K_2SO_4), which dissociates into K^+ and SO_4^{2-} ions in a 2:1 ratio.

Strong acids and bases and many salts ionize completely in dilute aqueous solution. (In fact, we may regard this as a definition of a strong acid or base—it is a substance that is completely dissociated into ions in solution.) In solutions of weak or less active acids and bases, a substantial fraction of the dissolved substance is present in molecular form. So although the total concentration may be high, the concentration of ions is low. Partial ionization (or dissociation) accounts for their lower conductivity. Salts that are only slightly soluble, even though they may be completely dissociated into ions in solution, have low electrical conductivity because of their limited solubility.

The role of the solvent in dissociation reactions

Most of the solution reactions we have studied have been in aqueous solution. However, it is important to remember that nonaqueous solvents are often used and that the dissociation of acids, bases, and salts depends very much on the properties of the solvent. For example, the dissociation of an acid involves the transfer of a proton from the acid to the solvent. Therefore, the extent of dissociation of an acid will depend on the intrinsic basicity of the solvent. Although acetic acid is only partially dissociated in water, it is completely dissociated in a more basic solvent, such as liquid ammonia.

$$HC_2H_3O_2 + H_2O \rightleftharpoons H_3O^+ + C_2H_3O_2^-$$
PARTIAL REACTION

$$HC_2H_3O_2 + NH_3 \longrightarrow NH_4^+ + C_2H_3O_2^-$$
COMPLETE REACTION

The dielectric constant of the solvent and the solvation energy of ions in a solvent are also important. If two ions of opposite charge are placed in a vacuum, the force between them is given by Coulomb's law,

$$force = \frac{kq_1q_2}{r^2}$$

where k is a constant, q_1 and q_2 are the charges on the ions, and r is the distance between the centers of the ions. When the two ions are immersed in a material medium, the force between them is reduced in inverse proportion to the *dielectric constant, ϵ,* of the medium, as expressed in the equation

$$force = \frac{kq_1q_2}{\epsilon r^2}$$

Therefore, the total energy expended in separating the ions will decrease as the dielectric constant increases. That means that a solvent with a large dielectric constant will tend to promote the dissociation of an ionic solute more than will a solvent with a small dielectric constant. The hydrocarbons (with $\epsilon < 5$) do not promote the dissociation of ionic solutes. Water (with $\epsilon \cong 80$) is a good solvent for ionic solutes.

Solvation energy is also an important factor. When an ionic solute dissociates in a solvent, the total energy of the reaction may be thought of as composed primarily of two terms: (1) the energy expended in the separation of the positive and negative ions of the crystal lattice (called the *lattice energy*), and (2) the *solvation energy* liberated by the association of the ions with solvent molecules. Solvation energy is the result of the interaction between the electric charges of the ions and the electric dipoles of the solvent molecules. When there is enough solvation energy to offset the energy expended in separating the positive and negative ions, the ionic solute will dissociate in the solvent.[2] A polar solvent, therefore, will tend to promote the dissociation of an ionic solute.

Ionic equations

In describing a reaction in which ions are either the reactants or products, we often focus our attention on the essential changes, ignoring "spectator" ions that are not participating in the chemical reaction. We will call this abbreviated description the *net ionic equation.*

For example, the reaction of aqueous NaOH with aqueous HNO_3 could be written in full as

$$Na^+ + OH^- + H_3O^+ + NO_3^- \rightarrow Na^+ + NO_3^- + 2\ H_2O$$

---

[2]A more careful examination of the factors that determine the solubility of a solute must include the effect of changes in the solvent structure that occur when ions are introduced, causing polar solvent molecules to cluster around the ions. The entropy change of the system is a measure of this effect, which can be strong enough that salts such as NaCl will dissolve even though the lattice energy is greater than the solvation energy.

which we will call the *total ionic equation.* Note that in this reaction, neither the Na^+ nor the NO_3^- has reacted; both appear as separate particles on each side of the equation. When these spectator ions are omitted, the equation that expresses the essential change is the *net ionic equation:*

$$OH^-(aq) + H_3O^+(aq) \rightarrow 2\ H_2O(l)$$

(Note that the net ionic equation is still a balanced equation with respect to charge and the number of each kind of atom. The spectator ions that are not shown are also still contributing to the electrical conductivity of the solution.)

Typical ionic reactions are those in which ions unite to form a weakly ionized or insoluble substance. For example, if a weakly acidic substance such as ammonium chloride, NH_4Cl, reacts with a base such as sodium hydroxide, NaOH, the net change is expressed by

$$NH_4^+(aq) + OH^-(aq) \rightarrow NH_3(aq) + H_2O(l)$$

A typical example of ions uniting to form an insoluble substance is expressed by the equation

$$Ca^{2+}(aq) + CO_3^{2-}(aq) \rightarrow CaCO_3(s)$$

It makes no difference whether the Ca^{2+} is derived from calcium chloride, calcium nitrate, or any other soluble calcium salt. Carbonate ion, CO_3^{2-}, could be obtained equally well from sodium carbonate, ammonium carbonate, or any soluble substantially ionized carbonate salt.

Often the net result of an ionic reaction is determined by a competition between two largely undissociated substances. For example, carbonic acid, H_2CO_3, is formed when a strong acid acts on a carbonate salt:

$$CaCO_3(s) + 2\ H_3O^+(aq) \rightarrow$$
$$Ca^{2+}(aq) + H_2CO_3(aq) + 2\ H_2O(l)$$

Carbonic acid readily dehydrates to CO_2 and water, so the net ionic reaction is

$$CaCO_3(s) + 2\ H_3O^+(aq) \rightarrow$$
$$Ca^{2+}(aq) + CO_2(g) + 3\ H_2O(l)$$

Thus, this reaction is driven to the right by two factors: (1) the production of a very weakly dissociated substance, H_2O, and (2) the escape of volatile CO_2 gas.

The experimental method: Electrical conductivity

In this experiment, we will measure the electrical conductivity of pure liquid substances and dissolved substances to determine whether there is any solvent-assisted ionic dissociation. We will also try to determine the net result of any reaction that may take place when solutions of ionic substances are mixed.

Solutions of ionic-type substances conduct electric current by the movement of their ions under the influence of an electric field. The faster the ion moves, the greater its ion mobility. At the solution–electrode interfaces, the circuit is completed by electron transfer reactions, because in metal wires the only charge carriers are electrons.[3] For a given applied voltage, the amount of current depends on the concentration of ions and, to a lesser extent, on differences in individual ion mobilities. (Hydrogen ion has an unusually high mobility—five to eight times that of many other ions. To better understand why, see the Consider This discussion at the end of the experiment.) Figure 18-1 shows a simple apparatus for the comparison of electrical conductivities.

We will compare the electrical conductivities of a number of solutions and of the products formed after some chemical reactions, then use the data we obtain to interpret the character of the solutions and the course of any reactions. We will interpret these results in terms of the net ionic equations for the reactions.

Experimental Procedure

Special Supplies: Conductivity apparatus. (Use a commercial conductivity meter or an apparatus constructed as described in the notes that follow and shown in Figure 18-1.)

Chemicals: Pure 17 M acetic acid; 6 M acetic acid; 0.1 M acetic acid; methanol (CH_3OH); 6 M HCl; 0.1 M HCl; toluene; anhydrous HCl in toluene produced by bubbling HCl gas through dry toluene (*if gaseous HCl is not available, you may substitute a solution prepared by dissolving pure acetic acid in toluene to make a 1 M solution*); 0.1 M $HgCl_2$; 0.1 M NaOH; 0.1 M NH_3; 0.1 M H_2SO_4; 0.1 M $Ba(OH)_2$; 0.1% thymol blue indicator solution; sucrose (s); NaCl(s); $CaCO_3$(s) (marble chips); zinc metal (mossy).

▨ NOTES TO INSTRUCTOR ▨

Notes on the Construction and Use of the Conductivity Apparatus Shown in Figure 18-1

1. Build the circuit in a small aluminum chassis box, with the power cord, fuse, and pilot light on the rear; the banana jacks on the side; and the push-button switch on top. Mount the chassis box on a

---

[3]With alternating current (AC), the net chemical reaction at the electrode surface is diminished; at high frequency, the chemical reaction is practically eliminated.

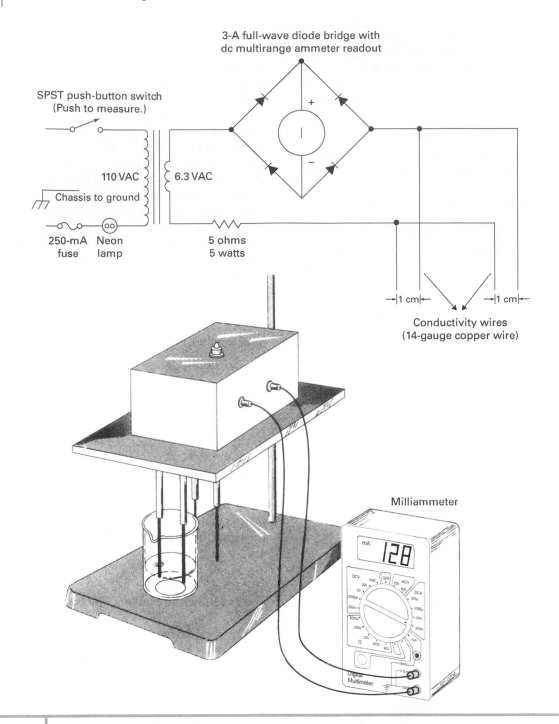

FIGURE 18-1 | **An apparatus for comparing the electrical conductivities of various solutions.**

piece of ³/₄-in. plywood through which holes have been drilled to allow two pairs of 14-gauge copper wire electrodes (connected in parallel) to protrude about 15 cm. Space the wires about 1 cm apart. All but the bottom 5 cm of the copper wire should be insulated. Fasten a length of ½-in.-OD aluminum rod on the underside of the plywood so that the apparatus can be supported by a clamp holder on a ring

stand. Connect the aluminum chassis to the green ground lead of the three-wire power supply cord.

2. The voltage across the electrodes is about 6 V AC rms (18 V peak-to-peak) when the push-button switch is depressed, so be careful not to touch the electrodes while the switch is depressed. The 5-ohm, 5-watt resistor limits the current to a safe value if the electrodes are inadvertently shorted.

3. An inexpensive digital multimeter with a direct-current (DC) multirange ammeter provides a suitable readout. Two to four current ranges covering 50 microamperes (μA) to 200 or 300 milliamperes (mA) will work well.

4. To make a measurement, adjust the ammeter to the highest current range (200–300 mA). Put enough of the test substance in a 30-mL beaker to give a depth of about 1 cm. Lower the electrodes into the beaker so that they are immersed to a depth of 1 cm.

Then depress the push-button switch to get a current reading. Switch the ammeter to the most sensitive current range that can be used without driving the meter off scale. The solution conductivity is directly proportional to the current. Record the current reading, and return the ammeter to the highest current range.

5. Between each measurement, rinse the electrodes with distilled water and dry them with tissue. As an extra precaution, unplug the apparatus while cleaning the electrodes. Periodically clean the electrodes with steel wool and rinse with water.

WASTE COLLECTION:
Mercuric chloride, $HgCl_2$, is toxic. After you have made measurements of the conductivity of 0.1 M $HgCl_2$, pour the solution into a waste bottle. *Do not pour it down the drain.* Likewise, toluene and solutions of HCl (or acetic acid) in toluene should be poured into waste containers, not down the drain.

1. Electrolytes and Nonelectrolytes If you will be using the conductivity apparatus shown in Figure 18-1, read over the preceding notes describing how to use the apparatus. If you will use a commercial conductivity meter, follow your instructor's directions. When you are confident that you understand how to use the conductivity meter, measure and record the conductivity of the following *substances and/or their solutions in water*. Determine whether they are largely, moderately, or poorly ionized, or whether they are essentially nonelectrolytes. (For an apparatus like that shown in Figure 18-1, a nonelectrolyte will give a very small current, less than a few microamperes. A completely dissociated strong electrolyte, such as 0.1 M HCl, will typically give a current of more than 100 mA.) Test the pure substance first; then add some water to it and retest. In each

case, the magnitude of the current will be a measure of the conductivity of the solid or solution, indicating whether or not the substance contains ions that are free to move.

(a) To begin, test distilled (or deionized) water and tap water. What causes the difference in conductivity? Then test the following substances and their aqueous solutions: methanol (CH_3OH); glacial acetic acid (CH_3COOH); sucrose (table sugar, $C_{12}H_{22}O_{11}$); sodium chloride (NaCl); and *dilute solutions only* (not the pure substances) of 0.1 M HCl, 0.1 M NaOH, and 0.1 M $HgCl_2$. Be sure to dispose of the mercuric chloride, $HgCl_2$, in the waste container reserved for this purpose. *Do not pour it down the drain—it is toxic.*

(b) Observe the *effect of the solvent* on the ionization of HCl in toluene and in water. First test a little pure toluene (C_7H_8) in a thoroughly dry beaker, and then a solution of HCl in toluene. (*Use the solution already on your reagent shelf,* prepared by bubbling anhydrous HCl through dry toluene—do *not* add aqueous concentrated HCl to toluene.) Then add 5 mL of deionized water to this solution, mix it well, and repeat the test with the electrodes immersed farther into the lower aqueous layer. (*If HCl in toluene is not available, you may substitute a 1 M acetic solution in toluene; the effect is less dramatic because acetic acid is not a strong acid in water like HCl.*) Be sure to dispose of the toluene solutions in the waste container reserved for this purpose.

(c) Compare the chemical behavior of 6 M HCl and 6 M acetic acid with your conductivity data. Put three or four marble chips into each of two clean beakers. Add 6 M HCl to one beaker and 6 M acetic acid to the other. Compare the rate of evolution of CO_2 gas. Put one or two pieces of mossy zinc metal into another pair of beakers, add 6 M HCl and 6 M acetic acid, and compare the rates of evolution of hydrogen (H_2) gas. Record your observations and write balanced chemical reactions for each of the four reactions you observed.

2. Typical Ionic Reactions You will experiment with some ionic reactions to determine the nature of their products. Through conductivity tests of the separate reactants and of the mixture after reaction, you will discover whether the acids, bases, and salts concerned are largely ionized (strong electrolytes) or only moderately ionized (weak electrolytes). For each reaction, write the total ionic equation and the net ionic equation that best describes the reaction taking place. This will require careful thinking on your part. Use the following procedures.

(a) *0.01 M HCl with 0.01 M NaOH.*[4] These solutions can be prepared by diluting 5 mL of each of the 0.1 M solutions to 50 mL. It is an advantage to use both pairs of electrodes connected in parallel (see Figure 18-1) by placing one pair of electrodes in 0.01 M HCl and simultaneously placing the other pair in an equal volume of 0.01 M NaOH contained in a separate beaker. Then mix the solutions, divide them into two equal portions, and again place the two solutions simultaneously in contact with the two electrode pairs. (This compensates for the dilution effect of mixing two solutions.) Why is the conductivity of the product solution lower than that of the reactant solutions?

(b) *0.1 M acetic acid, $HC_2H_3O_2$, with 0.1 M NII_3.*[5] As in part 2(a), test equal volumes of the separate solutions, mix them, divide the solution into two equal parts, then remeasure the conductivity. Record your measurements.

(c) *0.1 M H_2SO_4 with 0.1 M Ba(OH)_2.* Measure and record the conductivity of each solution by itself. Then add two or three drops of 0.1% thymol blue indicator solution to the sulfuric acid solution, and add 0.1 M $Ba(OH)_2$ solution one drop at a time to the sulfuric acid solution, stirring continuously. The sulfuric acid will be exactly neutralized when the indicator changes from its pink form through its yellow (acid) form to its blue (basic) form. If you add too much $Ba(OH)_2$, overshooting the end point, add 0.1 M H_2SO_4 drop by drop, stirring continuously, until the *exact* end point—shown by a sharp color change of the indicator—is reached. Now, measure and record the conductivity of this mixture. Write balanced ionic equations to describe the reactions that take place.

---

[4]Quite dilute 0.01 M solutions are suggested so that you can interpret moderate conductivity changes more easily.
[5]Solutions in water are sometimes formulated as NH_4OH, called *ammonium hydroxide.* Because the evidence for the existence of such a molecule in solution is ambiguous, we will simply refer to a solution of ammonia in water as "aqueous ammonia solution."

CONSIDER THIS

Solutions of strong bases like NaOH and strong acids like HCl have much higher conductivity than a solution of the same concentration of NaCl. By examining a number of examples of this, scientists have discovered that it is the hydronium ion (H_3O^+) and hydroxide ion (OH^-) that are responsible for the higher conductivity.

Conductivity is related to the mobility of an ion (how fast it moves in solution under the influence of an electric field). In order to get from point A to point B in solution, most ions have to physically move from point A to point B, dragging along the cluster of water molecules that surrounds each ion. However, for hydronium ions, the whole hydronium ion does not need to move in order to transfer a positive charge from point A to point B. The way this works is shown in the following diagram:

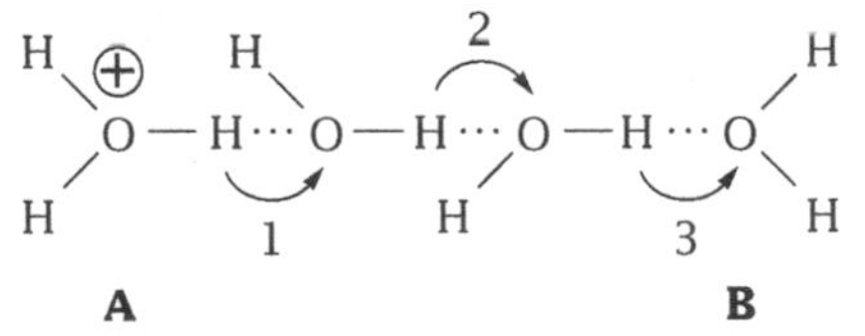

Imagine that there is a hydronium ion at point A on the left end of a chain of water molecules. The water molecules are loosely associated by hydrogen bonding, as shown by the dotted lines representing the hydrogen bonds. The positive charge can be transferred from the hydronium ion at point A to the water molecule at point B by having three protons hop from one molecule to the next, as indicated by the three arrows. After transfer of the three protons, the water molecule at point B would be transformed into a hydronium ion (H_3O^+), so the charge associated with a hydronium ion would have moved from point A to point B without physical movement of the whole hydronium ion originally at point A. This rapid proton hopping (or proton transfer) process allows charge to move along much faster than physical movement of the whole ion, such as a Na^+ or Cl^- ion.

Now, you draw a similar diagram, starting with a hydroxide ion on the right at point B, that shows how the −1 charge associated with a hydroxide ion could move from point B to point A, by shifting protons left to right from one water molecule to the next. (Hydronium ions and hydroxide ions under the influence of the same electric field would move in opposite directions.)

Ionic and Covalent Bonding

Conductivity of Solutions of Ionic and Covalent Compounds

Name _____

Date _____ Section _____

Locker _____ Instructor_____

Observations and Data

1. Electrolytes and nonelectrolytes

(a) Record the conductivity measurement of each *substance and solution* tested. Then, after noting the range of measurements made (about five orders of magnitude), classify the substances and solutions as containing essentially no ions (lowest values), a few ions (middle values), or many ions (highest values).

	Conductivity	Category		
		No ions	Few ions	Many ions
Distilled $H_2O(l)$				
Tap H_2O				
Methanol $CH_3OH(l)$				
$CH_3OH(aq)$				
Glacial acetic acid, $CH_3COOH(l)$				
Aqueous acetic acid, $CH_3COOH(aq)$				
Sucrose, $C_{12}H_{22}O_{11}(s)$				
Aqueous sucrose, $C_{12}H_{22}O_{11}(aq)$				
NaCl(s)				
NaCl(aq)				
0.1 M $HgCl_2(aq)$				
0.1 M HCl(aq)				
0.1 M NaOH(aq)				

(i) Write the equation for the reaction that forms the few (but important) ions in pure water.

(ii) Why is tap water more conductive than distilled water?

(iii) Is there any evidence that methanol forms ions either in the pure state or when dissolved in water? Would you classify methanol as a strong electrolyte, weak electrolyte, or nonelectrolyte?

(iv) Why does acetic acid form ions when it is dissolved in water but not when it is in the pure (glacial) form? Write an equation to describe the formation of ions in aqueous acetic acid.

(v) Compare your measurements for $CH_3OH(aq)$, $CH_3COOH(aq)$, and $HCl(aq)$. Would you classify aqueous acetic acid as a strong electrolyte, weak electrolyte, or nonelectrolyte? What species are present in aqueous acetic acid? In glacial acetic acid?

	Species present
$CH_3COOH(aq)$	
$CH_3COOH(l)$	

(vi) Does sucrose behave like a strong electrolyte, weak electrolyte, or nonelectrolyte when it is dissolved in water? What species are present in $C_{12}H_{22}O_{11}(aq)$?

(vii) The models of solid NaCl describe it as consisting of Na^+ cations and Cl^- anions. If this is the case, why is NaCl(s) not a conductor?

(viii) Explain the distinctly different behavior in the conductivity of NaCl(s) and NaCl(aq). What species are present in each of these substances?

	Species present
NaCl(aq)	
NaCl(s)	

(ix) Is $HgCl_2(aq)$ a strong electrolyte, weak electrolyte, or nonelectrolyte? What species are present in this solution? Is this consistent with the statement in the Pre-Lab Preparation that *"many salts ionize completely in dilute aqueous solution"*?

Name _____ Date _____

(x) What species are present in 0.1 M HCl and NaOH solutions? Write equations to describe the reactions that yield these species.

	Species present	Equation
HCl(aq)		
NaOH(aq)		

(b) The effect of the solvent

	Conductivity	Category		
		No ions	Few ions	Many ions
Toluene, C_7H_8(l)				
HCl (*in toluene*) (or CH_3COOH *in toluene*)				
Aqueous layer after mixing with HCl (*in toluene*) (or CH_3COOH *in toluene*)				

Why is the conductivity of HCl (or CH_3COOH) different in the two solvents? What causes this difference?

(c) Correlating chemical and conductivity behavior

(i) Compare the rates of reaction of $CaCO_3$(s) with 6 M acetic acid and 6 M hydrochloric acid. The reaction of $CaCO_3$ with an acid, HA, is

$$2\ HA(aq) + CaCO_3(s) \rightarrow Ca^{2+}(aq) + 2\ A^-(aq) + CO_2(g) + H_2O(l)$$

(ii) Compare the rates of reaction of Zn(s) with 6 M CH_3COOH and 6 M HCl. The reaction of zinc with an acid, HA, is

$$2\ HA(aq) + Zn(s) \rightarrow Zn^{2+}(aq) + 2\ A^-(aq) + H_2(g)$$

(iii) Is there a correlation between the rates of the chemical reactions of 6 M CH_3COOH and 6 M HCl and the conductivities observed for CH_3COOH(aq) and HCl(aq) in part 1(a)? Suggest an explanation.

2. Typical ionic reactions

(a) *0.01 M HCl with 0.01 M NaOH.* The relative conductivities of the solutions tested are as follows:

0.01 M HCl/0.01 M NaOH measured in parallel: _____

Mixture measured in parallel: _____

Do the reactants exist as molecules or ions? _____

Does the conductivity change indicate the creation of more ions or a decrease in the number of ions?

_____

The total ionic equation for the reaction is _____

The net ionic equation for the reaction is _____

Interpret any changes in the conductivity of the solutions, before and after mixing, in accordance with the preceding equations:

(b) *0.1 M CH₃COOH with 0.1 M NH₃.* The relative conductivities of the solutions tested are as follows:

0.1 M CH$_3$COOH/0.1 M NH$_3$ measured in parallel: _____

Mixture measured in parallel: _____

Do the reactants exist as molecules or ions? _____

Does the conductivity change indicate the creation of more ions or a decrease in the number of ions?

_____

The total ionic equation for the reaction is _____

The net ionic equation for the reaction is _____

Interpret any changes in the conductivity of the solutions, before and after mixing, in accordance with the preceding equations:

(c) *0.1 M H₂SO₄ with 0.1 M Ba(OH)₂.* The relative conductivities of the solutions tested are as follows:

0.1 M H$_2$SO$_4$: _____   0.1 M Ba(OH)$_2$: _____   Mixture: _____

The total ionic equation for the reaction is _____

The net ionic equation for the reaction is _____

Interpret any changes in the conductivity of the solutions, before and after mixing, in accordance with the preceding equations:

Name _____   Date _____

CONSIDER THIS

In the Consider This section at the end of the experiment, we discussed a way to explain the large mobility of hydronium ion (H_3O^+) and suggested that a similar proton transfer (or proton hopping) mechanism might explain the large mobility of hydroxide ion (OH^-).

Let's begin with a diagram showing a hydroxide ion at point B and a water molecule at point A:

$$
\begin{array}{c}
H \\
\backslash \\
O-H \cdots O-H \cdots O-H \cdots O^- \\
/ / \backslash \\
H H H
\end{array}
$$

$$\quad\quad \mathbf{A} \quad\quad\quad\quad\quad\quad \mathbf{B}$$

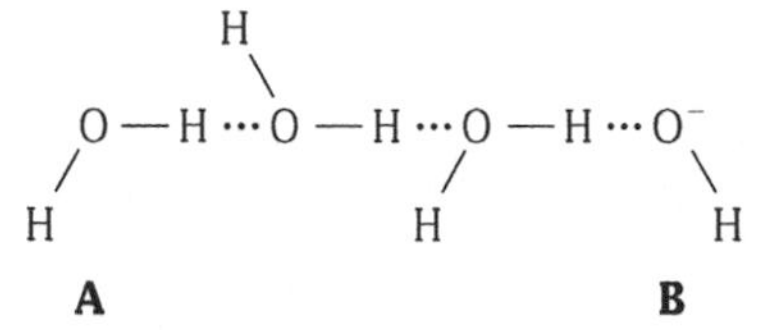

Add curved arrows to the diagram above to show how proton transfer from one water molecule to the next could lead to the transfer of (−) charge from point B to point A, effectively transforming the hydroxide ion at point B into a water molecule, and the water molecule at point A into a hydroxide ion.

Ions move, thereby transferring charge, under the influence of an electric field. Add lines at each end of the diagram to represent the electrodes in a conductivity apparatus.

Would the charge on the electrode at point A have to be positive or negative to attract a negatively charged hydroxide ion (OH^-)? _____

Would the charge on the electrode at point B have to be positive or negative to attract a positively charged hydronium ion (H_3O^+)? _____

What should the polarity of the electrodes be in order to induce proton transfer from one water molecule to the next, moving from left to right (point A to point B), considering that protons are positively charged particles?

      Charge on the electrode at point A (positive or negative?): _____

      Charge on the electrode at point B (positive or negative?): _____

An electric current is a flow or movement of charged particles. In a metal wire, electrons are the charge carriers, hopping from atom to atom. In solution, it is the movement of ions in solution that constitutes the electric current. Note that both positive and negative ions contribute to the electric current in solution.

Explain in simple language why ions move and in which direction they move in the presence of an electric field created between two electrodes having opposite charges.

Writing Lewis Structures

Purpose

• Learn how to draw Lewis structures of covalently bonded molecules and molecular ions.

• Derive from these diagrams the *bond order,* a measure of the strength of the bonds.

Introduction

The notion of the chemical bond as a shared-electron-pair (covalent) bond and the representation of these by diagrams, now called Lewis structures (or Lewis diagrams or simply electron dot diagrams), was presented in the years 1916–1919 by G. N. Lewis, a professor at the University of California, Berkeley, and by Irving Langmuir, a research scientist at the General Electric Company.

Lewis structures can be drawn using only paper and pencil and they provide a crude but still useful way to visualize the bonding in molecules. In presenting his ideas, Lewis did not use the ideas or language of quantum mechanics. Quantum mechanics has revolutionized our way of thinking about the behavior of small particles but did not come on the scene until after 1925.[1]

In his 1923 book, *Valence and the Structure of Atoms and Molecules,* summarizing his ideas, Lewis said, "*The chemical bond is at all times and in all molecules merely a pair of electrons held jointly by two atoms.*" Although this definition has since proved to be too restrictive and is not to be taken as the whole story, it's still a useful rule of thumb and will apply to most of the molecules you are likely to encounter in a general chemistry course. Important exceptions are the so-called electron-deficient boranes, such as diborane, which are compounds of boron and hydrogen.[2]

Like most chemists, Lewis thought with the aid of visual images. He visualized the arrangement of the electrons in the outer (valence) electron shell of an atom as a cube with places for electrons at the eight corners of the cube. For example, as shown in Figure B-1, two fluorine atoms, each with a vacancy in the valence shell, could fill the vacant positions in their cubes by sharing electrons. In the final structure, all the cube positions are filled, with the two atoms sharing a cube edge. Similarly, a double bond between two atoms could be represented by two cubes sharing a common face with four shared electrons. (A triple bond doesn't easily fit into a scheme where the electrons are arranged at the corners of a cube.) Lewis was able to draw structures for many more molecules known to him that were consistent with the notion of a covalent bond as an *electron pair* shared between two atoms.

Although the idea of a covalent shared-electron-pair bond undoubtedly arose from the static image of two cubes sharing an edge, a more profound explanation for the tendency of electrons to come in pairs appeared in 1925 after the discovery that electrons have a property called *spin.*[3]

---

[1]The qualitative concepts found in your general chemistry textbook that are borrowed from quantum mechanics include the notion of quantum numbers, orbitals, and the notation for the electron configuration of an atom designating the number of electrons occupying atomic orbitals of successively higher energy. For example, the electron configuration for a neutral argon atom is written as $1s^2 2s^2 2p^6 3s^2 3p^6$, where 1, 2, and 3 are the principal quantum numbers $n,$ corresponding to electron shells 1, 2, and 3. The term *orbital* was apparently first used by the American chemist Robert S. Mulliken in 1932 to mean a one-electron hydrogen-like wavefunction. In the interpretation of Max Born, the square of the amplitude of the wavefunction defines the probability per unit volume of finding an electron at a particular point in an atom.

---

[2]Diborane, B_2H_6, better represented as $H_2BH_2BH_2$, contains two bridging B–H–B bonds involving three atoms bonded by a single electron pair, now generically called *two-electron–three-center bonds.*

[3]To date, no simple and easy explanation has been given to explain the origin of the spin angular momentum of electrons and most other fundamental particles. Even neutrons with zero charge and photons with zero rest mass and zero charge have spin. In a magnetic field, electrons show just two spin states. We can call them *spin up* and *spin down.* There are two important consequences of electron spin—the tendency of electrons with opposite spin to pair and the exclusion of other electrons from any quantum state already occupied by another electron, called the *Pauli exclusion principle.* The exclusion effect is not a physical force. It's more like a statistical probability, but its effect is even stronger than the normal Coulomb repulsion between two negatively charged electrons. In the absence of the exclusion effect, we assume that atoms with more than one electron would show no shell structure because all electrons would occupy the same state; then, chemistry and life as we know it would not exist.

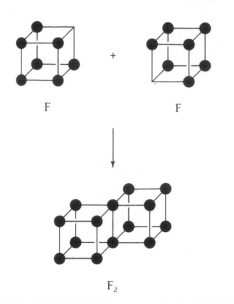

F   F

F_2

FIGURE B-1 G. N. Lewis imagined an octet of valence electrons occupying the eight corners of a cube. A fluorine atom with seven valence electrons has one electron position vacant. A bond between two fluorine atoms is formed by sharing two electrons so that the octets on both atoms become filled.

To summarize, Lewis had two ideas that are useful approximations for describing the covalent bonding in molecules and molecular ions involving hydrogen and the nonmetal elements:

• A covalent bond is an electron pair shared between two atoms.

• Bond formation proceeds until each atom has a share of eight electrons in its outermost (valence) shell. This is called the *octet rule.*

We emphasize again that these rules are approximate descriptions of chemical bonding, not laws of nature. We have already mentioned the boranes as exceptions to the first rule. Likewise, the *rule of eight,* as Lewis called it—later dubbed the octet rule by Langmuir[4]—applies most rigorously to molecules formed from atoms in Period 2 (particularly compounds containing C, N, O, and F) but has a number of exceptions. Hydrogen is unique. The hydrogen atom can share a maximum of two electrons because that is the maximum capacity of the valence shell of

hydrogen. (This corresponds to a filled $1s$ orbital and quantum number restrictions dictate that $1p$ orbitals do not exist.)

Besides recognizing the restriction on hydrogen, Lewis also recognized that compounds such as PCl_5, or any compound having an atom forming bonds to more than four other atoms, must be an exception to the octet rule. Thus, atoms of Period 3, such as P, S, and Cl, can share more than eight electrons. At the simplest level of explanation, their larger size allows more atoms to fit around them. This is also consistent with the orbital description of bonding. In atoms beyond Period 2, more than four bonds can be formed by the participation of vacant d orbitals in the bonding, which effectively enlarges the capacity of the valence shell.

When atoms of elements combine with one another to form molecules like water, H_2O, or molecular ions like nitrate, NO_3^-, or ammonium, NH_4^+, the bonding can be described as a bond in which a pair of electrons is simultaneously attracted to two or more atomic nuclei. This picture can be simplified by imagining that the filled inner electron shells of an atom neutralize much of the nuclear charge of an atom, producing an atom core with a residual effective nuclear charge. Then we can approximately describe the chemical bond as the result of mutual attraction between the outer valence electrons of the atoms and the effective nuclear charge of the atom cores (or kernels). Chemical bonding is almost entirely a phenomenon involving only the outer-shell valence electrons. The electron density of the core is not significantly changed when a bond is formed to another atom (excepting hydrogen atoms, which have only a valence shell with no inner core). This is why Lewis structures need show only the valence electrons of the atoms.

In order for there to be a strong mutual attraction between two atoms, there must be vacancies in the valence shell of at least one of the atoms. Strong covalent (electron-pair) bonds are not formed between two atoms if both have filled valence shells. When the outer orbitals (valence shells) of two atoms are completely filled, the electrons of one atom cannot penetrate closely enough to the nucleus of the other atom to be strongly attracted. Consequently, the interactions between the two atoms will be very weak. This is typical of the interatomic forces between atoms of a noble gas, such as between two argon atoms.

However, the heavier noble gases such as xenon will form covalent bonds with very electronegative atoms such as fluorine and oxygen that have vacancies in their valence shells. This can be explained in the context of the orbital model of the atom by assuming that the xenon atom can make use of

[4]Irving Langmuir also coined the term *covalent bond,* which replaced the cumbersome but more descriptive term, *shared-electron-pair bond.* He also was the first to assign *formal charges* to the atoms in a Lewis structure, which he called *residual atomic charge.* We will discuss the concept of formal charge later.

vacant *d* orbitals that lie at higher energy than the filled *s* and *p* orbitals. Promotion (or excitation) of xenon *p*-orbital electrons to the xenon *d* orbitals creates outer orbital vacancies on the xenon atom, expanding the capacity of the valence shell. These outer orbital vacancies allow the sharing of electrons and the formation of stable bonds with fluorine or oxygen. In this way of explaining bonding, the energy required for promotion is more than recovered when the stable bonds are formed.

Uses and Limitations of Lewis Diagrams

What are Lewis structures good for?

• They provide a way to represent covalent bonding with a simple visual image that shows how the atoms are linked together in the molecule.

• They provide the foundation for predicting the shape of a molecule using the simple rules of valence shell electron-pair repulsion (VSEPR) theory.

• They help to explain the bond-forming capacity of an element, often called the *valency,* in relation to the position of the element in the periodic table.

• They provide a simple way for estimating the relative strength of a bond between two atoms, called the *bond order* (or bond multiplicity). The bond order usually correlates well with experimentally measured bond strengths and bond lengths.

What are some of the limitations of Lewis structures?

• For some molecules, a Lewis structure that conforms to the simple rules for drawing Lewis structures cannot be drawn. For example, there are a number of exceptions to the octet rule.

• Lewis structures cannot always be relied on to give even a qualitatively correct representation of the charge distribution in a molecule or molecular ion, as we shall see later.

• Lewis structures are static images, so they convey no sense of the dynamic motions of the electrons and atoms in a molecule. This may give the incorrect impression that the nuclei and electron pairs binding them together sit motionless in fixed positions.[5]

---

[5]A detailed analysis of covalent bonding shows that the tendency of electrons to spread out in the region between two atoms rather than be confined to the orbitals of just one atom acts to lower the kinetic energy of the electrons and is an important factor in covalent bonding, as described by G. B. Bacskay, J. R. Reimers, and S. Nordholm, "The Mechanism of Covalent Bonding," *J. Chem. Educ.* **1997,** *74,* 1494–1502.

• Structures drawn on a two-dimensional sheet of paper may give a misleading impression of the true three-dimensional shape of the molecule.

The Basic Rules While learning a few simple rules for drawing Lewis structures, keep in mind the basic ideas of covalent bonding: Each bond is formed by sharing a pair of electrons between two atoms. Generally, enough electron-pair bonds are formed to fill the outer orbital vacancies of each atom.

Covalent bonds are formed mainly between the nonmetals. These are hydrogen and the elements in Groups 13–18 (main Groups III–VIII), often called the *p*-block elements because they are those elements where the *p* subshell is being filled. Elements in these groups have valence shells comprised of the outer *s* orbitals (holding two electrons) and the three *p* orbitals (holding six electrons). Therefore, the completely filled valence shells will accommodate eight electrons, giving rise to the octet rule, which states that each atom (except hydrogen) in a stable molecule will have a share of eight electrons, with an electron configuration like that of a noble gas. Note, however, that the Lewis theory of a covalent bond did not use concepts such as *s*, *p*, and *d* orbitals, which derive from quantum mechanics, only the notion that atoms had an outer valence shell of electrons, the number of electrons in the shell being determined by the length of a period in the periodic table.

The octet rule results not so much from any special stability of the noble gas electron configuration, nor from any special driving force to achieve a noble gas configuration, but because bond formation proceeds until all the vacancies in the valence shell are filled. Exceptions to the octet rule often involve the larger nonmetal atoms of Period 3 and beyond. Because they are larger, more atoms can be bonded to Period 3 atoms without overcrowding. Also, from the point of view of the orbital model, participation of *d* orbitals is energetically more favorable than for Period 2 elements.

A Method for Writing Lewis Structures

In this study assignment, we will first formulate some rules for drawing Lewis structures. We will then restate each rule and discuss details of its application and possible exceptions. We will work out several examples illustrating each rule, then work out sample structures by using the rules step by step. Finally, we will discuss some of the finer points in order to ensure that the structures we draw are reasonable representations of the bonding in the molecules. You should then be able to apply the rules to draw Lewis structures for the molecules and molecular ions listed at the end of this study assignment.

Rules for writing simple Lewis structures

1. Write the chemical formula for the molecule (or ion) and determine the total number of valence electrons in the molecule.

2. Draw the skeletal arrangement of the molecule showing single bonds connecting the atoms.

3. Assume that each bond in the skeleton requires two valence electrons (an electron-pair bond). After subtracting two electrons for each bond from the total number of valence electrons, assign the remaining electrons to give each atom an octet, or a share of eight electrons.

4. If, after each atom has been given a share of eight electrons, additional electrons remain, assign the extra electrons to the central atom of the molecule.

5. If there are not enough electrons to give each atom a share of eight electrons, then form multiple bonds between atoms by moving electron pairs to form double (or triple) bonds.

Rule 1:   Determine the total number of valence electrons

The number of valence electrons in a molecule is simply the sum of the valence electrons of the atoms that make up the molecule. The number of valence electrons for each atom can be determined either by noting the main group to which the atom belongs or by counting across the periodic table from the left (ignoring the transition metals). Thus, hydrogen in Group I has one electron, silicon in Group IV has four valence electrons, all of the halogens (F, Cl, Br, and I) in Group VII have seven, and the noble gases have eight valence electrons.

For each of the following molecules, the total number of valence electrons is shown.

H_2 $(2 \times 1e^-) = 2e^-$

PCl_5 $(5e^-) + (5 \times 7e^-) = 40e^-$

CH_4 $(4e^-) + (4 \times 1e^-) = 8e^-$

H_2O_2 $(2 \times 1e^-) + (2 \times 6e^-) = 14e^-$

CO_2 $(4e^-) + (2 \times 6e^-) = 16e^-$

N_2O_4 $(2 \times 5e^-) + (4 \times 6e^-) = 34e^-$

If the molecule is a molecular ion, then the charge on the ion modifies the electron count. Because elec-trons have a negative charge, one electron is added to the total for each negative charge on the ion and one is subtracted for each positive charge. For each of the following molecular ions, the total number of valence electrons is shown.

OH^- $(6e^-) + (1e^-) + (1e^-) = 8e^-$
(One electron has been added for the −1 charge.)

NH_4^+ $(5e^-) + (4 \times 1e^-) - (1e^-) = 8e^-$
(One electron has been subtracted for the +1 charge.)

NO_2^- $(5e^-) + (2 \times 6e^-) + (1e^-) = 18e^-$

SO_4^{2-} $(6e^-) + (4 \times 6e^-) + (2e^-) = 32e^-$

NO^+ $(5e^-) + (6e^-) - (1e^-) = 10e^-$

PO_3^{3-} $(5e^-) + (3 \times 6e^-) + (3e^-) = 26e^-$

$C_2H_3O_2^-$ $(2 \times 4e^-) + (3 \times 1e^-) + (2 \times 6e^-) + (1e^-) = 24e^-$

Rule 2:   Draw the bond skeleton of the molecule

Rule 2 is easy to apply if the structure consists of a central atom to which the surrounding atoms are bonded, because the central atom is customarily written first. Thus, the following molecules have the bond skeletons shown:

There are some exceptions to this rule, but they are easily recognized, and their structures are determined by noting a consequence of the rules for drawing Lewis structures: An atom will tend to form one bond for each electron it needs to fill its valence orbitals. Thus, hydrogen forms only one bond and is always found on the periphery of a molecule. Group VI atoms, such as oxygen, tend to form two bonds (two electrons are required to complete the octet). Group V atoms tend to form three bonds, and so on. The skeleton should be drawn so that each atom forms $8 - n$ bonds. Eight is the total capacity of the s and p orbitals, while n is the number of valence electrons that each atom has. The difference, $8 - n$, gives the number of vacancies in the valence orbitals of an atom. As we will see in the final Lewis structure, most of the time each atom will form $8 - n$ bonds.

In the following formulas, the central atom is not written first, but consideration of the number of bonds that an atom is able to form will often lead to the correct bond skeleton.

H_2O
WATER
H—O—H   (H CANNOT BE CENTRAL IF IT FORMS ONLY ONE BOND.)

N_2H_4
HYDRAZINE

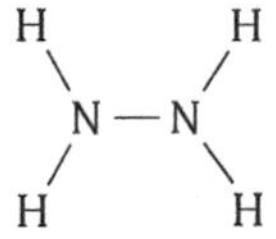

```
 H        H
  \      /
   N  —  N
  /      \
 H        H
```

H_2O_2
HYDROGEN PEROXIDE
H—O—O—H

HCN
HYDROGEN CYANIDE
H—C—N   or
H—N—C

H_2CO
FORMALDEHYDE

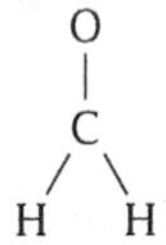

```
    O
    |
    C      or
   / \
  H   H
```

H—C—O—H

N_2O
DINITROGEN OXIDE
N—N—O   or
N—O—N

O_3
OZONE

```
O—O—O   or   O—O
              \ /
               O
```

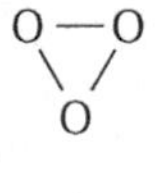

C_3H_6
PROPENE

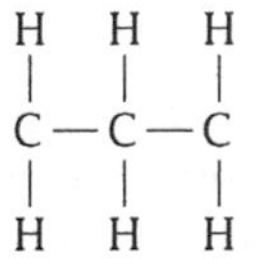

```
 H   H   H
 |   |   |
 C — C — C   or
 |   |   |
 H   H   H
```

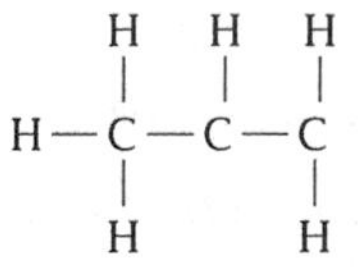

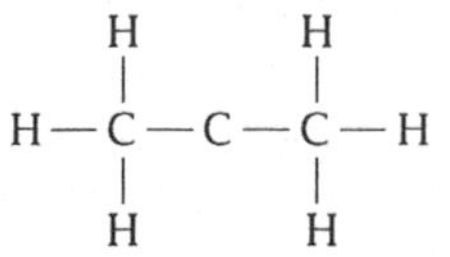

```
 H   H   H
 |   |   |
 H—C — C — C   or
 |   |
 H   H
```

```
   H       H
   |       |
 H—C — C — C—H   or
   |       |
   H       H
```

```
  H      H
  |      |
H—C——C—H
  |    |
  H    C
      / \
     H   H
```

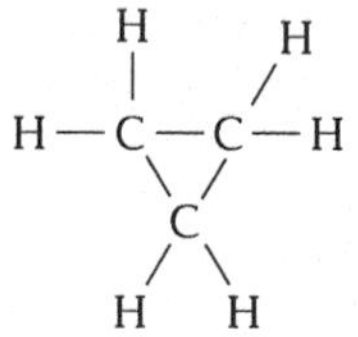

For molecules like HCN, H_2CO, N_2O, O_3, or C_3H_6, where more than one bond skeleton is possible, we will need more information about the structure of the molecule in order to draw the correct structure. (The bond skeleton of HCN is implied by the way the formula is written; for the other molecules, one of the structures may later emerge as the best one.)

Bond skeletons that have three-membered rings, such as

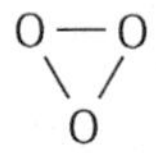

```
O — O                 C — C
 \ /        and        \ /
  O                      C
```

for ozone and propene, respectively, can often be discarded because they involve a great deal of strain or distortion of normal bond angles. In general, do not draw a structure that contains a three-atom ring unless you know that the molecule contains such a ring.

Rule 3: Assign the remaining electrons to give each atom an octet

Each bond that is drawn in a skeleton represents a pair of electrons. The number of valence electrons left to assign is the total number, determined by Rule 1, minus the number of electrons required to form the bonds in the skeleton, which is twice the number of bonds shown in the skeleton structure drawn according to Rule 2.

Remaining electrons = total valence electrons −
(2 × number of bonds)

It generally works best to assign the remaining electrons in pairs to each of the atoms in the skeleton structure, starting with the outer atoms and ending with the central atom. (Remember that a hydrogen atom can share no more than two electrons, not eight; see Table B-1.)

Rule 4: If there are additional electrons, place them on the central atom

If all of the electrons are assigned by Rule 3 and each atom (except hydrogen) has a share of eight electrons, then the Lewis structure is complete. In some instances, as in the case of SF_4 in Table B-1, there are still electrons to be assigned after all octets are complete. In these cases, the extra electrons are assigned

TABLE B-1 Where Possible, Assign Electrons so That Each Atom Has an Octet (a Share of Eight Electrons)

Molecule	Number of valence electrons	Skeleton	Remaining electrons	Complete octets
HF	8	H—F	6	H—$\ddot{\underset{..}{F}}$:
H_2O	8	H—O—H	4	H—$\ddot{O}$—H
CCl_4	32	 Cl Cl \ / C / \ Cl Cl	24	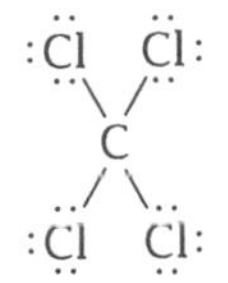 :Cl Cl: \ / C / \ :Cl Cl: (with lone pairs)
PF_3	26	F | P—F | F	20	:F: | :P—F: | :F:
SF_4	34	 F F \ / S / \ F F	26	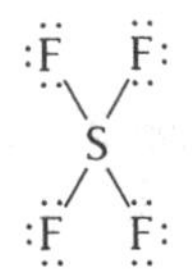 :F: F: \ / S / \ :F: F: (2 electrons unassigned)
O_2	12	O—O	10	:Ö—Ö: (insufficient electrons)

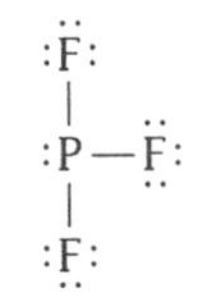

TABLE B-2 Put "Extra" Electrons on the Central Atom Giving It an Expanded Octet (More Than Eight Electrons)

Molecule	Number of valence electrons	Skeleton	Remaining electrons	Lewis structure
SF_4	34	 F F \ / S / \ F F	26	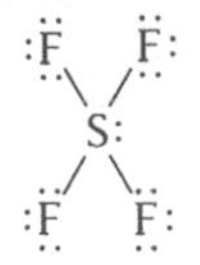 :F: F: \ / S: / \ :F: F:
XeF_2	22	F—Xe—F	18	:F—Xe—F:
ICl_4^-	36	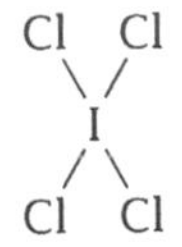 Cl Cl \ / I / \ Cl Cl	10	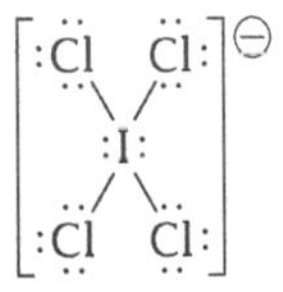 [:Cl Cl:]⊖ \ / :I: / \ :Cl Cl:

Rule 5: If there are not enough electrons to complete all of the octets, then move electron pairs to form multiple bonds between atoms

In the example of the O_2 molecule shown in Table B-1, there are insufficient electrons to complete each atom's octet. In such a situation, two atoms can share more than one pair of electrons between themselves, thus forming multiple bonds. If two pairs are shared between the same two atoms, a double bond is formed; three shared pairs between two atoms constitute a triple bond. Measurements of bond energies and bond lengths show that these multiple bonds are real. The bonds in a double or triple bond are stronger and shorter than in a single bond.

In practice, to complete the octets for the structures of H_2CO and HCN (shown in the discussion of Rule 2 and for O_2 shown in Table B-1), we must move some lone pairs on atoms into position between that atom and an adjacent atom with an incomplete octet. Thus, for O_2, H_2CO, and HCN, we complete the Lewis structures by forming double or triple bonds (see Table B-3).

Exceptions to the octet rule

Although the octet rule (based on filling each orbital vacancy so that each atom has a noble gas electron configuration) is the central rule in drawing Lewis structures, there are some instances where the best description of a molecule contains atoms that do not have an octet of electrons. A brief summary and

to the central atoms as lone pairs. This gives the central atom an *expanded octet,* as shown by the completed Lewis structure shown for SF_4 in Table B-2.

The explanation for this violation of the octet rule is that in these instances, the central atoms have a *d* subshell with empty orbitals that has the same principal quantum number as the *s* and *p* valence orbitals and is thus close in energy to the *s* and *p* valence-shell orbitals. The atomic orbitals from this *d* subshell can hold some of the electrons and thus participate in the bonding in the molecule. For molecules like PCl_5 and SF_6, the octet of the central atom must be expanded to contain all of the valence electrons of the atoms. (See also the examples for XeF_2 and ICl_4^- in Table B-2.)

TABLE B-3 **If There Are Not Enough Electrons to Complete All of the Octets, Then Move Lone Pairs to Form Multiple Bonds Between the Atoms Until Each Atom Has a Share of Eight Electrons**

Molecule	Number of valence electrons	Skeleton	Rule 3 structure	Lewis structure[a]
O_2	12	O—O	:Ö—Ö	Ö=Ö
H_2CO	12	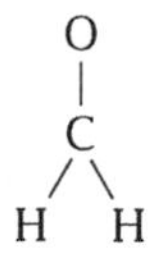 O	 :Ö:	 :O:
		(C with H, H)	(C with H, H)	(C with H, H)
HCN	10	H—C—N	H—C—N̈:	H—C≡N:

[a]Note that the application of the rules leads to a structure that implies that O_2 would have no unpaired electrons, whereas O_2 is known to be paramagnetic with two unpaired electrons.

explanation of these situations may help you to recognize circumstances when the octet rule is likely to be broken.

1. *Hydrogen.* For hydrogen atoms, the maximum valence orbital capacity is two electrons. This is because no $1p$ orbital exists, owing to quantum number restrictions.

2. *Beryllium and boron.* In compounds of beryllium and boron, experiments show that when these atoms form two and three bonds, respectively, the bonds are single bonds. Thus, BeF_2 and BCl_3 have the Lewis structures

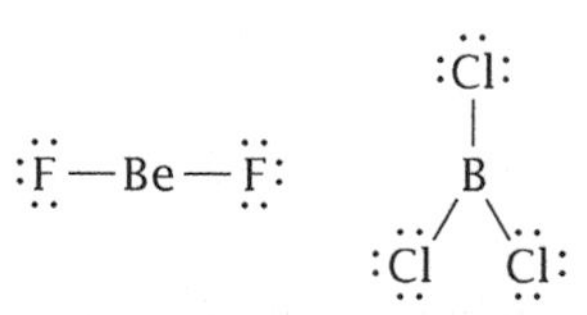

:F̈—Be—F̈:

:C̈l:
|
B
/ \
:Cl   Cl:

even though strict application of the rules would predict that they would have the structures

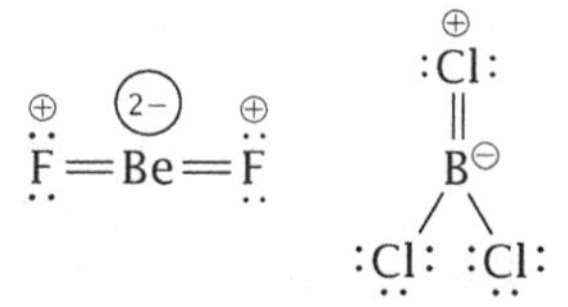

F̈=Be=F̈ (with ⊕, 2−, ⊕ charges)

⊕
:C̈l:
‖
B⊖
/ \
:Cl: :Cl:

These possible structures are improbable (high-energy) electron distributions because they put positive charge on the electronegative atoms F and Cl. (In the following subsection, we show how the formal charges indicated on the atoms are calculated.)

3. *Expanded octets.* The central atom of some molecules may have more than an octet of electrons. This occurs in cases where the central atom has more than four atoms bonded to it, as in PCl_5 and SF_6, or where there are more than enough electrons to satisfy the octets of all of the atoms, as in XeF_2. These extra electrons are assigned to the central atom as lone pairs (Rule 4). An expanded octet can occur only in atoms that have empty d orbitals in their valence shell. (Note that $2d$ valence orbitals do not exist, so a nitrogen atom cannot have an expanded octet, whereas a phosphorus atom may.)

4. *Odd-electron molecules.* There are a few molecules that contain an odd number of electrons; NO, NO_2, and CH_3 are examples. Molecules like these, which contain an unpaired electron, are known as *free radicals* and tend to be quite reactive. It is impossible to form octets from an odd number of electrons, so the Lewis structures of radicals contain an atom that has only seven electrons instead of an octet.

Formal charge

Assigning formal charges to atoms in the Lewis structure of a molecule or molecular ion is the equivalent of simple electron bookkeeping. As the term *formal charge* implies, it is determined by a formal convention or set of arbitrary rules. *It does not show in general the actual distribution of electric charge among the atoms in a molecule or molecular ion.* Sometimes we find that we can draw several different Lewis structures that conform to the octet rule; we use the formal charges on the atoms to help us choose the most plausible Lewis structure(s).

The formal charge on any atom is assigned by comparing the number of valence electrons of the atom with the number of electrons that belong to the atom in the Lewis structure. When counting electrons, we assume that bonding electrons are equally shared, so that half the bonding electrons belong to the atom, and that lone-pair electrons on the atom belong completely to the atom. So, to determine the formal charge of an atom in a Lewis structure:

• Count the bonds to other atoms (equal to half the bonding electrons).

• Add to the count the number of electrons on the atom in the form of lone pairs.

• Subtract this total electron count from the number of valence electrons for a neutral isolated atom of that element.

The formal charge of an atom may be zero, positive, or negative. In molecular ions like NH_4^+ and SO_4^{2-}, there will inevitably be at least one atom with a formal charge, but formal charges may also exist in neutral molecules. For a neutral molecule, the formal charges of all the atoms must sum to zero. For a molecular ion, the sum of the formal charges must equal the net charge of the ion.

As an example, let's look at the Lewis structure for the chlorate ion, ClO_3^-. It shows 26 valence electrons. [The Cl atom, belonging to main Group VII, has seven valence electrons; the O atoms, belonging to main Group VI, each have six valence electrons. We add one electron for the net (−1) charge.]

Chlorate ion

Expressed in equation form, the formal charge of an atom is given by

Formal charge = (no. of valence electrons of atom)
  − (no. of bonds)
  − (no. of lone pair electrons)

Formal charge on Cl = 7 − 3 − 2 = +2

Formal charge on O = 6 − 1 − 6 = −1

(We adopt the convention that the formal charges, to make them more visible, will be enclosed in a circle located near the atom and that formal charges of zero will not be shown.) Note that for chlorate ion the formal charges sum to −1, the net charge on the ion.

We will use formal charges to help us choose the best structure(s) among several different Lewis structures that involve different distributions of the electron-pair bonds. The preferred structures will be ones that have the following characteristics:

• The least charge separation (smallest formal charges of opposite sign on adjacent atoms)

• No adjacent atoms with formal charges of the same sign

• The negative formal charges on the most electronegative elements

As an example, consider the following structures for dinitrogen oxide, N_2O, usually called nitrous oxide.

Acceptable structure   Acceptable structure

Implausible structure: large charge separation; puts (+) charge on O atom; has adjacent atoms with same formal charge.

Structures D, E, and F are implausible because each structure has an atom with a share of only six electrons. Structures E and F also have implausible charge distributions.

Lewis structures A and B are the preferred structures. Structure C is a less plausible (higher energy) structure because of its large charge separation and because it puts a plus (+) charge on the O atom, which has the largest electronegativity and is therefore expected to have the most negative (−) charge.

Structures D, E, and F do not conform to the octet rule and would be less stable (higher energy) structures because they have only three covalent bonds compared to the four covalent bonds in the other structures.

As described earlier under Exceptions to the Octet Rule, molecules containing Group II or III atoms, such as Be or B, do not have enough valence electrons to give Lewis structures in which each atom has a shared octet without assigning a negative formal charge to the Be or B atom; this gives a structure with an implausible charge distribution. The most plausible Lewis structure for BF_3 leaves the B atom with a vacancy in the valence shell. This allows a place to form another bond with a species that can donate an electron pair. For example, the BF_3 molecule can add a fluoride ion, F^-, to form the tetrafluoroborate ion, BF_4^-.

Tetrafluoroborate ion

Conforms to the rules for drawing Lewis structures, but has an implausible charge distribution.

Although this structure conforms to our rules for drawing Lewis structures, the negative charge on B relative to the F atoms is known to be wrong. Accurate calculations show that the negative charge is mostly distributed over the four fluorine atoms and not centered on the B atom. We must conclude that the formal charges in Lewis structures do not always give a true picture of the charge distribution, reinforcing our earlier statement that Lewis structures are not infallible guides to the charge distribution.

Resonance

In applying the rules for drawing Lewis structures to nitrous oxide, N_2O, we saw that we could draw at least two plausible structures (A and B) with the same number of bonds but different bond distributions, implying the possible existence of two different kinds of N–O bonds in this molecule.

When this situation exists, it can be shown that a structure that is a combination of the different structures will have a lower energy, and will therefore be more stable, than either of the contributing structures A and B. These structures having comparable charge distributions but different distributions of bonds are called *resonance* structures, each contributing a certain amount to the true bond distribution.

Unfortunately, the term *resonance* may give you the impression that the molecules flip back and forth among the various resonance structures. This is decidedly not true. A molecule in its most stable energy state has just one stable charge and bond distribution.

When a single Lewis structure is inadequate to show the best charge and bond distribution, you might imagine drawing each resonance structure on a transparent sheet, like an overhead transparency, then superimposing them. The composite of the superimposed structures would give you a better picture of the charge and bond distribution.

Now let's consider a second example, the Lewis structure of the nitrate ion, NO_3^-.

Although this structure is consistent with the rules for drawing a Lewis structure, it gives the impression that there is one N–O double bond and two N–O single bonds, whereas we know experimentally that all three bonds are the same. To improve the picture, we draw the three equivalent resonance structures and imagine that the true structure is a combination of all three.

The best structure is a combination or superposition of these three structures with the electrons distributed equally in the three bonds. We will call these *equivalent* resonance structures because we could have drawn the first Lewis structure with the double bond located on any of the three O atoms. More formally, each of the three structures is related by a simple symmetry operation: Rotation by 120° about an axis passing through the N atom and perpendicular to the plane of the paper converts one structure into the next.

For equivalent Lewis structures, we adopt the convention of using an equals sign to indicate the equivalence of the resonance structures, avoiding the use of a double arrow, which tends to convey the impression that the molecule is flipping back and forth among the three equivalent structures.

The bonding in molecules for which we can draw several resonance structures of comparable charge distribution but with a different bond distribution can be described as a bond in which there is spreading out of the electron pairs over more than two atoms. This is called electron *delocalization*, which gives a more understandable interpretation to the term resonance: *Resonance implies electron delocalization.*

Bond order – questions

The Lewis bond order (BO) is an arbitrary measure of the strength of a covalent bond between two atoms. It is not a measurable physical property like bond length or bond angle but is derived directly from the Lewis structure model. In order to make valid comparisons of the bonding among different pairs of atoms, we preferably compare bonds having the same atom pair. For example, drawing conclusions by comparing the bond order of a C–C bond with a C–O or N–O bond must be done with considerable caution, because single bonds involving different atom pairs have intrinsic bond energies and bond lengths that are different. With this proviso in mind we can say that, although the Lewis model is very simple, the Lewis bond order usually correlates well with experimentally measured bond strengths (bond energies) and bond lengths.

Rules for calculating the bond order (BO)

• In any single Lewis structure, the bond order of the bond connecting any two atoms is equal to the

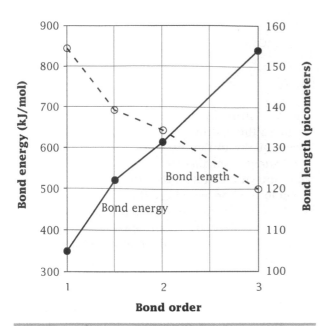

FIGURE B-2 | **Bond energy and bond length versus bond order for the C–C bonds in ethane, benzene, ethene, and ethyne. As the bond order increases, the bonds get stronger and shorter.**

number of electron pairs in the bond. In this case, the bond order will be an integer (BO = 1, 2, or 3) corresponding to one, two, or three electron pairs in the bond:

Ethane
C–C bond order = 1
C–H bond order = 1

Ethene
(ethylene)
C–C bond order = 2
C–H bond order = 1

Ethyne
(acetylene)
C–C bond order = 3
C–H bond order = 1

• If the Lewis structure has two or more resonance structures, the bond order is the average over all the resonance structures. In this case, the bond order may be fractional, that is, noninteger.

Benzene, C_6H_6, is known to have a ring structure. Furthermore, all of the C–C bonds are known to be of equal length—that is, they are all equivalent. The structures we draw that show electron pair bonds conforming to the octet rule give rise to resonance structures with alternating single and double bonds between the carbon atoms. Again we invoke the idea of representing the best structure as a combination (or superposition) of the resonance structures

$$\text{C-C bond order} = \frac{1+2}{2} = \frac{3}{2} = 1.5$$

C–H bond order = 1

The bond orders are the averages for the two (equivalent) resonance structures.

Figure B-2 shows how the bond order is correlated with bond strength (bond energy) and bond length for the C–C bonds in ethane, benzene, ethene (ethylene), and ethyne (acetylene). From the graph we may make the generalization that as bond order increases, bonds get stronger and shorter.

Next, we show how to derive the bond order in the carbonate ion, known to have three equivalent C–O bonds, by representing the true structure as a combination of the three equivalent resonance structures:

$$\text{C-O bond order} = \frac{1+2+1}{3} = \frac{4}{3} = 1.33$$

The C–O bond order is the average for the three equivalent resonance structures. The sum of the formal charges is equal to the net charge of the carbonate ion.

Perhaps you can now see the principle at work when calculating the bond order for molecules with several equivalent resonance structures:

• Count the number of bonds in a single resonance structure and divide this number by the number of resonance structures.

Finally, we dip our toes into the textbook controversy over the best way to represent with Lewis structures the bonding in molecules having multiple bonds to O atoms—for example, ozone, sulfur dioxide, and phosphate ion. The issue is this: Are the best structures the ones that conform to the octet rule wherever possible, or can one best represent the structure by minimizing the formal charge on P and S at the expense of the octet rule, allowing S and P to have a share of more than eight electrons, which they must certainly have in the Lewis structures for molecules such as SF_6 or PCl_5?

Ozone

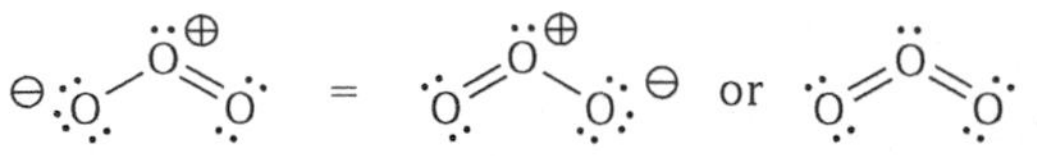

Preferred structures
The two equivalent resonance structures conform to the octet rule.
O–O bond order = 1.5

Not acceptable
Minimizes charge separation, but gives the central O atom a share of ten electrons, not allowed for a period 2 element.

Sulfur dioxide

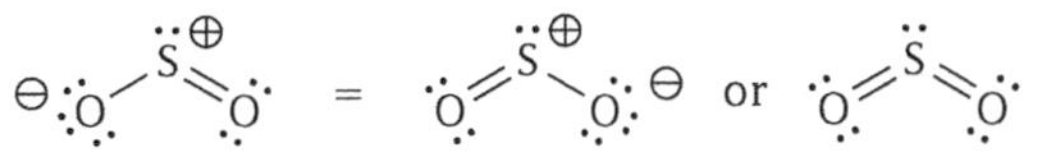

Acceptable structures
The two equivalent resonance structures conform to the octet rule.
S–O bond order = 1.5

Acceptable structure
Minimizes charge separation, but gives the central S atom a share of ten electrons, allowable for a period 3 element.
S–O bond order = 2

Phosphate ion

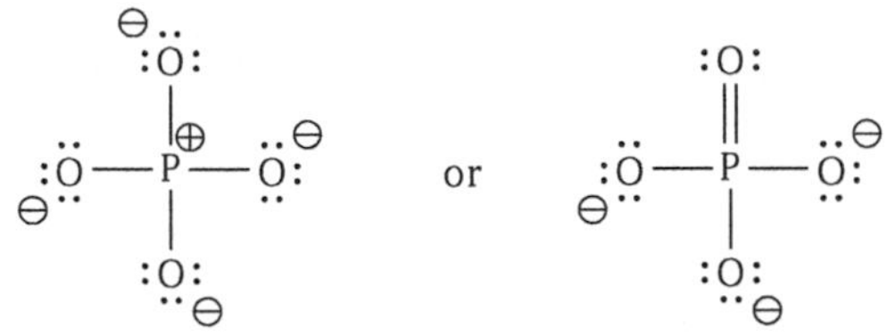

Acceptable structure
Conforms to the octet rule.
P–O bond order = 1

Acceptable structure
Minimizes charge separation, but gives the central P atom a share of ten electrons, allowable for a period 3 element.
Has a total of four equivalent resonance structures.
P–O bond order = 1.25

The relative merits of the structures for sulfur dioxide and phosphate ion and a review of the controversy are discussed by Purser (see the Bibliography). We consider that either of the structures is satisfactory for most purposes. Our position is that Lewis structures provide a crude but still useful model of bonding, but that we should not overinterpret their significance. Calculations of electron density based on quantum mechanics will give you a

much deeper insight into chemical bonding. To explore this further, read the article by the Shustermans and browse the monograph by Gillespie and Popelier listed in the Bibliography.

Summary

A systematic method for drawing Lewis structures is summarized by the following procedures.

1. Determine the total number of valence electrons in the molecule.

   (a) Sum the valence electrons of each atom in the molecule.

   (b) Add an electron for each negative charge or subtract one for each positive charge on a molecular ion.

2. Draw the bond skeleton of the molecule.

   (a) Often the central atom of the molecule is written first in the chemical formula; then all other atoms are bonded directly to it.

   (b) If the central atom is not written first, make use of the fact that atoms tend to form $8 - n$ bonds, where n is the number of valence electrons of the neutral atom.

3. Assign the remaining electrons to give each atom an octet of electrons. H, Be, and B do not form an octet.

4. If there are more than enough electrons to give each atom an octet, place the additional electrons on the central atom to form an expanded octet.

5. If there are not enough electrons to complete all of the octets, move lone pairs to form multiple bonds between atoms until each atom has a share of eight electrons.

6. Determine the formal charge of each atom in the resulting structures, and eliminate any structures with unlikely charge distributions (adjacent atoms with the same charge).

7. Look to see if there are possible equivalent resonance structures.

Bibliography

Gillespie, R. J.; Popelier, P. L. A. *Chemical Bonding and Molecular Geometry—From Lewis to Electron Densities,* Oxford University Press, New York, 2001.

Pauling, L. *The Nature of the Chemical Bond,* 3rd ed. Cornell University Press, Ithaca, NY, 1960.

Popelier, P. *Atoms in Molecules—An Introduction,* Prentice-Hall, an imprint of Pearson Education Limited, Essex, England, 2000.

Purser, G. H. "Lewis Structures Are Models for Predicting Molecular Structure, *Not* Electronic Structure," *J. Chem. Educ.* **2001,** *78,* 981–983.

Purser, G. H. "Lewis Structures in General Chemistry: Agreement between Electron Density Calculations and Lewis Structures," *J. Chem. Educ.* **2001,** *78,* 981–983.

Shusterman, G. P.; Shusterman, A. J. "Teaching Chemistry with Electron Density Models," *J. Chem. Educ.* **1997,** *74,* 771–776. The color figures with additional text accompanying this article can be viewed at http://academic.reed.edu/chemistry/alan/ED/JCE/figlist.html

Models of Molecular Shapes

VSEPR Theory and Orbital Hybridization

Purpose

• Build models of molecules of the type AX_mE_n, following the rules of the valence shell electron-pair repulsion (VSEPR) method. [A stands for the central atom, X represents an atom bonded to the central atom, and E represents the lone pair(s) of electrons on the central atom.]

• Use the models to help visualize the three-dimensional arrangement of the atoms in a molecule and to help predict if the molecule will have a dipole moment.

• Explore the connections among Lewis structures, VSEPR theory, and the orbital description of a covalent chemical bond in which the valence orbitals of the central atom are combined to form a set of hybrid orbitals.

Introduction

Chemists use symbols and models to help them think about the chemistry of atoms and molecules. For example, the symbols in a chemical equation, such as $2H_2 + O_2 \rightarrow 2H_2O$, provide a visual image of the different kinds of atoms that participate in a chemical reaction, making it easier to see at a glance the chemical composition of the reactants and products.

Chemists also use mathematical equations to model the functional relationships among several physical variables represented by symbols. For example, the ideal gas equation, $PV = nRT$, concisely models the relation between the pressure, volume, number of moles, and temperature of an ideal gas. Relationships between the variables P, V, T, and n may also be represented by pictures, called *graphs,* where values of one variable are plotted versus another.

We often draw pictures of orbitals to represent the distribution of electrons about the nucleus of a hydrogen atom. Physical models of atoms and molecules are also useful for visualizing how atoms are connected in three-dimensional space—what we call the *molecular geometry* of a molecule. In fact, using computers and computer graphics software, we can now perform accurate calculations of molecular geometry that can be translated into visual images on a video terminal. Such images can be rotated so that we can view them from any angle and are useful for understanding the three-dimensional shapes of molecules, particularly those that are large and complex, such as proteins.

Figure 19-1 presents two images of the molecule ammonia (NH_3) produced by molecular modeling programs. The ball-and-stick model clearly shows how the atoms are connected by chemical bonds and the molecule's overall geometry. The space-filling model provides a more accurate impression of the molecule's shape and how the molecule might look if we could magnify its size and see the electron clouds that define its size and shape. The surface plotted is called an *isodensity surface* because it represents a surface having a constant electronic charge density that includes about 98% of the molecule's electronic charge.

Let's review some of the models we use in talking about the electronic structure of atoms and the chemical bonds between atoms.

Electronic structure of atoms

Most general chemistry texts show two ways of representing the electronic structure of an atom, using the $1s^2 2s^2 2p^6$ . . . notation or using boxes or circles to represent hydrogen-atom-like orbitals. For example, the electronic structure of a carbon atom in its lowest energy state, called the *ground state,* can be represented by the notation $1s^2 2s^2 2p^2$ or by the orbital diagram

  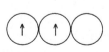

$1s$ $2s$ $2p$

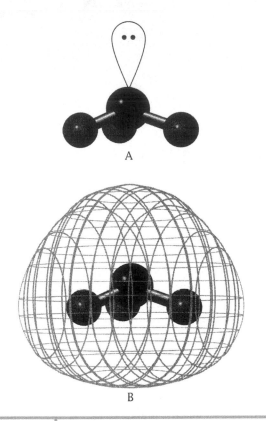

A

B

FIGURE 19-1 | **Two different models of the molecule ammonia, NH₃. In (A), the ball-and-stick model shows the geometric arrangement of the atoms and lone pair of electrons clearly and corresponds most closely to the way we draw the Lewis structure for the NH₃ molecule. The space-filling model shown in (B) gives a more accurate picture of the overall shape and distribution of electronic charge density in the molecule.**

Covalent chemical bonding

We saw in Study Assignment B that G. N. Lewis and Irving Langmuir were among the first to attempt a description of a chemical bond as a shared electron pair (covalent bond), which Lewis represented by diagrams now called Lewis diagrams. These molecular diagrams show the valence electrons of the atoms grouped in bonding pairs (single or multiple bonds) and lone pairs.

Later, quantum chemistry provided an interpretation or explanation of why electrons tend to group in pairs. This behavior is closely related to a property of electrons that we now call *spin*. Electrons with opposite spins tend to form pairs occupying a region of space that is defined by a mathematical function called a *wavefunction*. Electrons with the same spin are excluded from occupying the same region of space, as described by the Pauli exclusion principle. Roughly speaking, we may think of these groups of spin-paired electrons as impenetrable charge clouds

that occupy a certain volume of space, other electrons being excluded from this space as described by the Pauli principle (Pauli exclusion). Some authors use the term *electron-pair domains* or *electron charge clouds* to describe these regions of space occupied by two electrons with paired spins.

The orbital model of covalent bonding

In the orbital model of covalent bonding, a shared-electron-pair bond between two atoms, A—B, is imagined to be formed by overlapping an *orbital* on atom A containing one electron with an orbital on atom B, also containing one electron. These overlapping orbitals form a *bonding orbital* containing a shared pair of electrons.

Exactly what is an orbital? The word *orbital* was apparently first used in 1932 by the American chemical physicist R. S. Mulliken, who coined the term to stand for a hydrogen-like one-electron wavefunction and interpreted it as meaning "something like an orbit in which there is some high probability of locating an electron." So the images of orbitals in our textbooks are pictorial representations of mathematical functions that are solutions of the Schrödinger equation that describes the behavior of the single electron in a hydrogen atom.

Orbitals are not physical observables such as electron density or bond length or bond angle, which can all be determined by experiment. You cannot perform an experiment to "see" an orbital. It's necessary to make this point because the orbital concept permeates chemical thinking and the orbital model has proved very useful as a basis for making calculations of quantities that *are* physical observables. Furthermore, textbooks and molecular modeling software packages used in computational chemistry provide images of atomic and molecular orbitals, so it's easy to fall into the trap of believing that orbitals exist as physical objects that can be observed.

To better understand the significance of atomic orbitals, it should be noted that the electron density of any single gaseous atom has the shape of a ball with diffuse or fuzzy edges. In other words, the distribution of electron density is spherically symmetric with no bumps or bulges corresponding to identifiable orbitals, even if the atom has an electron configuration with incompletely filled or empty orbitals. It's as if an electron spends equal time in all of the orbitals so that, averaged out, the electron density is the same in any direction as you move away from the nucleus. This is also consistent with the mathematics of the wavefunctions. The square of the wavefunctions gives the probability density for electrons (the probability of finding an electron in a particular volume of space). Summing the squares of the three *p*-orbital wavefunctions or the five *d*-orbital wavefunctions gives mathematical functions with spherical symmetry, consistent with the spherical distribution of electrons around a single isolated atom.

Orbital Hybridization In the 1930s, Linus Pauling and others showed how the number of bonds and general shape of a molecule could be explained by combining the *s* and *p* orbitals (as well as *d* and *f* orbitals) on an atom to form what are called *hybrid orbitals.* Combining the *s* and *p* orbitals of a carbon atom, which means mathematically combining their wavefunctions, leads to improved overlap with an orbital on another atom. This improved overlap translates into a structure with lower energy; lower energy is always associated with an improved wavefunction corresponding to an atom configuration that has greater stability and is therefore more likely to be closer to the true structure.

The mathematical procedure for combining the wavefunctions is called *orbital hybridization.* In the case of a carbon atom, the combination of the *s* and three *p* orbitals on carbon gives rise to a set of four equivalent orbitals, called *sp³* hybrid orbitals, that can form four bonds having a tetrahedral spatial configuration in agreement with what is observed both chemically and by physical methods of structure determination. Does this mean that the tetrahedral structure was dictated by the spatial arrangement of the *sp³* hybrid orbitals? Not really. A better way of looking at it is to say that the orbital model provides

a *description* of the bonding that is determined by the interaction of the carbon atom with the four other atoms bonded to it.

Hybrid orbitals have a property similar to the unhybridized atomic orbitals. Adding together the squares of their wavefunctions gives a mathematical function with spherical symmetry. Therefore, on an isolated gaseous carbon atom you would *not* find four bulges of electron density directed toward the corners of a tetrahedron corresponding to four *sp³* hybrid orbitals. If we want to attribute human qualities to atoms, aside from our own existence, we could say that atoms "know" nothing about orbitals. The orbital model *describes* rather than determines the behavior of electrons.

Second, when hydrogen-like *s, p,* etc. orbitals are combined, the number of hybrid orbitals produced always equals the number of atomic orbitals combined. For example, combining an *s* orbital and three *p* orbitals for a carbon atom gives rise to four equivalent *sp³* hybrid orbitals. [The superscript 3 on the *p* simply tells you how many *p* orbitals have been used in forming the *sp³* hybrid orbitals. It doesn't specify the number of electrons, as do the superscripts used in writing the electron configuration $(1s^2 2s^2 2p^6 \ldots$ etc.), nor does it mean that the *p*-orbital mathematical wavefunction has been cubed.]

A pictorial way of representing the process of orbital hybridization in the formation of the six sulfur–fluorine bonds in the molecule SF_6 is shown in Figure 19-2. First, we should say that we might

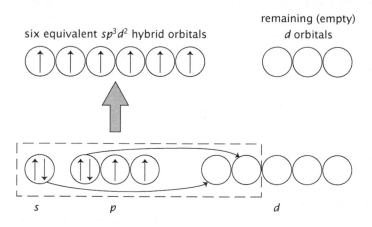

The ground state electron configuration of the electrons in the valence shell of a sulfur atom.

FIGURE 19-2 | **A pictorial representation of orbital hybridization of the sulfur atom in molecular SF_6, in which a set of six orbitals in the ground state of a sulfur atom (one *s* orbital, three *p* orbitals, and two *d* orbitals) are combined to form a set of six equivalent *sp³d²* hybrid orbitals.** **The set of six *sp³d²* hybrid orbitals has octahedral symmetry, identical to the octahedral symmetry of the AX_6 VSEPR structure predicted for the molecule SF_6, as shown at the bottom of Table 19.1.**

expect the ground-state electron configuration of a sulfur atom to allow formation of only two shared-electron-pair bonds, since the valence shell has two half-filled p orbitals. And in fact, hydrogen sulfide, H_2S, is an example of a molecule in which the sulfur atom forms just two bonds. The formation of the molecule SF_6 (having six sulfur–fluorine bonds) is rationalized by imagining the formation of a set of six hybrid orbitals formed by combining one s orbital, three p orbitals, and two d atomic orbitals on a sulfur atom.

Conceptually, we might imagine the overall process being broken down into two steps. We first imagine promoting two electrons (one from the sulfur s orbital and one from the sulfur p orbital) to each of two empty d orbitals. This gives a total of six half-filled orbitals (one s orbital, three p orbitals, and two d orbitals). Second, these orbitals are then combined to form a set of six equivalent hybrid orbitals, labeled as sp^3d^2 (or d^2sp^3) orbitals to indicate the atomic orbitals that were combined to form them. These six orbitals are directed toward the six vertices of a regular octahedron. (Recall that octa, meaning eight, comes from the fact that a regular octahedron has eight faces, each face being an equilateral triangle.) Overlap of these orbitals with the half-filled p orbitals on six fluorine atoms then allows formation of six sulfur–fluorine bonds, giving the molecule SF_6, which has the octahedral symmetry of an AX_6 molecule shown at the bottom of Table 19-1.

Overall, we can say that the process of bond formation is exothermic because the energy required to unpair the electrons in the s and p orbitals and promote them to the higher energy d orbitals (an energy-consuming process) is more than recovered by formation of the six sulfur–fluorine bonds, which is an energy-releasing (exothermic) process.

The Connection Between VSEPR Theory and Hybrid Orbitals VSEPR theory is used to describe or predict the shape of a molecule. Most of these molecular shapes are identical to or closely related to idealized structures that have a considerable amount of symmetry, that is, they possess symmetry elements such as planes of symmetry or rotational axes of symmetry.

• For each idealized structure found in VSEPR theory, there is a corresponding set of hybrid orbitals having the same symmetry.

So it's no accident that there is a connection between VSEPR theory and the hybrid orbital description of bonding. Hybrid orbital and molecular orbital descriptions of bonding are woven into the fabric of chem-

istry, so the time you spend learning about this will be useful to you if you continue your study of chemistry.

Finally, we should note that VSEPR theory is essentially a qualitative theory. It's useful for predicting shapes of molecules and correlating trends in their properties, but it is not used for calculating properties of molecules.

VSEPR theory and molecular shapes

In 1940, the British chemists Sidgwick and Powell surveyed the geometry of singly bonded AX_n molecules and showed that most of their structures could be rationalized by assuming that the geometry of an AX_n molecule was determined by the total number of bonding and nonbonding electron pairs in the valence shell of the central atom A. This idea languished, probably a victim of World War II, until it was revived and extended in 1957 by the British chemists Gillespie and Nyholm into the theory now known as the *valence shell electron-pair repulsion* (VSEPR) theory. It's essentially a qualitative theory that extends the idea of Lewis structures with a few simple rules to allow the prediction of the shapes of molecules, mainly those in which there is a central atom bonded to a surrounding cluster of other atoms.

VSEPR rules are very simple.

• The primary rule is that bonding electrons and lone pairs of electrons adopt a spatial arrangement that keeps them as far apart as possible.

In the words of Gillespie and Popelier, "They act *as if* they repel each other." This behavior is more a result of the Pauli principle (the mutual exclusion of electrons having the same spin) than of the electrostatic repulsion of electrons.[1]

• A secondary rule is that the magnitude of the repulsive interactions (whether Pauli exclusion or electrostatic repulsion) is assumed to be

Lone Pair–Lone Pair (LP–LP) >
Lone Pair–Bonding Pair (LP–BP) >
Bonding Pair–Bonding Pair (BP–BP)

This secondary rule can be used to rationalize the preferred geometry when there are lone pairs of electrons

[1]Pauli exclusion is not a physical force like electrostatic repulsion. It might be described as being more like a statistical probability effect resulting from electron spin. The physicist R. P. Feynman put it this way, referring to the consequences of electron spin: "It appears to be one of the few places in physics where there is a rule that can be stated very simply, but for which no one has found a simple and easy explanation. The explanation is deep down in relativistic quantum mechanics."

A useful discussion of the Pauli exclusion principle is given by Popelier and Gillespie in Chapter 3 of their book, referenced in the Bibliography.

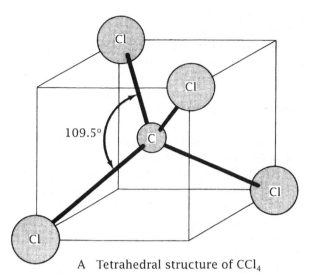

A Tetrahedral structure of CCl_4

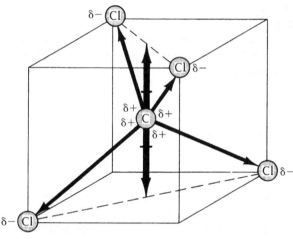

B Canceling bond dipoles in CCl_4

FIGURE 19-3 **The symmetry of the tetrahedral structure of CCl_4 shown in (A) causes the bond dipole moments to cancel one another as shown in (B). Thus, CCl_4 has a zero resultant dipole movement.**

1. The notion of an orbital in a many-electron atom is a somewhat artificial but useful concept based on the notion that electrons in an atom with many electrons behave somewhat like the single electron in a hydrogen atom. Therefore, atomic orbitals in atoms with many electrons are sometimes said to be *hydrogen-like* orbitals. Atomic orbitals in atoms with many electrons are constructed so that when we add up the electron density of the orbitals, we get the correct electron density for an atom or molecule as determined by comparison with the experiment.

2. The concept of orbital hybridization provides a way to explain the geometry (arrangement in three-

dimensional space) of the chemical bonds. Thus, the pictures we draw of covalent bonds in molecules are often based on hybrid orbitals. Why is this? In simple language, it is because our brains don't add or integrate the visual images of the electronic charge density of the pure atomic *s*, *p*, and *d* atomic orbitals very accurately. As a result, we often use hybrid orbitals (mixtures of *s*, *p*, and *d* atomic orbitals) rather than native (unhybridized) *s*, *p*, and *d* atomic orbitals, because hybrid orbitals help us draw pictures of the molecular geometry that are in better agreement with the facts. It's also true that the labels used to designate hybrid orbitals are firmly entrenched in the chemistry curriculum. So you will encounter the hybrid orbital labels in later courses, particularly in organic chemistry, which emphasizes the chemistry of carbon compounds.

3. Aside from the conceptual limitations of the orbital concept, the physical models of molecules you will build also have limitations; that is, there are many ways in which they fall short of adequately representing real molecules, particularly their dynamic character. As you build the models, think about how they differ from real molecules. You will be asked to note some of the limitations of the models in your conclusion to the experiment.

Experimental Procedure

Special Supplies: Molecular model kits (Prentice-Hall Molecular Model Set for General Chemistry, available from Sigma-Aldrich, www.sigma-aldrich.com, tel 800-558-9160, are a convenient size and are durable). Each student (or pair of students) should have a minimum of one 6-hole octahedral atom center, one 5-hole trigonal bipyramid atom center, and one 4-hole tetrahedral atom center to represent the central A atoms; six 1-hole white atoms to represent the bonded (X) atoms; and six medium-length gray plastic connectors that will represent the bonding and lone-pair electrons.

▓ NOTES TO INSTRUCTOR ▓

If you use Prentice-Hall molecular models, it should be noted that the 6-hole octahedral atom centers and 5-hole trigonal bipyramid atom centers are contained only in the General Chemistry model sets and are not contained in the Organic Chemistry model sets. However, it is useful to also have one Organic Chemistry model set to provide spare (white) 1-hole atoms, as they are easily lost. One general chemistry set can serve two students working individually or four students working in pairs.

Other activities combine well with model building. For example, we sometimes have students build VSEPR models only through the AX_4 family, view a film or video on infrared spectroscopy, take an infrared spectrum of a gas such as carbon dioxide or methane, and view animations of molecular vibrations using a molecular modeling program, such as Spartan (from Wavefunction, Inc.). These activities show students the dynamic character of molecules. The absorption of infrared light by molecules can be related to the greenhouse effect and to the notion that each molecule has a unique infrared spectrum that serves as an identifying "fingerprint." Or, if you have enough computers and molecular modeling programs available, students can build visual images of the molecules they are studying rather than working with physical models. Space-filling (isodensity-surface) renderings also give students a more realistic idea of molecular shapes.

The models are constructed by joining the atom centers with gray connectors. A gray connector without a (white) X atom joined to it will be considered to be a lone pair of electrons on the central A atom. Thus, the gray connectors that link atoms will represent either bonding electrons or lone-pair electrons on the central (A) atom, depending on whether or not a (white) X atom is connected to the central atom.

Following the directions given in the Report form, draw Lewis structures and build models of each assigned molecule. As you proceed, fill in the table on the first page of the Report form. The models can be grouped into five families, each family having a characteristic maximum number of electron pairs (bonding pairs + lone pairs). For each model, begin with the appropriate central atom.

Total number of bonds and lone pairs. The total number of bonds and lone pairs on the central atom determines the VSEPR family. This number also represents the maximum number of electron pairs on the central atom and the maximum number of atoms that can be bonded to the central atom. For each VSEPR family, the sum of the number of bonds and the number of lone pairs remains constant. Within each VSEPR family there are one or more VSEPR classes, in which atoms are replaced by lone pairs of electrons.

Orbital Hybridization Associated with each VSEPR family is an idealized geometry and orbital hybridization. The relationships among the five VSEPR families, total number of bonds and lone pairs, orbital hybridization, and idealized geometries are summarized in Table 19-1.

VSEPR Class, Molecular Geometry, and Dipole Moment For each VSEPR family, there will be one or more VSEPR classes given by the general formula AX_mE_n, in which atoms have been replaced by lone pairs of electrons. Each of the thirteen VSEPR classes shown in Table 19-1 has a molecular geometry with a particular name that describes the arrangement of the atoms. Note that the number of bonds and lone pairs determines the idealized geometry, but if lone pairs are present on the central atom, these are ignored when naming the molecular geometry. In other words, only the geometric arrangement of the atoms is considered when naming the molecular shape.

The overall symmetry of the molecule will determine whether or not the molecule has a permanent dipole moment.

CONSIDER THIS

Almost every model that chemists use has features designed to highlight certain aspects of the system being modeled. As a result, any model distorts other aspects of the system. Critique the models used in this experiment by answering these questions.

1. In Figure 19-1(A), how are bond pairs of electrons represented? What features of bond pairs are well represented by this "stick" model? Which features of electrons are greatly distorted here?

2. What features of lone-pair electrons are well represented in Figure 19-1(A)? What is meant by the dots in the figure?

3. Compare Figures 19-1(A) and 19-1(B). In what ways is (B) an improvement?

In your text, you can probably find the ammonia molecule represented by a structure such as

$$H-\overset{\displaystyle \cdot\cdot}{N}-H$$
$$|$$
$$H$$

What features of molecular structure does this represent? Which features are ignored by this structure?

Can you invent other ways to represent the structures of molecules that are more realistic or less confusing than the ones you see in your text?

There are a few molecules with coordination numbers beyond the six-coordinate examples used here, such as IF_7. Apply the principles that you learned about VSEPR and orbital hybridization to describe the geometry of this molecule and the bonding within it.

Bibliography

Gillespie, R. J.; Hargittai, I. *The VSEPR Model of Molecular Geometry,* Allyn & Bacon, 1991.

Gillespie, R. J.; Popelier, P. L. A. *Chemical Bonding and Molecular Geometry: From Lewis to Electron Densities,* Oxford University Press, New York, 2001.

Pauling, L. *The Nature of the Chemical Bond,* 3rd ed., Cornell University Press, Ithaca, NY, 1960, Chapter 4.

Name _____   Date _____

2. What features of lone-pair electrons are well represented in Figure 19-1(A)? What is meant by the dots in the figure?

3. Compare Figure 19-1(A) and 19-1(B). In what ways is (B) an improvement?

In your text, you can probably find the ammonia molecule represented by a structure such as

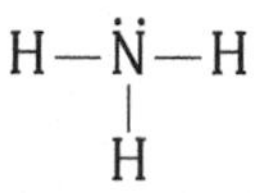

$$H - \overset{\cdot\cdot}{N} - H$$
$$|$$
$$H$$

What features of molecular structure does this represent? Which features are ignored by this structure?

Can you invent other ways to represent the structures of molecules that are more realistic or less confusing than the ones you see in your text?

There are a few molecules with coordination numbers beyond the six-coordinate examples used here, such as IF_7. Apply the principles that you learned about VSEPR and orbital hybridization to describe the geometry of this molecule and the bonding within it.

Intermolecular Forces

Slime Gel: Making and Killing Slime

Purpose

• Make "slime," a gel of poly(vinyl alcohol) and borax.

• See how the physical properties of linear polymers can be altered by forming intermolecular bonds between the chains.

Pre-Lab Preparation

If in some cataclysm, all of scientific knowledge were to be destroyed, and only one sentence passed on to the next generation of creatures, what statement would contain the most information in the fewest words? I believe it is the *atomic hypothesis* (or the atomic *fact,* or whatever you wish to call it) that *all things are made of atoms—little particles that move around in perpetual motion, attracting each other when they are a little distance apart, but repelling upon being squeezed into one another.* In that one sentence, you will see, there is an *enormous* amount of information about the world, if just a little imagination and thinking are applied.[1]

Richard P. Feynman

Feynman's description of two particles attracting each other when they are a little distance apart but repelling each other upon being squeezed into one another is graphically shown in Figure 20-1. One curve represents the *energy of interaction,* the other represents the *force of interaction* between the two particles. Net repulsive energy and force are taken as positive; net attractive force and energy are taken as negative. The zero value corresponds to separation of the two particles at a large (infinite) distance apart.

If we represent the potential energy of interaction by the symbol *U,* the interaction force is just the

[1]Richard P. Feynman, Robert B. Leighton, Matthew Sands, *The Feynman Lectures on Physics,* Vol I, Addison-Wesley Publishing Co., Inc., 1963, pp. 1–2.

(negative) derivative of the potential energy curve with respect to the interparticle distance, r—that is, force = $-dU/dr$. The interparticle distance is plotted as a normalized or dimensionless parameter, in which the interparticle distance, r, is divided by the equilibrium interparticle distance, $r_{equilibrium}$. At the point where $r/r_{equilibrium} = 1$, the energy is at its minimum value, corresponding to the trough of the energy curve, and the force is zero. The depth of the trough below the $U = 0$ line is a measure of the energy binding the two particles together (100 kJ/mol in this example). As r becomes smaller than $r_{equilibrium}$ ($r/r_{equilibrium} < 1$), the potential energy and repulsive force increase rapidly as the two particles are squeezed together (U and the force become more positive). As r becomes larger than $r_{equilibrium}$ ($r/r_{equilibrium} > 1$), the potential energy and the force both rapidly approach zero, illustrating that the intermolecular forces between neutral particles typically operate over a very short range.

What is an atom? A picture of an atom has emerged from early twentieth century investigations as having a small, unimaginably dense nucleus comprised of neutral particles (*neutrons*) and positively charged particles (*protons*). The positively charged nucleus is surrounded by an approximately spherical cloud of negatively charged particles with much smaller mass, called *electrons*. This electron cloud defines the volume of space that the atoms occupy when they are squeezed together and in effect determines the size of the atom.

When atoms get close enough together, they can form stable aggregates that we call *molecules.* If the forces holding the atoms together are sufficiently large, we say that the atoms have formed a chemical bond, in which the atoms are held together in a stable configuration that resists change.

When molecules are close to one another, they also tend to stick together, attracted by weak short-range forces. If there were no *intermolecular forces* between molecules, all molecules would be gases. Molecules cannot exist in a liquid or solid form

Potential Energy, *U* and Force vs. Interparticle Distance

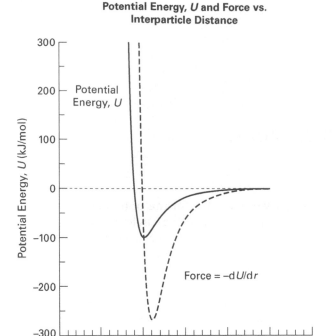

FIGURE 20-1 | **The potential energy, *U*, and interaction force between two particles as a function of the normalized distance between the particles, *r*/*r*$_{equilibrium}$. The potential energy is a minimum and the force is zero at the equilibrium separation, where *r*/*r*$_{equilibrium}$ = 1.00 and *r* = *r*$_{equilibrium}$. As the particles are squeezed together, both the potential energy and the interaction force approach large positive values, reflecting the repulsion between the particles. As the distance between the particles becomes large, both the potential energy and interaction force approach zero.**

unless there are intermolecular forces holding them together. Without these forces, there would be no liquid water or ice, no solid wood, nothing to hold the strands of protein together in the tissues and muscles of our bodies. In short, life as we know it could not exist.

At the root of atomic and molecular structure is the mutual attraction of particles with opposite charge—what are called *electrostatic* or *electrodynamic forces*, depending on whether the particles are stationary or moving. The fundamental kind of electric force binding two hydrogen atoms to an oxygen atom to form a water molecule is the same as the force that holds water molecules together to form liquid water. In this sense, there is no fundamental distinction between a chemical process, such as the formation of water by reaction of molecular hydro-

gen and oxygen, and a physical process, such as the evaporation of liquid water to form gaseous water. All can be viewed as resulting from the mutual interaction of positively charged nuclei and negatively charged electrons. So the distinction between a physical and chemical process is one of degree and not of kind. Physical processes involve weaker intermolecular (electric) forces and chemical processes involve stronger interatomic (electric) forces.

Weak intermolecular electric forces are responsible for many of the familiar bulk properties of liquids and solids, including viscosity (the resistance to flow), the volatility or vapor pressure of liquids and solids, the melting point and energy of fusion (melting), the boiling point and energy of vaporization, the solubility of one kind of molecule in another, the adsorption and catalytic activity of surfaces, adhesion (related to the properties of adhesives and glues), surface tension, friction and lubrication, the formation of monomolecular films and bilayer membranes, detergent action, the formation of emulsions and foams, gelation, hydrogen bonding, and specific interactions in synthetic or biological systems (often called *molecular recognition*). These properties are dependent on the shapes and charge distribution in interacting molecules. For example, enzymes in the body exert their catalytic effects on chemical reactions by binding a molecule in a way that facilitates or speeds up the reaction rate of the molecule. Perhaps you can use your own thinking and imagination to add to the list.

In this experiment, we will study how altering the intermolecular forces between polymer chains has a dramatic effect on the physical properties of the polymer solution. The term *polymer* comes from the Greek (*poly*, "many," + *meros*, "part") and refers to a large molecule made up of many parts. *Macromolecule* is a term synonymous with polymer. Polymers are made by linking together simple molecules called *monomers* ("single parts"). If the monomers add together so that the polymer contains all of the atoms of the monomer molecules, the reaction is called *addition polymerization*. If small by-product molecules such as water are formed during the polymerization process, the reaction is called *condensation polymerization*.

In this experiment, we will study a synthetic polymer, made by addition polymerization of the monomer vinyl acetate, followed by hydrolysis of the acetate group to produce an alcohol (—OH) group. The resulting polymer is called *poly(vinyl alcohol)*, or *PVA*. It is a water-soluble linear polymer with a carbon atom backbone having a predominant pattern of —OH groups on alternate carbon atoms, as shown in Figure 20-2.

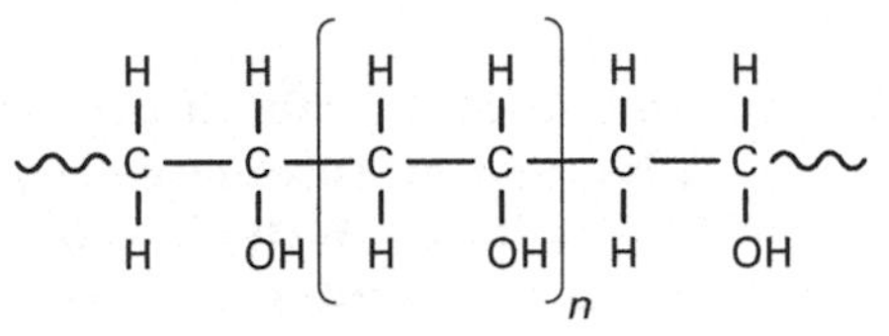

FIGURE 20-2 Poly(vinyl alcohol), or PVA, is a linear polymer in which the predominant structure is a carbon atom backbone with hydroxy (—OH) groups on alternate carbon atoms. The PVA polymer strands have an average molecular weight of about 78,000, corresponding to an average of about 1800 —[CH$_2$–CH–OH]— monomer units (MW = 44) in each strand.

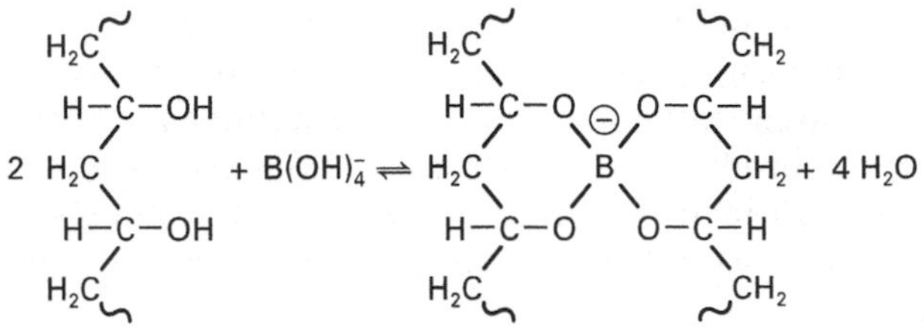

FIGURE 20-3 A tetrahydroxyborate ion, B(OH)$_4^-$, forms a cross-link between two strands of poly(vinyl alcohol). Each tetrahydroxyborate ion can react with four —OH groups. The B—O bonds are labile, rapidly and continuously breaking and reforming.

The gelation of poly(vinyl alcohol) with borax

Adding a molecule that produces cross-linking inter-actions between the chains of a linear polymer produces a structure with less flexibility. In the limit where the cross-linking bonds, once formed, are very strong and do not let go, a rigid three-dimensional structure will be formed. Examples of this would include epoxy resins and vulcanized rubber.

If the cross-link bonds are weaker or are not per-manently formed, something can form that is between a liquid and solid, having properties of a viscous gel. It can flow, but only slowly. Intermolecular links of this kind are formed when a solution of borax (sodium tetraborate decahydrate, Na$_2$B$_4$O$_7$·10H$_2$O, FW 381.4) is added to a solution of PVA. When dissolved in water, sodium tetraborate forms an equimolar mixture of boric acid and sodium tetrahydroxyborate ion, B(OH)$_4^-$:

$$Na_2B_4O_7(s) + 7\,H_2O \rightleftharpoons$$
$$2\,B(OH)_3(aq) + 2\,Na^+(aq) + 2\,B(OH)_4^-(aq) \quad (1)$$

The concentration of B(OH)$_4^-$ in solution can be determined by titration of B(OH)$_4^-$ with a strong acid such as sulfuric acid:

$$2\,Na^+ + 2\,B(OH)_4^- + H_2SO_4 \rightarrow$$
$$2\,B(OH)_3(aq) + 2\,Na^+ + SO_4^{2-} + 2\,H_2O \quad (2)$$

From the overall stoichiometry of Equations (1) and (2), you can see that one mole of sulfuric acid titrates the two B(OH)$_4^-$ ions produced by one mole of borax.

When a solution containing tetrahydroxyborate ion, B(OH)$_4^-$, is added to an aqueous solution of poly(vinyl alcohol), the borate ion forms cross-links between the alcohol groups on adjacent polymer chains, as shown in Figure 20-3. These look like covalent bonds, and in fact they are polar covalent bonds, but they have an additional important property—they are *labile*. This means that they are continually and rapidly breaking and reforming. This is very much like the bonding in water between hydrogen and oxygen atoms. The polar covalent bonds in water are fairly strong, but there is a tendency for water molecules to dissociate to a slight extent, forming H$_3$O$^+$ ions and OH$^-$ ions. This process of breaking and reforming bonds is so rapid that if you put a small amount of D$_2$O (where deuterium is the form of hydrogen with mass 2, having an extra neutron in the nucleus) into a large amount of ordinary water, the deuterium form of hydrogen is rapidly distributed throughout all of the water molecules so that most of the deuterium-labeled water will be in the form of HDO, with very few water molecules existing as D$_2$O. Similarly, when B(OH)$_3$ labeled with ^{18}O (a form of oxygen with two extra neutrons in the nucleus) is added to a solution containing B(OH)$_4^-$ ion, the ^{18}O label is rapidly and uniformly distributed among all of the boron-containing species.

When the cross-links are labile (rapidly breaking and reforming), the linear polymer chains can still move past one another but not as easily as when there is no cross-linking, so a viscous gel forms that still retains some liquid-like properties. However, if the gel is stirred very rapidly, the gel becomes more viscous—the harder you stir, the more viscous it becomes. In other words, it does not behave like a normal liquid. In fact, the gel has three distinct responses to shearing and stressing: With slow shearing (stirring), it flows like a highly viscous fluid, such as thick molasses; at somewhat faster shearing, it behaves like a piece of rubber; fast shearing causes it to fracture when it is pulled apart suddenly.

Natural polymers

Natural linear polysaccharide (polysugar) polymers can form similar gels with borax. A plaything called Slime®, marketed by the Mattel Corporation, is a gel formed by mixing a solution of the natural polysaccharide *guar* with borax. Guar is extracted from a leguminous plant native to India.

Other natural linear polysaccharides include agar and carrageenan (both extracted from seaweed) and starch (extracted from corn or potatoes). These natural products are widely used to thicken and give a smooth texture to foods and cosmetics and also find uses as adhesives and as sizing for silk and paper.

Natural polymers are of crucial importance to living organisms. Without natural structural polymers such as the *cellulose* in plants or the *proteins* that constitute the muscle and sinew of animals, no living plant or animal could stand upright against the pull of gravity. We would all be creatures of the sea, where our delicate membranes could be supported by floating in water. Plant and animal *enzymes*, which catalyze all vital biochemical reactions, are polymers of amino acids. Finally, deoxyribonucleic acid (DNA), the very stuff of the genetic inheritance of all plants and animals, is a polymer of nucleic acids.

> ☠ **At the end of the laboratory, wash your hands well to remove any residue of PVA or borax. Water can be used to clean up any residue that may accidentally get on your clothing or the benchtop. The gel may be discarded in the trash.**
>
> **If your instructor permits you to take the gel with you from the laboratory, put it in a plastic sandwich bag. Be careful not to create messes by getting it on clothing or furniture. Do not leave it where small children may play with it or try to eat it. Finally, note that handling the gel will eventually contaminate it, leading sometimes to the growth of molds. The gel may be disposed of by dissolving it in water or putting it in the trash.**

Experimental Procedure

Special Supplies: 1-mL microburet (see the Introduction to the lab manual); polyethylene transfer pipets with 0.5-mL and 1.0-mL calibration marks; wooden stirrers (tongue depressors or popsicle or craft sticks); plastic sandwich bags; Handi-Wrap film.

Chemicals: 4% (40 g/L) poly(vinyl alcohol) solution (PVA powder, J. T. Baker, 99.0–99.8% fully hydrolyzed, MW 77,000–79,000; guar gum is a suitable and less expensive substitute for PVA); 4% (40 g/L) borax, sodium tetraborate decahydrate ($Na_2B_4O_7 \cdot 10H_2O$) solution; 3.0 M H_2SO_4; 1.0 M NaOH; 0.1% bromocresol green, sodium salt indicator.

▨ NOTES TO INSTRUCTOR ▨

PVA powder dissolves slowly. It must be sprinkled in the water in small increments with vigorous stirring, sufficient to produce a small vortex; otherwise, lumps will form that are difficult to dissolve. Warming the water to 75–80 °C helps dissolve the powder faster. A large stirrer/hot plate, set to low heat with a large magnetic stirring bar, provides both vigorous stirring and warming. If the PVA solution is heated on a hot plate set on high with inadequate stirring, the solution is likely to scorch, producing a very disagreeable odor. Allow 24–48 h to prepare the PVA solution and get it fully hydrated and free of lumps.

Use the same precautions if guar gum is substituted for poly(vinyl alcohol).

1. Titration of Borax with Sulfuric Acid Borax, sodium borate decahydrate, is an efflorescent salt, meaning that it loses water when exposed to air of 50% or less relative humidity. Therefore, it is necessary to titrate the borax solution in order to determine its concentration accurately. Put 10 mL of the 4% borax solution into a 25-mL Erlenmeyer flask. (The 4% borax solution can be measured in a 10-mL graduated cylinder, or you can weigh out 10.0 g into the 25-mL flask, because 10.0 mL ≅ 10.0 g.)

Add 10 drops of 0.1% bromocresol green indicator and titrate the resulting blue solution with 3.0 M H_2SO_4 in a 1-mL microburet to the first appearance of a permanent yellow color. (See the Introduction to the lab manual for a description of the 1-mL microburet, made from a 1-mL tuberculin syringe.) Repeat the titration using a second 10-mL sample of 4% borax solution. From the volume and concentration of sulfuric acid, calculate the molar concentration of borax. [One mole of H_2SO_4 titrates one mole of borax, as shown by Equations (1) and (2)]. Compare with the expected theoretical molarity expected for 4.0% borax, 0.105 M.

2. Making "Slime," a Gel of Poly(Vinyl Alcohol) and Borax Place 50 mL of a 4% aqueous solution of poly(vinyl alcohol) in a 250-mL beaker. Add 10 drops of 0.1% bromocresol green indicator to the PVA solution. Stirring continuously, add 1.0 mL of 4% borax solution. It will be sufficiently accurate to measure the volume of borax solution using a polyethylene transfer pipet having a 1-mL calibration mark. Something stiff and wide, like a wooden popsicle or craft stick or a half-inch-wide stainless steel scoop, makes a better stirrer than a glass rod (recall mixing cans of paint). Continue vigorous stirring until the mixture forms a homogeneous gel. If the mixture separates into two phases (a gel and liquid), try scraping the gel off the stirrer and using a back and forth stirring motion straight across the beaker to chop up the gel, rather than a circular stirring motion. With time and sufficient stirring, the solution will form a homogeneous gel. The resulting gel has a pH of about 9, causing the bromocresol green indicator to appear blue.

Note and record the characteristics of the gel, such as its viscosity (resistance to flow). Try stirring it slowly, then as fast as you can. Is there a difference in the way the gel behaves? Can you pour the gel from one container to another? What happens to the gel when you place a lump of it in the center of a clean watch glass or clean piece of plastic film (such as Handi-Wrap)?

Now add, with stirring, a second 1.0-mL portion of 4% borax solution to the gel. Stir until the gel is homogeneous. Note and record any change in the properties of the gel. Is it more viscous? Or less? Is it still as easily stirred?

Finally, add with stirring a third 1.0-mL portion of 4% borax solution to the gel. As before, note and record any changes in the properties of the gel. After preparing the gel, observe its unusual viscoelastic properties. When completely gelled, the material can be handled with the bare hands. (This is the fun part.) Try rolling the gel into a ball, then seeing how long it takes for the ball to flatten out when placed on a clean watch glass or piece of Handi-Wrap film. Try pulling on the gel at different speeds, slowly then very rapidly. What happens when you try to pull the gel apart suddenly? Can you knead pieces of the gel together by working them with your fingers? Note and record how the gel behaves under these different conditions.

3. Killing and Resurrecting Slime Gel From the borax titration you did in part 1, you know the volume of 3.0 M H_2SO_4 that it took to neutralize 10 mL of borax solution. Therefore, it should take 0.30 times this volume to neutralize the total of 3.0 mL of borax added to the gel. Using your 1-mL microburet, add precisely this volume of 3.0 M H_2SO_4 to the gel. (It should be a small volume, in the 0.10–0.15 mL range.) In principle, this is precisely the amount of H_2SO_4 required to react with all of the $B(OH)_4^-$ ion that was present in the borax solution. If it is tetrahydroxyborate ion that forms the cross-links in the gel, what would you expect to happen if the ion is converted to $B(OH)_3$? Stir the gel until it is again homogeneous. If

you have added enough H_2SO_4 to neutralize the $B(OH)_4^-$, the bromocresol green indicator should be yellow, like the end point in the titration. During the intermediate stage of mixing, before the mixture is homogeneous, you may see lumps of blue gel in a yellow solution. If so, keep stirring and mixing until the solution is homogeneous.

Note and record any changes in the properties of the gel resulting from the addition of the H_2SO_4. Is the gelation process reversible? To test this, add 2.0 mL of 1.0 M NaOH, measured with a polyethylene transfer pipet having a 1-mL calibration mark. This is more than enough to neutralize the added H_2SO_4. The bromocresol green indicator should again be blue. Stir the gel again. Does the gel regain its high viscosity? Note and record your observations.

CONSIDER THIS

Let's try to imagine what a solution of poly(vinyl alcohol) or PVA looks like on a microscopic scale. First, show that the average strand of PVA (MW 78,000) has about 1800 monomer units. If each unit contains two carbon atoms, and if each carbon atom has a diameter of about 154 picometers (pm), how long would the PVA strand be if stretched out in a straight line, like a strand of uncooked spaghetti?

Next, if we had 1 L of PVA solution spread out in a sheet whose thickness was equal to the length of a PVA strand, what would be the area of the sheet? If we divide this area by the number of PVA strands in 1 L of 4% PVA solution (about 3×10^{20}), we get the average area per strand. If we think of this average area/strand as a little square containing, on average, the cross-section area of one strand of PVA, the square root of the area of this little square would give the edge length of the square. This edge length would be equal to the average distance between strands of PVA. (Imagine each little square containing a cross-section area of a PVA strand located at the center of the square.) Show that this edge length is about 2.4 nanometers (nm), or about ten times the diameter of a PVA strand assumed to be about 0.2–0.3 nm. So the strands of PVA must be fairly close together in solution but not as tightly packed as individual strands in a bowl of cooked spaghetti. It would be more like spaghetti strands being cooked in a pot of water, if the strands were stirred so that they were uniformly distributed throughout the pot.

Now imagine that tetrahydroxyborate ion forms labile bonds between the strands of PVA. We talked before about the lability of —OH bonds in water

[handwritten in left margin: 1.0–1.5 sulfuric & start stirring]

being based on the fast reaction that results in the ionization of water:

$$H_2O + H_2O \rightleftharpoons H_3O^+ + OH^- \quad (K_w = 10^{-14}) \quad (3)$$

An analogous equilibrium involving $B(OH)_3$ and $B(OH)_4^-$ ion would be the fast reaction that results in the ionization of $B(OH)_3$:

$$B(OH)_3 + H_2O \rightleftharpoons H_3O^+ + B(OH)_4^- \quad (K_a \cong 10^{-9}) \quad (4)$$

In the equilibrium of Equation (4), $B(OH)_3$ acts as a weak Lewis acid (an electron pair acceptor). The continual rapid breaking and reforming of B—OH bonds could account for the lability of the bonds between tetrahydroxyborate and PVA strands shown in the chemical equilibrium in Figure 20-3. These labile bonds tend to stick the PVA strands together, while still allowing them to slowly slide past one another.

Calculate the molar concentration of —OH groups in 4% PVA solution, assuming 40 g/L PVA and a monomer molecular weight of 44 g/mol. Show that one mole of borax provides enough $B(OH)_4^-$ to form bonds with eight moles of —OH groups. Considering this information, how many mL of 4% (0.1 M) borax solution would be required to form bonds with all of the —OH groups in 50 mL of 4% PVA solution?

Bibliography

Cassassa, E. Z.; Sarquis, A. M.; Van Dyke, C. H. "The Gelation of Polyvinyl Alcohol with Borax: A Novel Class Participation Experiment Involving the Preparation and Properties of a 'Slime'," *J. Chem. Educ.* **1986,** *63,* 57–60.

McLaughlin, K. W.; Wyffels, N. K.; Jentz, A. B.; Keenan, M. V. "The Gelation of Poly(Vinyl Alcohol) with $Na_2B_4O_7 \cdot 10H_2O$: Killing Slime," *J. Chem. Educ.* **1997,** *74,* 97–99.

Sarquis, A. M. "Dramatization of Polymeric Bonding Using Slime," *J. Chem. Educ.* **1986,** *63,* 60–61.

Sarquis, A. M. (ed.) *Chain Gang: The Chemistry of Polymers,* Terrific Science Press, Center for Chemical Education, Miami University Middletown, 4200 E. University Blvd., Middletown, Ohio 45042-3497, http://www.terrificscience.com/sciencestore/product.php?pid=28.

Walker, J. "Serious Fun with Polyox, Silly Putty, Slime, and Other Non-Newtonian Fluids," *Scientific American,* **1978,** *239*(5), 186.

Woodward, L. *Polymers All Around You,* Terrific Science Press, Center for Chemical Education, Miami University Middletown, 4200 E. University Blvd., Middletown, Ohio 45042-3497, http://www.terrificscience.com/sciencestore/product.php?pid=29.

Name _____   Date _____

CONSIDER THIS

Show that the average strand of PVA (MW 78,000) contains about 1800 monomer units.

Calculate the approximate length of a PVA strand, assuming that each monomer unit contributes two carbon atom diameters to the length (1 carbon atom diameter = 154 pm).

Show that 1.00 L of 4% (40 g/L) PVA contains about 3×10^{20} PVA strands.

Assuming that 1.00 L of 4% PVA solution is spread out to form a thin sheet of thickness equal to the average length of a PVA strand, calculate the total area of the liquid sheet. Then calculate the average area of the sheet of solution per strand of PVA. Assuming this area is a little square, the square root of this area gives the edge length of the square which is approximately the average distance between PVA strands. Show that this distance is about ten times the estimated diameter of a PVA strand (0.2–0.3 nm).

Calculate the molar concentration of —OH groups in 4% PVA solution, assuming 40 g/L PVA and a monomer molecular weight of 44 g/mol. Show that one mole of borax provides enough $B(OH)_4^-$ to form bonds with eight moles of —OH groups. Considering this information, how many mL of 4% (0.1 M) borax solution would be required to form bonds with all of the —OH groups in 50 mL of 4% PVA solution?

Liquids and Solids

The Vapor Pressure and Enthalpy of Vaporization of Water

The Enthalpy of Fusion of Water

Purpose

• Determine the vapor pressure of water as a function of temperature.

• Use the vapor pressure data to make a Clausius–Clapeyron plot [Ln(P) versus $1/T$].

• Determine the enthalpy of vaporization of water from the slope of the Ln(P) versus $1/T$ plot.

• Measure the enthalpy of fusion of ice using a Styrofoam-cup calorimeter.

Introduction

Heat and molecular motion

Our accumulated knowledge about the effect of heat provides a rather detailed picture of what happens when you heat (or transfer energy to) a substance. On an atomic or molecular scale, heating increases the motion of the atoms. If the atoms are linked by chemical bonds to form a molecule, we can focus our attention not on individual atoms but on the molecule as a whole.

If we heat a molecule, it moves about faster. This motion of the molecule's center of mass is called *translational* motion (meaning "motion from place to place"). Gaseous molecules have a substantial fraction of their total energy stored as kinetic energy of translation. They move about rapidly, bouncing off one another and off the walls of their container.

Molecules can also *rotate* about their centers of mass; as a gas is heated, the molecules rotate faster. Energy is stored in the rotating molecules, just as energy is stored in a rotating flywheel.

Individual atoms in molecules are linked by chemical bonds that are "springy," constraining the atoms that are bonded together to move like masses connected by springs. As the molecules are heated, the energies and amplitudes of their *vibrations* increase.

To summarize, the energy we put into a molecule is stored as *translational, rotational,* and *vibrational* motions of the molecules.

When we cool a substance, we remove energy from it, and molecular motions slow down. We reach the lowest temperature possible, called *absolute zero,* when we have removed as much energy as possible from the substance. At this temperature, molecular motion nearly ceases.

As we remove energy from gaseous molecules, at some temperature they will condense into a liquid and eventually into a solid, because there are attractive forces between molecules caused by dipole–dipole (or induced-dipole) attraction. To separate the molecules in a solid to the distances characteristic of those in a gas, we must do work against the opposing attractive forces. (Recall that *work = force × distance.*) This work is the energy required to melt the solid and vaporize the liquid.

Now imagine the following experiment. Suppose we begin adding energy to ice at a uniform rate while measuring the temperature. For water, a plot of temperature versus energy (or time) would look like Figure 21-1. As we warm the solid, the molecules begin to move. On average, the molecules in the solid are not going anywhere. They simply begin to bounce around in the cage formed by their surrounding neighbors. In effect, they are like gas molecules confined to a very small container not much larger than the molecule itself. At the same time, the amplitude

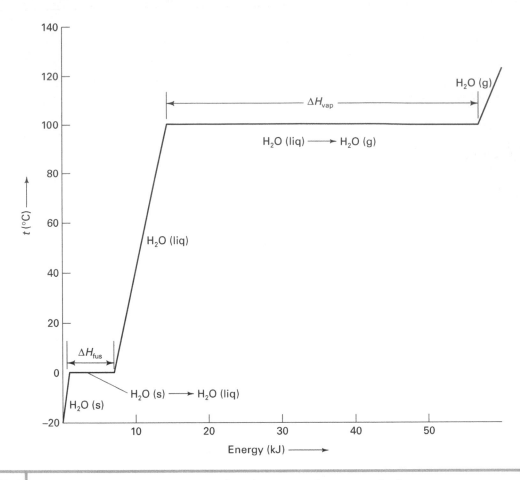

FIGURE 21-1 | The *temperature* versus *energy* heating curve for one mol of water.

of vibration of the atoms bonded together increases, and the molecules may begin to rotate in their small cages. A few molecules at the surface of the ice may acquire enough kinetic energy to break free from the solid to become gas molecules. The number of gas molecules may be large enough to produce a measurable pressure. This pressure is called the *vapor pressure* of the solid.

At the temperature we call the *melting point,* the amplitude of the motions increases to the point where the forces between neighboring molecules are no longer sufficient to hold them in fixed positions. The regular structure of the solid breaks down, forming "holes" (or vacancies) in the solid. The molecules can move more freely, jumping from hole to hole. This produces a phase change—the solid becomes a *liquid.* The liquid has a dramatically lower viscosity than the solid, so the material adopts the shape of the container.

On average, the molecules of the liquid have more energy than those of the solid. The energy required to change the solid into a liquid is called the *heat of fusion* or the *enthalpy of fusion.*

Because liquid molecules have a greater average energy than molecules in the solid, a greater number of them can escape into the gas phase, so the vapor pressure of the liquid is greater than that of the solid. As we continue to heat the liquid, a greater number of molecules acquire sufficient energy to escape from the liquid, so the vapor pressure increases as the temperature of the liquid increases. At the temperature we call the *normal boiling point,* where the vapor pressure is equal to 1 atm, all of the energy added goes into a second phase change, called *vaporization.* If the container is not closed, the vapor pressure cannot exceed 1 atm, so the liquid begins to boil, forming bubbles of vapor that escape from the liquid. If we keep adding energy, we eventually convert all of the liquid to the gas phase.

The data showing the gas, liquid, and solid equilibria can be displayed in a diagram called a *phase diagram.* The variables plotted are the pressure and temperature. The phase diagram for water between −20 and +20 °C is shown in Figure 21-2. The lines on the phase diagram represent the temperatures and pressures at which two phases can coexist in

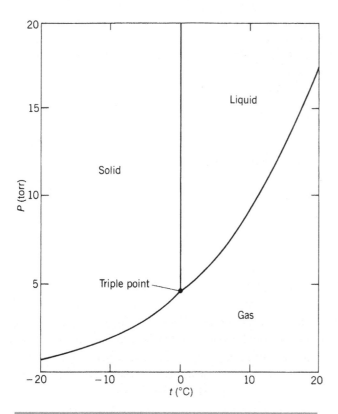

FIGURE 21-2 | The *pressure* versus *temperature* phase diagram for water in the range −20 °C to +20 °C. The lines show where two phases can coexist in equilibrium. At the triple point, all three phases coexist in equilibrium.

equilibrium. The unique point where all three phases (gas, liquid, solid) can coexist is called the *triple point*.

Now recall the experiment where we imagined that we began to heat a solid initially at a low temperature. Figure 21-1 shows how the temperature changes as we add energy to the substance, while Figure 21-2 shows how the vapor pressure of ice and liquid water changes with temperature.

Latent heat

When you heat a substance, will a thermometer immersed in the substance always indicate a temperature increase? Not necessarily. Figure 21-1 shows that while the solid is melting or the liquid is vaporizing, the temperature remains constant. Heat that produces a phase change such as the melting of ice (or the boiling of water) without producing a temperature change is called *latent* heat. A thermometer, therefore, can be used to measure temperature changes but cannot always be used to measure energy changes.

Vapor pressure and the Clausius–Clapeyron equation

Look again at Figure 21-2. Note that the vapor pressure of the ice and liquid water does not increase in a linear fashion with temperature. It increases in a more rapid, or *exponential,* fashion. An equation that fits quite well the vapor pressure data shown in Figure 21-2 for the solid–gas and liquid–gas equilibria is the *Clausius–Clapeyron* equation:

$$P = A \times e^{-\Delta H/RT} \qquad (1)$$

where ΔH is the enthalpy of *sublimation* (J/mol) for the *solid–gas* equilibrium or the enthalpy of *vaporization* for the *liquid–gas* equilibrium; R is the gas constant (8.314 J · mol^{-1} · K^{-1}); T is the Kelvin temperature (K); and A is a constant with the units of pressure.

By taking the natural logarithm of both sides of Equation (1), we can convert the equation to a linear form:

$$\text{Ln}(P) = \text{Ln}(A) - \frac{\Delta H}{R}\left(\frac{1}{T}\right) \qquad (2)$$

A plot of Ln(P) versus $1/T$ gives a straight line whose slope is equal to $-\Delta H/R$ (see Figure 21-3). Note that the slope for the ice vapor pressure line is steeper than that for the water vapor pressure line. This is because the enthalpy of sublimation of the solid is greater than the enthalpy of vaporization of the liquid by an amount of energy equal to the enthalpy of fusion:

$$\Delta H_{\text{sublimation}} = \Delta H_{\text{fusion}} + \Delta H_{\text{vaporization}} \qquad (3)$$

If we have vapor pressure data for the liquid, we can determine the enthalpy of vaporization of the liquid. If we have vapor pressure data for the solid, we can determine the enthalpy of sublimation of the solid. The difference between them gives the enthalpy of fusion of the solid, as shown by Equation (3).

The enthalpy of fusion of water

There is another way to determine the enthalpy of fusion of solid water (ice): making use of the temperature change of a known amount of liquid water when ice is added to it. This method requires a knowledge of the heat capacity of water. The experiment can be carried out by using a simple calorimeter made of Styrofoam coffee cups and a thermometer to measure the temperature changes.

It takes energy to melt ice which is initially at 0 °C and to warm the water from the melted ice to the final temperature of the water in the calorimeter. That energy comes from the warm water initially

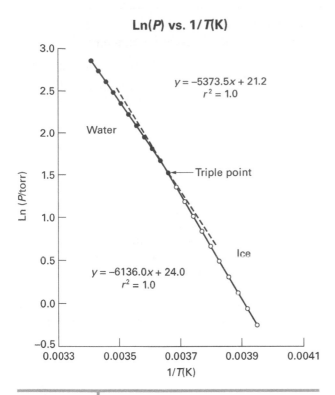

Ln(*P*) vs. 1/*T*(K)

$y = -5373.5x + 21.2$
$r^2 = 1.0$

Water

Triple point

$y = -6136.0x + 24.0$
$r^2 = 1.0$

Ice

FIGURE 21-3 A Ln(*P*) versus 1/*T* (Clausius–Clapeyron) plot for water in the range −20 °C to +20 °C. The slope for the water line segment is −5373 K. The slope for the ice line segment is −6136 K. The value of $r^2 = 1.0$ indicates a good fit to a straight line. The enthalpy of vaporization can be calculated from the slope of the water line segment and the enthalpy of sublimation from the slope of the ice line segment: $\Delta H = -R \times$ slope, where R is the gas constant in units of J/mol · K.

placed in the calorimeter. Using the conservation of energy principle, we can write

$$\Delta H_{\text{calorimeter}} + \Delta H_{\text{warm water}} + \Delta H_{\text{ice water}} + \Delta H_{\text{fusion}} = 0 \tag{4}$$

We want to calculate the enthalpy of fusion of the ice. Assume that the enthalpy change of the calorimeter (Styrofoam cups plus thermometer) is negligible (zero), and that the enthalpy change of the warm water in the calorimeter is given by

$$\Delta H_{\text{warm water}} = m\, C_{\text{p}}\, \Delta T \tag{5}$$

where m is the mass of the warm water put in the calorimeter; C_{p} is the heat capacity of water at constant pressure ($4.18\ \text{J} \cdot \text{g}^{-1} \cdot \text{K}^{-1}$); and ΔT is the temperature change of the warm water, $T_{\text{final}} - T_{\text{initial}}$.

The enthalpy change of the melted ice water is given by a similar expression,

$$\Delta H_{\text{ice water}} = m\, C_{\text{p}}\, \Delta T \tag{6}$$

where m is the mass of the added ice; C_{p} is the heat capacity of the melted ice water (as before); and ΔT is the temperature change of the melted ice water, $T_{\text{final}} - 0$ °C.

Thus, we can calculate each of the first three terms in Equation (4), which enables us to determine the remaining unknown term, ΔH_{fusion}.

Experimental Procedure

Special Supplies: If available, an electric stirrer/hot plate and magnetic stirring bar (egg shaped, 40-mm length); 7 × 300 mm stirring rods, if hand stirring is to be used; 1 × 0.01 mL disposable borosilicate glass serological pipets (Corning 7079-1N or Kimble 72120-11100); 1/8-in. ID × 1/32-in. wall thickness silicone rubber tubing; rubber septa (size 4 natural rubber septa available from the Aldrich Chemical Co. as catalog number Z10,073-0); 5-mL disposable plastic Luer-tip syringes to use as filling devices; 1-L tall-form Berzelius beakers (Corning 1060-1L or Kimble 14030-1000); 3/4-in. width water-resistant labeling tape (TimeMed tape or equivalent, or electrician's tape); crushed ice; 6-oz Styrofoam cups; 0–110 °C thermometer (or, if available, a digital thermometer reading to 0.1 °C, such as the Hanna Instruments Inc. CheckTemp 1 HI 98509); string or 22-gauge copper wire.

> **⚠ SAFETY PRECAUTIONS:**
> **Determining the enthalpy of vaporization of water involves heating one liter of water in a beaker. If the beaker is supported on an iron ring over a burner, make sure that a second ring or beaker clamp is used to secure the beaker so that it cannot tip over.**

▮ NOTES TO INSTRUCTOR ▮

Only students working very efficiently can complete all three parts of the experiment and the report in a 3-hour lab period. We suggest three possible options: allow two lab periods to complete the whole experiment; assign students parts 1 and 2 only; or have students omit part 1 (the gathering of vapor pressure data for water) and do parts 2 and 3, using vapor pressure data for water in the 50 to 75 °C range from Table 3 of the Appendix.

A stirrer/hot plate and digital thermometer are conveniences. If a stirrer/hot plate is not available, the solution can be heated on a simple hot plate or over a Bunsen burner while stirring by hand with a stirring rod. Likewise, a digital thermometer allows more accurate and convenient temperature measurements, but an ordinary thermometer can be used, estimating the temperature to the nearest 0.5 °C.

If the serological pipets are precut and lightly fire polished beforehand, students can move ahead more quickly to the vapor pressure measurements. (Any cotton plugs in the pipets should be removed. This is most easily done by connecting them to a water tap and forcing them out by water pressure.) In our experience, polystyrene serological pipets are troublesome when used with water, because water does not wet the polystyrene, making it very difficult to remove trapped air bubbles.

1. The Vapor Pressure of Water Assemble the microscale vapor pressure measuring device as shown in Figure 21-4. You will need two 1-mL borosilicate glass serological pipets (with 0.01-mL graduations), a size 4 rubber septum (or stopple), a 6-cm length of silicone rubber tubing (⅛ in.-ID × ¹⁄₃₂-in. wall thickness), and two 10-cm lengths of tape (TimeMed or electrician's tape).

Take one of the 1-mL serological pipets and, using a glass knife or glass tubing cutter or triangular file, score it precisely at the 0 mark and the 0.85-mL mark. In turn, moisten the score, and grasp the tubing with both thumbs opposite the score. Apply pressure with your forefingers pushing against your thumbs. The glass pipet will normally break cleanly with only light pressure. Then lightly fire polish each end where the pipet has been cut, just enough so that the sharp edge is just slightly rounded so that it will not cut the rubber tubing. (200-grit sandpaper can also be used to take off the sharp edge of the glass.) Take the second pipet and make a single score at the 0.85-mL mark. Snap off the bottom end of the pipet, and lightly fire polish the end of the pipet.

Next, cut a 6-cm length of silicone rubber tubing and push each end onto the lower ends of the pipets to make a connecting loop. The ends of the rubber tubing should come up to about the 0.82-mL mark. When you face the assembly, make sure that you can easily read the graduation numbers on the pipets.

Now cut a 10-cm length of TimeMed or electrician's tape and lay it on the bench top with the adhesive side up. Lay the connecting rubber loop in the middle of the tape, with the two pipets parallel and spaced about 3 cm apart and with the top edge of the tape just even with the ends of the rubber tubing.

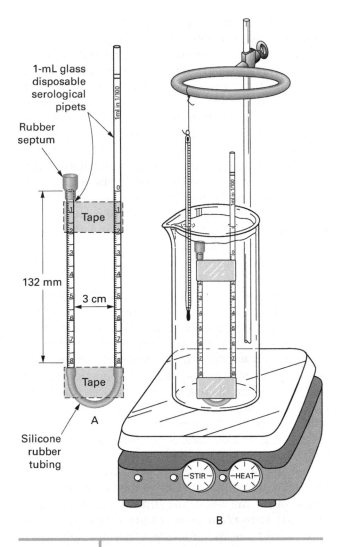

FIGURE 21-4 (A) Enlarged view of the construction of the microscale apparatus for measuring vapor pressure. (B) Overall view of the apparatus.

Then fold over the ends of the tape, so that the adhesive sides of the tape are sealed to each other to make a semirigid connection at the bottom of the pipets.

Next, turn the assembly upside down and insert the untaped end of the shorter pipet into a beaker containing deionized water. Apply suction to draw the water up into the two pipets. (This can be done with a 5-mL Luer-tip syringe and short length of silicone rubber tubing connected to the end of the longer pipet.) When the assembly is turned upright, the water level should be somewhere between the 0.35- and 0.40-mL marks in both pipets. Add or remove water as necessary. Take care to eliminate all bubbles in the pipets and connecting rubber tubing loop before proceeding to the next step.

Next, get a rubber septum and force the smaller diameter end (3.4-mm ID) of the septum onto the unsealed end of the shorter pipet. This makes a tight seal, trapping a small volume of air and water vapor. Make sure that the septum is forced on as far as it will go, so that it seals off the end of the pipet. (The small end of the septum will come down to about the 0.04-mL mark.) Then take another 10-cm piece of tape, adhesive side up as before, and lay the pipets down on the tape with the two pipets parallel and the top edge of the tape just below the rubber septum, as shown in Figure 21-4(A). Fold over the ends of the tape to make a second connection. Now you can use the longer pipet end as a handle, and the whole assembly will be semirigid.

Fill a tall-form 1-L beaker nearly to the brim with water. If available, a stirrer/hot plate is convenient for making measurements, along with a Teflon-covered magnetic stirring bar. (We find that an egg-shaped bar gives good stirring and is less prone to flopping around.) If you are going to heat the beaker over a gas burner and stir by hand, place an iron ring or beaker clamp around the upper portion of the beaker to reduce the risk of it tipping over. (**Caution:** Hot water can produce very painful burns.) Use a sturdy stirring rod, about 300 mm in length, to do the stirring, and be careful not to strike the fragile bulb of the thermometer.

Put the microscale vapor pressure assembly in the beaker up against the side of the beaker so that it will be out of the way of the stirrer if you are using a magnetic stirrer/hot plate. (A piece of tape can be used to hold it upright in place.) Thorough stirring is necessary to ensure constant temperature throughout the beaker before each volume and temperature reading is made. Using string or copper wire, suspend the thermometer so that the bulb is located approximately at the interface between the water and trapped air in the pipet, and so you can easily see the temperature graduations. (A digital thermometer is even more convenient.)

Stir the solution until the temperature reading is constant; then record the initial volume of air and water vapor to the nearest 0.005 mL. (Read the volume at the bottom of the curved meniscus, as you do when reading the volume in a graduated cylinder or buret.) Rapidly heat the water until the volume of trapped gas is about 0.8 mL (this typically requires a temperature of about 80 °C if the initial volume at room temperature is about 0.4 mL). Then turn down the heat (or turn it off) and stir the solution nearly continuously (if you are stirring by hand), pausing only to take nearly simultaneous readings of the volume of trapped gas and temperature. Don't allow the

water bath to cool too rapidly—about a degree per minute is satisfactory. Take readings about every 5 °C as the water bath cools to about 50 °C, stirring when you are not taking a reading. Cold water or small amounts of ice may be added, followed by thorough stirring, to obtain temperatures below 60 °C more quickly. You can use a small beaker to dip out water so that the 1-L beaker does not overflow.

When the temperature reaches 50 °C, record the last readings of temperature and volume. Then wrap a towel around the beaker to protect your hands, and carry it to a sink. Without disturbing the vapor pressure apparatus, cautiously add cold tap water, allowing the water in the beaker to overflow into the sink. After a minute or so, carefully lift out the vapor pressure apparatus, taking care to keep it upright, and fill the 1-L beaker with a slurry of crushed ice and water. Place the vapor pressure apparatus in the ice bath, and stir until the temperature is below 3 °C (preferably less than 1 °C). When the temperature is nearly constant, take readings of the temperature and the volume of trapped gas. Be careful to take this reading accurately because it corresponds to the volume of air in the apparatus, a crucial quantity in subsequent calculations.

Read and record the barometric pressure and temperature of the barometer.

Calculations: If we assume that the vapor pressure of water at 3 °C is negligible in comparison to atmospheric pressure, we can calculate the number of moles of air trapped in the pipet, n_{air}, by assuming that air behaves as an ideal gas:

$$n_{air} = \frac{PV_{3\,°C}}{RT} \tag{7}$$

where P is the atmospheric pressure (in torr); $V_{3\,°C}$ is the volume (in milliliters) measured at the lowest temperature (<3 °C); R is the gas constant (6.237×10^4 mL • torr • mol^{-1} • K^{-1}), and T is the Kelvin temperature of the ice water.

For each of the temperature–volume measurements made between 75 °C and 50 °C, calculate the partial pressure (in torr) of air in the pipet from the equation

$$P_{air} = n_{air} \times \frac{RT}{V} \tag{8}$$

For each temperature, we obtain the vapor pressure of water from Dalton's law of partial pressures:

$$P_{water} = P_{total} - P_{air} \tag{9}$$

where P_{total} is equal to the barometric pressure, ignoring the small pressure exerted on the column of air by the weight of water corresponding to the difference in the water heights in the two arms of the apparatus as the gas expands. (If desired, a correction for this effect can be easily made, since it is a linear function of the volume reading of the gas.)

2. The Enthalpy of Vaporization of Water Using the vapor pressure data obtained in part 1, calculate the natural logarithm of the pressure, Ln(P) (torr), and $1/T$ (K) for each data point, retaining four significant figures. Plot Ln(P) versus $1/T$. Draw the best straight line through the points. Calculate the slope of the plot, using two points on the line that are far apart. (A spreadsheet program, such as Microsoft Excel or another graphing program, also may be used to make the plot and to determine the best-fit straight line and slope, if you have access to the program and know how to use it. Figure 21-3 is an example of a plot produced using Excel.)

$$\text{slope} = \frac{\text{Ln}(P)_2 - \text{Ln}(P)_1}{(1/T)_2 - (1/T)_1} \quad (10)$$

(Note that the slope is a negative number, with units of Kelvins.) Calculate the enthalpy of vaporization from the relation

$$\Delta H_{vap} \text{ (J/mol)} = -R \times \text{slope } (K) \quad (11)$$

taking $R = 8.314$ J • mol^{-1} • K^{-1}.

3. The Enthalpy of Fusion of Ice Weigh to the nearest 0.1 g and record the mass of a pair of nested 6-oz Styrofoam cups to be used as a calorimeter. Place 100 mL of water in a 250-mL beaker and warm it to about 10 °C above room temperature. Carefully pour all of the warm water into the Styrofoam-cup calorimeter. Reweigh and record the mass to the nearest 0.1 g. Put a digital thermometer in the calorimeter or suspend a 0 to 100 °C thermometer in the calorimeter by means of a string or copper wire from an iron ring on a ring stand to reduce the risk of breaking the thermometer (see Figure 21-5). Stir the warm water with the thermometer until the reading is constant. Record the temperature to the nearest 0.2 °C.

Blot about 25 g of crushed ice in a triple layer of absorbent paper towels to remove as much of the adhering water as possible. Using a stirring rod, brush the ice into the calorimeter. Stir until all of the ice has melted, noting the approximate temperature. If needed, add more (blotted) ice until the temperature is about 10 °C below room temperature. Carefully note the lowest temperature attained, and record it to the nearest 0.2 °C. To get the mass of ice added, reweigh the calorimeter and record the mass to the nearest 0.1 g.

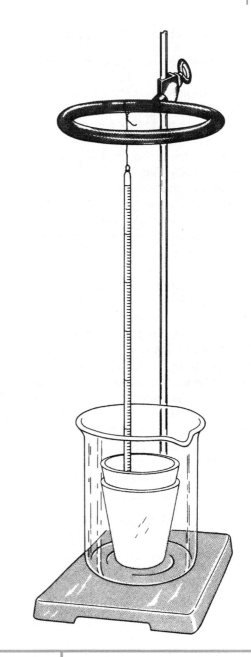

FIGURE 21-5 A Styrofoam-cup calorimeter. The beaker provides extra stability so that the nested cups won't tip over.

Calculations: Calculate the enthalpy of fusion of ice (in joules per gram) from your data, using Equations (4) to (6). Also calculate the enthalpy of fusion in units of joules per mole and compare your results with the accepted experimental value.

CONSIDER THIS

You can use the microscale vapor pressure apparatus to determine the vapor pressure of other pure volatile liquids, such as alcohols (methanol, ethanol, 2-propanol, etc.), or of solutions, where the effect on the vapor pressure of adding a nonvolatile solute to a volatile solvent can be observed.

Water has a considerably higher boiling point and lower vapor pressure than other molecules of similar or larger molar mass. For example, dinitrogen (N_2), methane (CH_4), methanol (CH_3OH), acetone [$O{=}C(CH_3)_2$], and dimethyl ether (CH_3OCH_3) all have lower boiling points and higher vapor pressures than water at the same temperature. How might this be explained?

If the intermolecular forces between molecules of a substance were very small, what effect would you expect this to have on the boiling point and vapor pressure of the substance? Give one or two specific examples of molecules where you might expect the intermolecular forces to be very small. What would you look for? Should the atoms of the molecules be large or small? Why might this matter? Should the electronegativity difference of the atoms bonded to one another be large or small? Why might this matter?

Use the information presented in Figure 21-3 to calculate a value for the enthalpy of fusion of water. Compare it to the literature value of 6009 J/mol.

Bibliography

De Muro, J. C.; Margarian, H.; Mikhikian, A.; No, K. H.; Peterson, A. R. "An Inexpensive Microscale Method for Measuring Vapor Pressure, Associated Thermodynamic Variables, and Molecular Weights," *J. Chem. Educ.* **1999,** *76,* 1113–1116.

Levinson, G. S. "A Simple Experiment for Determining Vapor Pressure and Enthalpy of Vaporization of Water," *J. Chem. Educ.* **1982,** *59,* 337.

Liquids and Solids

The Vapor Pressure and Enthalpy of Vaporization of Water

The Enthalpy of Fusion of Water

Name _____

Date _____ Section _____

Locker _____ Instructor_____

Data and Calculations

1. The vapor pressure of water

Data

V(mL)	t(°C)	Barometric pressure (torr)
	(t < 3 °C)	

Calculations

Calculate the moles of air in the graduated cylinder, using the temperature–volume data at the lowest temperature (t < 3 °C).

$$n_{air} = \frac{PV}{RT}$$

where P is the barometric pressure (in torr), V is the volume (in milliliters) at ice-water temperature, $R = 6.237 \times 10^4$ mL • torr • mol^{-1} • K^{-1}, and T is the Kelvin temperature of the ice water.

n_{air}: _____ mol

Calculate each entry in the following table, using your volume–temperature data and the equations

$$T(K) = t\,(°C) + 273.2$$

$$P_{air} = \frac{n_{air}RT}{V}$$

(Calculate P_{air} in units of torr, using the value of n_{air} from the previous step.) Then calculate P_{water} using Dalton's law,

$$P_{water} = P_{total} - P_{air}$$

where P_{total} is the barometric pressure. (Calculate P_{water} in units of torr.)

V(mL)	T(K)	P_{air}(torr)	P_{water}(torr)

2. The enthalpy of vaporization of water

Data

Use your data from part 1. (If you did not do part 1, your instructor will direct you to a source of vapor pressure–temperature data.)

Calculations

Write the values of the vapor pressure of water and the temperature from the table in part 1 (or another source, if your instructor so directs) in the following table, and calculate for each pair of pressure–temperature values the log of the pressure and the reciprocal of the temperature, retaining four significant figures.

P(torr)	T(K)	Ln(P)	1/T, (K^{-1})

Plot Ln P(torr) versus $1/T$(K) and draw the best straight line through the points; that is, draw the line so that as many points as possible lie near the line and so that there are about as many positive as negative deviations for individual points. Or better, if you have access to a computer with a spreadsheet program such as Excel, use the graphing capabilities of the spreadsheet to make a plot of your data points and add a best-fit linear trendline.

Name _____ Date _____

Ln (*P*/torr) for water versus $1/T$ (in reciprocal kelvins, K^{-1}) in the range 50–75 °C

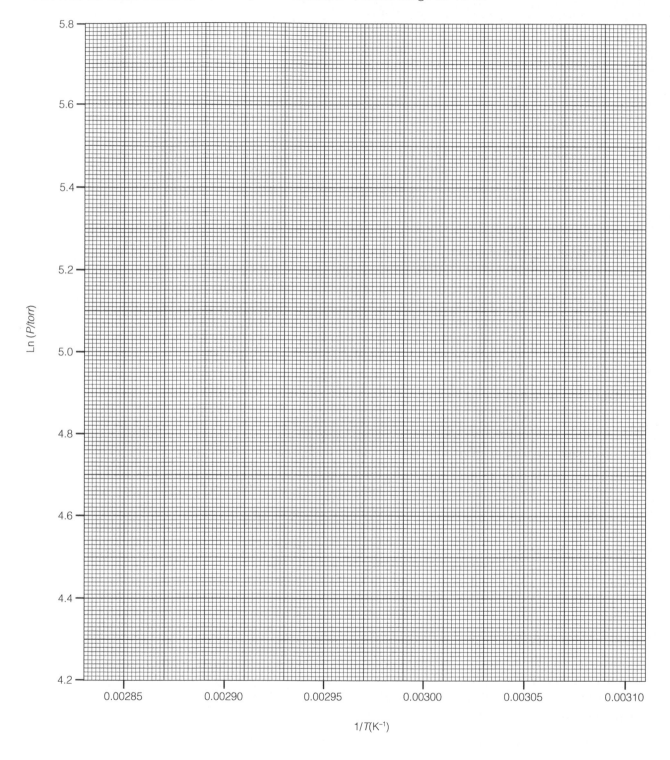

Calculate the enthalpy of vaporization from the slope of the Clausius–Clapeyron plot. Choose two points that lie on the straight line you have drawn through your data points on the Ln(P) versus 1/T plot, preferably points that are near each end of the line. Calculate the slope of the line from the equation

$$\text{slope} = \frac{\text{Ln}(P)_2 - \text{Ln}(P)_1}{(1/T)_2 - (1/T)_1}$$

Calculate the enthalpy of vaporization from the equation

$$\Delta H_{vap} = -R \times \text{slope}$$

_____ J/mol

The accepted value for ΔH_{vap} is 42.6 kJ/mol. (The value given in most textbooks of 40.7 kJ/mol is the enthalpy of vaporization at 100 °C. The value given here is the average value for the temperature range 50–75 °C.)

$$\text{percentage difference} = \frac{\text{your value} - \text{accepted value}}{\text{accepted value}} \times 100\% = $$

_____ %

3. The enthalpy of fusion of water

Calculations

Mass of empty Styrofoam cups: _____ g

Mass of Styrofoam cups + 100 mL warm water: _____ g

Initial temperature of warm water: _____ °C

Mass of cups + warm water + ice: _____ g

Final temperature of water + melted ice: _____ °C

Data

From your data, calculate the enthalpy of fusion of ice using Equations (4) to (6).

$$\Delta H_{calorimeter} + \Delta H_{warm\ water} + \Delta H_{ice\ water} + \Delta H_{fusion} = 0$$

$$0 + m_{ww}\, C_P\, \Delta T_{ww} + m_{iw}\, C_P\, \Delta T_{iw} + \Delta H_{fusion} = 0$$

where m_{ww} is the mass of the warm water, ΔT_{ww} the temperature change of the warm water [see Equation (5)], m_{iw} the mass of the ice water (mass of ice added), and ΔT_{iw} the temperature change of the ice water [see Equation (6)]. Show your calculation.

Calculated heat of fusion of ice in joules per _gram_ of water, ΔH_{fusion}: _____ J/g

Calculate the heat of fusion of ice in joules per _mole_ of water (i.e., for 18.0 g of water).

Calculated molar heat of fusion of ice, ΔH_{fusion}: _____ J/mol

The accepted value of $\Delta H_{fusion} = 6.01$ kJ/mol.

Calculate the percentage difference between your value and the accepted value.

$$\text{percentage difference} = \frac{\text{your value} - \text{accepted value}}{\text{accepted value}} \times 100\% = $$

_____ %

Name _____   Date _____

CONSIDER THIS

1. Water has a considerably higher boiling point and lower vapor pressure than other molecules of similar or larger molar mass. For example, dinitrogen (N_2), methane (CH_4), methanol (CH_3OH), acetone [$O=C(CH_3)_2$], and dimethyl ether (CH_3OCH_3) all have lower boiling points and higher vapor pressures than water at the same temperature. How might this be explained?

2. If the intermolecular forces between molecules of a substance were very small, what effect would you expect this to have on the boiling point and vapor pressure of the substance? What properties would you look for in a molecule in order to have a low boiling point and high vapor pressure? Should the atoms of the molecules be large or small? Why might this matter? Should the electronegativity difference of the atoms bonded to one another be large or small? Why might this matter? Give two or three examples of molecules where you might expect the intermolecular forces to be very small.

Questions

3. Use the information presented in Figure 21-3 to calculate a value for the enthalpy of fusion of water. Compare it to the literature value of 6009 J/mol.

4. Why would contact with steam at 100 °C produce a more severe burn than contact with liquid water at the same temperature?

5. Orange growers often spray water on their trees to protect the fruit in freezing weather. Explain how the energy of the water → ice phase transition could provide protection from freezing weather.

Colligative Properties

The Molar Mass of a Soluble Substance by Freezing-Point Depression

Purpose

• Explore one of the colligative properties of a solution, the freezing-point depression.

• Use this property to determine the molar mass of an unknown substance.

Pre-Lab Preparation

When a nonvolatile solute is dissolved in a solvent, the solution properties change from those of the pure solvent. Relative to the pure solvent, the solution has a lower vapor pressure, a lower freezing point, and a higher boiling point. These effects are called *colligative* properties of the solution because they are linked together by a common feature: They all depend primarily on the ratio of the number of solute particles to the number of solvent particles. The size of the particles or whether they are molecules or ions makes little difference; it is the *relative number of particles* that is important.[1]

Vapor-pressure lowering, freezing-point depression, and boiling-point elevation—all of these characteristics are observed in both aqueous and nonaqueous solutions. For example, a solution of a nonvolatile solute like sucrose or ethylene glycol in water shows the same kind of effects as a solution of aspirin dissolved in cyclohexanol or sulfur in naphthalene.

We make practical use of the colligative properties of an aqueous solution of ethylene glycol by putting it in the cooling system of an automobile. Because antifreeze—an aqueous solution of ethylene glycol—has both a lower freezing point and a higher boiling point than pure water, the solution protects the car's cooling system against both freezing and boiling over. Boil-over protection is increased by using a spring-loaded pressure cap, which allows the system to operate above atmospheric pressure. Increases in pressure further raise the boiling point of the solution.

The freezing-point depression and boiling-point elevation of an aqueous solution can in principle be observed with the setup shown in Figure 22-1. The phase diagram shown in Figure 22-2 describes what takes place in the apparatus. This diagram shows that solvent vapor pressure is lowered when a nonvolatile solute is dissolved in the solvent. Note that the vapor-pressure curve for the solution (shown by the dotted curve) is lower than the vapor-pressure curve for pure water (shown by the solid curve). Recall that the boiling point is the temperature at which the equilibrium vapor pressure equals the atmospheric pressure. The equilibrium vapor pressure of the solvent over the solution is *less* than that of the pure solvent at the same temperature. Therefore, the solution must be heated to a higher temperature than the pure solvent in order to reach atmospheric pressure. That means the boiling point of the solution is higher than that of the pure solvent. Increasing the pressure above atmospheric pressure, as is done when you use a pressure cap on your auto radiator, raises the boiling point even more. (If you drew a horizontal line at $P = 2$ atm, it would intersect the vapor pressure curve at a higher temperature.)

When the temperature of most solutions is lowered to the freezing point, the solid that separates out is mainly pure solvent. Lowering the vapor

---

[1] These statements are quantitatively true for ideal solutions, in which the forces between molecules are small, and for molecules having moderate molecular weights (<500 g/mol). For real solutions, except at great dilution, there is always some deviation. For polar or ionic substances or for solutions of high-molecular-weight polymers, this deviation may be considerable. In particular, ions cannot move independently because of the strong electric forces between them. When measuring the colligative properties, we therefore try to use as dilute a solution as will allow satisfactory experimental precision.

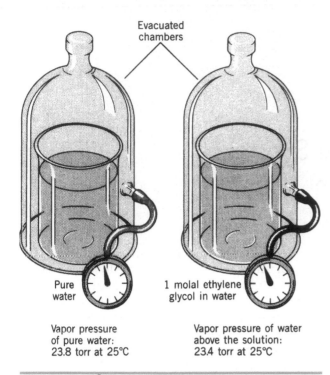

Evacuated chambers

Pure water

1 molal ethylene glycol in water

Vapor pressure of pure water: 23.8 torr at 25°C

Vapor pressure of water above the solution: 23.4 torr at 25°C

FIGURE 22-1 | **An illustration of the phenomenon of vapor pressure. If a beaker of a volatile liquid such as water is placed in an evacuated chamber, some of the liquid evaporates to fill the chamber with vapor. The pressure of this vapor is determined solely by the temperature of the liquid. When a nonvolatile solvent, such as ethylene glycol, is added to the pure solvent, the resulting solution has a vapor pressure less than the pure solvent at each temperature.**

pressure of a solvent by adding a nonvolatile solute causes the equilibrium vapor-pressure curve to intersect the sublimation-pressure curve at a lower temperature than the curve for the pure solvent (see Figure 22-2). The freezing-point curves for the solution and the pure solvent begin at the intersections of the sublimation-pressure and vapor-pressure curves, that is, at the *triple points* (shown in Figure 22-2), where all three phases (solid, liquid, and gas) simultaneously coexist. (The liquid phase is either the pure solvent or the solution containing added solute.) Therefore, the freezing-point curve for the solution lies at a *lower* temperature than the freezing-point curve of the pure solvent.

The freezing-point-depression effect has several important scientific uses. It allows us to compare the freezing-point depressions of a known and unknown substance in the same solvent in order to determine the molar mass of the unknown. This type of comparison is called *cryoscopy,* the stem *cryo-* coming from the Greek word meaning "icy" or "cold." In this

experiment, you will determine the molecular weight of an unknown dissolved in cyclohexane by the cryoscopic method.

Cryoscopy is also used in the clinical laboratory to measure the total concentration of solutes in urine, making use of the fact that the freezing-point depression is proportional to the sum of the concentrations of all dissolved particles. The urine that initially is formed within the kidney has a much lower concentration of solutes than the urine that leaves the kidney. This is because the kidney conserves body water by recovering much of the water in urine as it passes through the kidney tubules. The kidney's ability to concentrate urine is one of this organ's most important functions, and when the renal tubules are damaged, this ability is one of the first functions to be lost. Thus, a cryoscopic measurement of the solute content of urine is an indication of kidney function.

Raoult's law

The equation that quantitatively describes vapor-pressure lowering is called *Raoult's law.* It is expressed as:

$$P_1 = X_1 P_1^0 \qquad (1)$$

where P_1 is the vapor pressure of the solvent above the solution, P_1^0 is the vapor pressure of the pure solvent, and X_1 is the mole fraction of the solvent in the solution. The mole fraction is a dimensionless concentration unit that expresses the fraction of the total number of molecules (or moles) that are solvent molecules:

$$X_1 = \frac{n_1}{n_1 + n_2} \qquad (2)$$

where n_1 is the number of moles of solvent and n_2 is the number of moles of solute. This justifies the earlier statement that the colligative properties depend only on the relative numbers of solvent and solute molecules.

It is convenient in working with the colligative properties of solutions to introduce the *molality concentration scale* for solutes. We define the molality, m, of a solute as the number of moles of the solute per 1000 g of solvent:

$$\text{molality} = \frac{\text{moles of solute}}{1000\,\text{g of solvent}} = \frac{\text{moles of solute}}{\text{kilogram of solvent}} \qquad (3)$$

For example, a solution prepared by dissolving 1.00 mol of KCl (74.55 g) in 1.00 kg of water is 1.00 molal in KCl. The "particle molality" would be

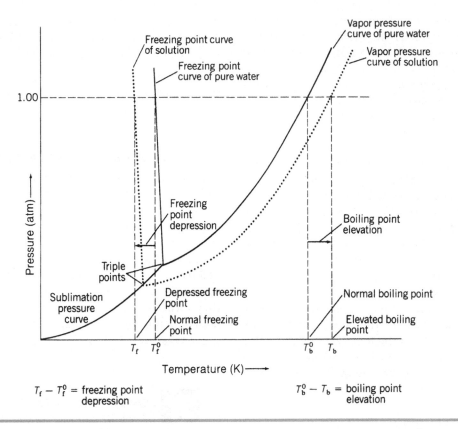

FIGURE 22-2 Phase diagrams for pure water (solid line), and for water containing a nonvolatile solute (dotted line). The presence of the solute lowers the vapor pressure of the solvent. The reduced vapor pressure of the solvent gives rise to an *increase* in the boiling point of the solution relative to pure water, $T_b - T_b^0 > 0$, and to a *decrease* in the freezing point of the solution relative to pure water, $T_f - T_f^0 < 0$. (When the temperature of an aqueous solution is lowered to the freezing point, the solid that separates out is almost always pure ice, so the sublimation-pressure curve is assumed to remain unchanged.) The extent of vapor-pressure lowering is exaggerated in the diagram to show the effects more clearly.

2.00 molal, because KCl dissociates into two ions, K^+ and Cl^-.

For dilute solutions, the molality is proportional to the mole fraction. Both molality and mole fraction are temperature-independent concentration scales, in contrast to the molar concentration, which changes with temperature because of expansion or contraction of the solution volume.

The freezing-point depression

The equation describing the freezing-point depression is[2]

$$\Delta T_f = K_f m$$

The symbol ΔT_f represents the freezing-point depression: $\Delta T_f = T_f^0 - T_f$, where T_f^0 is the freezing point of the pure solvent and T_f is the freezing point of the solution. This quantity is directly proportional to the concentration of particles expressed as the molality, m, of the particles in the solution. The proportionality constant, K_f, is called the *molal freezing-point-depression constant*. By using a solute of known molar mass, we can determine K_f for a particular solvent. The molality (and molar mass) of an unknown compound can then be determined by measuring the freezing-point depression of a solution of the unknown in the same solvent.

Note the phrase *particle molality* used in discussing the molality of an ionic solute, KCl. If the solute does not dissociate into two or more particles, the particle molality equals the molality of the solute. This will be true for all the solutions we will study in this experiment, solutions of *p*-dichlorobenzene and

[2]Some texts write the freezing-point-depression equation with a minus sign, $-\Delta T_f = K_f m$, because ΔT_f defined as $\Delta T_f = T_f - T_f^0$ is a negative quantity, so the minus signs will cancel. For simplicity, we are ignoring the minus signs.

organic compounds dissolved in cyclohexane. The distinction between particle molality and solute molality is important to keep in mind when dealing with aqueous solutions. A 0.1 *m* KCl solution would lower the freezing point of water approximately twice as much as a 0.1 *m* sucrose solution. (Sucrose, ordinary table sugar, does not dissociate in water.) Does this help to explain why $CaCl_2$ is often used in northern climates to melt ice on roads and sidewalks?

The freezing-point depression for a 1 *m* solution in water (the molal freezing-point constant) is 1.86 °C, and the corresponding boiling-point rise (the *molal boiling-point constant*) is 0.52 °C. The same principles apply for solutions in other solvents. For cyclohexane, which we will use in this experiment, the freezing point is 6.5 °C, with an ideal molal freezing-point constant of 20.2 °C/*m*.

These constants apply to measurements in quite dilute solution. For the more concentrated solutions (0.5 *m*) that we must use, the freezing-point constant may deviate slightly from the ideal value. This is the reason we will first determine the constant under the conditions of the experiment, and then use this value for determining molar mass.

Calculations of molar mass

Either boiling point or freezing point may be used to determine the molar mass of a soluble substance. For the latter, it is necessary only to determine the freezing point of a solution containing a known mass of the solute in a known mass of the solvent and to compare this with the freezing point of the pure solvent, as follows.

1. Use the measured freezing-point depression and the molal freezing-point constant in Equation (1) to determine the molality of the unknown cyclohexane solution.

2. This molality is equal to the moles of unknown divided by the kilograms of the solvent, cyclohexane; so if we multiply the molality by the mass of cyclohexane (in kilograms), we will know the number of moles of unknown that the solution contains.

3. The molar mass is the number of grams per mole, so dividing the mass of the unknown by the number of moles (which we just determined in step 2) gives the molar mass of the unknown.

Experimental Procedure

Special Supplies: Thermometer (1 °C divisions are suitable; electronic thermometers with 0.1 °C resolution are more convenient to read and give better precision);[3]

27 cm of ¹⁄₁₆-in. diameter brass rod (brazing rod available from welding supply companies is ideal); 20 × 150 mm test tube; No. 2 solid stopper; No. 2 one-hole stopper.

Chemicals: Cyclohexane; *p*-dichlorobenzene; solid organic compounds for use as unknowns for molar mass determination; salt (NaCl; see following note); ice.

▮▮ NOTES TO INSTRUCTOR ▮▮

If an electronic thermometer with 0.1 °C resolution is used, a smaller 15 × 125 mm test tube may be used, omitting the rubber stopper, and the amount of cyclohexane may be decreased to 5 mL and the solute mass to 0.25–0.30 g.

Some grades of salt (NaCl) contain anticaking additives that cloud the ice–water solution, making it more difficult to see the crystals of cyclohexane as they form. It is best to use a grade of salt that does not contain these additives.

⚠ SAFETY PRECAUTIONS: Cyclohexane is flammable. No open flames should be in the laboratory while this experiment is in progress. Keep the cyclohexane containers stoppered as much as possible, or use the fume hood to minimize odors.

☠ WASTE COLLECTION: The cyclohexane solutions should be disposed of in a properly labeled container. This container should be in a fume hood if possible.

1. The Freezing Point of Cyclohexane Figure 22-3 illustrates the simple apparatus required. Using a sharp razor blade, a 45° pie-shaped sector is cut longitudinally from a No. 2 one-hole rubber stopper.

[3]Moderately priced electronic digital thermometers (approximately $30) with a 4-in. stainless steel temperature sensing probe and 1-m cable, a range of −50 to +150 °C, and temperature resolution of 0.1 °C, are available from Hanna Instruments Inc. (Model Checktemp 1C, HI 98509) or Flinn Scientific (Catalog No. AP8559).

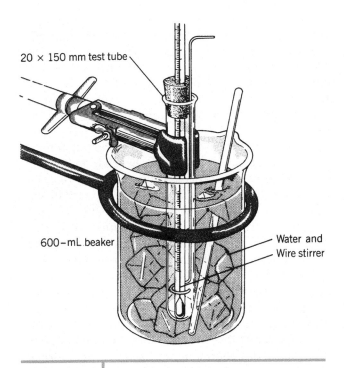

FIGURE 22-3 | **Freezing point apparatus for determination of molar mass.**

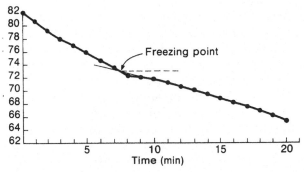

FIGURE 22-4 | **A typical freezing-point crystallization curve.**

The split stopper easily holds a thermometer in position without danger of breakage. A circular stirrer, fashioned from 1/16-in. diameter brazing rod (or 14-gauge copper wire), is fitted through the notch at the side of the stopper.

Weigh the empty 20 × 150 mm test tube with the solid stopper to ±0.01 g on a balance. (You can stand the test tube up in a beaker.) Add about 10 mL of cyclohexane (C_6H_{12}) and weigh the test tube apparatus again. Accurately weigh (to the nearest milligram) about 0.5 to 0.6 g of p-dichlorobenzene, $C_6H_4Cl_2$ (molar mass 147.0), on a creased sheet of glazed weighing paper.

Clamp the 20 × 150 mm test tube with the cyclohexane in a water bath (550 mL of ice and water in a 600-mL beaker). The bottom of the test tube should be about 2 cm from the bottom of the beaker. Place the thermometer and stirrer in position, with the thermometer bulb about 2 to 3 mm above the bottom of the test tube, threaded through the ring of the stirrer. This arrangement ensures good stirring around the thermometer bulb. Stir the cyclohexane briskly and continuously. When crystals of solid cyclohexane begin to form and the temperature is constant, read the thermometer to the nearest 0.1 C° and record this value. You may need to wipe condensation from the outside of the beaker in order to read the thermometer clearly. Warm the cyclohexane to room temperature; then repeat the cooling operation

until you are certain of the freezing point (as accurately as your thermometer can be read). When reading the thermometer, be sure to have your eye on a level with the mercury meniscus, to avoid parallax in the reading.

2. The Molal Freezing Point Constant of Cyclohexane Remove the stopper and attached thermometer and stirrer (avoiding loss of cyclohexane), just far enough to add the sample of p-dichlorobenzene. Weigh the glazed weighing paper to obtain, by difference, the weight of the p-dichlorobenzene added. Warm the mixture and stir it to dissolve the solute completely. Pour most of the water out of the 600-mL beaker, refill the beaker with ice, add 50 mL (approximately 65 g) of solid NaCl, and refill the beaker with water, leaving enough room for the test tube. Stir the mixture of salt and ice water to dissolve as much salt as possible, then lower the test tube into the ice–salt-water mixture and clamp it in place.

Now stir briskly as before and observe as accurately as possible the steady temperature at which a *small* amount of solid solvent is in equilibrium with the solution. There may be a small amount of supercooling before crystallization. Study Figure 22-4, noting especially that when a larger amount of solvent is frozen out, the remaining solution is more concentrated and therefore gives a declining freezing point. This is why readings must be taken as soon as the first crystals appear in the solution, when only a very small amount of the solvent has frozen solid.[4]

Repeat this freezing-point measurement several times by warming the solution and allowing it to cool

[4]For precise determination, readings are taken every 30 to 60 s during the cooling process. A curve similar to Figure 22-4 is plotted, and the exact freezing point determined by extrapolation, as illustrated.

until you are certain of this equilibrium temperature, to within 0.1 °C if possible. When you have finished the measurements, pour the solution into the waste cyclohexane bottle provided. *Do not pour it into the sink.* Remove the last trace of solution by rinsing the tube with cyclohexane in the fume hood and drying the tube with a gentle stream of air in the hood.

From these data, calculate the molal freezing-point constant for cyclohexane—that is, the change in freezing point caused by 1 mol of *p*-dichlorobenzene per kilogram of cyclohexane. Compare your value with the literature value of 20.2 °C/*m*. (The literature value is an idealized value, extrapolated to zero concentration of solute.) If sufficient time is available, repeat the determination with a fresh cyclohexane sample and another weighed sample of *p*-dichlorobenzene.

3. The Molar Mass of an Unknown Solid Obtain from your instructor a sample of an unknown organic material. The freezing-point measurements for pure cyclohexane and for the unknown dissolved in the cyclohexane are made exactly as in part 1, using a 0.5-g sample of unknown. From these data, calculate the molar mass of the unknown, using the molal freezing-point constant for cyclohexane determined in part 2. Be sure to dispose of the cyclohexane solution in the proper container when you are finished. If your instructor so directs, repeat the measurement with a second sample of unknown.

CONSIDER THIS

Is the freezing-point depression a linear function of molality? Make a series of different concentrations of *p*-dichlorobenzene in cyclohexane and graph the freezing-point temperature versus concentration.

Why is the molality concentration scale used in the context of colligative properties and not the more common molarity scale? Use the data in the *CRC Handbook of Chemistry and Physics* (in a section entitled "Concentrative Properties of Aqueous Solutions: Conversion Tables") to plot ΔT_f versus molality and versus molarity in order to explore these relationships.

Most everyone has participated in the fun of making homemade ice cream. (If you haven't, do it soon!) A key feature of the process takes advantage of a colligative property. Describe why it is necessary to add salt to the ice that surrounds the ice-cream container in order to freeze the ice cream inside. This is an interesting system to explore. Place a quantity of ice in a Styrofoam-cup calorimeter (see Experiment 14 or 15). Monitor the temperature as you add salt to the system. How can the temperature drop if no heat can escape the system? How low can you get the temperature to go? Convert this temperature to °F. (*Note:* This was one of Fahrenheit's calibration points as he developed the thermometer. Why is the normal freezing point of pure water a better calibration point?)

Cool a clear plastic (*not glass*) bottle of a soft drink in the freezer until it just begins to freeze. Take the cap off the bottle. (Over the sink is a good place to do this.) What happens? Explain in terms of the colligative properties that you explored in this experiment. Can you design an experiment to determine the carbon dioxide concentration in a soft drink using the freezing-point-depression property?

What has a lower freezing point, a diet soda or a regular version of the same soda? Why? Can you use the calorie content listed on the can to estimate the sugar concentration in the soda, then calculate an expected freezing point for the soda? What complications come into play?

REPORT **22**

Colligative Properties

The Molar Mass of a Soluble Substance by Freezing Point Depression

Name _____

Date _____ Section _____

Locker _____ Instructor_____

1. Experimental data

Sample or unknown no. _____

	p-dichlorobenzene		Unknown	
	Sample 1	Sample 2	Sample 1	Sample 2
Mass of test tube + beaker	g	g	g	g
Mass of cyclohexane + test tube + beaker	g	g	g	g
Mass of cyclohexane	g	g	g	g
Mass of sample + weighing paper	g	g	g	g
Mass of weighing paper	g	g	g	g
Mass of unknown	g	g	g	g
Freezing point of cyclohexane	°C	°C	°C	°C
Freezing point of solution	°C	°C	°C	°C

Use the graph below to plot the data from one of your cooling curves if so directed by your instructor.

2. Freezing point constant (Show a sample calculation.)

		p-dichlorobenzene	
	Calculations	Sample 1	Sample 2
Freezing point depression $\Delta T_f = T_f^0 - T_f$		°C	°C
Moles of solute, *p*-dichlorobenzene		mol	mol
Molality of solution = $\dfrac{\text{moles of solute}}{\text{kilograms of solvent}}$		mol/kg	mol/kg
Molal freezing-point constant, from $\Delta T_f = K_f m$		°C · kg/mol	°C · kg/mol

3. The molar mass of an unknown solid (Show a sample calculation.)

		Unknown	
	Calculations	Sample 1	Sample 2
Freezing point depression $\Delta T_f = T_f^0 - T_f$		°C	°C
Molality of unknown solution, from $\Delta T_f = K_f m$		mol/kg	mol/kg
Moles of unknown solute, from molality of solution = $\dfrac{\text{moles of solute}}{\text{kilograms of solvent}}$		mol	mol
Molar mass of unknown = $\dfrac{\text{grams of unknown}}{\text{moles of unknown}}$		g/mol	g/mol
Percentage error = $\dfrac{\text{your value} - \text{actual value}}{\text{actual value}} \times 100\%$		%	%

Approval of Instructor: Your unknown is _____

Name _____   Date _____

Questions on Experimental Error

1. (a) Assume that you have a solution of an unknown solute in cyclohexane. If the solution has a freezing-point depression of 9.50 °C, what is the molality of this solution? (The molal freezing-point constant of cyclohexane is 20.2 °C/m.)

_____ m

(b) Since the thermometer used has markings for each Celsius degree, assume that you can be certain of the freezing-point depression only to within ±0.2 °C. If the freezing-point depression is measured as 9.7 °C, then what molality would be reported for this solution?

_____ m

(c) What is the relative error in the molality that is introduced by this limitation in the thermometer? (*Hint:* The relative error in the molality will be the same as the relative error in the freezing-point depression.)

$$\left(\text{relative error} = \frac{\text{measured value} - \text{actual value}}{\text{actual value}} \times 100\%\right)$$

_____ %

2. The relative error can be reduced if a more precise thermometer is used. Assume that with the more precise thermometer you can determine the temperature to within ±0.02 °C.

(a) If the measured freezing-point depression with such a thermometer is 9.52 °C, what molality is determined for this solution?

_____ m

(b) What is the relative error in the molality as determined with this more precise thermometer?

_____ %

Problems

1. Cyclohexanol, $C_6H_{11}OH$, is sometimes used as the solvent in molecular weight determinations by freezing-point depression. If 0.253 g of benzoic acid, C_6H_5COOH, dissolved in 12.45 g of cyclohexanol, lowered the freezing-point of pure cyclohexanol by 6.55 °C, what is the molal freezing-point constant of this solvent?

2. Since the freezing point of a solution depends on the relative number of *particles*, what would you calculate to be the freezing point of 0.1 *m* solutions in water of (a) NaCl, (b) BaCl$_2$? Assume that these salts are 100% ionized in solution. (Compare your answers with the actual respective freezing points: −0.348 °C and −0.470 °C. The difference is due to the decreased activity of the ions. Because of the attractive forces between the positively and negatively charged ions, they do not move completely independently of one another.)

(a) _____

(b) _____

3. How many grams of each of the following per kilogram of water in your car radiator are needed to give equal protection against freezing down to −10 °C? (a) Methyl alcohol, CH$_3$OH, b.p. 64.6 °C. (b) Ethylene glycol, C$_2$H$_4$(OH)$_2$, b.p. 197.2 °C. In spite of higher cost, what advantage does ethylene glycol possess over methyl alcohol as a winter antifreeze and/or summer coolant?

(a) _____

(b) _____

CONSIDER THIS

Is the freezing-point depression a linear function of molality? Make a series of different concentrations of *p*-dichlorobenzene in cyclohexane and graph the freezing-point temperature versus concentration.

Why is the molality concentration scale used in the context of colligative properties and not the more common molarity scale? Use the data in the *CRC Handbook of Chemistry and Physics* (in a section entitled "Concentrative Properties of Aqueous Solutions: Conversion Tables") to plot ΔT_f versus molality and versus molarity in order to explore these relationships.

Name _____   **Date** _____

Most everyone has participated in the fun of making homemade ice cream. (If you haven't, do it soon!) A key feature of the process takes advantage of a colligative property. Describe why it is necessary to add salt to the ice that surrounds the ice-cream container in order to freeze the ice cream inside. This is an interesting system to explore. Place a quantity of ice in a Styrofoam-cup calorimeter (see Experiment 14 or 15). Monitor the temperature as you add salt to the system. How can the temperature drop if no heat can escape the system? How low can you get the temperature to go? Convert this temperature to °F. (Note: This was one of Fahrenheit's calibration points as he developed the thermometer. Why is the normal freezing point of pure water a better calibration point?)

Cool a clear plastic (*not glass*) bottle of a soft drink in the freezer until it just begins to freeze. Take the cap off the bottle. (Over the sink is a good place to do this.) What happens? Explain in terms of the colligative properties that you explored in this experiment. Can you design an experiment to determine the carbon dioxide concentration in a soft drink using the freezing-point depression property? Do not use a glass bottle of soft drink. It might explode, possibly causing injury.

What has a lower freezing point, a diet soda or a regular version of the same soda? Why? Can you use the calorie content listed on the can to estimate the sugar concentration in the soda, then calculate an expected freezing point for the soda? What complications come into play?

Some Examples of Chemical Equilibria

Le Châtelier's Principle

Purpose

• Observe a number of interesting and colorful chemical reactions that are examples of chemical systems at equilibrium.

• See how these systems respond to changes in the concentrations of reactants or products or to changes in temperature.

• Learn how the shift in equilibrium tends to at least partially offset the change in conditions, a principle first clearly stated by Le Châtelier.

Pre-Lab Preparation

Any chemical equation that describes an actual chemical reaction sums up the result of an experiment (more often, several experiments). Someone had to measure out the reactants, carry out the reaction, identify the products, and measure quantitatively the number of moles of products produced per mole of reactant consumed. The last step is the writing of a balanced chemical equation that concisely summarizes the experimental observations.

Many chemical reactions go essentially to completion, but some do not, stopping at some point of equilibrium that lies between no reaction and an essentially complete reaction. A state of equilibrium is the point of balance where the rate of the forward reaction equals the rate of the backward reaction. For these reactions, the same point of equilibrium can be reached from either end of the reaction, by mixing together either the reactants or the products, clearly indicating that chemical reactions can go either forward or backward.

In principle, every chemical reaction is a two-way reaction, but if the point of equilibrium greatly favors the reactants, we say that there is *no reaction.* If the point of equilibrium greatly favors the products, we say the reaction is *complete.* For either of these two extremes it is difficult to measure experimentally the concentrations of all the reactants and products at equilibrium. In the first case (reactants greatly favored), the concentration of products is practically zero. In the second case (products greatly favored), no significant quantity of the reactants will be produced when the products are mixed together. Thus, when the equilibrium greatly favors the products, the reaction appears to go in only one direction: *reactants → products.*

Nearly every chemical reaction consumes or releases energy. (We call these, respectively, *endothermic* or *exothermic* reactions.) For these reactions, we can regard energy as if it were a reactant or product; by adding energy to a system (by heating it) or removing energy from a system (by cooling it), we can produce a shift in the point of chemical equilibrium. The concentrations of reactants and products change to reflect the new equilibrium.

There is an added complication that must be considered. Not all chemical reactions are rapid. In fact, some are so slow that you might be fooled into thinking that no reaction occurs, whereas the equilibrium point, when it is finally reached, might greatly favor the products. The rate at which a chemical reaction reaches equilibrium will be taken up when you study the kinetics of chemical reactions.

For reactions that rapidly reach equilibrium, the point of equilibrium may be approached from either the side of the reactants or the side of the products, which emphasizes the dynamic nature of chemical reactions. The reactions you will study in this experiment are all rapid reactions, so you will be able to observe almost immediately the effects of changing the concentration of either reactants or products. In particular, you will observe that when you change the concentration of a reactant or product, the point of equilibrium shifts in a direction that tends to offset the change. This behavior can be summarized in a general principle that was first fully stated in 1884 by the French chemist Henri Louis Le Châtelier: *A chemical reaction that is displaced from equilibrium by a change in conditions (concentrations, temperature, pressure, volume) proceeds toward a new equilibrium state in the direction that at least partially offsets the change in conditions.*

This introductory experiment is designed to show qualitatively several important features of chemical equilibria. In subsequent experiments, you will learn about the quantitative aspects of chemical equilibria, measuring the equilibrium constants for a variety of reactions, and making calculations of equilibrium concentrations from previously measured equilibrium constants.

Experimental Procedure

Special Supplies: Electric hot plate; crushed ice; polyethylene transfer pipets (25 drops/mL).

Chemicals: 1 M K_2CrO_4; 3 M H_2SO_4; 6 M NaOH; 0.1% methyl orange indicator; 0.1% phenolphthalein indicator; 6 M HCl; 0.1 M CH_3COOH (acetic acid); 1 M $NaCH_3COO$ (sodium acetate); 0.1 M NH_3; 1 M NH_4Cl; 0.1 M $Fe(NO_3)_3$; 0.1 M KSCN (potassium thiocyanate); 0.15 M anhydrous $CoCl_2$, in methanol (20 g anhydrous $CoCl_2$ per liter of methanol; if only the hydrated salt $CoCl_2 \cdot 6H_2O$ is available, it must be dried before use); concentrated (12 M) HCl; saturated (5.4 M) NaCl; 0.1 M $BaCl_2$; 0.1 M $CaCl_2$; 0.5 M $H_2C_2O_4$ (oxalic acid); 0.5 M $K_2C_2O_4$ (potassium oxalate); 6 M NH_3.

> **⚠ SAFETY PRECAUTIONS:**
> **Be careful when handling concentrated (12 M) HCl and 6 M NaOH. They cause skin and eye burns. Wear eye protection.**
>
> **Ba^{2+} is toxic, and CrO_4^{2-} and $Cr_2O_7^{2-}$ are known to be carcinogenic. Rinse them off immediately if they contact your skin. Dispose of them in waste containers. Do not pour them down the drain.**
>
> **Take care when heating the methanol solution of $CoCl_2$. Methanol vapors are toxic and flammable, so the heating must be done by placing the test tube in a beaker of hot water at 65 to 70 °C on a hot plate in the fume hood.**

1. The Shifting of Equilibria in Acid–Base Reactions; The Common Ion Effect

(a) *The Chromate Ion–Dichromate Ion Equilibrium.* Yellow chromate ion reacts with hydrogen ion first to form hydrogen chromate ion and then, by condensation and loss of H_2O, orange dichromate ion:

$$2\,CrO_4^{2-} + 2\,H_3O^+ \rightleftharpoons$$
$$2\,HCrO_4^- + 2\,H_2O \qquad (1)$$
$$\rightleftharpoons Cr_2O_7^{2-} + 3\,H_2O$$

At present, we need consider only the overall reaction,

$$\underset{\text{yellow}}{2\,CrO_4^{2-}} + 2\,H_3O^+ \rightleftharpoons \underset{\text{orange}}{Cr_2O_7^{2-}} + 3\,H_2O \qquad (2)$$

To 3 mL of 1 M K_2CrO_4 in a test tube, add several drops of 3 M H_2SO_4. Mix this and observe any change. Now add several drops of 6 M NaOH, stirring continuously, until a change occurs. Again add H_2SO_4. Interpret the observed changes. How did the equilibrium shift in response to the added reagents—toward the formation of yellow chromate ion or toward the formation of orange dichromate ion? Explain how the addition of hydroxide ion exerts an effect, even though it doesn't appear in the overall equation.

(b) *Weak Acid–Base Indicator Equilibria.* In Experiment 18, you compared the relative concentrations of molecules and ions in weak acids and weak bases. Let us now see how the chemical equilibria shift when the concentrations of these ions or molecules are changed. First we will observe the effects of strong acid and strong base on indicators, which are themselves weak acids and bases. The chemical equation for the dissociation of the indicator methyl orange can be written as

$$\underset{\text{red}}{HIn} + H_2O \rightleftharpoons H_3O^+ + \underset{\text{yellow–orange}}{In^-} \qquad (3)$$

Protonated HIn form
(red)

Deprotonated In⁻ form
(yellow)

where HIn represents the protonated (acid) form of the indicator (red in color) and In^- represents the deprotonated (or base) form of the indicator (yellow-orange in color). Methyl orange indicator changes color around pH 4, which corresponds to a hydronium ion concentration of 10^{-4} M. (For a fuller discussion of indicators and the pH scale, see your text.)

Other acid–base indicators change color at different hydronium ion concentrations. Phenolphthalein, whose acid form is colorless, changes to the pink base form around pH 9, which corresponds to a hydronium ion concentration of 10^{-9} M.

First, in order to observe the effects of acid and base on indicators, add a drop of methyl orange to 3 mL of water. Then add 2 drops of 6 M HCl, followed by 4 drops of 6 M NaOH. Repeat the experiment, using phenolphthalein indicator in place of the methyl orange. Record your observations.

(c) *Weak Acid–Weak Base Equilibria.* Now we will use the indicators to observe changes involving weak acids and bases that are themselves colorless. To each of two 3-mL samples of 0.1 M CH_3COOH (acetic acid), add a drop of methyl orange. To one of the samples, add 1 M $NaCH_3COO$ (sodium acetate), a few drops at a time, with mixing. Compare, then record the colors in the two test tubes. The added salt, sodium acetate, has an ion in common with acetic acid, a weak acid that dissociates in water to give acetate ion:

$$CH_3COOH + H_2O \rightleftharpoons H_3O^+ + CH_3COO^- \qquad (4)$$

Note that adding acetate ion produces a change in the color of the indicator. The color change indicates that the indicator has been changed from its acid to its base form, which in turn must mean that the hydrogen ion concentration became smaller when the sodium acetate was added. Explain your observations in terms of the equilibria shown in Equations (3) and (4). The effect on the dissociation of acetic acid that is produced by adding sodium acetate is called the *common ion effect.*

To each of two 3-mL samples of 0.1 M NH_3, add a drop of phenolphthalein indicator. Note and record the color. To one sample, add 1 M NH_4Cl, a few drops at a time, with mixing. To the other add 6 M HCl, a drop at a time, with mixing. In each case, note any changes in the solution's color and odor.

Write the equation for the reaction of NH_3 with water to form NH_4^+ and OH^-. Interpret the results in terms of the changes of H_3O^+ concentration (shown by the color change of the indicator) and the equilibrium for the reaction of NH_3

with water. Explain clearly how the equilibria shift when NH_4^+ ions (from NH_4Cl) and H_3O^+ (from HCl) are added. Write the net ionic equation for the reaction of NH_3 with HCl.

2. Complex Ion Equilibria It is common for cations (especially those with +2 or +3 charge) to attract negatively charged ions (or neutral molecules with lone pairs of electrons) to form aggregates called *complexes.* If the resulting aggregate has a net charge, it is called a *complex ion.* The composition of these complexes may vary with the proportion and concentration of reactants.

(a) *The Thiocyanatoiron(III) Complex Ion.* This ion, sometimes called the *ferric thiocyanate* complex ion, is formed as a blood-red substance described by the equilibrium equation[1]

$$Fe^{3+} + SCN^- \rightleftharpoons Fe(SCN)^{2+} \qquad (5)$$

In a 100-mL beaker, add 3 mL of 0.1 M $Fe(NO_3)_3$ to 3 mL of 0.1 M KSCN. Dilute by adding 50 to 60 mL of water until the deep-red color is reduced in intensity, making further changes (either increases or decreases in color) easy to see. Put 5 mL of this solution in a test tube; add 1 mL (about 25 drops) of 0.1 M $Fe(NO_3)_3$. To a 5-mL portion in a second test tube, add 1 mL of 0.1 M KSCN. To a third 5-mL portion, add 5 to 6 drops of 6 M NaOH. [$Fe(OH)_3$ is very insoluble.] Put a fourth 5-mL portion in a tube to serve as a control. Compare the relative intensity of the red color of the thiocyanato complex in each of the first three test tubes to that of the original solution in the fourth tube. Interpret your observations, using Le Châtelier's principle and considering the equilibrium shown in Equation (5).

(b) *The Temperature-Dependent Equilibrium of Co(II) Complex Ions.* The chloro complex of cobalt(II), $CoCl_4^{2-}$, is tetrahedral and has a blue color. The aquo complex of cobalt(II), $Co(H_2O)_6^{2+}$, is octahedral and has a pink color. (Figure 23-1 shows the geometry of the tetrahedral and octahedral complexes.) There is an equilibrium between the two forms in aqueous solution, and because the conversion of one form to another

---

[1]For simplicity, we use the unhydrated formulas. This makes no difference in our equilibrium consideration, however, since water is always present at high constant concentration. With the addition of SCN^- to hydrated iron(III) ion, $Fe(H_2O)_6^{3+}$, a substitution of SCN^- for H_2O occurs, with possible formulas such as $Fe(H_2O)_5(SCN)^{2+}$, $Fe(H_2O)_4(SCN)_2^+$, $Fe(H_2O)_3(SCN)_3$, . . . , $Fe(SCN)_6^{3-}$.

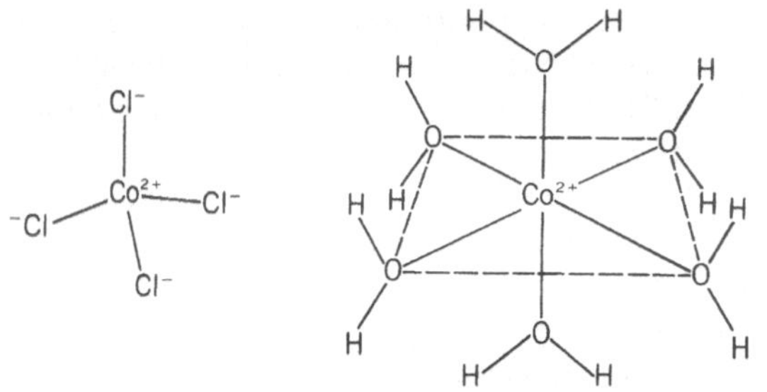

tetrahedral $CoCl_4^{2-}$
blue

octahedral $Co(H_2O)_6^{2+}$
pink

FIGURE 23-1 | **The cobalt(II) complexes with chloride ion and water have different molecular geometries and different colors.**

involves a considerable energy change, the equilibrium is temperature dependent:

$$CoCl_4^{2-} + 6\,H_2O \rightleftharpoons$$
$$4\,Cl^- + Co(H_2O)_6^{2+} + energy \quad (6)$$

Le Châtelier's principle applied to Equation (6) predicts that, if energy is removed (by cooling the system), the equilibrium tends to shift toward the aquo complex, because a shift in this direction produces some energy, thus partly offsetting the change.

Put 3 mL of 0.15 M $CoCl_2$ (in methanol) into a 13 × 100 mm test tube. (Methanol is used as a solvent so that you can observe the effects of adding water.) Using a dropper, add *just enough* water to the blue methanol solution to change the color to that of the pink aquo complex. Divide the pink solution into two equal portions in two 13 × 100 mm test tubes. Add concentrated (12 M) HCl one drop at a time to one test tube until you observe a color change. Record your observations. Heat the test tube containing the other portion of the pink solution in a beaker of hot water (65 to 70 °C) on a hot plate in a fume hood. [**Caution:** Methanol vapors are toxic and flammable. If a fume hood is not available, use an inverted funnel connected to an aspirator (as shown in Figure 5-1) to remove the methanol vapor.] You should note a color change. (If you do not, you probably added too much water to the original methanol solution. Try again.) The color change is reversible. Cooling the solution in an ice bath will restore the original pink color. Repeat the cycle of heating and cooling to verify this. Record your observations, and interpret them by applying Le Châtelier's principle to the equilibrium shown in Equation (6).

3. The Equilibria of Saturated Solutions

(a) *Saturated Sodium Chloride.* To 4 mL of saturated (5.4 M) sodium chloride in a 13 × 100 mm test tube, add 2 mL of concentrated (12 M) HCl. Mix, observe, and record your observations. The Cl^- concentration in the original solution is 5.4 M (the same as the NaCl concentration). Calculate the total Cl^- concentration that the solution would have after adding 2 mL of 12 M HCl, and explain what you observed in terms of the saturated solution equilibrium.

(b) *Saturated Barium Chromate.* To 3 mL of 0.1 M $BaCl_2$, add 5 to 6 drops of 1 M K_2CrO_4, stir, and then add 10 to 12 drops of 6 M HCl. Record your observations. Write an equation for the reaction of Ba^{2+} with CrO_4^{2-}. Using the equation you have written, explain how the addition of HCl shifts the equilibrium of the reaction. [Recall your observations in part 1 as summarized by Equation (2).]

4. Application of the Law of Chemical Equilibrium to Solubility Equilibria

In qualitative analysis, Ca^{2+} is usually precipitated as insoluble calcium oxalate:

$$Ca^{2+} + C_2O_4^{2-} \rightleftharpoons CaC_2O_4(s) \quad (7)$$

What conditions will make the precipitation as complete as possible? Since Ca^{2+} is the unknown ion, it is desirable to drive the reaction as far as possible to the right by obtaining the maximum $C_2O_4^{2-}$ concentration possible. Would it be better to use a substance that completely dissociates, such as the soluble salt $K_2C_2O_4$, or a substance that forms only a slight amount of oxalate ion, such as the weak acid $H_2C_2O_4$? The dissociation equilibria are

$$K_2C_2O_4 \rightleftharpoons 2\,K^+ + C_2O_4^{2-} \quad (8)$$

and

$$H_2C_2O_4 + H_2O \rightleftharpoons H_3O^+ + HC_2O_4^-$$
$$+$$
$$H_2O$$
$$\updownarrow$$
$$H_3O^+ + C_2O_4^{2-} \quad (9)$$

Would it be better to make the solution acidic or basic to achieve the maximum $C_2O_4^{2-}$ concentration? Test your reasoning by the following experiments.

Mix 4 mL of 0.1 M $CaCl_2$ with 4 mL of deionized water and place equal portions in three 13 × 100 mm test tubes. To tube 1, add 0.3 mL (6 to 7 drops) of 0.5 M $H_2C_2O_4$; to tube 2, add 0.3 mL (6 to 7 drops) of 0.5 M $K_2C_2O_4$. Compare the results. In your records, refer to these experiments as solution 1 and solution 2, respectively.

To tube 1, add 0.5 mL (10 drops) of 6 M HCl and mix. (For purposes of your records, refer to this new mixture as solution 3.) Explain the results. Now, again to the solution in tube 1, add a slight excess of 6 M NH_3, and mix. (Refer to this mixture as solution 4.) Is the precipitate that forms $Ca_2C_2O_4$ or $Ca(OH)_2$? [$Ca(OH)_2$ could be a possible product because the reaction of NH_3 with water generates some OH^- ion.] To find out if the precipitate is $Ca(OH)_2$, add a few drops of 6 M NH_3 to the third tube that you prepared (the one that contains diluted 0.1 M $CaCl_2$ solution). (For reference, call this solution 5.) Does a precipitate form? From this result, what conclusion can you draw about the identity of the precipitate in solution 4? Explain all of your results in terms of the equilibria shown in Equations (7), (8), and (9) and the reaction with acid of NH_3:

$$NH_3 + H_3O^+ \rightleftharpoons NH_4^+ + H_2O \qquad (10)$$

Would the acidity of the solution be more important in the precipitation of the salt of a strong acid or in the precipitation of the salt of a weak acid? To put this question another way, is H_3O^+ more effective in competing with the metal ion for the *anion of a strong acid* or the *anion of a weak acid?*

CONSIDER THIS

When two chemical equilibria have a species in common, we say that they are coupled. Changes affecting one equilibrium will also affect the other, if the equilibria are coupled. Give an example of a system in which two coupled equilibria coexist. Next, give an example of a chemical system in which two independent equilibria, not coupled, simultaneously coexist.

Bibliography

Berger, T. G.; Mellon, E. K.; Dreisbach, J. "A Multistep Equilibria-Redox-Complexation Demonstration to Illustrate Le Châtelier's Principle," *J. Chem. Educ.* **1996,** *73,* 783.

Bodner, G. M. "On the Misuse of Le Châtelier's Principle for the Prediction of the Temperature Dependence of the Solubility of Salts," *J. Chem. Educ.* **1980,** *57,* 117–119.

Jensen, W. B. "Le Châtelier's Principle, Solubility, and Déjà Vu," *J. Chem. Educ.* **1987,** *64,* 287–288 (Letter to the editor).

Some Examples of Chemical Equilibria

Le Châtelier's Principle

Name _____

Date _____ Section _____

Locker _____ Instructor_____

Observations and Data

1. The shifting of equilibria in acid–base reactions; the common ion effect

(a) *The Chromate Ion–Dichromate Ion Equilibrium*

Rewrite the equation for this equilibrium.

Describe the color changes on the addition of

H_2SO_4

NaOH

Interpret these observations in terms of the equilibrium described by Equation (2) and Le Châtelier's principle.

How does OH^- exert an effect?

(b) *Weak Acid–Base Indicator Equilibria*

Describe the effects of adding HCl and NaOH to

methyl orange

phenolphthalein

Interpret these effects in terms of Equation (3) and Le Châtelier's principle.

(c) *Weak Acid–Weak Base Equilibria*

Complete the equation for the dissociation equilibrium of acetic acid in water.

$$CH_3COOH + H_2O \rightleftharpoons$$

Explain the observed changes when 1 M sodium acetate is added to 0.1 M acetic acid (+ methyl orange).

Complete the equation for the reaction of ammonia with water:

$$NH_3 + H_2O \rightleftharpoons$$

Describe any observed odor or color changes when you add the following to 0.1 M NH_3 (+ phenolphthalein):

NH_4Cl

HCl

In which direction, left or right, does each reagent above shift the equilibrium for the reaction of NH_3 with water? Explain fully.

Write the net ionic equation for the reaction of NH_3 with HCl.

2. Complex ion equilibria

(a) *The Thiocyanatoiron(III) Complex Ion*

Write the equation for this equilibrium [Equation (5)].

Compare the intensity of the red color on addition of each of the following with the intensity of the color in tube 4. Interpret your observations in terms of the equation describing the equilibrium and Le Châtelier's principle.

$Fe(NO_3)_3$

KSCN

NaOH [Assume that the brown precipitate is formed by the reaction: $Fe^{3+} + 3\ OH^- \rightarrow Fe(OH)_3(s)$.]

(b) *The Temperature-Dependent Equilibrium of Co(II) Complex Ions*

Write the equation for this equilibrium [Equation (6)].

Describe what happens when you add 12 M HCl to the pink (aquo) complex.

Describe what happens when you heat the pink (aquo) complex and when you cool it.

Interpret these results in terms of the equation describing the equilibrium and Le Châtelier's principle.

3. The equilibria of saturated solutions

(a) *Saturated Sodium Chloride*

Describe what happens when 12 M HCl is added to saturated (5.4 M) NaCl.

What is the Cl^- concentration

    (i) in the original solution?

    (ii) after addition of the 12 M HCl (before any precipitate forms)?

    Explain this behavior in terms of the Cl^- concentrations in the two solutions, and in terms of the equilibrium equation $NaCl(s) \rightleftharpoons Na^+(aq) + Cl^-(aq)$.

(b) *Saturated Barium Chromate*

Write the equation for the equilibrium established when K_2CrO_4 and $BaCl_2$ solutions are mixed. Explain the changes observed when this solution is treated with 6 M HCl.

4. Application of the law of chemical equilibrium to solubility equilibria

Solutions mixed	Relative amounts of precipitate, if any
(1) $CaCl_2(aq) + H_2C_2O_4(aq)$	
(2) $CaCl_2(aq) + K_2C_2O_4(aq)$	
(3) Results of (1) + 6 M HCl	
(4) Results of (3) + 6 M NH_3	
(5) $CaCl_2(aq)$ + 6 M NH_3	

Answer questions (a) and (d), relating these data on the relative amounts of $CaC_2O_4(s)$ formed, or dissolved, under various conditions to the following coupled equilibrium equations. (The equilibria are said to be coupled because each equilibrium shares at least one species in common with another equilibrium.)

$$Ca^{2+} + C_2O_4^{2-} \overset{\boxed{1}}{\rightleftharpoons} CaC_2O_4(s)$$
$$+$$
$$H_3O^+ + NH_3 \overset{\boxed{2}}{\rightleftharpoons} NH_4^+ + H_2O$$
$$\boxed{3} \big\Vert$$
$$HC_2O_4^- + H_3O^+ \overset{\boxed{4}}{\rightleftharpoons} H_2C_2O_4 + H_2O$$
$$+$$
$$H_2O$$

(a) Account for the difference in behavior of $H_2C_2O_4$ and $K_2C_2O_4$ in tubes 1 and 2. Specifically, why does one solution form so much more $CaC_2O_4(s)$?

(b) From this array of coupled equilibria, explain the effect of adding HCl on these equilibria. (Note that H_3O^+ participates in equilibria 2, 3, and 4.) Why does the addition of HCl change the amount of $CaC_2O_4(s)$?

(c) Consider solutions 4 and 5, where NH_3 was added. Does $Ca(OH)_2(s)$ form when NH_3 is added to a Ca^{2+} solution (solution 5)? How does ammonia affect the amount of $CaC_2O_4(s)$ present in solution 4?

(d) Why does the H_3O^+ concentration have more influence on the precipitation of the salt of a weak acid, such as $CaC_2O_4(s)$, than on the precipitation of the salt of a strong acid, such as AgCl(s) or $CaSO_4(s)$?

████████   **CONSIDER THIS**   ████████

When two chemical equilibria have a species in common, we say that they are coupled. Changes affecting one equilibrium will also affect the other, if they are coupled. Give an example of a system in which two coupled equilibria coexist. Next, give an example of a chemical system in which two independent equilibria, not coupled, simultaneously coexist.

Determination of an Equilibrium Constant by Spectrophotometry

The Iron(III)–Thiocyanate Complex Ion

Purpose

• Confirm, by using the method of continuous variation, that the iron(III) ion forms a 1:1 complex with thiocyanate ion.

• Use a spectrophotometer to determine the amount of complex at equilibrium.

• Calculate the value of the equilibrium constant for the formation reaction of the complex.

Pre-Lab Preparation

When two or more substances react to form a particular set of products, we generally want to know two things about the chemical reaction: (1) the chemical composition of the product(s) and (2) the extent of reaction.

The extent of reaction is measured by the equilibrium constant for the reaction. If the equilibrium constant is very small (or zero), we say that essentially there is no reaction. If the equilibrium constant is very large, we say that the reaction goes to completion.

For many chemical reactions, the point of equilibrium lies between the two extremes of no reaction and complete reaction. In these circumstances, it is an interesting challenge to determine the composition of the equilibrium mixture of reactants and products and to calculate the equilibrium constant for the reaction.

The reaction we will consider is the formation of a complex ion by reaction of Fe(III) ion (ferric ion, Fe^{3+}) with SCN^- (thiocyanate ion) to form a red-colored complex ion:

$$Fe^{3+} + SCN^- \rightleftharpoons Fe(SCN)^{2+} \qquad (1)$$

$$\text{metal} + \text{ligand} \quad \text{metal–ligand}$$
$$\text{ion} \qquad\qquad \text{complex ion}$$

We will use the method of continuous variation to confirm that one ferric ion reacts with one thiocyanate ion to form a single product, called a *complex ion*. In this context, the term *complex ion* does not imply that the product is especially complicated. Rather, it means that the reaction product is an aggregate (or complex) of an Fe^{3+} ion and a SCN^- ion.

There is a universal tendency for positively charged metal ions to form complexes with negatively charged ions (for example, SCN^-) and with neutral molecules (for example, water and ammonia) that have molecular dipoles (those in which one side of the molecule is negative and the other side positive). This tendency can be understood as the result of an attractive force between the positively charged metal ions and negatively charged ions (or the negative end of a molecular dipole). Therefore, when you dissolve a metal salt in water, the metal ions will form complexes either with the water, with the anions of the salt, or with any other anions or neutral molecules having lone-pair electrons that may be added. The ions or neutral molecules bound to the metal ion are called *ligands* (from the Latin word *ligare,* "to bind").

When it has a net charge, the complex often remains in solution as a complex ion. The $Fe(SCN)^{2+}$ complex ion we are studying behaves in this way. The equilibrium is called a *homogeneous* equilibrium because all of the reactants and products are present together in *one phase,* the solution.

Some neutral complexes are very insoluble and precipitate out of solution. An example of this kind of behavior is the precipitation of insoluble solid $BaSO_4$ when a solution of Na_2SO_4 is added to a solution of $BaCl_2$. The equilibrium between a solid insoluble salt and its constituent ions is called a *heterogeneous* equilibrium because the system involves two phases: a solid phase, which consists of the pure insoluble salt, and a liquid phase, which contains the constituent ions of the salt in solution.

We make the distinction between the two kinds of equilibria, *complex formation* and *precipitation reactions,* on the basis of whether the equilibrium is homogeneous or heterogeneous. The driving force for both these kinds of reactions is the same, namely, the attraction between a positively charged metal ion and a negatively charged anion (or a neutral molecule with lone-pair electrons).

The method of continuous variation

Determining the stoichiometry of a chemical reaction by the method of continuous variation is based on a simple idea. Keep the total number of moles of reactants constant and vary their ratio, while measuring the number of moles of product produced by the reaction. The maximum amount will be formed when the mole ratio of the reactants is precisely the same as the ratio of the coefficients in the equation that describes the net chemical reaction.

We can use a simple analogy to explain how this works. It's based on the notion of a *limiting reactant,* a concept that you have already studied. Suppose that you are preparing hot dogs by putting frankfurters in buns. Each hot dog requires one frankfurter and one bun, so we can write the "reaction" for the formation of a hot dog in the following way:

$$\text{frankfurter} + \text{bun} \rightarrow \text{hot dog} \qquad (2)$$

Now suppose that the sum of the number of frankfurters and buns is constant, for example, 10. How many hot dogs could we make if we had 2 frankfurters and 8 buns? Just 2 hot dogs with 6 buns left over; the frankfurters represent the limiting reactant. When would we be able to produce the greatest number of hot dogs? If we had 5 frankfurters and 5 buns, we could make 5 hot dogs with no frankfurters or buns left over.

There is a special way of graphing or plotting the data to determine the ratio that represents the "stoichiometry" of our reaction, a 1:1 ratio of frankfurters and buns. This graph is constructed by plotting the number of hot dogs that we can make against the fraction of frankfurters, where we define the fraction of frankfurters, *F*, as

$$F = \frac{\text{number of frankfurters}}{\text{number of frankfurters} + \text{number of buns}} \qquad (3)$$

Defined this way, the fraction of frankfurters ranges from 0 to 1. In Figure 24-1, we have plotted the number of hot dogs that we can make versus the fraction of frankfurters for the case where the number of frankfurters plus the number of buns equals 10. We can make the greatest number of hot dogs

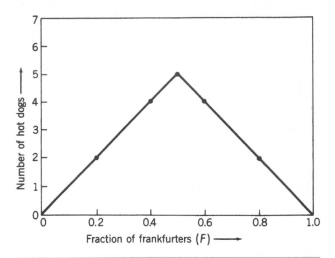

FIGURE 24-1 | **A graph of the number of hot dogs that can be made versus the fraction of frankfurters for the case where the number of frankfurters plus the number of buns equals 10. The maximum occurs when the fraction of frankfurters equals 0.5. At this point, the number of frankfurters and buns are both equal to 5 and the frankfurter-to-bun ratio is 1.0.**

when the fraction of frankfurters equals 0.5. This ratio corresponds to an equal number of frankfurters and buns, 5 in this example. So from the plot we can determine the "stoichiometry" of our "reaction." To make a hot dog, we require one frankfurter for each bun.

We will use the method of continuous variation just described to confirm the 1:1 stoichiometry of the $Fe(SCN)^{2+}$ complex. While holding the sum of the moles of Fe(III) and moles of SCN^- constant, we will vary the mole fraction of Fe(III) as we measure the amount of $Fe(SCN)^{2+}$ complex formed.

There are two further points to consider. First, what if the reaction does not go to completion when we mix the reactants? It turns out that we still find that the maximum yield of products comes when the reactants are mixed in the ratio corresponding to the stoichiometry of the complex, provided that only one chemical reaction exists at equilibrium in the range of concentrations being studied. [In the system we are studying, it has been found that only one complex is formed under the conditions we will use. However, if we increase the ratio of SCN^- to Fe(III) much above 4:1, a second complex, having two SCN^- ions to each Fe(III), begins to form.] If the reaction does not go to completion, we will obtain a plot that is rounded at the top rather than sharply peaked like a triangle. This rounding occurs because at equilibrium we have a significant amount of both reactants and products present together, so less product is formed than if the reaction were 100% complete.

The second point that needs to be made is that we must have some way of measuring the amount of product formed. We will do this by spectrophotometry, making use of the fact that the Fe(III)–thiocyanate complex formed is red and absorbs visible light, whereas the reactants, Fe^{3+} and SCN^- ions, are essentially colorless and therefore do not absorb light in the region of the spectrum where the complex does.

Why are transition metal ion complexes often colored?

The absorption of light is something like a chemical reaction in the sense that it involves an encounter (or collision) between two things: the absorbing molecule (or ion) and a packet of electromagnetic energy that we call a *photon*. Associated with the energy of the photon is a characteristic wavelength, λ, and frequency, ν. The product of the frequency times the wavelength of the photon is equal to the speed at which the photon travels, the speed of light, c:

$$\lambda\nu = c \tag{4}$$

The *energy* of the photon is related to the frequency and wavelength of the photon by the equation

$$E = h\nu = \frac{hc}{\lambda} \tag{5}$$

where E is the energy of the photon and h is Planck's constant.

In order for there to be a high probability that a photon will be absorbed, the photon's energy must match the difference between two energy states of the molecule. For visible light, the energy states correspond to those of the electrons in the molecule—in particular, to the energy difference between the energy states of the outermost valence electrons (*d*-orbital energy states) belonging to the Fe(III) ion in the complex ion.

We may be used to thinking that the five $3d$ orbitals of an Fe atom (or ion) all have the same energy, but that is true only for a free atom (or ion) in the gaseous state. The presence of any ion or molecule bound to an Fe^{3+} ion alters the energy of the d orbitals, producing energy differences between them. Relative to the free ion, electrons in d orbitals whose lobes of electron density point toward the ligands along the x, y, or z axes are generally raised in energy, while electrons in d orbitals pointing between the ligands are generally lowered in energy. This creates an energy difference between the d orbitals, as shown in Figure 24-2. So transition

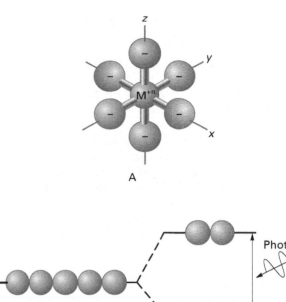

FIGURE 24-2 (A) Six negative charges arranged octahedrally along a set of *x*, *y*, and *z* Cartesian reference axes create an electrostatic field that splits the M^{+n} metal ion *d*-orbital energy levels. (B) An energy level diagram showing how the five *d* orbitals are split into two groups by an electrostatic field with octahedral symmetry. When a photon is absorbed, an electron is promoted from the lower to the upper *d*-orbital energy level.

metal ions such as Fe(III), which have partially filled d orbitals, can absorb photon energy if the photon's energy approximately matches the difference in energy of the d orbitals. In the process, an electron absorbs the photon energy, going from a filled or half-filled d-orbital energy state to a higher energy empty or half-filled d-orbital energy state. Such an electronic transition is not possible if all the d orbitals are filled with pairs of electrons having opposite spins, as in the Zn^{2+} ion. Thus, simple Zn^{2+} salts are usually colorless, whereas many transition metal ions having partially filled d orbitals absorb visible light and therefore display a variety of colors.

Group I, II, and III metal ions, such as Na^+, Ca^{2+}, and Al^{3+}, do not absorb visible light because the spacing between their electronic energy levels is much larger than the spacing between the d-orbital energy states in transition metal ions. As a result, the absorption occurs in the ultraviolet (which our eyes cannot see).

Measurement of the complex concentration by spectrophotometry

An instrument designed to measure the amount of light absorbed at different wavelengths is called a *spectrophotometer*. Figure 24-3 is a diagram of a widely used spectrophotometer that consists of a light source, an entrance slit, collimating lenses that form a parallel beam of light, a reflection grating to disperse (spread out) the light into its component wavelengths, an exit slit to isolate a narrow band of wavelengths of the dispersed light, a sample cell holder, and a detector to measure the amount of light that passes through the sample.

The light detectors used in most spectrophotometers provide an electrical signal that is proportional to the radiant power, P. We may think of the radiant power as a measure of light intensity proportional to the number of photons arriving per second at the detector.

In the Spectronic 20, which has only a single beam, the measurement of the light absorption by the sample is made by positioning the grating to give the desired wavelength of light. The phototube current is read on a meter that is first adjusted to read 0 when no light is falling on the detector. Then a tube containing only the reference solution (the solvent medium or water) is inserted in the sample cell, and the light control knob is adjusted so that the meter reads its maximum value (100% transmittance of the light). Then a tube containing the sample solution is placed in the cell holder. If the sample solution absorbs some of the light, less light will arrive at the detector, and the meter will read some number less than 100% transmittance. The transmittance, T (which ranges between 0 and 1), is defined by the equation

$$T = \frac{P}{P_0} \tag{6}$$

where P_0 is proportional to the number of photons per second passing through the reference tube, and P is proportional to the number of photons per second passing through the sample tube, as shown schematically in Figure 24-3. The percentage transmittance (which ranges between 0 and 100) is defined by the equation

$$\%T = 100T \tag{7}$$

Therefore, the transmittance, T, represents the fraction of photons that pass through the sample molecules without being absorbed, as shown in Figure 24-4.

The probability that a photon will be absorbed at a particular point in the cell is proportional to both the concentration of absorbing molecules and the number of photons arriving at the point. The absorption of photons causes the radiant power, P, to decline exponentially as it passes through the absorbing sample:

$$T = \frac{P}{P_0} = 10^{-\epsilon bc} \tag{8}$$

We have previously defined P_0 and P (see Figure 24-4). In Equation (8), the constant ϵ is called the *molar absorptivity* (it has units of liters per mole per centimeter), b is the cell *path length* in centimeters, and c is the *concentration* of the absorbing sample in moles per liter. Taking the logarithm (to the base 10) of both sides, we obtain a quantity, called the

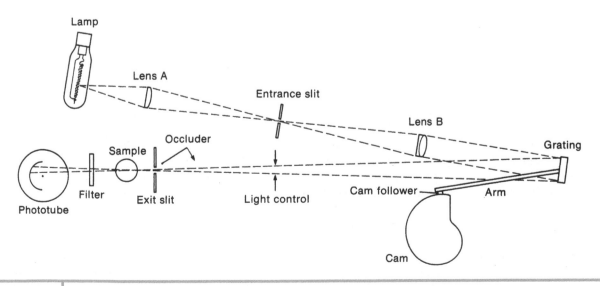

FIGURE 24-3 Schematic optical diagram of the Bausch & Lomb Spectronic 20.

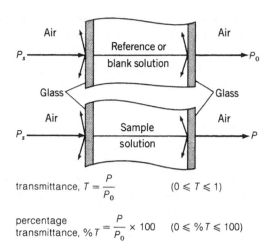

transcribance, $T = \dfrac{P}{P_0}$ $(0 \leqslant T \leqslant 1)$

percentage transmittance, $\%T = \dfrac{P}{P_0} \times 100$ $(0 \leqslant \%T \leqslant 100)$

FIGURE 24-4 | **The measurement of transmittance (or percentage transmittance). The radiant power from the source is P_s. When the reference or blank solution is in the cell, the radiant power that emerges from the cell is P_0. When the sample solution is in the cell, the radiant power that emerges is P. The use of a reference or blank solution compensates for reflection losses at the air–glass and glass–solution interfaces as well as absorption by the glass and solvent, to ensure that you measure only absorption due to the sample.**

absorbance, A, that is proportional to the concentration of the sample:

$$A = \log_{10}\left(\frac{P_0}{P}\right) = \epsilon bc \tag{9}$$

Spectrophotometers like the Spectronic 20 have both a linear $\%T$ scale and a nonlinear (logarithmic) absorbance scale. The relationship between these scales is shown in Figure 24-5.

We will determine the concentration of the absorbing $Fe(SCN)^{2+}$ complex by measuring the absorbance of the complex, either reading the $\%T$ or reading the absorbance directly. If you measure the $\%T$, it is a simple matter to compute the absorbance from the measured $\%T$ readings by using a calculator:

$$A = \log_{10}\left(\frac{P_0}{P}\right) = \log_{10}\left(\frac{100}{\%T}\right) \tag{10}$$

Knowing the value of the molar absorptivity and the cell path length b (or their product, ϵb), we can calculate the concentration (in moles per liter) of the absorbing complex from the simple relation

$$c = \frac{A}{\epsilon b} \quad \leftarrow 6120 \text{ L/mol} \tag{11}$$

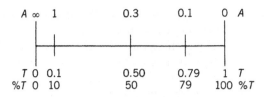

FIGURE 24-5 | **The relationship between absorbance (A), transmittance (T), and percentage transmittance ($\%T$). The absorbance scale is a nonlinear (logarithmic) scale. The transmittance scales are linear.**

Experimental Procedure

Special Supplies: Spectrophotometer; spectrophotometer cells (15 × 125 mm test tubes for the Spectronic 20); two burets (or 10-mL graduated cylinders); plastic wrap.

Chemicals: 0.00200 M KSCN in 0.50 M HNO_3 solution + 0.5 g/L sulfamic acid; 0.00200 M $Fe(NO_3)_3$ in 0.5 M HNO_3 solution + 0.5 g/L sulfamic acid;[1] $NaHCO_3$(s).

> **SAFETY PRECAUTIONS:**
> The stock solutions of KSCN and $Fe(NO_3)_3$ contain 0.50 M HNO_3, which is corrosive. Wipe up any spills immediately, and if you get any solution on your skin, immediately rinse it off with water.

> **WASTE COLLECTION:**
> When you have finished making absorbance measurements, empty all of the solutions into an 800-mL beaker, neutralize with sodium bicarbonate until the solution is just basic to litmus or pH paper, and discard the neutralized solution down the drain.

---

[1]When the KSCN and $Fe(NO_3)_3$ solutions are made up, add 0.5 g/L of sulfamic acid ($^+H_3NSO_3^-$) to each solution to eliminate nitrogen oxides from the nitric acid, which might oxidize thiocyanate ion. [The sulfamic acid does not interfere with the $Fe(III)/SCN^-$ reaction.] In each stock solution, the 0.50 M HNO_3 is present to suppress hydrolysis of the Fe^{3+} ion and to maintain a solution of constant ionic strength so that the equilibrium constant will not change. At low ionic strengths, the value of the equilibrium constant is strongly dependent on the ionic strength of the solution.

1. Absorbance Measurement Obtain seven 15 × 125 mm test tubes. Label the tubes from 1 to 7. In each test tube, put the quantities of 0.00200 M KSCN in 0.5 M HNO_3 solution and 0.00200 M $Fe(NO_3)_3$ in 0.5 M HNO_3 solution shown in Table 24-1. *The solution volumes must be measured as accurately as possible.* It is convenient to dispense the two stock solutions from burets. If burets are not available, use two 10-mL graduated cylinders, adjusting the volumes precisely with a dropper. Note that in each tube, the sum of the concentrations of the thiocyanate ion and the Fe(III) ion is constant and is equal to 0.00200 M. This fulfills the condition required for using the method of continuous variation discussed earlier.

Thoroughly mix the solution in each tube by covering the mouth of the tube with a small piece of plastic wrap to protect your thumb and inverting the tube several times. If you are using a Spectronic 20, adjust the spectrophotometer wavelength to 450 nm and set the 100% T value with a tube containing pure water. Measure the absorbance (or %T) of each solution at 450 nm, following the procedures given by your instructor for the particular instrument used in your laboratory. [If your instrument measures only %T, you can calculate the absorbance of each solution by using Equation (10).] Save the seven solutions you have prepared until you have calculated and plotted the absorbance values as described in part 2.

TABLE 24-1 **In the Method of Continuous Variation, the Total Solution Volume Is Kept Constant**

Tube Number	0.00200 M Solution (mL)	
	KSCN	Fe(NO$_3$)$_3$
1	10 5~	0 ~
2	8 4	2 1
3	6 3	4 2
✗ 4	5 3	5 3
5	4 2	6 3
6	2 1	8 4
7	0	10 5~

[2]The mole fraction of Fe(III) is fundamentally defined as mol Fe(III)/[mol Fe(III) + mol KSCN]. But because the number of moles of each substance is equal to the volume times the molarity of the solution, and because the molarities of the solutions are equal, the mole fraction is equal to the volume relationship of Equation (12).

2. Confirming the Mole Ratio of the Complex Determine the ratio of Fe^{3+} to SCN^- in the complex by plotting the absorbance of the seven solutions you measured versus the mole fraction of Fe(III), where the mole fraction of Fe(III) in each solution is calculated from the relation[2]

$$\text{mole fraction Fe(III)} = \frac{\text{mL Fe(III) sol'n}}{\text{mL Fe(III) sol'n} + \text{mL KSCN sol'n}} \quad (12)$$

Draw a smooth curve through the points you have plotted. Show the plot to your instructor. If it is satisfactory, determine the mole fraction of Fe(III) where the maximum value of absorbance occurs. What mole ratio of iron(III) to thiocyanate ion does this mole fraction of Fe(III) correspond to?

When you have completed all of your absorbance measurements and plotted the data points, you may discard the solutions into an 800-mL beaker. Neutralize the solutions with sodium bicarbonate (approximately 5 g) and discard them down the drain.

3. Calculating the Equilibrium Constant for the Reaction Three unknown concentrations appear in the equilibrium constant expression for Equation (1), so three independent equations are required to determine these unknown concentrations. The three equations we will use are the equilibrium constant expression

$$K = \frac{[Fe(SCN)^{2+}]}{[Fe^{3+}][SCN^-]} \quad (13)$$

and two material balance equations, which account for all the Fe(III)- and thiocyanate-containing species in the solution:

$$c_{Fe} = [Fe^{3+}] + [Fe(SCN)^{2+}] \quad (14)$$

$$c_{SCN} = [SCN^-] + [Fe(SCN)^{2+}] \quad (15)$$

The material balance equations state that all of the Fe(III)-containing species and all of the thiocyanate-containing species must add up to the amounts of Fe(III) and thiocyanate ion originally put into the solutions. For each solution, c_{Fe} and c_{SCN} are given by the expressions

$$c_{Fe} = 0.00200\,M \times \frac{\text{mL Fe(NO}_3)_3\text{ sol'n}}{10.0\,\text{mL}} \quad (16)$$

and

$$c_{SCN} = 0.00200\,M \times \frac{\text{mL KSCN sol'n}}{10.0\,\text{mL}} \quad (17)$$

We will calculate values of the equilibrium constant for just three solutions, those in tubes 2, 4, and 6. We start by using Equation (11) to calculate the concentration of the $Fe(SCN)^{2+}$ complex in each of the three solutions from the measured absorbances of the solutions.

> **NOTE** If you use conventional cells of square cross section with 1-cm path length, assume that $\epsilon = 4700$ L·mol^{-1}·cm^{-1} and $b = 1.00$ cm. If you use a Spectronic 20 with 15×125 mm test tubes as sample cells, take the product ϵb in Equation (11) to be equal to 6120 L·mol^{-1} [Frank and Oswalt (1947) and authors' values].

Next, for each solution, calculate the total concentration of Fe(III), c_{Fe}, from Equation (16) and the total concentration of thiocyanate ion, c_{SCN}, from Equation (17). Now, for each solution you know the concentration of the $Fe(SCN)^{2+}$ complex and the total concentrations, c_{Fe} and c_{SCN}.

Next, for each solution, calculate the equilibrium concentration of iron(III) ion, $[Fe^{3+}]$, using Equation (14) and the equilibrium concentration of thiocyanate ion, $[SCN^-]$, using Equation (15). This gives for each solution the three equilibrium concentrations you need to calculate a value of the equilibrium constant K from Equation (13).

Finally, calculate the average value of the equilibrium constants you calculated for the three solutions in tubes 2, 4, and 6.

CONSIDER THIS

Suppose that you took the contents of tube number 4, containing equal amounts of 0.0020 M Fe(III) and 0.0020 M KSCN solution, and diluted it by a factor of 2 by adding an equal volume of 0.5 M HNO$_3$. [It's important to add 0.5 M HNO$_3$ rather than water, in order to keep the ionic strength constant. Otherwise, the value of the equilibrium constant would change because its value depends on the ionic strength, as described in the articles by Frank and Oswalt (1947) and Cobb and Love (1998) listed in the Bibliography.]

When the solution has been diluted by a factor of 2, what would you expect to happen to the concentration of $Fe(SCN)2^+$ as measured by the absorbance at 450 nm? Would it decrease by a factor of 2? Try it and see. Can you explain why it decreases by a factor of approximately 4, rather than a factor of 2?

Thiocyanate is found in human saliva. You could determine the thiocyanate of your own saliva following the detailed procedure described by Lahti et al. (1999), which involves using a spectrophotometric measurement of the red color of the iron(III)–thiocyanate complex formed when Fe(III) is mixed with saliva in acid solution.

Bibliography

Cobb, C. L.; Love, G. A. "Iron(III) Thiocyanate Revisited: A Physical Chemistry Equilibrium Lab Incorporating Ionic Strength Effects," *J. Chem. Educ.* **1998**, *75*, 90–92.

Frank, H. S.; Oswalt, R. L. "The Stability and Light Absorption of the Complex Ion FeSCN^{++}," *J. Am. Chem. Soc.* **1947**, *69*, 1321–1325.

Lahti, M.; Vilpo, J.; Hovinen, J. "Spectrophotomorphic Determination of Thiocyanate in Human Saliva," *J. Chem. Educ.* **1999**, *76*, 1281–1282.

Ramette, R. W. "Formation of Monothiocyana-toiron(III): A Photometric Equilibrium Study," *J. Chem. Educ.* **1963**, *40*, 71–72.

Name _____   Date _____

CONSIDER THIS

Suppose that you took the contents of tube number 4, containing equal amounts of 0.00200 M Fe(III) and 0.00200 M KSCN solutions, and diluted it by a factor of 2 by adding an equal volume of 0.5 M HNO_3. [It's important to add 0.5 M HNO_3 rather than water, in order to keep the ionic strength constant. Otherwise, the value of the equilibrium constant would change because its value depends on the ionic strength, as described in the articles by Frank and Oswalt (1947) and Cobb and Love (1998) listed in the Bibliography.]

When the solution has been diluted by a factor of 2, what would you expect to happen to the concentration of $Fe(SCN)^{2+}$ as measured by the absorbance at 450 nm? Would it decrease by a factor of 2? Try it and see. Can you explain why it decreases by a factor of approximately 4, rather than a factor of 2?

Factors Affecting the Rates of Chemical Reactions

Purpose

• Study a number of factors that influence chemical reaction rates, including:

1. Concentration

2. The nature of the chemical reactants

3. The surface area in a heterogeneous reaction

4. The temperature of the reacting system

5. The presence of catalysts

6. The effects of diffusion in a poorly mixed solution of chemical reactants

Pre-Lab Preparation

At present, it is safe to say that we do not yet understand in detail every factor that influences the rate of a chemical reaction. Moreover, we cannot explore in two or three hours of experimentation everything that is known about the subject, so you should regard this experiment as an introduction to several of the more important factors that influence the rates of chemical reactions.

We have omitted from this experiment the study of some factors that are known to influence reaction rates. For example, we will not discuss the effect of light on chemical reactions. (The interaction of light photons with molecules is an important subdiscipline of chemistry called *photochemistry.*) We might also discuss the effects of molecular geometry and *stereochemistry,* for it is well known that the shapes of molecules influence reaction rates. Reaction rates in solution also are affected by the nature of the solvent. These are all subjects you will encounter in more advanced courses in chemistry.

Measuring the rate of a chemical reaction

Let's begin by defining what we mean by the rate of a chemical reaction. The simplest and most fundamental definition of reaction rate is either the number of moles of reactants that disappear per unit time or the number of moles of products that appear per unit time. According to this definition, the units of reaction rate would be moles per second (mol/s).

If the volume of the reacting system is constant (or nearly constant), it is convenient to define the reaction rate in moles per unit volume per unit time—for example, $mol \cdot L^{-1} \cdot s^{-1}$. This approach allows us to measure reaction rates as changes in concentration per unit time, which is convenient when we study reactions in solution (where volume changes are usually small) or in the gas phase (where the reactants and products are confined to a reaction flask of fixed volume).

To measure the rate of a chemical reaction, we must make use of chemical or physical properties of the system, properties that allow us to determine changes in the number of moles (or concentration) per unit time. As an example, the rate of a reaction that evolves a gas can be measured by measuring the volume of gas produced per second. If a reactant or product has a characteristic color, we can measure the intensity of the color (proportional to concentration) at a particular wavelength in an instrument called a *spectrophotometer.*

Another method, one that we will use in this experiment, requires us to measure the time required for a fixed amount of a reactant to disappear. This is called a *clock reaction,* and we will use this method to study the reaction of hydrogen peroxide with iodide ion:

$$H_2O_2 + 2\,I^- + 2\,H_3O^+ \rightarrow I_2 + 4\,H_2O$$

The method works like this: A fixed amount of sodium thiosulfate ($Na_2S_2O_3$) is added to the reaction mixture, along with H_2O_2, KI, starch, and an acetic acid–sodium acetate buffer. (The buffer keeps the hydrogen ion concentration nearly constant.) As the hydrogen peroxide reacts with iodide ion, iodine (I_2) is produced. As the I_2 is formed, it reacts very rapidly with thiosulfate ion to reform iodide ion (thus keeping the iodide ion concentration constant):

$$2\,S_2O_3{}^{2-} + I_2 \rightarrow 2\,I^- + S_4O_6{}^{2-}$$

When the last bit of thiosulfate ion is used up, the I_2 concentration increases. In the presence of iodide ion, molecular I_2 forms the triiodide complex ion

$$I_2 + I^- \leftrightarrows I_3^-$$

which in turn forms a dark blue-black complex with the added starch. The appearance of the blue color signals that all of the thiosulfate (and a corresponding amount of H_2O_2) has been used up.

The time required for a fixed amount of thiosulfate (and H_2O_2) to react is measured. The faster the reaction, the shorter the reaction time, so the reaction rate and the reaction time are inversely related. The amount of thiosulfate is deliberately made small so that not more than 10% of the H_2O_2 reacts. Thus, during the course of the reaction, the concentrations of H_2O_2, I^-, and H_3O^+ remain nearly constant, approximately equal to their initial values. This allows us to study the effects of concentration by varying the initial concentration of one or more reactants.

Effects of concentration

In order to react, molecules or ions must encounter one another. These encounters (or collisions) are often violent enough to produce a rearrangement of the atoms and the formation of new products. The more molecules there are per unit volume, the more likely they are to collide with one another, so increasing the concentrations of the reactants generally increases the rate of reaction. However, the dependence of reaction rate on the concentrations of the reactants is intimately linked to the reaction pathway; in some instances, an increase in the concentration of a reactant has no effect (or even a slowing effect) on the rate of reaction. Experiment 26 involves a quantitative study of the dependence of reaction rate on the concentrations of the reactants.

Influence of the nature of the reactants

The alkali metals all react with water. The net reaction is the same for each metal:

$$M + H_2O \rightarrow M^+(aq) + OH^-(aq) + \tfrac{1}{2}H_2(g)$$

where M represents Li, Na, K, Rb, or Cs. The rate of reaction increases noticeably in the sequence lithium → cesium. Lithium reacts briskly, but cesium reacts violently, producing enough heat to ignite the hydrogen produced. The metals that react most rapidly have the lowest melting points and the smallest ionization energies.

The reactions of the halogens with hydrogen show the opposite progression. The lightest element, fluorine, reacts explosively with hydrogen at room temperature, whereas iodine reacts slowly.

Effect of surface area in a heterogeneous reaction

In a reaction between a solid and a gas (or liquid), the reactants can get together only at the surface of the solid. The larger the surface area of the solid, the greater the number of atoms available to react. Unmilled wheat doesn't burn very well, but when ground to fine flour and dispersed in air, it forms an explosive mixture. When this mixture explodes, it produces so much force that it can destroy huge concrete grain elevators.

Effect of temperature

Chemical reactions involve the rearrangement of atoms into new configurations. Chemical bonds are broken, and new bonds are formed. This process necessarily involves the stretching of bonds and the distortion of bond angles—which is exactly what happens when you put more energy into a molecular system. There is an increase in the amplitude of the vibrations of atom–atom stretching and bending, putting the atoms into configurations where they can more easily react. In addition, the kinetic energy of the molecules increases. They move about faster, and thus collide with greater force, producing the bond-length and bond-angle distortions characteristic of reacting molecules. So a molecule at a higher temperature has higher energy and a higher probability of reacting.

Dramatic temperature effects are often observed with reactions that are exothermic. The reaction of paper with the oxygen in air is extremely slow at room temperature, but as the paper is heated it suddenly bursts into flame. The temperature at which this happens is called the *ignition temperature*, a point where the energy produced by the chemical reaction cannot be dissipated rapidly enough to keep the reactant from heating up. This self-heating process results in a dramatic rise in temperature and rate of reaction—a characteristic of combustions and chemical explosions.

Effect of catalysts

A catalyst increases the rate of a chemical reaction but is not permanently altered in the net chemical reaction. Catalysts generally function by interacting with one or more reactants to form an intermediate that is more reactive. In effect, catalysts change the

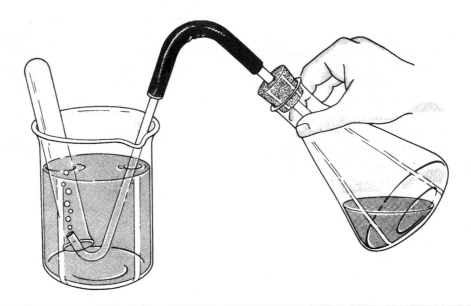

FIGURE 25-1 A simple apparatus for measuring the rate of evolution of a gas. Invert a test tube filled with water over the gas delivery tube in the beaker; then measure the time required to fill the test tube with gas.

reaction pathway so that less energy is required for the reaction to take place. Many enzyme catalyst molecules are very specific, catalyzing the reaction of a single kind of substrate molecule. Other catalysts are more general. For example, finely divided palladium metal catalyzes the rather nonspecific addition of H_2 to C=C bonds found in different kinds of molecules.

Effects of diffusion and mixing

In the gas phase, molecules are moving quite rapidly. Nitrogen molecules at room temperature have an average speed of about 300 m/s (about 650 mph). But nitrogen molecules released in a corner of a room wouldn't get to the other side in a fraction of a second. This is because they collide so often with other molecules that they must jostle their way to the other side, like a person fighting through a crowd at a football game.

If it's difficult to travel rapidly in the gas phase, imagine how hard it is in a liquid, where each molecule is surrounded by a cage of molecules. Just to move a distance roughly equal to the size of a molecule requires that the molecule push and shove its neighbor out of the way. So if two different reactant solutions are not thoroughly mixed, the reaction time can be extended by minutes until the random motions of the slowly diffusing molecules produce a

more uniform distribution of reacting molecules. This is an important consideration to keep in mind when studying chemical reactions in solution (or when making up a reagent solution). Thorough mixing is essential.

Experimental Procedure

Special Supplies: Mortar and pestle; apparatus for collecting gases shown in Figure 25-1; 2.5×2.5 cm squares of cotton cloth (muslin).

Chemicals: 3% (0.9 M) H_2O_2; solution A (containing in the same solution 0.2 g/L starch,[1] 0.5 M CH_3COOH, 0.05 M $NaCH_3COO$, 0.30 M KI, and 0.030 M $Na_2S_2O_3$); Mg metal; turnings; Ca metal turnings; 0.1% phenolphthalein indicator; marble chips; 1 M HCl; steel wool; iron nails (approx. 4 cm long); methanol; 50:50 (v:v) 2-propanol–water; 1 M $CuCl_2$; 3 M $CuCl_2$; 3 M $FeCl_3$; $KMnO_4$(s); 0.005 M $KMnO_4$; 0.5 M $H_2C_2O_4$ (oxalic acid); 6 M H_2SO_4.

---

[1]Put 0.2 g starch in 200 mL of boiling water and heat for 5 min. Dilute the solution to about 800 mL, add the other reagents in the order given, allow to cool, and dilute to 1 L.

SAFETY PRECAUTIONS:
Part 4 of this experiment involves igniting small quantities of flammable solvents. If you follow the directions, this is no more hazardous than lighting a Bunsen burner. It is recommended that the flammable solvents be kept in the fume hood and that the requisite amounts be dispensed to you there. *There must be no open flames in the hood where the solvents are stored.* **After obtaining your small amounts of solvent, you should take them to the lab bench where the experiment is to be carried out.** *The containers of flammable solvents must never be taken to the lab bench.* **If a fume hood is not available, the instructor should demonstrate the experiments in parts 4b and 4c.**

1. Effects of Concentration on Reaction Rates

Place 5 mL of solution A in a 50-mL beaker. Add 5 mL of 3% H_2O_2, mix, and note the number of seconds required for the solution to turn blue. Record the reaction time.

Repeat the procedure using 5 mL of solution A, 4 mL of water, and 1 mL of 3% H_2O_2. Record the reaction time. Is it longer? Which reactant concentrations have been changed? What is the relationship between the reaction time and the initial concentration of H_2O_2?

2. The Influence of the Nature of the Reactants

Obtain three or four small pieces of magnesium metal turnings and about the same amount of calcium metal turnings. Add enough water to each of two 15 × 150 mm test tubes to fill them about one-third full. Place the magnesium turnings in one test tube and the calcium turnings in the other. Both of these metals are in the same chemical family in the periodic table. Which metal is more reactive with water?

To determine what gas is produced, invert an empty test tube for about 15–30 s over the mouth of the tube in which the metal is reacting more vigorously. Quickly move the mouth of the inverted tube near the flame of a Bunsen burner. Is the gas flammable? What is the chemical composition of the gas?

Look closely at the slower-reacting metal as you agitate the test tube. Is any gas being produced? Add a drop of phenolphthalein indicator solution to each

test tube containing the metals. What other product is formed besides a gas? Write chemical equations describing the reactions of the metals with water.

Is there a correlation of the reaction rates with the ionization potentials of these metals (is it easier to remove electrons from magnesium or calcium)?

3. The Influence of Surface Area in a Heterogeneous Reaction

(a) Set up the apparatus shown in Figure 25-1, using a 125-mL Erlenmeyer flask, a 13 × 100 mm test tube to collect the gas, and a large beaker or plastic basin. Fill the test tube with tap water and, holding your thumb or finger over the mouth, invert it in the beaker of water over the end of the bent gas delivery tube. Place 3 g of coarse marble chips ($CaCO_3$) in the 125-mL Erlenmeyer flask. Add 25 mL of water; then add 25 mL of 1 M HCl. Immediately insert the one-hole stopper in the flask and record the time required to fill the test tube with gas. What gas is produced?

Rinse the Erlenmeyer flask with tap water and place the used marble chips in the waste receptacle provided. *Do not discard the marble chips in the sink.* Obtain another 3 g of marble chips and, using a mortar and pestle, pulverize them to the size of grains of sugar. Using a rolled-up sheet of paper as a funnel, transfer the granular material to the 125-mL Erlenmeyer flask. Fill a test tube with water and invert it over the delivery tube as before. Add 25 mL of water to the flask, followed by 25 mL of 1 M HCl, and stopper immediately. Record the time required to fill the test tube with gas. Save this apparatus for use in part 4. Is there a correlation between the surface area of the marble and the reaction rate? Write an equation describing the chemical reaction between $CaCO_3$ and HCl.

(b) Place a small iron nail (cleaned so that it is free of rust) in a 20-mm-diameter test tube. In a second 20-mm test tube, place a 0.6-g ball of steel wool, using a glass rod to push it to the bottom of the test tube. Add 10 mL of 1 M $CuCl_2$ to each test tube and watch them closely. Observe and record any color changes and note in which tube they occur more rapidly. Is a new solid substance forming? What do you think it is? Agitate each test tube and feel the lower portion of the tubes. Is one tube warmer than the other? What happened to the steel wool? Would the same thing eventually happen to the nail? Write an equation describing the chemical reaction that takes place.

4. The Influence of Temperature

(a) Measure 5 mL of solution A (used in part 1) into a test tube. Measure 4 mL of water and 1 mL of 3% H_2O_2 into a second test tube. Place both test tubes in a 250-mL beaker of warm water (50 °C). After about 3 min, pour solution A and then the H_2O_2 solution into a 50-mL beaker, stir, and note the time. When the blue color appears, note and record the elapsed time. Compare the reaction time with that observed for the second reaction mixture you studied in part 1.

(b) *Ignition Temperature.* Review the Safety Precautions. In a fume hood, place 10 to 20 drops of methanol in a 30-mL beaker. Remove the beaker to your lab bench. Heat the end of a glass stirring rod at the tip of the inner blue cone of a Bunsen burner until the glass starts to soften. Quickly touch the heated end to the methanol. A chemical reaction should begin; if it does not, reheat the rod and try again. *Stay at arm's length from the beaker. Do not touch the beaker until it has cooled.* Describe the reaction that takes place. Write a balanced chemical equation for the reaction. Why doesn't methanol burst into flames at room temperature when exposed to air?

(c) Obtain a small square of cotton cloth (muslin) and a book of matches. (The cotton cloth should be tightly woven, about 2 to 3 cm on a side, and cleanly cut, so there are no ragged edges.) Working in a hood, place about 10 mL of 50:50 (v:v) 2-propanol–water in a 50- or 100-mL beaker. Using crucible tongs, grasp the square by one corner and dip it into the 2-propanol–water solution. Lift the cloth and allow the excess solution to drain until no more solution drips off the cotton. Place a small watch glass over the solution in the beaker. Carry the cotton, still held in the tongs, to your lab bench. Have a lab partner light a match and bring it just near enough to the square of cotton (held in the tongs) to ignite the 2-propanol-soaked cotton. Make certain that no other person is within arm's length of the flaming cloth. When the flame dies out, carefully note the condition of the cloth. How can the 2-propanol burn without igniting the cotton?

Demo #### 5. Effect of Catalysts on the Rate of Decomposition of H_2O_2 Using the same apparatus as in part 3a, place 20 mL of 3% H_2O_2 in the flask. Fill a 13×100 mm test tube with water and invert it over the gas delivery tube. Then add 20 drops of 3 M $CuCl_2$ to the flask, replace the stopper, continuously swirl the contents, and note and record how long it takes to fill

the test tube with the evolved gas. The gas coming over first will contain some air that was present in the flask. After 2–3 min, test the gas in the Erlenmeyer flask by inserting a glowing splint into the flask. Does the gas support combustion? What is the identity of the gas?

Rinse the flask and repeat the experiment, adding only the H_2O_2 solution to the flask. Continuously swirl the contents of the flask. Does the gas evolve more slowly? After a few minutes' wait, you should be able to tell whether the $CuCl_2$ catalyzes the reaction.

Now fill another test tube with water and place it over the gas delivery tube. Remove the stopper just long enough to add 2 drops of 3 M $FeCl_3$. Insert the stopper and continuously swirl the solution. Note and record the time required to fill the test tube with gas.

Rinse the flask. Add to it 20 mL of 3% H_2O_2, 1 drop of 3 M $CuCl_2$, and 1 drop of 3 M $FeCl_3$, in that order. Quickly replace the stopper, continuously swirl the solution in the flask, and record the time required to fill a 13×100 mm test tube with gas.

Repeat the procedure using 20 mL of 3% H_2O_2, 5 drops of 3 M $CuCl_2$, and 5 drops of 3 M $FeCl_3$. Quickly stopper the flask and continuously swirl the contents of the flask. Note and record the time required to fill a 13×100 mm test tube with gas.

6. The Importance of Mixing Reactants

(a) Half-fill two 13×100 mm test tubes with water. Drop a small crystal of solid $KMnO_4$ into each tube. Put one tube aside in a test tube rack or beaker and let it remain undisturbed. Swirl the second tube to dissolve the crystal of $KMnO_4$. Observe the crystal in the undisturbed tube every few minutes, noting how long it takes for the crystal to dissolve and then to spread (diffuse). Record your observations.

(b) Add 10 mL of 0.50 M $H_2C_2O_4$ (oxalic acid) to 10 mL of 6 M H_2SO_4 and mix thoroughly. Place 8 mL of this solution into each of two 15×125 mm test tubes. Measure out two separate 8-mL portions of 0.005 M $KMnO_4$. Add one portion of $KMnO_4$ solution to one of the test tubes containing oxalic acid, holding the test tube at a 45° angle and carefully pouring the $KMnO_4$ down the side of the test tube to avoid mixing the two solutions. Note the time. Pour another 8 mL of $KMnO_4$ into the second test tube, stopper the tube, mix by inverting the tube several times, and note the time. As the reactions proceed, MnO_4^- will oxidize the oxalic acid to CO_2 and will itself be reduced to colorless Mn^{2+} ion. Record the times required for the solutions in the two test tubes to become completely colorless.

Sit and observe a campfire. How are the factors discussed in this experiment evident in the burning of a wood fire? (You should be able to find examples of each factor, with the possible exception of catalysis.)

The process of cooking food involves chemical reactions (charring and denaturing of proteins are typical examples); physical processes (water vaporization); and the energy transfer issues of radiation, conduction, and convection. In a typical kitchen, find examples that illustrate the factors listed in this experiment.

The growth of a plant is a series of chemical reactions. Are these reactions temperature dependent? Design an experiment to test this factor. Why does the growth rate decrease above certain temperatures?

Is the rate of plant growth dependent on concentration? Design experiments to test the fertilizer-concentration dependence. Why does an excess of fertilizer cause a decrease in the growth rate? (You may need to consult Experiment 22 for hints.)

Can you measure the CO_2 dependence of the growth rate for plants? This factor is a significant component of the discussion about global warming. Why?

Design an experiment to measure the rate of the chemical reaction known as rusting,

$$4\ Fe(s) + 3\ O_2(g) + n\ H_2O \rightarrow 2\ Fe_2O_3 \bullet n\ H_2O(s)$$

Can you determine the temperature dependence? The oxygen concentration dependence?

Name _____   Date _____

2. A stoichiometric mixture of two volumes of H_2 and one volume of O_2 in a flask will not react if undisturbed, but any spark or small flame will cause a violent explosion, shattering the flask. How do you interpret this rate of behavior?

CONSIDER THIS

Sit and observe a campfire. How are the factors discussed in this experiment evident in the burning of a wood fire? (You should be able to find examples of each factor, with the possible exception of catalysis.)

The process of cooking food involves chemical reactions (charring and denaturing of proteins are typical examples); physical processes (water vaporization); and the energy transfer issues of radiation, conduction, and convection. In a typical kitchen, find examples that illustrate the factors listed in this experiment.

The growth of a plant is a series of chemical reactions. Are these reactions temperature dependent? Design an experiment to test this factor. Why does the growth rate decrease above certain temperatures?

Is the rate of plant growth dependent on concentration? Design experiments to test the fertilizer-concentration dependence. Why does an excess of fertilizer cause a decrease in the growth rate? (You may need to consult Experiment 22 for hints.)

Can you measure the CO_2 dependence of the growth rate for plants? This factor is a significant component of the discussion about global warming. Why?

Design an experiment to measure the rate of the chemical reaction known as rusting:

$$4\ Fe(s) + 3\ O_2(g) + n\ H_2O \rightarrow 2\ Fe_2O_3 \cdot n\ H_2O(s)$$

Can you determine the temperature dependence? The oxygen concentration dependence?

The Rate of a Chemical Reaction

Chemical Kinetics

Purpose

- Determine the rate law for a chemical reaction, specifically

$$H_2O_2 + 2\,I^- + 2\,H_3O^+ \rightarrow 4\,H_2O + I_2$$

- Observe how changing the concentrations of H_2O_2, I^-, and H_3O^+ affects the rate of reaction.

- Observe the effect of a temperature increase on the reaction rate.

- Observe the effect on the reaction rate of adding a catalyst, in this case molybdate ion.

Pre-Lab Preparation

Two important questions may be asked about a chemical reaction: (1) *How far* or how completely do the reactants interact to yield products, and (2) *how fast* is the reaction? "How far" is a question of chemical equilibrium. The study of chemical systems at equilibrium is the realm of chemical energetics, more commonly called *chemical thermodynamics*. The second question—"How fast"—is the realm of *chemical kinetics*, the subject of this experiment. For elementary reversible reactions, there is a direct relation between the equilibrium constant and the rate constants for the forward and reverse reactions. Furthermore, a study of the factors affecting the rate often reveals important information about the reaction pathway or the mechanism of the chemical reaction.

In this experiment, we will see how changing the concentrations of each of the reactants affects the rate and yields the rate law for the reaction. We will measure the effect of a catalyst and of temperature on the rate of the reaction. This should give you a better understanding of the particular reaction being studied and increase your understanding of the more complete discussion of these problems in your text. Let us first discuss some of the questions to be considered.

1. *How does the concentration of each of the reacting substances affect the rate?* For a homogeneous[1] chemical reaction, it is usually possible to express the rate as an algebraic function of the concentrations of the reactants. For many chemical reactions, this algebraic function is simply the product of the concentrations of the reactants, each concentration being raised to a power called the *order* of the reaction with respect to that reactant. As an example, suppose we have a reaction whose overall stoichiometry is

$$2\,A + B \rightarrow P \qquad (1)$$

and whose *rate law* can be expressed by an algebraic function of the type

$$\text{rate} = k[A]^x[B]^y \qquad (2)$$

The proportionality constant k, called the *specific rate constant*, is the rate when the concentrations of A and B are both 1.0 mol/L. Brackets around the symbols A and B mean "molar concentrations of" A and B.

Right at the beginning, we emphasize that the values of the exponents x and y that appear in the rate expression cannot be predicted from the overall stoichiometry and must be determined by experiment. In other words, *rate expressions based on the stoichiometric equation are seldom correct*. This is because the reaction pathway (or reaction mechanism) seldom involves the simultaneous collision of all reactants to yield products in a one-step reaction, which we will

---

[1]A *homogeneous* reaction is one occurring uniformly throughout *one phase* (e.g., a solution); a *heterogeneous* reaction is one occurring at the interface of *two phases* (e.g., a solid and a liquid, or a gas and a solid). The latter reaction rate depends on the extent of subdivision and surface area of the phases and on adsorption and other phenomena.

TABLE 26-1 Rate Expressions for Different Reaction Paths for the Overall Reaction $2A + B \rightarrow$ Products

Reaction Pathway[a]	Rate Law or Rate Expression	Reaction Order with Respect to Reactant A	Reaction Order with Respect to Reactant B	Total Reaction Order $(x + y)$
Pathway 1: $2A + B \rightarrow P$	rate $= k[A]^2[B]$	2	1	3
Pathway 2: $2A \rightarrow C$ (*slow*) $B + C \rightarrow P$ (*fast*)	rate $= k[A]^2$	2	0	2
Pathway 3: $A + B \rightarrow D + E$ (*slow*) $A + D \rightarrow E$ (*fast*) $2E \rightarrow P$ (*fast*)	rate $= k[A][B]$	1	1	2

[a]The symbols A and B represent reactants, and P represents the products of the stoichiometric reaction. The symbols C, D, and E represent chemical *intermediates* that are formed in the elementary reactions. Intermediates have only a fleeting existence and do not appear as final products of the reaction.

call an *elementary reaction*.[2] The stoichiometric equation more often represents the sum of several elementary reactions, at least one of which may be distinctly slower than the others and is therefore the rate-controlling reaction. The rate expression thus corresponds to the rate for the slow step in the total reaction. Also keep in mind that in writing a rate expression like that of Equation (2), we have assumed that the rate of the backward reaction is negligible.

To illustrate how the rate law depends on the reaction pathway, Table 26-1 presents three different hypothetical reaction mechanisms, each having the overall stoichiometry $2A + B \rightarrow P$, where A and B represent the reactant molecules and P represents the products. The rate expression for each reaction pathway is shown, together with the reaction orders with respect to A and B. The terminology here uses the designations x and y for the order of the reaction with respect to the reactant concentrations [A] and [B]. The *total reaction order* is the sum of the individual reaction orders, $x + y$. Note that for each reaction pathway, the sum of all the elementary reactions gives the stoichiometric equation, but that the reaction orders *do not* necessarily correspond to the stoi-

chiometric coefficients. For the first reaction pathway, where the elementary reaction is identical to the stoichiometric reaction, doubling the concentrations of A and B would increase the rate by a factor of 8 ($2^2 \times 2^1 = 8$); for the second and third pathways, the rate would be increased by a factor of 4.

Determining the rate law is often one of the first steps in a kinetic investigation. This task can often be simplified by using a small concentration of one reactant in the presence of 20- to 100-fold larger concentrations of all the other reactants. Under these conditions, the reactants that constitute the larger concentration remain essentially constant, so that the total order of the reaction is reduced to the order of the reactant that constitutes the small concentration. For example, the first reaction mechanism shown in Table 26-1 has the rate law given by rate $= k[A]^2[B]$. If the concentration of A is made 50 times larger than the concentration of B, the rate expression is given by rate $= k'[B]$, where $k' = k[A]^2$, the concentration of A being treated as a constant because it changes only 4% during the course of the reaction. The total third-order reaction can now be treated as a *pseudo first-order* reaction with respect to the reactants. This is sometimes called the "flooding" technique. A related technique, called the *initial rate* method, is to measure the rate of a reaction over a period of time short enough so that less than 5 to 10% of the reactants is consumed. The reactant concentrations can be treated approximately as constants, and the reaction order with respect to each reactant can be determined by varying the concentration of only one constituent at a time and measuring the corresponding rate change.

---

[2]The *reaction pathway* (or *reaction mechanism*) of a chemical reaction is composed of a set of molecular reactions, which we call *elementary reactions* or *steps*, and which are written just like overall chemical reactions. You must be careful to distinguish between an elementary reaction and the stoichiometric chemical equation. Only for a single-step reaction, in which the elementary reaction is identical to the overall stoichiometric reaction, will the exponents appearing in the rate expression be the same as the coefficients appearing in the stoichiometric equation.

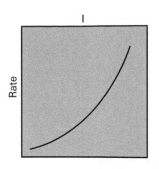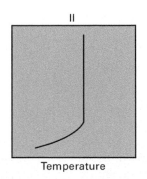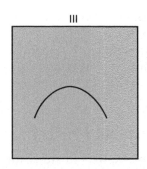

FIGURE 26-1 Effect of temperature on reaction rate. Class I: exponential (Arrhenius) behavior typical of most simple inorganic and organic reactions. Class II: typical of explosion reactions. Class III: typical of enzyme-catalyzed reactions. [After A. A. Frost and R. G. Pearson, *Kinetics and Mechanism*, 2nd ed., Wiley, New York, Chapter 2, Figure 7, Copyright © 1961.]

2. *How does the temperature at which a reaction occurs affect the rate?* Figure 26-1 shows a plot of reaction rate versus temperature for three different classes of reactions. Reactions in Class I (most ordinary chemical reactions in the gas phase and in solution) show a regular exponential increase over a temperature range of 50 to 100 °C. Class II behavior is typical of a combustion or explosion reaction. Class III behavior is characteristic of many enzyme-catalyzed reactions: At lower temperatures, the reaction rate increases with temperature, but at higher temperatures it decreases as the configuration of the enzyme catalyst is altered and enzyme denaturation occurs. The reaction we will study in this experiment belongs to Class I, often called *Arrhenius-type behavior* after the chemist who proposed an explanation to account for it.

When we compare measured rates of chemical reactions in the gas phase with the collision rates calculated from the kinetic gas theory, it becomes clear that only a small proportion of the many molecular collisions result in reaction.[3] In most collisions, the molecules simply rebound like billiard balls, without change. Only molecules with the proper geometric orientation and sufficient relative momentum collide with sufficient energy to "break and make" chemical bonds, momentarily forming an "activated complex" and eventually forming stable products. The proportion of molecules that have a given energy can be calculated from the Maxwell–Boltzmann distribution law; this proportion increases

at higher temperatures, as shown in Figure 26-2. We will call the minimum net energy that the colliding molecules must possess in order to react the *activation energy, E*. The situation is somewhat analogous to an attempt to roll a ball up from a mountain valley, over a pass, and down to another valley. The energy necessary to get the ball up to the top of the pass is the activation energy.

3. *How does a catalyst affect a reaction?* A catalyst is a substance that produces a new reaction pathway with a lower activation energy than the original uncatalyzed reaction pathway. A lower activation energy for the catalyzed reaction means that at a given temperature, a larger fraction of the molecules will possess enough energy to react; thus, the catalyst increases the rates of *both* the forward and backward reactions without affecting the equilibrium

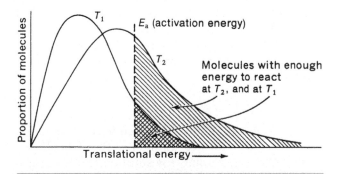

FIGURE 26-2 The Maxwell–Boltzmann distribution law. At a higher temperature (curve T_2), the total number of molecules (shaded area under this curve) with energies greater than E_a, the critical energy necessary for reaction, is much larger than at the lower temperature (curve T_1).

---

[3]We assume in this experiment that the arguments based on kinetic gas theory also apply qualitatively to chemical reactions in solution.

constant for the reaction. Although the catalyst is not consumed in the overall reaction, it may participate in the formation of intermediates and then be reformed in subsequent steps to produce the final products.

The Experimental Method

In this experiment, we will study the rate of oxidation of iodide ion by hydrogen peroxide, which proceeds according to the net stoichiometric reaction

$$H_2O_2 + 2 I^- + 2 H_3O^+ \rightarrow I_2 + 4 H_2O \qquad (3)$$

By varying the concentrations of each of the three reactants (H_2O_2, I^-, and H_3O^+), we will determine the order of the reaction with respect to each reactant and the rate law of the reaction, which is of the form

$$\text{rate} = \frac{-\Delta[H_2O_2]}{\Delta t} = k[H_2O_2]^x[I^-]^y[H_3O^+]^z \qquad (4)$$

In this formula, x, y, and z must be experimentally determined and are most often integers (0, 1, 2, etc.) or half-integers. We will use an acetic acid–acetate ion buffer to control the H_3O^+ concentration at a small constant value, around 10^{-5} M. By using a solution containing known concentrations of acetic acid and acetate ion, we can readily calculate the H_3O^+ concentration from the equilibrium constant expression for the dissociation of acetic acid, given by

$$K_a = \frac{[H_3O^+][C_2H_3O_2^-]}{[HC_2H_3O_2]} \qquad (5)$$

We will also examine the effect of temperature on the reaction rate and the effect of a catalyst [3.3×10^{-4} M Mo(VI)].

To measure the reaction rate, we will employ a clever variation of the initial rate method. A small amount of sodium thiosulfate and starch indicator is added to the reaction mixture. Thiosulfate ion does not react at an appreciable rate with any of the reactants, but it does react rapidly with iodine, according to the equation

$$2 S_2O_3^{2-} + I_2 \rightarrow 2 I^- + S_4O_6^{2-} \qquad (6)$$

As a result, iodine is reduced back to iodide ion as fast as it is formed until all of the thiosulfate is used up. At this point, the solution suddenly turns blue because the iodine concentration rapidly increases to the point where the I_2 forms an intense blue complex with the starch indicator. The time required to consume a fixed amount of thiosulfate is very reproducible, and since we measure this time, this type of reaction is often called a *clock reaction*.

While thiosulfate ion is present in solution, both Reactions (3) and (6) are simultaneously taking place; by adding them together, we see that the net result in solution is given by

$$H_2O_2 + 2 S_2O_3^{2-} + 2 H_3O^+ \rightarrow 4 H_2O + S_4O_6^{2-} \qquad (7)$$

You should be careful to note that Equation (7) expresses only the net result of the two reactions. Hydrogen peroxide is not reacting directly with $S_2O_3^{2-}$, but the result is the same as if it did, and two moles of thiosulfate are consumed per mole of hydrogen peroxide reacted. Knowing the amount of thiosulfate we put in the reaction mixture, we can calculate the reaction rate from equations that can be approximated by

$$\text{rate} \cong \frac{-\Delta[H_2O_2]}{\Delta t} = -\frac{1}{2}\frac{\Delta[S_2O_3^{2-}]}{\Delta t} \qquad (8)$$

and

$$\text{rate} \cong \frac{\frac{1}{2}[VM_{\text{thiosulfate}}/V_{\text{tot}}]}{\Delta t} \qquad (9)$$

where $VM_{\text{thiosulfate}}$ is the number of millimoles of thiosulfate added ($V_{\text{mL}} \times M_{\text{mol/L}}$), V_{tot} is the total volume (in milliliters) of the reaction mixture, and Δt is the time required for all of the thiosulfate to react.

To make the expression for the rate of reaction a positive quantity, we insert the minus sign, because $\Delta[H_2O_2]/\Delta t$ is a negative quantity due to the fact that the hydrogen peroxide concentration is decreasing with time. The factor of $1/2$ is inserted because hydrogen peroxide disappears half as fast as does thiosulfate, as shown by the stoichiometry of Equation (7).

Equation (8) makes the assumption that only a small amount of hydrogen peroxide is consumed (say, less than 3% of the total). Our rate measurement will be a very close approximation to the exact expression and will accurately measure the initial rate of reaction of the hydrogen peroxide.

Now we are in a position to understand how the thiosulfate plays an important role in the rate measurement. First, it allows us to make a simple measurement of the reaction rate by measuring the time required for a given amount of thiosulfate to be consumed under conditions where only a small fraction of the hydrogen peroxide is consumed. It also reacts rapidly with the reaction product I_2 as it forms, regenerating I^-. This keeps the I^- concentration constant and suppresses any backward reaction [the reverse of Equation (3)], so that we can be confident that we are measuring only the rate of the forward reaction.

Experimental Procedure

Special Supplies: Buret; 5-mL graduated pipets; 10-mL graduated pipets; thermometer; timer (a digital watch or a clock with a sweep second hand is suitable). The optional standardization procedure requires the foregoing and a 1-mL pipet.

> **NOTE** All solutions for this experiment should be made up in high-quality distilled or deionized water. Traces of metal ions, particularly copper or iron salts, catalyze the decomposition of H_2O_2.

Chemicals: 0.8 to 0.9 M H_2O_2 containing 0.001 M H_2SO_4 (the small amount of acid stabilizes the H_2O_2 solution and has a negligible effect on the rate measurements; a fresh 3% H_2O_2 solution sold in drugstores is about 0.9 M and can be used for the H_2O_2 solution, which must be standardized, preferably on the day the solution is to be used); 0.050 M sodium thiosulfate ($Na_2S_2O_3$); 0.050 M KI; 0.050 M acetic acid–sodium acetate buffer (0.05 M in $HC_2H_3O_2$ and 0.05 M in $NaC_2H_3O_2$); 0.30 M acetic acid ($HC_2H_3O_2$); 0.1% starch solution (freshly made); 0.010 M Mo(VI) catalyst [containing 1.76 g/L of ammonium hepta-molybdate $(NH_4)_6Mo_7O_{24} \cdot 4H_2O$]. The H_2O_2 standardization procedure requires, in addition, KI(s), 2 M H_2SO_4, and 3% ammonium heptamolybdate catalyst solution in a dropping bottle. If a 1.00-mL microburet is used for the standardization titration, a 2.50 M $Na_2S_2O_3$ solution will also be required.

▌ NOTES TO INSTRUCTOR ▐

If students are expected to complete the experiment and write-up in a three-hour laboratory period, they should be provided with standardized hydrogen peroxide solution and omit part A of the experiment. Alternatively, you may wish to have students do part A and replicate each reaction mixture measurement, allotting a second laboratory period for the write-up and a discussion of the results.

If students make only a single reaction time measurement for each reaction mixture, a "wild" result will spoil the calculations. To avoid this, we suggest that you gather the reaction time data (or have the students write it on the board) so that a wild result can be spotted and the measurement repeated if necessary.

Two students may work together advantageously in this experiment, one student mixing the solutions and the other noting the time to the exact second. *However, each student should record the data completely* before leaving the laboratory.

A. Standardization of the H_2O_2 solution

The instructor will tell you if you are to perform the standardization. For best results, the H_2O_2 solution should be standardized the same day it is to be used. The concentration should be in the 0.8 to 0.9 M range in order to yield conveniently measurable reaction times.

To perform the standardization, first fill a 50-mL buret with 0.050 M $Na_2S_2O_3$. Then pipet exactly 1.00 mL of the approximately 0.8 M H_2O_2 solution into a 125-mL Erlenmeyer flask containing 25 mL of water.

Add 10 mL of 2 M H_2SO_4, 1 g of solid KI, and 3 drops of 3% ammonium molybdate catalyst.

Swirl until the KI dissolves, then immediately titrate the brown iodine solution that forms with 0.050 M sodium thiosulfate until the brown color begins to fade to yellow.

At this point, add 2 mL of 0.1% starch indicator and titrate to the disappearance of the blue color of the starch–iodine complex. The end point is very sharp and should require 32 to 36 mL of 0.050 M $Na_2S_2O_3$ for the conditions specified.

If a 1.00-mL microburet is used, the concentration of sodium thiosulfate titrant should be increased to 2.50 M, with the titration requiring 0.64–0.76 mL for the conditions specified.

Calculate the exact concentration of the hydrogen peroxide stock solution, which should be in the 0.8 to 0.9 M range.

B. Reaction rate measurements

Six reaction mixtures will provide the information necessary to determine the effects of the concentrations of H_2O_2, I^-, and H_3O^+ on the rate of the reaction, as well as the effect of temperature and a catalyst. Table 26-2 specifies the temperature and reagent volumes to be used for each reaction mixture.

Make up each reaction mixture in a 250-mL beaker, adding the reactants for each reaction mixture in the order in which they appear in Table 26-2 (reading from left to right). Read and record the temperature to the nearest 0.2 °C. Graduated cylinders (100, 50, and 10 mL) can be used to measure the volumes of water, buffer, KI, starch, acetic acid, sodium acetate, and catalyst. Special care should be taken to measure the volumes as accurately as possible. If you performed the standardization, the 0.050 M $Na_2S_2O_3$ solution can be conveniently dispensed from a buret; otherwise, use a 5-mL graduated pipet. The H_2O_2 is conveniently dispensed from a 10-mL graduated pipet (or 10-mL graduated cylinder).

TABLE 26-2 Composition of the Reaction Mixtures

Reaction Mixture	Temperature (°C)	Water (mL)	0.05 M $HC_2H_3O_2$– $NaC_2H_3O_2$ buffer (mL)	0.3 M $HC_2H_3O_2$ (mL)	0.05 M KI (mL)	0.1% Starch (mL)	0.05 M $Na_2S_2O_3$ (mL)	0.01 M Mo(VI) Catalyst (mL)	0.8 M H_2O_2 (mL)
1	R.T.[a]	75	30	0	25	5	5	0	10
2	R.T.	80	30	0	25	5	5	0	5
3	R.T.	50	30	0	50	5	5	0	10
4	R.T.	30	30	45	25	5	5	0	10
5	R.T. + 10	75	30	0	25	5	5	0	10
6	R.T.	70	30	0	25	5	5	5	10

[a]R.T. stands for room temperature.

To avoid contamination of the reagents, it is best to dispense each solution from a separate pipet or graduated cylinder. Put labels on them so you don't confuse them. You must be especially careful not to get any contaminant in your hydrogen peroxide stock solution. Iodide ion catalyzes the decomposition of H_2O_2 into oxygen and water, and saliva contains substances that catalyze decomposition of H_2O_2, so do not pipet any solutions by mouth. If you use pipets, draw up the solutions with a rubber bulb. *Do not return any reagent solutions to the stock bottles.*

The first step in each reaction measurement is to place in a clean 250-mL beaker the specified volumes of each reagent listed to the left of the vertical line in Table 26-2. (The beakers *must* be clean. If they are not, wash them with detergent and rinse them thoroughly with tap water; then give them a final rinse with deionized water and shake off the excess. Do not attempt to dry them with a paper towel; it may leave contamination.) The beaker now contains all the reactants except H_2O_2. The timer or stopwatch is started when the hydrogen peroxide is added (with vigorous stirring) and stopped when the solution turns blue. The final volume of each reaction mixture is 150 mL. Knowing this fact and the amount of each reactant added, you can easily compute the initial concentration of each reactant.

You will run a duplicate measurement on the first reaction mixture in order to check your technique. If time permits (and your instructor so directs), make duplicate measurements for each reaction mixture.

Inaccuracies most often arise when mistakes are made in measuring out the reactants (or forgetting to add a reactant!) or when the solution is inadequately stirred while the hydrogen peroxide is being added. You must also be careful not to get any contaminant in your hydrogen peroxide stock solution, since many substances catalyze its decomposition into water and oxygen.

Mixture 1 Place all reactants except the H_2O_2 in a 250-mL beaker. Note and record the temperature of the solution. Arrange your watch or have your timer ready.

Then, using a pipet, add exactly 10 mL of H_2O_2 solution and stir continuously. Start the timer when half of the H_2O_2 solution has drained from the pipet; continue stirring until you are certain the solution is thoroughly mixed.

Watch the solution carefully for the sudden appearance of the blue color and stop the timer when it appears. (The total elapsed time should be on the order of 2 to 4 min.)

Discard the solution and rinse the beaker.

Repeat the measurement using a fresh solution, if your instructor so directs. The times should agree within a few seconds.

Mixture 2 Prepare solution 2 and note and record the temperature, which should be within 0.5 °C of the temperature recorded for mixture 1. (The temperature may be adjusted if necessary by cooling or heating in a water bath contained in a larger beaker.)

When the timer is ready, add 5 mL of H_2O_2 solution. Since the concentration of H_2O_2 is less than in mixture 1, you may expect that the reaction time will be at least as great as that for mixture 1.

Mixture 3 Prepare solution 3. Note and record the temperature, which should be the same as for mixtures 1 and 2 (adjust if necessary). Note that the concentration of KI has been doubled over that contained in mixture 1, so the reaction time may be shorter.

Carry out the measurement as described for mixture 1.

Mixture 4 Carry out the measurement as described for mixture 1. This solution contains ten times as much acetic acid, therefore the H_3O^+ concentration is ten times larger. If the reaction time is the same as for mixture 1, what does this suggest about the reaction order with respect to H_3O^+?

Mixture 5 Prepare solution 5, which has the same composition as solution 1, and place it in a beaker of hot water.

Stir until the temperature is about 10 to 12 °C above that of mixture 1; then add 10 mL of H_2O_2 solution, start the timer, and note and record the temperature.

When the solution turns blue, stop the timer, and again note and record the temperature. (The average of the two readings is taken as the temperature of the reaction mixture.)

Mixture 6 This solution has the same composition as mixture 1, except that it contains 5.0 mL of 0.010 M Mo(VI) catalyst. Mixture 6 is adjusted to the same temperature as mixture 1, 10 mL of H_2O_2 is added, and the reaction time is recorded.

C. Interpretation of the data

1. Calculating the Reaction Orders When we want to measure the order of a reaction with respect to a reactant, we vary the initial concentration of that reactant and keep constant the temperature and the initial concentrations of all other reactants.

For example, note that in reaction mixtures 1 and 2, only the concentration of H_2O_2 has been changed (see Table 26-2). To get the reaction order with respect to H_2O_2, we first note that the reaction rate is inversely proportional to the reaction time, Δt, as shown by Equation (9). Because we put the same amount of thiosulfate ion in each reaction mixture, the concentration terms cancel out when we take the ratio. Therefore, the ratio of the reaction rates for any two reaction mixtures will be equal to the inverse of the ratio of the reaction times.

$$\frac{\text{rate}_1}{\text{rate}_2} = \frac{\Delta t_2}{\Delta t_1} \qquad (10)$$

Starting with Equation (4), we can write expressions for the rates of reaction in reaction mixture 1:

$$\text{rate}_1 = k[H_2O_2]_1^x [I^-]^y [H_3O^+]^z \qquad (11)$$

and reaction mixture 2

$$\text{rate}_2 = k[H_2O_2]_2^x [I^-]^y [H_3O^+]^z \qquad (12)$$

Dividing Equation (11) by Equation (12), we obtain an expression for the ratio $\text{rate}_1/\text{rate}_2$.

$$\frac{\text{rate}_1}{\text{rate}_2} = \frac{\Delta t_2}{\Delta t_1} = \left(\frac{[H_2O_2]_1}{[H_2O_2]_2}\right)^x \qquad (13)$$

Note that when we write the equation for the ratio of the rates, all of the concentrations that have been held constant in the two reaction mixtures cancel out, leaving only the ratio of the concentrations of H_2O_2 (raised to the power of the reaction order, x). In addition, if the temperature is constant, the specific rate constant, k, that appears in Equations (11) and (12) will also cancel out.

Because we know the reaction times and the concentrations of H_2O_2 in reaction mixtures 1 and 2, we can calculate the unknown reaction order, x, with respect to H_2O_2. The nearest-integer value of the reaction order can usually be determined by inspection, but if the logarithms of both sides of Equation (10) are taken, we can calculate x from the resulting equation.

$$x = \frac{\log(\Delta t_2/\Delta t_1)}{\log([H_2O_2]_1/[H_2O_2]_2)} \qquad (14)$$

The same method is used to obtain the reaction order with respect to I^- (using the data for reaction mixtures 3 and 1) and the reaction order with respect to H_3O^+ (using the data for reaction mixtures 4 and 1).

2. Effect of Temperature In reaction mixtures 5 and 1, the concentrations of all the reactants are the same, but the temperature of mixture 5 is greater than that of mixture 1. As before, the ratio of the rates is given by the inverse ratio of the reaction times. Because all the concentration terms are the same, they cancel out; but the specific rate constants do not cancel because the specific rate constant changes with temperature. Therefore, we have

$$\frac{\text{rate}_5}{\text{rate}_1} = \frac{\Delta t_1}{\Delta t_5} = \frac{k_{T_5}}{k_{T_1}} \qquad (15)$$

where k_{T_5} and k_{T_1} are the specific rate constants at the temperatures T_5 and T_1 of reaction mixtures 5 and 1.

3. Effect of a Catalyst Using the measured reaction times, we can calculate the ratio of the reaction rates from the data for reaction mixtures 6 and 1.

$$\frac{\text{rate}_6(\text{catalyzed})}{\text{rate}_1(\text{uncatalyzed})} = \frac{\Delta t_1}{\Delta t_6} \qquad (16)$$

How many times faster is the catalyzed reaction rate? What is the molar concentration of the Mo(VI) catalyst in the reaction mixture?

CONSIDER THIS

It takes longer to soft boil an egg in Denver, Colorado (elevation approximately 1 mile), than at sea level. The average barometric pressure at the altitude of Denver is about 0.83 atm (630 torr). Assuming that the denaturation of the protein as the egg white cooks is a chemical reaction (or a set of chemical reactions), explain why it would be expected to take longer to soft boil an egg as the altitude increases. Experiment 21 and Table 3 of the Appendix provide some hints.

In this experiment, you studied the following net chemical reaction that we will call Reaction (1):

$$H_2O_2 + 2 H_3O^+ + 2 I^- \rightarrow I_2 + 4 H_2O \qquad (1)$$

with the following proposed three-step reaction mechanism:

Step 1: (slow) $H_2O_2 + I^- \rightarrow OH^- + HOI$

Step 2: (very fast) $OH^- + H_3O^+ \rightarrow 2 H_2O$

Step 3: (fast) $HOI + H_3O^+ + I^- \rightarrow I_2 + 2 H_2O$

In neutral solution containing only hydrogen peroxide and iodide ion, a different net chemical reaction occurs:

$$2 H_2O_2 \rightarrow 2 H_2O + O_2(g) \qquad (2)$$

The following three-step reaction mechanism has been proposed to explain Reaction (2):

Step 1: (slow) $H_2O_2 + I^- \rightarrow OH^- + HOI$

Step 2a: (very fast) $OH^- + HOI \rightarrow IO^- + H_2O$

Step 3a: (fast) $H_2O_2 + IO^- \rightarrow I^- + H_2O + O_2(g)$

Show that when you sum the three reaction steps, the net result is Reaction (2).

What would be the expected rate law for the reaction mechanism, assuming that the rate of Reaction (2) is limited by the slow step 1?

Is the expected rate law for Reaction (2) the same as or different from the rate law you determined for Reaction (1)?

To explain the different set of products produced in net Reactions (1) and (2), we must suppose that the concentration of H_3O^+ plays an important role in determining the outcome of the competition between step 3 of Reaction (1) and step 3a of Reaction (2). Which of the step 3 reactions is favored (made faster) by a higher H_3O^+ concentration—step 3 of Reaction (1) or step 3a of Reaction (2)?

In a solution that is slightly basic, say pH 8, which net reaction would you expect to be favored—Reaction (1) or (2)? Describe an experiment to test your prediction.

Bibliography

Bamford, C. H.; Tipper, C. F. H. (eds.) *Comprehensive Chemical Kinetics*, Vol. 6, Elsevier Publishing, New York, 1972, p. 406. A summary discussion with references of the reaction of hydrogen peroxide with halide ions.

Hansen, J. C. "The Iodide-Catalyzed Decomposition of Hydrogen Peroxide: A Simple Computer-Interfaced Kinetics Experiment for General Chemistry," *J. Chem. Educ.* **1996,** *73,* 728–732.

King, E. L. *How Chemical Reactions Occur,* W. A. Benjamin, Menlo Park, CA, 1963, pp. 36–38, 80–82.

The Rate of a Chemical Reaction

Chemical Kinetics

Name _____

Date _____ Section _____

Locker _____ Instructor_____

A. Standardization of the H_2O_2 Solution

If your instructor so directs, carry out the standardization procedure in duplicate.

Data	Trial 1	Trial 2
Buret final reading	mL	mL
Initial reading	mL	mL
Net volume of $Na_2S_2O_3$ (0.050 M for a 50-mL buret) (2.50 M for a 1.00-mL buret)	mL	mL

Calculate the H_2O_2 concentration, which should be in the range of 0.8 to 0.9 M. If it is outside this range, consult your instructor.

B. Reaction Rate Measurements

Make a single measurement on each solution, unless your instructor directs otherwise. If a measurement is spoiled for any reason, the procedure should be repeated.

Reaction mixture	Trial 1		Trial 2	
	Time (s)	Temperature (°C)	Time (s)	Temperature (°C)
1				
2				
3				
4				
5				
6				

C. Interpretation of the Data

1. Calculating the reaction orders

(a) Order of the reaction with respect to hydrogen peroxide, H_2O_2

Reaction mixture	Time (s)	H_2O_2 concentration in the reaction mixture (mol/L)
1		
2		

Using Equation (14), calculate the reaction order and round off to the nearest integer.

(b) Order of the reaction with respect to iodide ion, I^-

Reaction mixture	Time (s)	I^- concentration in the reaction mixture (mol/L)
1		
3		

Using Equation (14), calculate the reaction order and round off to the nearest integer.

(c) Order of the reaction with respect to H_3O^+

Taking K_a for acetic acid as 1.8×10^{-5} M, calculate the H_3O^+ concentration from Equation (5), using the known concentrations of acetic acid and sodium acetate contained in reaction mixtures 1 and 4. (In mixture 4, be sure that you account for the acetic acid contained in both the buffer and the added 0.3 M acetic acid.)

Reaction mixture	Time (s)	Acetic acid concentration (mol/L)	Sodium acetate concentration (mol/L)	Calculated H_3O^+ concentration (mol/L)
1				
4				

Show your calculations of the H_3O^+ concentration.

Name _____   Date _____

Using Equation (14), calculate the reaction order and round off to the nearest integer.

2. Effect of temperature

Reaction mixture	Time (s)	Temperature (°C)
1		
5		

Using Equation (15), calculate the reaction rate ratio for reaction mixtures 5 and 1. When the temperature is increased, does the reaction rate decrease or increase?

$$\frac{\text{rate}_5}{\text{rate}_1} =$$

3. Effect of a catalyst

Reaction mixture	Time (s)	Mo(VI) concentration in the reaction mixture (mol/L)
1		
6		

Using Equation (16), calculate the ratio of the rates of the catalyzed and uncatalyzed reactions.

$$\frac{\text{rate (catalyzed)}}{\text{rate (uncatalyzed)}} =$$

Questions

1. Calculating the reaction orders

(a) From your measurements, what are the reaction orders with respect to each of the three reactants H_2O_2, I^-, and H_3O^+?

H_2O_2 _____ I^- _____ H_3O^+ _____

(b) What is the rate expression (rate law) for the uncatalyzed reaction?

(c) What is the total reaction order?

(d) Is the rate law consistent with the idea that the mechanism of the reaction is $H_2O_2 + 2\,I^- + 2\,H_3O^+ \rightarrow I_2 + 4\,H_2O$ (all in one step)? Explain why or why not.

(e) The following reaction pathway has been proposed for the pH range 3.5 to 7:

$$\text{Step 1: (slow)} \quad H_2O_2 + I^- \rightarrow OH^- + HOI$$

$$\text{Step 2: (fast)} \quad H_3O^+ + OH^- \rightarrow 2\,H_2O$$

$$\text{Step 3: (fast)} \quad HOI + H_3O^+ + I^- \rightarrow I_2 + 2\,H_2O$$

Add the equations above. Do they give the overall reaction?

According to your results, which of the steps would be the rate-determining step?

If the proposed mechanism agrees with the experimental rate law, does this prove that the mechanism is correct and that no other pathway is possible? Defend your answer.

Name _____   Date _____

2. Effect of temperature

Do your measurements for part C(2) confirm the oft-quoted rule of thumb that "the rate of a chemical reaction approximately doubles for every 10 °C temperature increase"?

Problems

1. Considering the volumes and the concentrations used in mixture 1, what percentage of the moles of H_2O_2 present have been consumed during the timed reaction? Calculate the number of moles of H_2O_2 present initially and determine how many have reacted with iodide when the thiosulfate runs out. How does the I^- concentration change during this same time? Why?

2. Using Equations (9) and (11), calculate the specific rate constant, k, from your data for reaction mixture 1. [Use the integer reaction orders you determined for x, y, and z in the rate expression of Equation (11).] Compare your value with the reported value of $0.0115 \ L \cdot mol^{-1} \cdot s^{-1}$ at 25 °C for the uncatalyzed reaction.

$$k = \text{_____}$$

3. It has been found that in acid solution (greater than 1 M H_3O^+) the rate expression for the reaction of H_2O_2 with I^- is of the form: rate $= k_2[H_2O_2][I^-][H_3O^+]$, where $k_2 = 0.25 \ L^2 \cdot mol^{-2} \cdot s^{-1}$ at 25 °C. Therefore, the rate law over the whole range of acidity from very acid solutions to a neutral solution is given by a *two-term* rate expression:

$$\text{rate} = k_1[H_2O_2][I^-] + k_2[H_2O_2][I^-][H_3O^+]$$

where k_1 has the value of $0.0115 \ L \cdot mol^{-1} \cdot s^{-1}$ at 25 °C. Compare the relative magnitudes of these two terms under the conditions employed in your rate measurements for mixture 1. (You should find that the second term makes a negligible contribution to the reaction rate at H_3O^+ concentrations smaller than 3×10^{-4}.)

It takes longer to soft boil an egg in Denver, Colorado (elevation approximately 1 mile), than at sea level. The average barometric pressure at the altitude of Denver is about 0.83 atm (630 torr). Assuming that the denaturation of the protein as the egg white cooks is a chemical reaction (or a set of chemical reactions), explain why it would be expected to take longer to soft boil an egg as the altitude increases. Experiment 21 and Table 3 of the Appendix provide some hints.

In this experiment, you studied the following net chemical reaction that we will call Reaction (1):

$$H_2O_2 + 2\,H_3O^+ + 2\,I^- \rightarrow I_2 + 4\,H_2O \tag{1}$$

with the following proposed three-step reaction mechanism:

Step 1: (slow) $H_2O_2 + I^- \rightarrow OH^- + HOI$

Step 2: (very fast) $OH^- + H_3O^+ \rightarrow 2\,H_2O$

Step 3: (fast) $HOI + H_3O^+ + I^- \rightarrow I_2 + 2\,H_2O$

In neutral solution containing only hydrogen peroxide and iodide ion, a different net chemical reaction occurs:

$$2\,H_2O_2 \rightarrow 2\,H_2O + O_2(g) \tag{2}$$

The following three-step reaction mechanism has been proposed to explain Reaction (2):

Step 1: (slow) $H_2O_2 + I^- \rightarrow OH^- + HOI$

Step 2a: (very fast) $OH^- + HOI \rightarrow IO^- + H_2O$

Step 3a: (fast) $H_2O_2 + IO^- \rightarrow I^- + H_2O + O_2(g)$

Show that when you sum the three reaction steps, the net result is Reaction (2):

What would be the expected rate law for the reaction mechanism, assuming that the rate of Reaction (2) is limited by the slow step 1?

Is the expected rate law for Reaction (2) the same as or different from the rate law you determined for Reaction (1)?

To explain the different set of products produced in net Reactions (1) and (2), we must suppose that the concentration of H_3O^+ plays an important role in determining the outcome of the competition between step 3 of Reaction (1) and step 3a of Reaction (2). Which of the step 3 reactions is favored (made faster) by a higher H_3O^+ concentration—step 3 of Reaction (1) or step 3a of Reaction (2)?

In a solution that is slightly basic, say pH 8, which net reaction would you expect to be favored—Reaction (1) or (2)? Describe an experiment to test your prediction.

The pH Scale and Acid–Base Titrations

Standardization of NaOH Solution
The Titration of Vinegar and Stomach Antacids

Purpose

- Understand the meaning of pH.

- Differentiate between the H_3O^+ concentration and the total acidity of an acid.

- Learn the technique of acid–base titration by using a buret to measure volume and an acid–base indicator to determine the end point of the titration.

- Quantitatively determine the neutralization capacity of various commercial antacid tablets.

Pre-Lab Preparation

A quick look in the cabinets of the kitchen, laundry, or bathroom of the average home would turn up a number of substances that are acids or bases. Acidic substances we might find include citrus juices, vinegar, carbonated soft drinks, and toilet bowl cleaners. We might also find such bases as baking soda (sodium bicarbonate), detergents, household ammonia, chlorine-based bleaches, drain cleaners, and stomach antacid tablets.

Studies of our own body fluids would also reveal that some are acidic and others are nearly neutral or slightly basic. The contents of the stomach can be very acidic—the cells lining the walls of the stomach secrete hydrochloric acid. Human urine can be slightly acidic or basic, depending on what you may have last eaten or the state of your health. Human blood, tears, and saliva are usually just slightly basic.

Definitions of acids and bases

In about 1887, Svante Arrhenius proposed that substances that produce hydrogen ion in water be called

acids and substances that produce hydroxide ion in water be called *bases*. Since that time, the concept of an acid or base has been extended and redefined to explain more fully the terms *acidity* and *basicity* and the role of the solvent in acid–base equilibria.

In 1923, J. N. Brønsted and T. M. Lowry independently proposed the definition of an acid as a proton donor and of a base as a proton acceptor. This definition has proved useful because it explicitly points out the role played by the solvent (often water) in acid–base chemistry. The dissociation of water is described by the equilibrium

$$H_2O + H_2O \rightleftharpoons H_3O^+ + OH^- \tag{1}$$

$$K_W = [H_3O^+][OH^-] = 1.0 \times 10^{-14} \text{ at } 25\,°C \tag{2}$$

In this equilibrium, water plays the role of both an acid and a base (such substances are often called *amphiprotic*). Water can act as both an acid and a base because it can form two Brønsted acid–base pairs: H_3O^+/H_2O and H_2O/OH^-.

In pure water, the concentrations of H_3O^+ and OH^- are equal:

$$[H_3O^+] = [OH^-] = \sqrt{K_W} = 1.0 \times 10^{-7}\,M \text{ at } 25\,°C \tag{3}$$

When $[H_3O^+] = [OH^-]$, we say the solution is neutral (neither acidic nor basic).

What happens when we add a little vinegar (whose active ingredient is acetic acid) to pure water? The acetic acid donates a proton to a water molecule, thus creating additional H_3O^+ ions:

$$\underset{(HA)}{CH_3\overset{\overset{O}{\|}}{C}OH} + H_2O \rightleftharpoons H_3O^+ + \underset{(A^-)}{CH_3\overset{\overset{O}{\|}}{C}O^-} \tag{4}$$

The equilibrium constant expression for this reaction is

$$K_a = \frac{[H_3O^+][A^-]}{[HA]} = 1.76 \times 10^{-5} \qquad (5)$$

where K_a represents the acid dissociation constant for acetic acid, a weak monoprotic acid represented by the symbol HA. The anion of the acid, acetate ion, is represented by the symbol A^-. So the acetic acid acts as a proton donor, and the water acts as a proton acceptor. Since the H_3O^+ concentration increases, the OH^- concentration must decrease, because their concentrations are linked by the equilibrium constant expression shown in Equation (2).

Suppose that we add a little household ammonia to pure water. The ammonia reacts with water as a Brønsted base, accepting a proton from the water and leaving OH^- ions in solution:

$$NH_3(aq) + H_2O \rightleftharpoons NH_4{}^+(aq) + OH^-(aq) \qquad (6)$$

The equilibrium constant expression for this reaction is

$$K_b = \frac{[NH_4{}^+][OH^-]}{[NH_3]} = 1.8 \times 10^{-5} \qquad (7)$$

Note that water acts as a base (proton acceptor) with respect to acids that are stronger acids than water, but it acts as an acid (proton donor) with bases that are stronger bases than water. The equilibria shown in Equations (1), (4), and (6) are examples of the amphiprotic behavior of water.

The pH scale

Typical H_3O^+ and OH^- concentrations in solution range from large to very small values (from approximately 10 M to 10^{-14} M). It would be impossible to represent such a range of concentrations on a linear graph because the concentrations would either have to be compressed into an impossibly tiny length on the low end—or because we would have to have an astronomically long piece of graph paper. In a situation like this, we resort to using a logarithmic scale, which assigns an equal length on the scale to each power of 10 (or decade) of the H_3O^+ concentration.

There is another important reason why a logarithmic scale is used. A hydrogen ion–sensing device called a *pH meter* (which employs an electrochemical cell to measure the $[H_3O^+]$) gives a voltage output that depends on the logarithm of the hydrogen ion concentration. So the digital readout (or deflection of the meter needle of the pH meter) is proportional to $\log[H_3O^+]$.

Therefore, for both practical and theoretical reasons, it is useful to define a variable called the *pH*, where

$$pH = -\log[H_3O^+] \qquad (8)$$

and the inverse relation

$$[H_3O^+] = 10^{-pH} \qquad (9)$$

We take the symbol "p" to mean *take the negative logarithm of the quantity that immediately follows*. So pK_a means $-\log K_a$, and so on. Including the minus sign in the definition is a convenience. Most of the concentrations of interest are less than 1 M, so if the choice had been made to define pH as $+\log[H_3O^+]$, we would be required always to carry along a minus sign with the numerical values. However, including the minus sign in the definition of pH has one adverse consequence that is sometimes confusing. We often call the pH scale an acidity scale, but as the acidity of a solution increases, the pH decreases.

The neutral point on the aqueous pH scale (pH = 7.00 at 25 °C) is a consequence of the magnitude of the equilibrium constant for the dissociation of water. (At temperatures other than 25 °C in water or in other solvents, the neutral point would not be at pH 7.00.)

The distinction between pH and total acidity

If an acid completely dissociates in solution, the hydrogen ion concentration (and the pH) will be a measure of the total amount of acid in the solution. For a strong acid

$$[H_3O^+] = C \qquad (10)$$

$$pH = -\log C = pC \qquad (11)$$

where C is the molar concentration of strong acid in solution. This relation will be true for common strong acids such as $HClO_4$, HCl, and HNO_3.

But if the acid is a weak monoprotic acid, it will be incompletely dissociated, and an approximation of the hydrogen ion concentration (or pH) of the solution is given by the expressions

$$[H_3O^+] = \sqrt{K_a C} \qquad (12)$$

$$pH = -\log \sqrt{K_a C} = \frac{1}{2}pK_a + \frac{1}{2}pC \qquad (13)$$

where K_a is the acid dissociation constant of the weak acid, and C is the molar concentration of the acid. [Take pencil and paper and derive Equations (12) and (13), consulting your text as necessary.] Equations (12) and (13) show that in a weak monoprotic acid, the hydrogen ion concentration is not equal to C and the pH is not equal to pC. For a weak acid, K_a is generally $\ll 1$, so $[H_3O^+] \ll C$. Contrast these results with the simple expressions shown in Equations (10) and (11) for a strong acid.

Acid–base titrations

We can't directly determine the concentration, C, of a weak acid just by measuring the pH. Equation (13) shows that we also need to know the acid's dissociation constant, K_a. But we can determine C (which we call the *total acidity* or the *stoichiometric concentration*) by measuring the amount of a strong base (such as NaOH) required to react completely with a sample of the acid. This is called an *acid–base titration*. The NaOH is dispensed from a buret, so if we know the concentration and the volume of NaOH required to titrate the acid, we can accurately calculate the total amount of weak acid in the solution.

At the end point in the titration of a weak acid with a base, such as acetic acid (CH_3COOH, which we will abbreviate as HA) with NaOH, the following relationship holds:

$$\text{number of moles of acid} = \text{number of moles of base} \tag{14}$$

This is because the net ionic equation describing the reaction is

$$HA + OH^- \rightarrow A^- + H_2O \tag{15}$$

and the equilibrium constant of this neutralization reaction is extremely large ($K = K_a/K_W \cong 10^9$), indicating that it proceeds essentially to completion. Because one mole of NaOH reacts with one mole of HA, we can calculate the moles of HA from the relation

$$\text{moles HA} = \text{moles NaOH} = M_{NaOH} \times V_{NaOH} \tag{16}$$

So we can determine the concentration of acetic acid in a sample of vinegar by titration with a NaOH solution of known concentration.

A plot of the pH of the solution versus the volume of NaOH solution added is called a *titration curve*. An example of a titration curve for a sample of vinegar (a dilute solution of acetic acid) titrated with NaOH is shown in Figure 27-1. Note that when the last bit of acid is titrated, the pH *increases* dramatically, indicating a rapid *decrease* in the hydrogen ion concentration. This rapid change signals the end point of the titration.

Either a pH meter or an acid–base indicator can be used to detect the end point of the titration. An acid–base indicator is an organic dye molecule that is itself a weak acid. The acid and base forms of the indicator have different colors, so when it goes from its acid form to its base form, the indicator shows a color change. The pH range where this color change takes place is governed by the dissociation constant

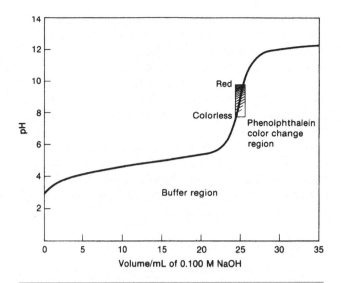

FIGURE 27-1 | **Titration curve for approximately 3 mL of vinegar diluted to 25 mL in water and titrated with 0.1 M NaOH. Phenolphthalein (or thymol blue) is a suitable indicator for detecting the end point of the titration.**

(K_a) of the acid indicator molecule. If we represent the protonated indicator molecule by the symbol HIn, we can write the equation for the dissociation of the indicator as

$$HIn + H_2O \rightleftharpoons H_3O^+ + In^- \tag{17}$$

$$K_a = \frac{[H_3O^+][In^-]}{[HIn]} \tag{18}$$

If we take the logarithm of both sides of Equation (18), we obtain

$$\log K_a = \log[H_3O^+] + \log\frac{[In^-]}{[HIn]} \tag{19}$$

which we can rearrange to the form

$$pH = pK_a + \log\frac{[In^-]}{[HIn]} \tag{20}$$

Now, take pencil and paper and prove to yourself that Equation (20) can be obtained from Equations (18) and (19). Then prove to yourself that when $pH = pK_a - 1$, $[In^-]/[HIn] = 0.1$, and that when $pH = pK_a + 1$, $[In^-]/[HIn] = 10$.

The pencil and paper exercise shows us that the indicator changes from being mostly in its acid form (HIn) to being mostly in its base form (In^-) in an interval of about 2 pH units, centered on the pK_a of the indicator. This is called *the pH interval for the color change of the indicator*. (Table 5 of the Appendix gives the colors and pH intervals of a number of indicators.)

In order to have the color change of the indicator accurately signal the end point, we must choose an indicator whose pK_a is about the same as the pH at the end point of the titration, as shown in Figure 27-1. Also note in Figure 27-1 that at the end point for the titration, the solution has a composition and pH that are identical to those of a sodium acetate solution having the same concentration.

Acid–base indicators have such an intense color that only a small amount of the indicator is necessary to produce a readily visible color. So the contribution of the indicator acid, HIn, to the total acidity of the solution is negligible.

Standardization of the NaOH titrant

The acid–base titration procedure we have just described requires that we know the concentration of the NaOH titrant in order to determine the concentration of the acid that we are titrating. The accurate determination of the concentration of the NaOH titrant is called a *standardization* procedure. It works like this: We weigh out a sample of an acid available in very pure form, called a *primary standard* grade of acid, dissolve it in water, and titrate the acid with our NaOH titrant. Knowing the mass of the acid used, its formula weight, and the volume of NaOH required to neutralize it, we can determine the concentration of the NaOH titrant. We will use a pure grade of potassium hydrogen phthalate, KHP, whose formula and structure are

Potassium hydrogenphthalate (KHP)

$KC_8H_5O_4$, FW = 204.23 g/mol

KHP is a weak carboxylic acid, like acetic acid, so we can use the same indicator to titrate KHP that we use for the titration of acetic acid in vinegar.

Titration of stomach antacids

In this experiment, we will determine the neutralization capacity of commercial stomach antacid tablets. These substances are bases, so direct titration would require an acid titrant. However, many of them do not readily dissolve in water and react rather slowly with acids. To circumvent this difficulty, we will use a back-titration procedure. An excess of a strong acid (HCl) is added to the antacid tablet, and the mixture is heated to ensure complete reaction. Some of the HCl is consumed by reaction with the antacid tablet. The antacid and HCl solution is then titrated with

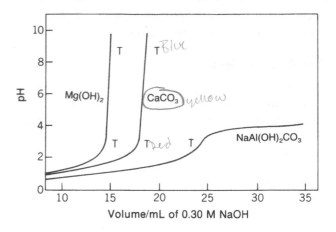

FIGURE 27-2 | **Back titration of unreacted HCl with standardized 0.3 M NaOH after addition of 5.0 mL of 3.0 M HCl to antacid tablets of various compositions. The indicator thymol blue has two color transitions, at the pHs shown approximately by the Ts.**

NaOH to determine how much HCl remains unreacted. Subtracting the amount of HCl remaining from that originally added gives the amount of HCl that the antacid tablet consumed. This gives an accurate measurement of the tablet's neutralization capacity. It also resembles the way the tablet reacts with HCl in the stomach. Typical titration curves for the back titration of antacid and HCl solutions with NaOH are shown in Figure 27-2. The indicator thymol blue (which is diprotic and therefore has two color transition regions) is a good choice for most antacids.

Antacid tablets are probably one of the most widely used self-prescribed medicines. They are taken to relieve the medically undefined conditions of heartburn or acid indigestion and sour stomach. Although such gastric distress is often attributed to excess production of HCl, sometimes the latter is not responsible for the symptoms. For instance, gastric HCl production may be less than normal in diseases such as superficial chronic gastritis and carcinoma of the stomach. The common condition popularly called heartburn is thought to be caused by regurgitation of the gastric contents into the esophagus, and acid indigestion may be merely due to overeating or to an irritating food.

Television commercials stress the competition among various brands of antacids to see which has the greatest neutralization capacity, but in actuality, if the pH in the stomach rises too high, the entire digestive process may be hindered. The digestion of proteins is catalyzed by the enzyme pepsin. Its proteolytic activity is inhibited if the pH of the stomach contents rises any higher than the natural buffering of the proteins in the food itself would allow it to go (around pH 4). Therefore, an antacid should only neutralize an amount of HCl in excess of what the body of a healthy individual normally secretes following a meal.

Studies show that HCl produced by gastric stimulation varies widely from individual to individual, ranging from about 1 to 22 mmol of HCl per hour, but the average figure for "excess" production is only about 10 mmol of HCl per hour greater than that of normal HCl production. Therefore, one dose of an antacid product should react with no more than about 10 mmol of HCl per hour. For many people the upper limit should be less. This raises the question of whether people who regularly consume the doses recommended on the labels of various antacid products are in fact interfering with normal digestion. The doses neutralize much more than 10 mmol of HCl, so the body's natural response raises the pH of the stomach contents above pH 4.

Experimental Procedure

Special Supplies: *Parts 1, 2, and 3: Microscale:* 1-mL buret; *Macroscale:* 25- or 50-mL buret.
Part 2: Distilled white vinegar (5% by weight acetic acid). Dark-colored vinegars are difficult to titrate because their color obscures the indicator end point.
Part 3: Stomach antacids (Tums Antacid Chewable Tablets, Regular; Phillips Milk of Magnesia tablets; etc.); 5-mL pipet; polyethylene transfer pipet.

Chemicals: *Parts 1, 2, and 3:* 0.1% thymol blue indicator.
Part 1 (optional standardization): Pure, dry potassium hydrogen phthalate (KHP).
Parts 1 and 2 (standardization and vinegar titration): Microscale: 1.00 M NaOH; *Macroscale:* 0.1 M NaOH.
Part 3 (antacid titration): 3.00 M HCl; *Microscale:* 6.00 M NaOH; *Macroscale:* 0.30 M NaOH.

NOTES TO INSTRUCTOR

If you prefer to use 25- or 50-mL burets instead of 1-mL burets, in part 2 use a 2.50-mL (or 2.50-g) sample of vinegar titrated with 0.100 M NaOH. Also in part 2, back-titrate the sample with 0.30 M NaOH in a 25- or 50-mL buret in place of 6.0 M NaOH in a 1-mL buret.

In part 3, it is preferable to use antacid tablets that will neutralize 10 to 11 mmol of HCl. Tablets with a larger amount of active ingredient will require a proportionate increase in the volume of added 3.00 M HCl. Use a volume of 3.00 M HCl that will leave about 4 to 5 mmol excess HCl after reaction with the tablet.

SAFETY PRECAUTIONS:
Parts 1 and 2: Sodium hydroxide is corrosive. Wear eye protection. Do not touch or handle the chemicals with your bare hands. Clean up any spills immediately.
Part 3: If available, use a hot plate to heat the flask of HCl containing an antacid tablet. If you use a Bunsen burner, be sure that the flask is clamped so that it cannot be overturned. Boiling hot solution that splashes on you from the bench top can inflict serious burns.

WASTE COLLECTION:
Collect all wastes in a single beaker, neutralize with sodium bicarbonate, and flush down the drain.

1. Standardization of NaOH with Potassium Hydrogen Phthalate (Optional) Your instructor will tell you whether you will be using microscale or macroscale burets. Read the section of the Introduction describing how to use burets.

Microscale. Obtain three clean 50-mL Erlenmeyer flasks and label them 1, 2, and 3. Tare each flask in turn and weigh into each flask about 0.160 g of pure, dry potassium hydrogen phthalate (KHP), recording the mass of each sample to the nearest milligram. Add 20 mL of deionized water to each flask and swirl to dissolve the KHP. Add 5 drops of 0.1% thymol blue indicator to the dissolved KHP.

Fill a 1-mL microburet (see the Introduction at the front of the lab manual) with approximately 1.0 M NaOH, taking care to expel all bubbles from the microburet before beginning the titration. Record the initial volume reading of the microburet and titrate the first solution to the appearance of the first permanent blue color. Then record the final volume reading of the buret. Repeat the titration with the remaining two samples of KHP.

Macroscale. Obtain three clean 250-mL Erlenmeyer flasks and label them 1, 2, and 3. Tare each flask in turn and weigh into each flask about 0.400 g of pure, dry potassium hydrogen phthalate (KHP), recording the mass of each sample to the nearest milligram. Add 100 mL of deionized water to each flask and swirl to dissolve the KHP. Add 5 drops of 0.1% thymol blue indicator to the dissolved KHP.

Fill a 25- or 50-mL buret (see the Introduction at the front of the lab manual) with approximately

0.10 M NaOH, taking care to expel all bubbles from the buret tip before beginning the titration. Record the initial volume reading of the buret and titrate the first solution to the appearance of the first permanent blue color. Then record the final volume reading of the buret. Repeat the titration with the remaining two samples of KHP.

Calculations: From the mass and formula weight of the KHP, calculate the number of millimoles of KHP in the sample. This will be equal to the number of millimoles of NaOH titrant. Divide the number of millimoles of NaOH by the volume of titrant added, in milliliters, to get the molar concentration of the NaOH titrant in units of moles/liter. Average the three values, which should be in good agreement (relative standard deviation < 5%).

2. Titration of Vinegar Record in your notebook the brand name and percentage of acetic acid in the vinegar that is specified on the manufacturer's label.

~~Microscale.~~ Obtain 5 mL of standardized 1.00 M NaOH and 5 mL of vinegar in two clean, dry, stoppered 13 × 100 mm test tubes. Obtain three clean 25-mL Erlenmeyer flasks to contain three vinegar samples and label them 1, 2, and 3. Using a 1.00-mL serological pipet and pipet-filling device, transfer exactly 1.00 mL of vinegar to each of the three Erlenmeyer flasks.

Alternatively, put the empty flask on the pan of the balance and push the tare bar so that the balance reads zero (or record the mass of the empty flask). Using a polyethylene pipet filled with vinegar, drop about 1.00 g of vinegar into the Erlenmeyer flask. Record the exact mass of the sample to the nearest 0.001 g. Repeat until you have three weighed samples of vinegar.

Add 10 mL of deionized water and 3 drops of 0.1% thymol blue indicator to each of the vinegar samples. Fill the 1-mL microburet with 1.00 M NaOH, taking care to remove any bubbles in the buret. (**Caution:** If you are using a 1-mL syringe as a microburet, take special care when filling the syringe or expelling bubbles not to get titrant on your skin or squirt yourself or your neighbor with the NaOH in the syringe.)

Record the initial volume reading of the buret. Titrate the sample of vinegar, swirling the flask to stir the sample while adding NaOH drop by drop from the microburet. When the solution begins to turn blue where the NaOH titrant hits the solution, you are nearing the end point. Titrate to the first appearance of a permanent blue color; then read and record the final volume of NaOH in the microburet. Discard the blue solution and rinse the flask with deionized water. Repeat the titration for the remaining two samples of vinegar.

Macroscale. In a clean, dry ~~flask~~ beaker, put 100 mL of standardized 0.10 M NaOH. ~~In a second flask, put about 15 mL of vinegar.~~ Next, obtain three clean —label 250-mL Erlenmeyer flasks to contain three vinegar samples. Using a 5-mL serological pipet and pipet-filling device, transfer exactly 2.50 mL of vinegar to each of the three flasks. (Alternatively, you may put the flask on an electronic balance, tare to zero, and transfer 2.50 g of vinegar to the flask, using a polyethylene transfer pipet. Record to the nearest milligram the exact mass of vinegar weighed into each flask.)

Add 100 mL of deionized water and 5 drops of 0.1% thymol blue indicator to each sample of vinegar. Rinse a 50-mL buret with standardized 0.10 M NaOH and fill the buret, taking care to expel all bubbles from the tip of the buret. Record the initial volume of the buret to the nearest 0.02 mL. Titrate the first sample of vinegar to the appearance of a permanent blue color. Record the volume reading of the buret. Repeat the titration for the remaining two samples of vinegar.

(Keep your buret, but rinse and refill it with 0.30 M NaOH if you are going to do part 3, the titration of antacid tablets.)

Calculations: From the concentration and added volume of the NaOH titrant, calculate for each titration the millimoles of NaOH added. This will equal the millimoles of acetic acid contained in the vinegar sample. Then calculate the mass of acetic acid contained in the sample; take as a given that the molar mass of acetic acid is 60.0 g/mol. If you measured the vinegar by volume to obtain the mass of vinegar, assume that the density of vinegar is about 1.005 g/mL (at 20 °C). Finally, calculate the mass percent of acetic acid in the sample as (g acetic acid/g of vinegar) × 100%.

3. Titration of Antacid Tablets Various brands of antacid tablets will be available, such as Tums Antacid Chewable Tablets, Regular ($CaCO_3$); Phillips Milk of Magnesia tablets [$Mg(OH)_2$]. The identity of the active ingredient and the mass of active ingredient per tablet will be found on the manufacturer's label. Record this information in your report.

Analyze one or more brands of tablets as your instructor directs, using the following procedure. Put an antacid tablet in a small beaker and weigh the beaker and tablet on the analytical balance to the nearest milligram. Record the mass. Then transfer the tablet to a clean 250-mL Erlenmeyer flask. Reweigh the empty beaker and record the mass so that you can obtain the mass of the tablet.

Put 5.00 mL of 3.00 M HCl in the flask containing the tablet, using a pipet or buret to accurately measure the volume of HCl solution.[1] Add approximately 50 mL of deionized water. Heat the flask to boiling on a hot plate or over a Bunsen burner. (**Caution:** If you use a Bunsen burner, clamp the flask so that it cannot be overturned.) Gently boil the solution for about 5 min. This expels any carbon dioxide produced by reaction with HCl. Use a glass stirring rod, if necessary, to help break up and disperse the tablet. (**Caution:** Do not use so much force that you risk breaking a hole in the flask with the glass stirring rod.)

Microscale: While the solution is boiling, rinse and fill a 1-mL microburet with standardized 6.0 M NaOH and record the initial buret reading.

Macroscale: Use the same 50-mL buret you used in part 2, but rinse and refill it with 0.30 M NaOH. Record the initial volume reading of the buret.

Cool the solution containing the antacid to room temperature by immersing it in a container of tap water. When the solution has cooled, add 10 drops of 0.1% thymol blue indicator. The solution should appear red, indicating that the pH is less than 2. Then titrate the solution with 6.00 M NaOH (*microscale 1 mL buret*) or 0.30 M NaOH (*macroscale buret*) until you get a color change from red to yellow. Record the buret reading.

> **NOTE** If you are titrating antacids that contain calcium carbonate or magnesium hydroxide as the active ingredient, the steep rise in pH at the end point, shown in Figure 27-2, means that the color transition from yellow to blue, which takes place in the pH range 8.0 to 9.6, will quickly follow the transition from red to yellow, which takes place in the pH range 1.2 to 2.8. If you are titrating too rapidly and overrun the end point slightly, you might not even see the red to yellow transition. For these samples, you can without serious error take the first appearance of the blue color as the end point of the titration.
>
> Repeat the titration with a second sample or with another brand of antacid, if your instructor directs.

Calculations: Calculate the millimoles of HCl added, the millimoles of NaOH added, the millimoles of HCl that were neutralized by the antacid, and the millimoles and milligrams of the active ingredient in the tablet.

Write a balanced chemical equation for the reaction of HCl with the active ingredient in the antacid tablet. From your titration data, calculate the mass of the active ingredient and compare this with the value given on the manufacturer's label. Compare the mass of the active ingredient you found with the mass of the tablet. The difference represents the mass of inert binders used in making the tablet. Calculate the milligrams of inert binder per tablet and the mass percentage of inert binder.

▰ CONSIDER THIS ▰

Using a pH meter, the pH of a cranberry juice cocktail was found to be 2.46, which indicates that the $[H_3O^+] = 3.47 \times 10^{-3}$ M. Titration of a 1.0-mL sample of the juice to approximately pH 7 required 0.94 mL of 0.10 M NaOH, indicating a total acid content of 0.094 M. Give a plausible explanation for why these two numbers don't agree.

Carbonated beverages like Coke, Pepsi, and 7-Up contain acids, usually phosphoric or citric acid. Propose a procedure to remove the dissolved carbon dioxide from these beverages and determine their total acid content.

In an early TV commercial for Rolaids antacid tablets, the manufacturer claimed that the tablet would neutralize 47 times its own weight of stomach acid. The tablet weighed 1.4 g and contained 0.334 g of the active ingredient, $NaAl(OH)_2CO_3$. Assume that the active ingredient reacts with HCl according to the reaction

$$NaAl(OH)_2CO_3 + 4\ HCl \rightarrow$$
$$NaCl + AlCl_3 + 3\ H_2O + CO_2$$

How many moles of HCl would one Rolaids tablet neutralize?

Stomach acid is about 0.14 M HCl and has a density about the same as water. Did a Rolaids tablet neutralize 47 times its own weight of HCl or 47 times its own weight of 0.14 M HCl solution (HCl + water)?

---

[1]This amount of HCl is calculated to neutralize a tablet that will consume 10 mmol of HCl, leaving about 5 mmol of unreacted HCl. If the tablet will consume more than 10 mmol of HCl, the volume of 3.00 M HCl added should be increased proportionately.

Bibliography

Batson, W. B.; Laswick, P. H. "Pepsin and Antacid Therapy: A Dilemma," *J. Chem. Educ.* **1979,** *56,* 484–486.

Breedlove, C. H., Jr. "Quantitative Analysis of Antacid Tablets," *Chemistry* **1972,** *45,* 27–28.

Fuchsman, W. H.; Garg, S. "Acid Content of Beverages: A General Chemistry Experiment Involving the Resolution of Apparent Anomalies," *J. Chem. Educ.* **1990,** *67,* 67–68. Titration of a variety of beverages makes clear the distinction between the [H_3O^+] and the total acid content of a beverage.

Hem, S. L. "Physicochemical Properties of Antacids," *J. Chem. Educ.* **1975,** *52,* 383–385.

Lieu, V. T.; Kalbus, G. "Potentiometric Titration of Acidic and Basic Compounds in Household Cleaners," *J. Chem. Educ.* **1988,** *65,* 184–185. (Note that the two titration curves in Figures 1 and 2 should be interchanged.)

Murphy, J. "Determination of Phosphoric Acid in Cola Beverages: A Colorimetric and pH Titration Experiment for General Chemistry," *J. Chem. Educ.* **1983,** *60,* 420–421.

Determination of the Molar Mass and Ionization Constant of a Weak Acid

Purpose

• Determine the molar mass and acid dissociation constant of acetic acid (a weak monoprotic acid) by titration and pH measurement of solutions containing a known concentration of acetic acid.

• Determine the molar mass and acid dissociation constant of an unknown weak acid.

• Learn how a pH meter works and how to measure the pH of a solution.

Pre-Lab Preparation

Acids and bases are important classes of chemical compounds. They control the pH of living systems and of many reactions carried out in the chemical laboratory. If the pH of the blood of most animals shifts by as little as 0.3 units above or below the normal range of 7.3 to 7.5, severe illness results. Therefore, if we understand the principles of how acids and bases function, we will be better informed about the functioning of biological systems, as well as of purely chemical systems.

In this experiment, we will study the properties of weak acids that can ionize (dissociate) by reaction with water to yield one hydronium ion (a protonated water molecule, H_3O^+). Such acids are called *weak monoprotic acids*. Acetic acid, the active ingredient in vinegar, is a common weak acid belonging to this class. We will determine its molar mass by titration and its ionization constant by making measurements of pure acetic acid solutions and a half-neutralized solution of acetic acid. We can also apply exactly the same principles to determine the molar mass and ionization constant of an unknown weak monoprotic acid.

Acetic acid has the molecular formula

The molecular formula for acetic acid is generally shortened to read CH_3COOH or $HC_2H_3O_2$. These chemical formulas are often further abbreviated to HA, where HA represents acetic acid (or any generic weak monoprotic acid). Then the symbol A^- stands for acetate ion, $CH_3CO_2^-$, the anion of acetic acid (or the anion of any weak monoprotic acid, HA).

The proton that dissociates in acetic acid is the one attached to the oxygen atom. The O—H bond in the acetic acid molecule is appreciably weaker than the C—H bonds, and the polarity of the O—H bond makes it easier for the proton to transfer to a water molecule. In contrast to the behavior of hydrogen atoms bonded to oxygen, there is no significant ionization of the hydrogen atoms bonded to carbon in aqueous solution.

Acetic acid ionizes according to the reaction

$$HA + H_2O \rightleftharpoons H_3O^+ + A^- \tag{1}$$

The equilibrium constant expression is given by

$$K_a = \frac{[H_3O^+][A^-]}{[HA]} \tag{2}$$

The literature value of the equilibrium constant for this reaction is 1.76×10^{-5} at 25 °C.[1] You are to measure the value of this constant in this experiment and compare your value with the literature value.

A straightforward way to obtain the value of K_a is to measure each of the quantities on the right-hand side of Equation (2) and carry out the indicated arithmetic. For solutions containing only the weak acid HA, the value of K_a can be determined by examining the ionization reaction shown in Reaction (1) and noting that the concentration of A^- will be equal to that of the H_3O^+ (ignoring the insignificant amount of H_3O^+ that comes from the ionization of water). The HA

---

[1]This is the value of K_a obtained by extrapolating to zero ionic strength. In any real solution of acetic acid, the ionic strength is greater than zero, and the dissociation constant is somewhat larger because of electrostatic interactions between the ions in solution.

concentration in the solution is equal to the *stoichiometric* concentration of HA in the solution minus the amount of HA lost by ionization.[2] For acetic acid and most other weak acids, the amount of HA lost by ionization is much smaller than the amount put into solution; thus, the equilibrium concentration of HA is only slightly smaller than the stoichiometric concentration.

In solutions that contain both acetic acid (HA) and sodium acetate (Na^+A^-), we will show that the equilibrium concentrations of HA and A^- are approximately equal to their stoichiometric concentrations. (Similar conclusions will apply to any solution containing a mixture of a weak monoprotic acid, HA, and its salt, Na^+A^-.) The stoichiometric concentrations can be readily determined by titration or by knowledge of the quantities of HA and Na^+A^- used to prepare the solutions. If we know the stoichiometric concentrations of HA and A^-, and if the equilibrium concentrations are equal to the stoichiometric concentrations, in order to determine K_a, all that remains is to calculate the $[H_3O^+]$ from the measured pH.

$$pH = -\log[H_3O^+]; \qquad [H_3O^+] = 10^{-pH} \qquad (3)$$

Measuring the total acid concentration by titration

To determine the molar mass of acetic acid, one of the things we need to know is the total amount of acid present in the original sample solution. This can be readily determined by titration of the acetic acid with a strong base of known concentration, such as sodium hydroxide. An acid–base indicator or a pH meter can be used to determine the end point of the titration. Figure 28-1 helps to explain how the titration works, illustrating how the pH of the solution changes as a strong base, NaOH, is added to the solution containing the weak monoprotic acid.

After an initial small rise in pH, the pH does not change very rapidly as NaOH is added. In the middle of the titration, there is an appreciable concentration of both HA and A^- present, so adding a small amount of base (or acid) does not change the pH very much. For this reason, this is called the *buffering* region.

When an amount of NaOH equivalent to the amount of acid has been added, the concentration of H_3O^+ decreases very sharply with a corresponding

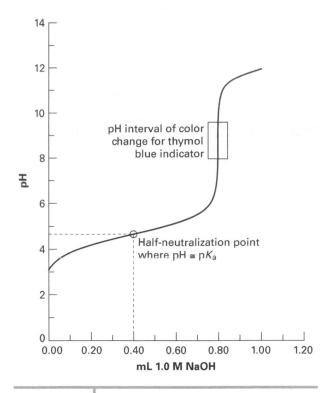

FIGURE 28-1 | **Titration of a weak acid with a strong base. The figure shows the titration of 20 mL of 0.04 M acetic acid with 1.0 M NaOH. Thymol blue is a satisfactory acid–base indicator for this titration.**

sharp increase in the pH. This is the end point of the titration, where the indicator changes color. The pH interval of the color change of the indicator must be a good match to the pH at the end point, as shown in Figure 28-1.

Determining K_a by the method of half-neutralization

At exactly the midpoint of the titration, when half of the acid has been neutralized, there will be equal amounts of HA and A^- in solution. This relationship is a good approximation, subject to the condition that [HA] and $[A^-]$ are both greater than $100K_a$, as discussed in greater detail later. From Equation (2) we see that when [HA] = $[A^-]$, the two terms will cancel, so that $K_a = [H_3O^+]$ and $pK_a = pH$ at the midpoint of the titration. This provides a simple way to measure the pK_a of a weak acid. If we have a half-neutralized solution of a weak acid, the pH of the solution will be equal to the pK_a of the acid. K_a and pK_a are related like $[H_3O^+]$ and pH:

$$pK_a = -\log(K_a); \qquad K_a = 10^{-pK_a} \qquad (4)$$

---

[2]The stoichiometric concentrations of HA and A^- are determined from the amounts of each substance that are put into the solution. However, HA and A^- can each react with water: HA + HOH $\rightleftharpoons$ HA + OH^- and A^- + HOH $\rightleftharpoons$ HA + OH^-. Therefore, in general, the actual concentrations of HA and A^- at equilibrium (which we will call the equilibrium concentrations) may not be exactly equal to the stoichiometric concentrations.

Determining the molar mass of a weak monoprotic acid

If we know the mass concentration of acid in units of g HA/L, it is a straightforward procedure to determine the molar mass of the acid (g HA/mol). We know that the molar mass of HA has units of g/mol. If we know the concentration of HA in g/L, then clearly we have to divide this quantity by mol HA/L (or mmol HA/mL) to get the molar mass.

If we take a known volume of HA solution, we can determine the total amount of HA in the sample by titration with a strong base of known concentration. Then, knowing the volume of the original sample (L) and the total amount of HA (mol), we have the necessary information to complete the calculation of the molar mass:

$$\text{molar mass} = \frac{\text{g HA}}{\text{L}} \times \frac{\text{L HA}}{\text{mol HA}} = \frac{\text{g HA}}{\text{mol}} \quad (5)$$

Theorem: In a solution of HA and Na^+A^-, the equilibrium concentrations of HA and A^- are approximately equal to their stoichiometric concentrations.

By adding sodium hydroxide to a larger amount of a weak acid, we can form a mixture of the weak acid, HA, and its salt, Na^+A^-. To determine the K_a of the weak acid, we must be able to calculate the equilibrium concentrations of HA, A^-, and H_3O^+. We will now show how to calculate the stoichiometric concentrations of HA and A^-. We will then show that the equilibrium concentrations are usually equal to the stoichiometric concentrations.

To see how we can determine the stoichiometric concentrations of the acid HA and the base A^-, let's suppose that we add a known amount of NaOH to a solution containing a larger (and known) amount of HA. The reaction is

$$HA + Na^+ + OH^- \rightarrow Na^+ + A^- + H_2O \quad (6)$$

[Na^+ ion, which appears on both sides of Reaction (6), is simply a spectator ion.]

The *stoichiometric* amount of sodium acetate (Na^+A^-) formed is equivalent to the amount of NaOH added:

$$\text{mol } Na^+A^-_{\text{formed}} = \text{mol NaOH}_{\text{added}} \quad (7)$$

The stoichiometric amount of HA left in solution is obtained easily by subtracting the amount of NaOH added from the original amount of HA present in the solution:

$$\text{mol } HA_{\text{final}} = \text{mol } HA_{\text{original}} - \text{mol NaOH}_{\text{added}} \quad (8)$$

The stoichiometric composition of the solution is precisely the same as would be obtained if we dissolved the same number of moles of acetic acid and sodium acetate as calculated from Equations (7) and (8) and in the same total volume. This is why these concentrations are called *stoichiometric concentrations.*

Now let's see how the equilibrium concentrations of HA and A^- are related to their stoichiometric concentrations. We will represent these stoichiometric concentrations by a subscript zero: $[HA]_0$ and $[A^-]_0$. We will make use of two equations, one based on a *material balance* and the other based on a *charge balance* condition.

Material and Charge Balance First we write the *material balance* equation that expresses the condition that the sum of the stoichiometric concentrations must equal the sum of the equilibrium concentrations:

$$[HA]_0 + [A^-]_0 = [HA] + [A^-] \quad (9)$$

This equation says that if we put known stoichiometric amounts of HA and A^- in solution, the sum of the stoichiometric concentrations will be exactly equal to the sum of the equilibrium concentrations. This is true because any HA that reacts with water produces A^- and vice versa, as shown by the equations in footnote 2.

The *charge balance* equation expresses the condition that the sum of the positive charges must equal the sum of the negative charges. This is true because the solution as a whole is electrically neutral. (The charge balance equation includes all of the ions in solution, including those that come from the ionization of water.)[3] The charge balance equation for a solution containing HA and Na^+A^- dissolved in water is written as

$$[Na^+] + [H_3O^+] = [A^-] + [OH^-] \quad (10)$$

The stoichiometric concentration of A^- is exactly equal to the concentration of Na^+:

$$[A^-]_0 = [Na^+] \quad (11)$$

---

[3]In the charge balance equation, the concentrations of H_3O^+ and OH^- include the contributions from all sources, both from the ionization of water and the reaction of HA and A^- with water. However, we don't need to worry about keeping separate account of ions coming from water and ions coming from the acid and its salt if they are the same kind of ions.) There is only one equilibrium concentration of H_3O^+ and one equilibrium concentration of OH^- in solution, no matter where they came from.

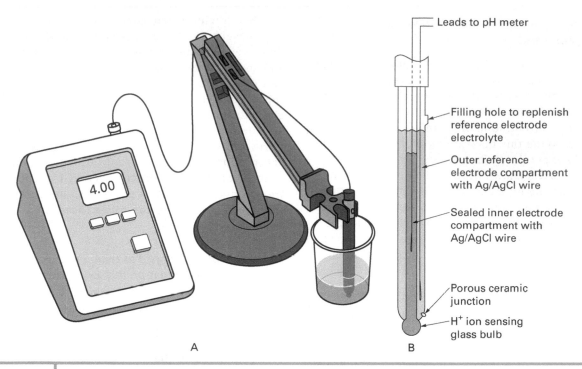

Leads to pH meter

Filling hole to replenish reference electrode electrolyte

Outer reference electrode compartment with Ag/AgCl wire

Sealed inner electrode compartment with Ag/AgCl wire

Porous ceramic junction

H^+ ion sensing glass bulb

A

B

FIGURE 28-2 The pH meter and its combination pH electrode. (A) The complete apparatus. (B) Detail of the combination pH electrode, which combines a hydrogen ion–sensing glass electrode and reference electrode in a single probe, forming an electrochemical cell. The voltage of the cell is proportional to the logarithm of the ratio of hydrogen ion activities inside and outside the glass bulb.

This is the same as saying that each mole of NaOH added to a solution containing a larger amount of HA produces a stoichiometric amount of sodium acetate given by Equation (7). Substituting $[A^-]_0$ for $[Na^+]$ in Equation (10) gives

$$[A^-]_0 + [H_3O^+] = [A^-] + [OH^-] \qquad (12)$$

This result shows that the equilibrium concentration of A^- is equal to the stoichiometric concentration of A^-, provided that the concentrations of H_3O^+ and OH^- are negligible compared to the concentration of A^-. From Equation (9) we see that if $[A^-]_0$ and $[A^-]$ are equal, $[HA]_0$ must equal $[HA]$. This illustrates the general rule that for a solution containing a mixture of HA and Na^+A^-, the equilibrium concentrations of A^- and HA will be equal to the stoichiometric amounts of A^- and HA calculated from Equations (7) and (8) (and divided by the total solution volume) *provided that HA is not too strong an acid* or *its concentration too dilute*. As a good rule of thumb, this rule will apply whenever the stoichiometric concentrations of HA and A^- are both greater than $100K_a$.

Using a pH meter to measure pH

All pH meters operate by measuring the voltage of an electrochemical cell. The cell is comprised of two half-cell electrodes: a glass-membrane electrode sensitive to the hydrogen ion concentration and a reference electrode.[4] (See Figure 28-2.) The reference electrode is often a silver–silver chloride electrode. Usually, the two half-cell electrodes are combined in a single probe unit called a *combination electrode*. The principle of operation, however, is the same for either type. The silver–silver chloride reference electrode supplies a constant potential ($E° = +0.20$ V versus the standard hydrogen electrode) determined by the half-reaction

$$AgCl(s) + e^- \rightleftharpoons Ag(s) + Cl^-(3.5 \text{ M}) \qquad (13)$$

The $[H_3O^+]$ in the solution determines the potential of the glass electrode. The potential (or voltage) developed across the glass membrane is proportional to the logarithm of the H_3O^+ concentration ratio inside and outside the glass electrode. The pH meter measures the total cell potential across the two half-cell electrodes and displays this measurement on the scale calibrated in pH units [a logarith-

---

[4]Some newer types of pH meters employ a solid state hydrogen ion sensor rather than one based on a glass membrane. The solid state electrodes have the advantage of being more rugged.

mic scale, as shown by Equation (3)]. When properly standardized with buffers of known pH, a pH meter provides an accurate and simple method for determining the pH of a solution. In Figure 28-2, note the construction details of the combination glass electrode. The thin glass bulb of the glass electrode is very fragile and is often protected with a special plastic guard. Be extremely careful not to hit the glass bulb against the side or bottom of a beaker.

Experimental Procedure

Special Supplies: pH meter; pH 4 and 7 buffers for standardizing the pH meter; 30-mL polyethylene beakers for measuring the pH of solutions; polyethylene transfer pipets.
Microscale: 1-mL microburet (Air-Tite 1 × 0.01 mL all-polypropylene syringe fitted with a fine tip cut from a fine-tipped polyethylene transfer pipet) or other form of microburet (see the Introduction to the lab manual for a description).
Macroscale: 25- or 50-mL buret.

Chemicals: 2.40 g/L (0.04 M) acetic acid; approximately equivalent molar concentrations of unknown weak monoprotic acids, labeled with their mass concentration (g/L); 0.1% thymol blue indicator (or phenolphthalein).
Microscale: Standardized 1.00 M NaOH.
Macroscale: Standardized 0.100 M NaOH.

1. Measurement of pH and Titration of Acetic Acid Solution

Your instructor will demonstrate the use of the pH meter. Use 30-mL polyethylene beakers, if available, for holding the samples. Use a wash bottle to rinse the electrode(s) between samples. Excess water can be removed by blotting the bulb with a tissue. (*Do not rub the bulb.*)

Microscale Procedure. Obtain 50 mL of 2.40 g/L acetic acid, 6–7 mL of standardized 1.00 M NaOH in a 13 × 100 mm test tube with size 00 rubber stopper, and a 1-mL microburet. Fill the microburet with 1.00 M NaOH and expel any air bubbles. Record the initial volume.

Pour 20 mL of the 2.4 g/L acetic acid solution into a clean 30-mL polyethylene beaker. Measure and record the pH. Then transfer the solution to a tared (or weighed) 50-mL Erlenmeyer flask. Using a polyethylene transfer pipet, adjust the mass of acetic acid solution in the flask to 20.0±0.1 g. Record the mass of the sample. Add 5 drops of 0.1% thymol blue indicator. Titrate the sample of acid to the first appearance of a permanent blue color. Record the final volume reading of the microburet.

Repeat the pH measurement and titration with a second 20.0-g sample of 2.4 g/L acetic acid. For each sample, calculate the mmol of acid/mL sample. (Assume that each 1.00 g of sample = 1.00 mL.) Average the values for the two samples. If they do not agree within 5%, titrate a third sample and average all three samples. Also average the initial pH values measured for each sample before they were titrated.

Macroscale Procedure. Obtain 80 mL of 2.4 g/L acetic acid, 75 mL of standardized 0.10 M NaOH, and a 25- or 50-mL buret. Rinse and fill the buret with 0.10 M NaOH, taking care to expel any bubbles from the tip of the buret. Record the initial volume reading.

Pour 20 mL of the 2.4 g/L acetic acid solution into a clean 30-mL polyethylene beaker. Measure and record the pH. Then transfer the solution to a tared (or weighed) 125-mL Erlenmeyer flask. Add more 2.4 g/L acetic acid and, using a polyethylene transfer pipet, adjust the mass of solution in the flask to 30.0±0.1 g. Record the mass of the sample. Add 5 drops of 0.1% thymol blue indicator. Titrate the sample of acid to the first appearance of a permanent blue color. Record the final volume reading of the buret.

Repeat the pH measurement and titration with a second 30.0-g sample of 2.4 g/L acetic acid. For each sample, calculate the mmol of acid/mL sample. (Assume that each 1.00 g of sample = 1.00 mL.) Average the values for the two samples. If they do not agree within 5%, titrate a third sample and average all three samples. Also average the pH values measured for each sample before they were titrated.

Calculations: Calculate the molar concentration of the 2.4 g/L acetic acid (mmol acid/mL of acid) from the average volume of NaOH required to titrate the samples of acetic acid. Calculate the molar mass of acetic acid (g/mol) using Equation (5). Compare with the expected molar mass for acetic acid whose molecular formula is CH_3COOH.

Using the average molar concentration and average pH, calculate the ionization constant for acetic acid. Compare this value with the literature value and with the value measured by the half-neutralization method.

2. Measurement of the pK_a of Acetic Acid by the Half-Neutralization Method

Weigh into a clean, tared Erlenmeyer flask the same size sample of 2.4 g/L acetic acid that you used in part 1 (*20.0 g for microscale; 30.0 g for macroscale*). Add to the flask exactly half of the average volume of NaOH required to titrate the sample, following the procedure you used in part 1. Mix thoroughly. This will produce a solution of acetic acid that is exactly half-neutralized.

Measure and record the pH of this solution. If the pH meter is properly calibrated, the pH will be equal, to a good approximation, to the pK_a of the acid, as was shown in the Pre-Lab Preparation.

3. The Molar Mass and Ionization Constant of an Unknown Weak Monoprotic Acid You will be provided with aqueous solutions of an unknown acid, identified only by its mass concentration (g/L). Following the same procedure used in parts 1 and 2, determine the molar mass and ionization constant for the weak monoprotic acid. Assuming the acid is a carboxylic acid of the general formula RCOOH, try to identify the acid from the list of acids given in Table 6 of the Appendix.

CONSIDER THIS

In making the comparison of the ionization constants that you measured with literature values, you should be aware that the *interionic attraction effect* causes the value of K_a for the solutions used in this experiment to increase (and pK_a to decrease) as the concentration of the ions in solution increases. In brief, the electrostatic attraction of oppositely charged ions acts to decrease the effective concentration (called the *activity*) of the ions in solution. This causes the ionization constants for weak monoprotic acids of the type we are studying in this experiment to increase by about 15% at an ionic strength of 0.02 M. Values of ionization constants tabulated in the literature have usually been determined by extrapolating to zero ionic strength, where interionic attraction effects are negligible.

In the method of half-neutralization for the measurement of the pK_a of an acid, we assume that the equilibrium concentrations of A^- and HA are equal to their stoichiometric concentrations. In order for this to be true, $[H_3O^+]$ and $[OH^-]$ must be negligible compared to $[A^-]$. Show that this condition is fulfilled in the measurements you made in part 2.

By taking the logarithm (base 10) of both sides of Equation (2) and multiplying all of the terms by -1, we can obtain the following equation:

$$-\log(K_a) = -\log[H_3O^+] - \log\frac{[A^-]}{[HA]} \quad (14)$$

which is equivalent to

$$pH = pK_a + \log\frac{[A^-]}{[HA]} \quad (15)$$

Now if we added 5% more than the calculated volume of NaOH required to half-neutralize an acid, the ratio of $[A^-]/[HA]$ would be approximately equal to 1.05/0.95. Show from Equation (15) that this would cause the measured pH to be 0.04 unit larger than if we had added the exact amount required. This is probably a fairly realistic estimate of the uncertainty in our measured value of pK_a.

The OH^- concentration of 0.500 M ammonia, NH_3, is 9.5×10^{-4} M.

(a) Write the chemical equation for the dissociation of NH_3 in water.

(b) Write the equilibrium constant expression for the reaction of NH_3 with water.

(c) Calculate the value of K_b using the information given.

Calculate the H_3O^+ concentration and pH of a carbonated beverage that is 0.10 M in dissolved CO_2. (Essentially all of the H_3O^+ comes from the first stage of dissociation, $CO_2(aq) + 2 H_2O \rightleftharpoons H_3O^+ + HCO_3^-$, for which $K_1 = 4.4 \times 10^{-7}$.)

Determination of the Molar Mass and Ionization Constant of a Weak Monoprotic Acid

Name _____

Date _____ Section _____

Locker _____ Instructor_____

Data and Calculations

1. Measurement of pH and titration of acetic acid solution

Concentration of standardized NaOH titrant _____ mol/L

Mass concentration of acetic acid _____ g/L

	Trial 1	Trial 2	Trial 3
Measured pH of the acetic acid solution			
Mass of acetic acid solution taken for titration	g	g	g
Initial buret reading of NaOH titrant	mL	mL	mL
Final buret reading of NaOH titrant	mL	mL	mL
Net volume of NaOH	mL	mL	mL
Millimoles of NaOH to end point of titration	mmol	mmol	mmol
Millimoles of acetic acid in sample	mmol	mmol	mmol
Molar concentration of acetic acid solution	mol/L	mol/L	mol/L
Calculated molar mass of acetic acid	g/mol	g/mol	g/mol

Calculation of the ionization constant for acetic acid from the measured pH of the acetic acid samples. (Average the pH and molar concentration values for the samples of acetic acid you titrated; use the average values in the calculation of the ionization constant.)

Average molar concentration of acetic acid for the samples you titrated: _____ mol/L

Average pH _____

Calculate the corresponding $[H_3O^+]$, compute the concentration of A^- and HA, and calculate the dissociation constant, K_a.

_____ M H_3O^+ _____ M A^- _____ M HA

$$K_a = \frac{[H_3O^+][A^-]}{[HA]} = \underline{\hspace{3cm}} ; \quad pK_a = \underline{\hspace{3cm}}$$

2. Measurement of the pK_a of acetic acid by the half-neutralization method

Concentration of standardized NaOH titrant _____ mol/L

	Trial 1	Trial 2	Trial 3
Mass of acetic acid solution	g	g	g
Volume of NaOH added to half-neutralize the acetic acid	mL	mL	mL
Measured pH of half-neutralized solution			

Average pH value _____

pK_a = _____; K_a = _____

Calculate the $[H_3O^+]$ and the concentrations of A^- and HA, in the half-neutralized solution.

(*Note*: In this solution, $[H_3O^+] \neq [A^-]$.)

_____ M H_3O^+ _____ M A^- _____ M HA

3. The molar mass and ionization constant of an unknown weak monoprotic acid

Concentration of standardized NaOH titrant _____ mol/L

Mass concentration of unknown weak monoprotic acid _____ g/L

	Trial 1	Trial 2
Measured pH of the unknown acid solution		
Mass of unknown acid solution taken for titration	g	g
Initial buret reading of NaOH titrant	mL	mL
Final buret reading of NaOH titrant	mL	mL
Net volume of NaOH	mL	mL
Millimoles of NaOH to end point of titration	mmol	mmol
Millimoles of unknown acid in sample	mmol	mmol
Molar concentration of unknown acid solution	mol/L	mol/L
Calculated molar mass of unknown acid	g/mol	g/mol

Calculation of the ionization constant for an unknown acid from the measured pH of the unknown acid samples. (Average the pH and molar concentration values for the samples of unknown acid you titrated; use the average values in the calculation of the ionization constant.)

Average molar concentration of unknown for the samples you titrated _____ mol/L

Average pH _____

Calculate the corresponding $[H_3O^+]$, compute the concentration of A^- and HA, and calculate the dissociation constant, K_a.

_____ M H_3O^+ _____ M A^- _____ M HA

$$K_a = \frac{[H_3O^+][A^-]}{[HA]} = \underline{\qquad\qquad}; \quad pK_a = \underline{\qquad\qquad}$$

Name _____ Date _____

Measurement of the pK_a of an unknown acid by the half-neutralization method

Concentration of standardized NaOH titrant _____ mol/L

	Trial 1	Trial 2
Mass of unknown acid solution	g	g
Volume of NaOH added to half-neutralize the unknown acid	mL	mL
Measured pH of half-neutralized solution		

Average pH value _____

pK_a = _____ ; K_a = _____

Calculate the $[H_3O^+]$ and the concentration of A^- and HA, in the half-neutralized solution.

(*Note*: In this solution, $[H_3O^+] \neq [A^-]$.)

_____ M H_3O^+ _____ M A^- _____ M HA

Comparing both the calculated molar mass and the ionization constant, which acid in Table 6 of the Appendix is the unknown acid most likely to be?

Give the molar mass and pK_a for the unknown acid that you measured:

Give the molar mass and K_a for the acid listed in Table 6 of the Appendix that you think is the same as your unknown acid.

▬▬▬   **CONSIDER THIS**   ▬▬▬

In the method of half-neutralization for the measurement of the pK_a of an acid, we assume that the equilibrium concentrations of A^- and HA are equal to their stoichiometric concentrations. In order for this to be true, $[H_3O^+]$ and $[OH^-]$ must be negligible compared to $[A^-]$. Show that this condition is fulfilled in the measurements you made in part 2.

By taking the logarithm (base 10) of both sides of Equation (2) and multiplying all of the terms by −1 we can obtain the following equation:

$$-\log(K_a) = -\log[H_3O^+] - \log\frac{[A^-]}{[HA]} \tag{14}$$

which is equivalent to

$$pH = pK_a + \log\frac{[A^-]}{[HA]} \tag{15}$$

Now if we added 5% more than the calculated volume of NaOH required to half-neutralize an acid, the ratio of $[A^-]/[HA]$ would be approximately equal to 1.05/0.95. Show from Equation (15) that this would cause the measured pH to be 0.04 unit larger than if we had added the exact amount required. This is probably a fairly realistic estimate of the uncertainty in our measured value of pK_a.

The OH^- concentration of 0.050 M ammonia, NH_3, is 9.5×10^{-4} M.

(a) Write the chemical equation for the dissociation of NH_3 in water.

(b) Write the equilibrium constant expression for the reaction of NH_3 with water.

(c) Calculate the value of K_b using the information given.

Calculate the H_3O^+ concentration and pH of a carbonated beverage that is 0.10 M in dissolved CO_2. (Essentially all of the H_3O^+ comes from the first stage of dissociation, $CO_2(aq) + 2\ H_2O \rightleftharpoons H_3O^+ + HCO_3^-$, for which $K_1 = 4.4 \times 10^{-7}$.)

_____ M

_____ pH

What's in Your Drinking Water?

Comparing Tap Water and Bottled Water

Purpose

• Show how the chemical composition of municipal drinking water is largely determined by geochemistry.

• Determine the concentrations of the two most important cations found in tap water and bottled water—calcium and magnesium—that are mainly responsible for water hardness.

• Determine the alkalinity of tap water and bottled water, due mainly to the presence of hydrogen carbonate ion, HCO_3^-.

• Make semi-quantitative tests for the presence of sulfate and chloride ions in tap water and bottled water.

• Compare the ion concentrations found in tap water and bottled water.

Introduction

It would be hard to imagine life without water. It is the medium in which all of the biological processes of living cells take place. Water is so essential that it can justifiably be called the most valuable of all our resources. Because we breathe oxygen, we might think that it is more essential to life, but that is a human-centered view. Many simple organisms can live without oxygen, but we know of no organism that can live and reproduce without water.

There is a lot of water on Earth (1.4×10^{21} L estimated), but it is mostly seawater. An estimated 97.2% is in the oceans and about 2.1% is trapped in the polar icecaps and glaciers, leaving about 0.7% of the water in the hydrosphere as the fraction on which we must rely. Our supply of fresh water would soon be exhausted except for the continual renewal of this resource by the solar energy–driven cycle of evaporation, condensation, and precipitation as rain and snow.

Ninety percent of Americans have access to municipal water utilities that provide drinking water simply by turning on the tap. Although municipal drinking water is generally regarded as safe, nearly all municipal drinking water contains low levels of contamination. The Environmental Working Group (EWG), a nongovernmental organization, has compiled the National Tap Water Quality Database (www.ewg.org/tapwater) by acquiring federally mandated testing data from state water offices. From these reports they have identified about 260 contaminants in the nation's water supply. About 140 of these have no enforceable safety limits, meaning that public health officials have not set safety standards for these contaminants, which include both inorganic and organic chemicals. The current EWG database contains data from 39,751 water utilities in 42 states. There is a good chance that your hometown is contained in the database, so you can see what contaminants are in your tap water and how that information stacks up against other cities and towns throughout the United States.

According to a 2005 report of the World Health Organization (WHO), an agency of the United Nations (UN), about 1.1 billion people lacked access to safe drinking water; about 2.8 billion, or over 40% of the total population of the Earth, lacked access to basic sanitation. Recognizing this, the UN declared the period from 2005 to 2015 the "International Decade for Action—Water for Life"; the goal is to halve, by 2015, the proportion of people without sustainable access to safe drinking water and basic sanitation (www.un.org/waterforlifedecade/factsheet.html).

The Chemical Composition of Natural Waters

In an area such as the Amazon River basin in South America, far removed from volcanic activity and the smokestacks of industry, rainwater can be very pure, containing a few dissolved substances— mainly carbon dioxide, traces of nitrogen oxides produced by lightning, and traces of NaCl. Near the sea, the rainfall will contain more NaCl. Although NaCl is not volatile, the interaction of wind and waves produces tiny droplets of seawater (called an *aerosol*) that are carried inland by the wind.

In areas near active volcanoes and in the industrialized countries, the precipitation contains appreciable concentrations of dissolved acids. In industrialized areas, these arise mostly from the combustion of fossil fuels—for heating, cooking, transportation, industrial processing, and generating electricity. Nitrogen oxides and sulfur dioxide (mainly from burning coal that contains sulfur) are produced in the primary combustion process. These are further oxidized to nitric and sulfuric acids in a complex series of reactions that take place in the atmosphere.

Table 29-1 shows the average composition of rainfall in the northeastern United States for weighted ion concentrations from 17 sampling sites over a period of 88 months. These data clearly show the presence of substantial quantities of sulfuric and nitric acids. In a region where the soils lack alkaline components that can neutralize the acids, this acid rain produces destructive ecological effects,

killing fish and other aquatic life and damaging vegetation.

The composition of river water is mainly determined by the reactions of rainwater (and water from melting snow) with rocks and soil as the water makes its way downhill to the sea. *Gypsum* ($CaSO_4 \cdot 2\ H_2O$) and *calcite* ($CaCO_3$) are commonly found in sedimentary deposits at the Earth's surface. Another less stable form of calcium carbonate, *aragonite*, is found in sediments that consist mainly of the mineral skeletons of marine organisms. A relatively common mineral in ancient rocks is *dolomite*, $MgCa(CO_3)_2$. Gypsum is slightly soluble in water and all of the carbonates can be readily dissolved by water containing traces of acid. Unless the rainfall is acidic from industrial emissions, the main acid present in rainwater is dissolved CO_2, which forms a slightly acidic solution.

$$CO_2(g) + 2\ H_2O(liq) \rightleftharpoons H_3O^+(aq) + HCO_3^-(aq) \quad (1)$$

In turn, the acid formed reacts with rocks containing calcium (and magnesium) carbonate.

$$H_3O^+(aq) + CaCO_3(s) \rightleftharpoons$$
$$Ca^{2+}(aq) + HCO_3^-(aq) + H_2O(liq) \quad (2)$$

TABLE 29-1 **The Average Chemical Composition of Natural Waters**[a]

	Rainfall, Northeastern United States (mmol/L)	Rivers of North America (mmol/L)	Seawater (mmol/L)
Cations			
Ca^{2+}	—	0.53	10.6
Mg^{2+}	—	0.21	54.5
Na^+	0.009	0.39	479.
K^+	—	0.036	10.2
H^+	0.072	—	—
NH_4^+	0.016	—	—
Anions			
HCO_3^-	—	1.1	2.3
SO_4^{2-}	0.028	0.21	28.9
Cl^-	0.012	0.23	546.
Br^-	—	—	0.85
F^-	—	0.008	0.07
NO_3^-	0.026	0.017	0.0001
$H_2SiO_4^{2-}$	—	0.15	—

[a]Quoted by J. N Butler, *Carbon Dioxide Equilibria and Their Applications*, Addison Wesley, Reading, MA, 1982, Chapter 5.

The net reaction is the dissolution of calcium carbonate as given by Equation (3), which is the sum of Equations (1) and (2).

$$CO_2(g) + H_2O(liq) + CaCO_3(s) \rightleftharpoons$$
$$Ca^{2+}(aq) + 2\ HCO_3^-(aq) \quad (3)$$

These geochemical processes largely account for the fact that calcium and magnesium ions are the most abundant cations and that bicarbonate ion is the most abundant anion in river water, as shown in Table 29-1.

By far the most abundant ionic solute in seawater is sodium chloride (see Table 29-1). This should not be too surprising, because sodium chloride and other sodium salts are both abundant and very soluble. The natural hydrologic cycle, operating over eons of time, has by now washed into the sea most of the soluble sodium salts that lie near the surface of the Earth.

In most regions of the United States, water for domestic use comes mainly from rivers and lakes (called *surface water*) and from wells (called *ground water*); it contains some dissolved salts, mostly calcium bicarbonate resulting from the reaction shown in Equation (3), with lesser amounts of magnesium, sodium, sulfate, and chloride ions.

To be suitable for use in drinking, cooking, bathing, and watering plants, water should not contain more than 200–300 parts per million (ppm) of dissolved solids, where 1 ppm = 1 mg/kg. Stated in words, 1 ppm is equal to 1 milligram (mg) of dissolved solids per million milligrams (1 kg) of water. (Because the densities of tap and bottled water are close to 1.00 g/mL, we can without serious error say that 1 ppm $\cong$ 1 mg/L.)

The total dissolved solids can be measured by evaporating a known volume of water to dryness and weighing the residue. (Or an approximate value can be obtained more quickly by measuring the electrical conductivity of the water because most of the dissolved solids are ionic.)

Water that contains Ca^{2+} and Mg^{2+} ions is called *hard water* because these ions form scale (mineral deposits) in boilers and plumbing and react with soaps (the carboxylate anions of long-chain fatty acids) to form an insoluble curdy scum. This precipitate reduces the cleaning efficiency of soap (and leaves a ring of scum on the tub when you take a bath).

If the water hardness is greater than 200 mg/L (expressed as the equivalent amount of $CaCO_3$ in mg/L), it may be desirable to soften the water used for laundry and washing by decreasing its content of calcium and magnesium ions. (Drinking soft water is not recommended for two reasons: It deprives your body of calcium and magnesium ions essential for good health, and it is more corrosive than hard water, able to leach potentially harmful trace metals from the pipes carrying the water.) Water softening can be done in several ways; before the advent of synthetic detergents (which do not form insoluble calcium and magnesium salts), washing soda ($Na_2CO_3 \cdot 10\ H_2O$) was often added to laundry water to remove calcium ions by the reaction

$$Ca^{2+}(aq) + CO_3^{2-}(aq) \rightarrow CaCO_3(s) \quad (4)$$

Another method, called *ion exchange*, replaces calcium and magnesium ions with sodium ions by passing the water through a tank containing small beads of natural or synthetic solid ion exchanger in the sodium ion form.

Bottled water and its critics

In the United States, the Federal Drug Administration (FDA) regulates bottled water as food, while the Environmental Protection Agency (EPA) regulates tap water that comes from municipal water supplies. The FDA defines several categories of bottled water, including mineral water, purified water, sparkling bottled water containing dissolved carbon dioxide, spring water, and artesian water. By FDA regulations, mineral water must contain no less than 250 mg/L total dissolved solids and come from a geologically and physically protected underground water source.

Americans seem to have a taste for water with a lower content of dissolved solids; several widely sold brands are therefore produced by purifying water from public water sources, then adding back small amounts of various salts commonly present in natural waters to produce a desirable taste. (Pure water containing no dissolved solids has a flat, unappetizing taste.) These brands of bottled water are labeled as "purified."

Is it Better to Drink Bottled Water than Tap Water? There is no simple yes/no answer to this question. There are cultural traditions, cost considerations, environmental and resource issues, and health and safety issues to think about. We will briefly discuss each of these in turn, and you can go to the sources cited in the bibliography for more information.

For decades, Europeans have had the custom of drinking bottled water, mainly with meals, in the belief that it aids digestion. This water often comes from local springs and water sources and is bottled and served in the local area. Various health spas located in areas where there are mineral springs have been (and still are) popular with wealthier Europeans, the rationale being that drinking and bathing in mineral waters supposedly has health benefits and curative powers.

Drinking bottled water has surged in popularity in the United States over the last 30 years. In 1976, annual consumption was less than 2 gal for every man, woman, and child. Americans are now drinking, on average, about 30 gal of bottled water a year, with sales recently growing by 10% per year—far faster than any other beverage. There is little doubt that drinking bottled water leads to better health and nutrition than drinking sugared and diet colas, especially for children.

The cost of bottled water. The critics of bottled water start with its cost. Bottled water typically costs 1000 times more per liter than high-quality municipal tap water. Its average retail cost is about the same price as gasoline at $3/gal. By comparison, tap water costs less than 1¢/gal and is delivered through an energy-efficient infrastructure.

Bottled water and resources. Critics also argue that bottled water is wasteful of resources and that the money spent on bottled water could be put to better use by spending it on improvements to municipal water supplies. Bottled water is usually served in single-use plastic bottles—about 30 billion of them made per year—with the petroleum-derived plastic requiring the equivalent of about 15 million barrels of crude oil.

Much bottled water now crosses state and national borders with brands from remote regions often having a special appeal for consumers. The Pacific Institute has estimated that the energy costs for the pumping, bottling, transportation, and refrigeration of bottled water requires the energy equivalent of another 50 million barrels of oil.

Nationwide, less than 20% of plastic bottles are recycled, leading to mountains of solid waste and litter. A few U.S. states have passed bottle deposit laws. Michigan's bottle deposit law was passed in 1976; at 10¢ per bottle, it is the highest in the country—and so is its 97% redemption rate. Environmental groups say that we need a federal bottle deposit law that would apply to all states.

Drinking water and health. Bottled water provides a convenient carry-along source of clean water, and it's likely that the great increase in the consumption of bottled water means that people are, on average, drinking more water.

Although most people have access to clean and safe municipal water, some prefer the taste of bottled water to their tap water; others wish to avoid drinking municipal tap water because it may contain low levels of substances that are known to be harmful. Two such harmful substances are nitrate ion and perchlorate ion, which pose a risk for infants. Municipal water is commonly treated with chlorine to kill harmful bacteria. Besides the residual chlorine, chlorination produces low levels of carcinogenic trihalomethanes if the water contains dissolved organic matter.

There have also been episodes of contamination of municipal water supplies by microorganisms, such as the massive 1993 cryptosporidium outbreak in Milwaukee, Wisconsin, that sickened 400,000 people over a period of two weeks. So it would not be surprising that people who read or hear about contamination of municipal water supplies prefer to minimize even low-level risk by drinking bottled water. However, analysis of bottled waters supported by the Natural Resources Defense Council showed that some samples of bottled water were not entirely free of microbial contamination, low-level chemical contamination, and materials leached from the plastic of the bottles.

Studies of the diet and nutrition of U.S. citizens show that many have less than optimal intakes of calcium and—more critically—of magnesium, which is an essential nutrient for good cardiovascular health. The U.S. National Academy of Sciences published a study in 1977 estimating that drinking water containing magnesium ion could make a significant contribution to cardiovascular health. Such epidemiological studies show an inverse relation between cardiovascular health and the calcium and magnesium content of the water—the softer the water, the higher the incidence of cardiovascular events, including premature death. For this reason, some public health officials and critics of bottled water have urged purveyors of bottled water to boost the magnesium ion content of their water to as high as 100 mg/L. Drinking a liter of water daily that contains this amount of magnesium ion would supply about 25% of the daily recommended intake of ~400 mg of magnesium.

In 2007, a public campaign was launched by activists, influential restaurateurs, and several U.S. cities to persuade consumers to choose tap water over bottled water. They argued that bottled water is costly, wasteful of resources, and is not superior to tap water in health benefits.

The U.S. Department of Agriculture (USDA) Agricultural Research Service maintains a Web site, called the Community Nutrition Mapping Project, that displays state-by-state information on the essential nutrient intake of the general population in each state.

It is suggested that three or four different brands of bottled water be made available in the laboratory, or students can be asked to bring their own favorite brand of bottled water. Each student is expected to gather data for local tap water and for one brand of bottled water.

The EDTA and alkalinity titrations of parts 2 and 3 could require as many as 18 titrations (3 analytes × 2 kinds of water × 3 replicate titrations = 18). For this reason, it is suggested that students work in pairs, dividing the responsibility for the titrations in parts 2 and 3.

No macroscale titration procedure is given for parts 2 and 3 because macroscale titrations take more time, require at least tenfold larger amounts of buffer and titrant, and give visual indicator end points that are less sharp than microscale titrations.

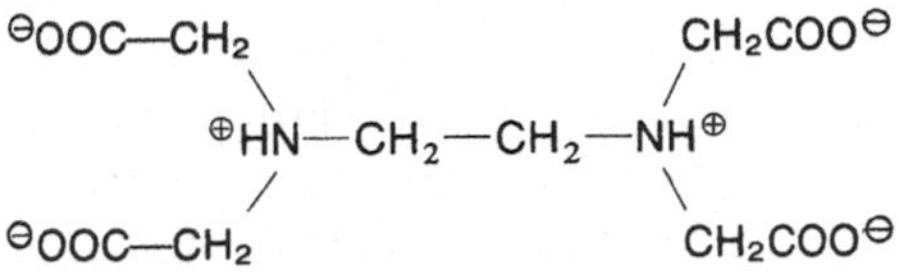

Pre-Lab Preparation

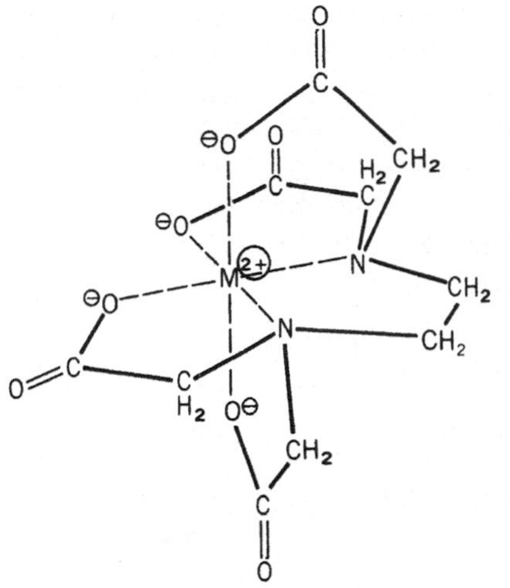

Determining Ca^{2+} and Mg^{2+} by EDTA Titration In this experiment, you will measure the concentrations of Ca^{2+} and Mg^{2+} by two sets of titrations, using ethylenediaminetetraacetic acid (EDTA), a chelating agent that forms 1:1 complexes with many metal ions having a +2 charge (M^{2+}).

The first set of titrations, done in pH 10 buffer using calmagite indicator, gives the sum of the concentrations of Mg^{2+} + Ca^{2+}. We call this sum the *water hardness*.

The second set, done at pH 12.5 using calcon indicator, gives the concentration of Ca^{2+} only, because Mg^{2+} is precipitated as Mg(OH)$_2$(s) at pH 12.5; therefore, it is not titrated by the EDTA. The difference between the two sets of titrations gives the concentration of Mg^{2+}:

$$\frac{mmol}{L}Mg^{2+} = \frac{mmol}{L}(Mg^{2+}+Ca^{2+}) - \frac{mmol}{L}(Ca^{2+}) \quad (5)$$
$$\text{(pH 10)} \qquad\qquad \text{(pH 12.5)}$$

Next, we will explain some of the chemistry behind the EDTA titration, using as our example the first titration, in which we determine water hardness as the sum of the concentrations of calcium and magnesium ions.

Ethylenediaminetetraacetic acid is a polyaminocarboxylic acid. This name is quite a mouthful, so it is usually abbreviated to EDTA. Fully protonated EDTA is a hexaprotic acid that we will represent by the symbol H$_6$Y^{2+}. The form of EDTA most commonly employed to make titrant solutions is the soluble

FIGURE 29-1 Structures of ethylenediaminetetraacetic acid (EDTA) and metal–EDTA chelates: (A) structure of the dianion of EDTA, H$_2$Y^{2-}; (B) structure of an M^{2+}–EDTA chelate, MY^{2-}.

disodium salt dihydrate, Na$_2$H$_2$Y · 2 H$_2$O, which is available in high purity. The diprotonated anion H$_2$Y^{2-} reacts[1] with Ca^{2+} (or Mg^{2+} or any metal ion with charge +2 or greater) to form a complex ion, CaY^{2-}:

$$H_2Y^{2-} + Ca^{2+} \rightleftharpoons CaY^{2-} + 2 H^+ \quad (6)$$

The structures of the anion H$_2$Y^{2-} and the EDTA complex with a metal ion, M^{2+}, are shown in Figure 29-1. Think of the EDTA anion as surrounding a metal ion like the large claw of a crab grasping its prey. The technical term for such a complex is *chelate*, which comes from the Greek word meaning "a pincer-like claw." In the chelate complex ion, negatively charged carboxylate groups and lone pair electrons on the nitrogen atoms are attracted to the positively charged metal ion.

To determine the end point of the titration, an indicator dye is added that forms a colored complex

---

[1]When Ca^{2+} reacts with H$_2$Y^{2-}, the two liberated H$^+$ ions are neutralized by the basic NH$_3$ component of the buffer, which acts to keep the pH essentially constant.

ion with M^{2+} ions. When added to a solution containing both Ca^{2+} and Mg^{2+} ions at pH 10, the triprotic calmagite indicator (added as a solution of $Na^+H_2In^-$) preferentially forms the red $MgIn^-$ complex ion because the equilibrium constant for formation of the indicator–magnesium ion complex is about 100 times larger than the equilibrium constant for formation of the indicator complex with calcium ion.

In a pH 10 buffer, any EDTA uncomplexed with metal ions exists mainly as HY^{3-}. The equilibrium constant for the formation of the EDTA–calcium ion complex is about 100 times larger than for the EDTA–magnesium complex; so as EDTA titrant is added to the solution, the following sequence of reactions takes place:

pH 10

$HY^{3-} + Ca^{2+} \rightleftharpoons CaY^{2-} + H^+$

$Mg^{2+}(free) + HY^{3-} \rightleftharpoons MgY^{2-} + H^+$

$HY^{3-} + MgIn^- \rightleftharpoons MgY^{2-} + HIn^{2-}$

RED   BLUE

As EDTA is added, the HY^{3-} reacts mainly with the Ca^{2+} ion, then with Mg^{2+} ion uncomplexed with indicator that we call the *free* Mg^{2+}. When enough EDTA has been added to react with all of the calcium and magnesium ions, the last bit of added EDTA displaces the indicator from the $MgIn^-$ complex ion, causing a color change from red to pure blue.

In this titration, if there is no Mg^{2+} ion in solution to initially form the red $MgIn^-$ complex, the end point will not be as sharp. Therefore, we deliberately add some Mg–EDTA solution (in the form of Na_2MgY) to make sure that some Mg^{2+} ion will be present. By adding it in the form of the Mg–EDTA complex, the volume of EDTA solution required to titrate the ions originally present is not significantly changed.

Water quality engineers typically express water hardness in units equivalent to parts per million (ppm), or mg/L, of $CaCO_3$, in effect pretending that the ions titrated by EDTA came from an equivalent amount of dissolved $CaCO_3$. (1 ppm $CaCO_3$ = 1 mg/kg water $\cong$ 1 mg/L $CaCO_3$ because the mass of a liter of tap or bottled water is approximately 1 kg, or 1 million mg.)

In most natural waters, some magnesium ion is present, so expressing water hardness in units of mg/L of $CaCO_3$ is an arbitrary convention that has been adopted for the sake of convenience (or tradition). The conversion from mmol/L of $Ca^{2+} + Mg^{2+}$ is made using the following equation:

$$\frac{mg}{L}CaCO_3 = \frac{mmol}{L}(Ca^{2+}+Mg^{2+}) \times$$

$$\frac{1 \text{ mol } CaCO_3}{1 \text{ mol } (Ca^{2+}+Mg^{2+})} \times \frac{100 \text{ g } CaCO_3}{mol \text{ } CaCO_3} \quad (7)$$

In effect, we see that the water hardness in units of mg/L of $CaCO_3$ is just equal to the total concentration of $Ca^{2+} + Mg^{2+}$ in units of mmol/L, multiplied by a factor of 100 (the molar mass of $CaCO_3$).

Alkalinity The alkalinity of tap or bottled water is the sum of all the bases in solution that can be titrated to a specified pH by a strong acid such as HCl. The most common base found in natural waters is hydrogen carbonate ion, HCO_3^-, formed when water flows over carbonate-containing rocks, as shown by the net reaction of Equation (3). Lesser amounts of alkaline silicate $SiO(OH)_3^-$, borate $B(OH)_4^-$, phosphate HPO_4^{2-}, and sulfate SO_4^{2-} ions also may be present. Hydrogen carbonate ion HCO_3^- acts as a buffer and tends to stabilize the pH of the water so that when quantities of acid or base smaller than the amount of HCO_3^- present are added to the water, the pH of the water does not change much.

As shown in Figure 29-2, when a dilute solution of hydrogen carbonate ion is titrated with HCl to form H_2CO_3 ($H_2O + CO_2$), the end point comes at a pH of about 4.6; this is usually the end point pH specified for determining alkalinity. The procedure you will use calls for titrating samples of tap water and

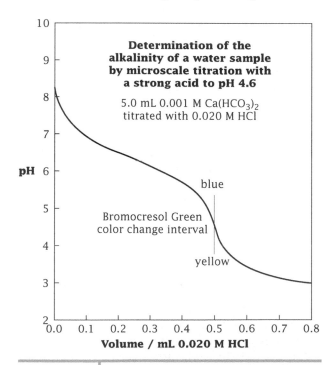

FIGURE 29-2 | Determining the alkalinity of a water sample by titration with a strong acid to pH 4.6. The alkalinity of natural waters is mainly caused by the presence of the hydrogen carbonate ion, HCO_3^-. The figure shows the titration of 5 mL of 0.001 M $Ca(HCO_3)_2$ with 0.02 M HCl. Bromocresol green is a satisfactory acid–base indicator for this titration. Calcium ion, Ca^{2+}, plays no significant role in this titration.

bottled water to the bromocresol green indicator end point, whose color change takes place in the pH range 5.4 to 3.8. The midpoint of this range corresponds to pH 4.6, as shown in Figure 29-2. (At the pH of the end point, sulfate ion SO_4^{2-} is too weak a base to be titrated to HSO_4^-, so any sulfate ion present will not make a significant contribution to the measured alkalinity.)

Alkalinity, like water hardness, is traditionally expressed in terms of the mg/L (ppm) of $CaCO_3$ equivalent to the total base present that is neutralized by the HCl titrant. The mmol HCl required to titrate the water sample is given by

$$mmol\ HCl = V_{HCl}\ (mL) \times M_{HCl}\ (mol/L) \qquad (8)$$

and the alkalinity expressed as the equivalent mg/L of $CaCO_3$ will be given by

$$alkalinity = \frac{mg}{L}\ CaCO_3 =$$

$$\frac{mmol\ HCl}{volume\ (L)\ of\ water\ sample} \times$$

$$\frac{1\ mol\ CaCO_3}{2\ mol\ HCl} \times \frac{100\ g\ CaCO_3}{mol\ CaCO_3} \qquad (9)$$

Note that if the alkalinity actually existed in the form of $CaCO_3$, it would take 2 mol of HCl to titrate 1 mol of $CaCO_3$ to the CO_2 end point:

$$CaCO_3 + 2\ HCl \rightarrow CaCl_2 + H_2O + CO_2 \qquad (10)$$

So each mmol of HCl titrant is formally equivalent to 0.5 mmol (50 mg) of $CaCO_3$. Finally, note that for a 1.00 mmol/L solution of $Ca(HCO_3)_2$, both the water hardness and the alkalinity would be equivalent to 100 mg/L (or 100 ppm) of $CaCO_3$.

Experimental Procedure

Special Supplies and Equipment: *Part 1:* pH meter and calibration buffers; conductivity meter and a conductivity calibration or test solution. Pocket conductivity testers having a range of ~2000 microSiemens/cm (microMhos/cm), equivalent to approximately 1000 TDS (total dissolved solids) in units of mg/L, are sufficiently accurate for this experiment. Most tap and bottled waters will have conductivities in the range 100–1000 microSiemens/cm.

Parts 2 and 3: 25-mL Erlenmeyer flask; 1-mL microburet (1-mL polypropylene Norm-Ject Luer Slip Syringe, Air-Tite Products Co., Inc., www.air-tite-shop.com, fitted with a fine tip cut from a 50 drops/mL polyethylene transfer pipet, as described in the Introduction of this laboratory manual); magnetic stirrer and small stirring bar (optional); 1-mL graduated polyethylene transfer pipets.

Part 4: Centrifuge; 13 × 100 mm culture tubes (rimless test tubes); test tube rack for holding the test tubes; 1-mL graduated polyethylene transfer pipets.

Chemicals: *Part 1:* pH calibration buffers; conductivity test solution (0.144 g/L NaCl has a conductivity of 300 microSiemens/cm at 25 °C.)

Part 2: 0.0100 M $Na_2H_2EDTA \cdot 2\ H_2O$ (3.722 g/L) 99+% ethylenediaminetetraacetic acid; disodium salt dihydrate, FW 372.24 (Sigma-Aldrich catalog no. 25,235-2 or 10,631-3); 1.5 M NH_3/0.3 M NH_4Cl buffer [100 mL of conc. (15 M) NH_3 and 16.0 g NH_4Cl per liter of solution]; 0.010 M $Na_2MgEDTA$/0.1 M NH_3 solution [6.7 mL of conc. (15 M) NH_3 and 3.6 g $Na_2MgEDTA$ per liter of solution]; 0.10 M (4.0 g/L) NaOH; 0.1% calmagite indicator (0.10 g/100 mL deionized water), Aldrich catalog no. C20-4, FW 358.37; 0.4% calcon indicator (0.2 g in 50 mL ethanol or methanol), Sigma Catalog No. H 1878, approximately 50% dye content; sodium bicarbonate, $NaHCO_3$(s).

Part 3: 0.020 M HCl; 0.1% bromocresol green sodium salt (100 mg/100 mL water).

Part 4: 1.0 M nitric acid (HNO_3); 0.05 M $BaCl_2$; 0.10 M $AgNO_3$; 30 mg/L SO_4^{2-} solution (0.044 g/L Na_2SO_4); 30 mg/L Cl^- solution (0.049 g/L NaCl).

☠ **WASTE COLLECTION AND CLEAN-UP:** Collect all of the titration solutions for part 2 (EDTA titrations) and part 3 (alkalinity titrations) in a single 400-mL beaker. Add about 8 g of sodium bicarbonate ($NaHCO_3$) and stir until all the added $NaHCO_3$ has dissolved. Flush the neutralized solutions down the drain.

For part 4, after making your observations on the relative amounts of precipitate in the test tubes, stir the solutions to redisperse the centrifuged precipitates. Then dispose of the solutions containing barium sulfate and barium ion (tubes 1–3) and silver chloride and silver ion (tubes 4–6) in separate waste collection bottles labeled as "barium waste" and "silver waste."

1. pH and Conductivity of Tap Water and Bottled Water Measure and record the pH of samples of tap water and bottled water, using a pH meter properly calibrated with standard pH buffers.

Read the directions for use of the conductivity meter or receive directions from the instructor. Some meters may require calibration with conductivity standards having particular values. Others may not read in preferred units of microSiemens/cm but are designed to read total dissolved solids (TDS) in units of mg/L. The TDS reading, which cannot be exact because the true TDS value depends on the unknown chemical composition of the water sample, is typically about half the reading of the conductivity in microSiemens/cm. As you make the conductivity readings, note and record the temperature of the samples.

To check that the conductivity meter is working normally and to give you confidence that you understand the conductivity units displayed on the meter, first make a reading on a conductivity test solution available in the laboratory (0.144 g/L NaCl solution has a conductivity of ~300 microSiemens/cm at 25 °C). Some meters have built-in temperature compensation—others don't. Conductivity typically has a positive temperature coefficient, increasing about 2% per °C.

Between each reading, rinse the electrode with some of the next sample contained in a beaker. Discard this rinsing, then make a reading on a fresh portion of the sample. Continue in this way until you have made and recorded readings of conductivity and temperature for samples of the conductivity standard, tap water, and bottled water.

Estimating the sample size for part 2a. You will use the conductivity reading to estimate a value of the sample size for the titration in part 2a, the next part of the experiment. The reasoning behind the estimate goes like this: $Ca(HCO_3)_2$ is used as a proxy for the sum of the concentrations of Ca^{2+} + Mg^{2+} because most natural waters contain more Ca^{2+} than Mg^{2+} and HCO_3^- is usually the predominant anion. A 0.0010 M solution of $Ca(HCO_3)_2$ has a conductivity of about 208 microSiemens/cm at 25 °C. We will aim to use at least 0.5 mL of EDTA titrant—half the capacity of our 1-mL microburet—to titrate the Ca^{2+} + Mg^{2+} in the sample. Assuming our EDTA solution is 0.0100 M, we estimate that a sample size of 5 mL would be about right:

$$mmol\ EDTA = mmol\ Ca^{2+}$$

$$0.5\ mL\ EDTA \times 0.0100\ \frac{mol}{L}\ EDTA = 0.005\ mmol =$$

$$5\ mL \times 0.001\ \frac{mol\ Ca^{2+}}{L}$$

So now we can establish an approximate working relationship between the conductivity—proportional to the $Ca(HCO_3)_2$ concentration—and the sample size:

$$V_{sample} \cong 5\ mL \times$$

$$\frac{208\ microSiemens/cm}{sample\ conductivity\ (microSiemens/cm)} \quad (11)$$

For convenience, you will round the calculated sample volume to the nearest mL (but you must be sure to measure the sample volume for all titrations to an accuracy of ±0.1 mL).

The estimated volume is approximate because there may be cations besides Ca^{2+} and Mg^{2+} in the tap water or bottled water samples that also contribute to the conductivity. For example, some bottled waters, particularly mineral waters with a high content of dissolved solids, often contain significant amounts of Na^+ or K^+ that contribute to the conductivity. For such waters, the concentration of $Ca(HCO_3)_2$ would be smaller than estimated from the conductivity. Then our sample size would have to be made larger—but you wouldn't know this until you make a first trial titration.

2a. EDTA Titration of Ca^{2+} + Mg^{2+} Ions at pH 10 Using Calmagite Indicator *Procedure:* Fill a 1-mL microburet with 0.0100 M Na_2H_2EDTA. (Read the solution labels carefully so that you do not mistakenly fill the microburet with 0.010 M $Na_2MgEDTA$ solution.) Then put into a 25-mL Erlenmeyer flask a volume of tap water calculated using Equation (11), measured with an accuracy of ±0.1 mL (after rounding the calculated value to the nearest mL). Add 5.0 mL of 1.5 M NH_3/0.3 M NH_4Cl buffer. Using a 1.0-mL graduated polyethylene transfer pipet, add ~0.5 mL of 0.010 M $Na_2MgEDTA$. Add enough deionized water to bring the total volume to about 10 mL. Finally, add 4 drops of 0.1% calmagite indicator.

If a magnetic stirrer is available, add a small stirring bar to the flask and place the flask on a magnetic stirrer; otherwise, swirl the flask by hand while adding the titrant. Put a sheet of white paper under the flask to make the color change more easily visible. Titrate the buffered tap water sample with 0.0100 M Na_2H_2EDTA from pink to a pure blue end point.

Repeat the titration, adjusting the volume of the sample if necessary so that the titration volume is between 0.5 and 1.0 mL. The ratio *mL 0.0100 M EDTA/mL water sample* should agree within 10% for the two titrations. If it does, move on to the titration of samples of bottled water; otherwise, titrate a third sample of tap water.

Repeat the entire titration procedure for at least two samples of bottled water, using a sample volume estimated from the conductivity of the bottled water.

Calculations: Calculate for each sample the total concentration of Ca^{2+} + Mg^{2+} in units of mmol/L, which will be equal to the mmol of EDTA added, then divide by the sample volume in units of liters. (The concentrations of other M^{2+} ions, such as Fe^{2+} or Mn^{2+}, in tap water are usually negligible compared to the total concentration of Ca^{2+} + Mg^{2+}.)

**2b. EDTA Titration of Ca²⁺ at pH 12.5 Using
Calcon Indicator** *Overview of the procedure:* Calcium is titrated with Na_2H_2EDTA at pH 12.5, where
magnesium ion is precipitated as $Mg(OH)_2(s)$. The
end point is easily detected as a color change from
pink to pure blue. To avoid using diethylamine, as
called for in the procedure originally described by
Hildebrand and Reilley (see the Bibliography), we
will obtain a solution pH of 12.5 by addition of
NaOH.

Procedure: Use the same 1-mL microburet
filled with 0.0100 M Na_2H_2EDTA that was used in
part 2a. Put into a 25-mL Erlenmeyer flask a volume of tap water equal to that used in part 2a,
measured to an accuracy of ±0.1 mL. Add 5.0 mL
of 0.10 M NaOH and enough deionized water to
give a total volume of about 15 mL. Stir the solution, then add 5 drops of 0.4% calcon indicator.
Immediately titrate with 0.0100 M Na_2H_2EDTA contained in the 1-mL microburet. (The indicator is
somewhat unstable at high pH.) The color change
at the end point is from pink to pure blue. Repeat
until you have completed at least two titrations
that agree to within 10%.

Repeat the entire titration procedure for at least
two samples of bottled water, using the same sample
volume that you used for bottled water in part 2a,
again measured to an accuracy of ±0.1 mL.

Calculations: Using the data of part 2a, calculate the total (sum) of the Ca^{2+} and Mg^{2+} concentrations in units of millimoles per liter (mmol/L).

Using the data of part 2b, calculate the Ca^{2+} concentration for the samples titrated and calculate the
average and standard deviation of the calculated
concentrations, in units of mmol/L. Once we have
the concentration of a particular ion in units of
mmol/L, conversion to mass units of mg/L is
straightforward—just multiply by the molar mass of
the ion. Thus, we calculate the concentration of Ca^{2+}
in units of mg/L of Ca^{2+} (*not* mg/L of $CaCO_3$) by multiplying the concentration in mmol/L by the molar
mass of Ca (40.1 g/mol). (Because electrons have
such a small mass, the difference between the
mass of a mole of Ca and a mole of its ion, Ca^{2+}, is
negligible.)

The Mg^{2+} ion concentration can be obtained in
units of mmol/L by taking the difference between
the titrations in part 2a and 2b:

$$Mg^{2+} \text{ mmol/L} = (Mg^{2+} + Ca^{2+}) \text{ mmol/L} - (Ca^{2+}) \text{ mmol/L}$$
$$\text{(from part 2a)} \qquad \text{(from part 2b)}$$

[Why can't you use mass concentration units (mg/L)
when calculating the concentration of Mg^{2+} by taking
the difference?] After calculating the concentration

of Mg^{2+} in units of mmol/L, convert to units of mg/L
of Mg^{2+} by multiplying by the molar mass of Mg.

**3. Determining Alkalinity (HCO_3^- concentration) by Titration with HCl Using Bromocresol
Green Indicator** *Procedure:* Use the sample volumes that were found satisfactory for the EDTA
titrations of tap water and bottled water in part 2a.
Using a graduated cylinder or graduated pipet,
place a sample of tap water in a 25-mL Erlenmeyer
flask. (Record the sample volume, accurate to
±0.1 mL.) Add 2–3 drops of 0.1% bromocresol green
indicator. If a magnetic stirrer is available, place the
flask on the stirrer over a sheet of white paper to
better see the color change. Otherwise, swirl the
flask by hand while adding the titrant. Using a 1-mL
microburet filled to the 1.00-mL mark with 0.020 M
HCl, titrate the sample to the color change from
blue to a greenish-yellow. (The equivalence point
comes at approximately pH 4.6.) If the sample
requires less than 0.5 mL of 0.020 M HCl, increase
the sample size for the second trial. Repeat the
titration on a second sample. The ratio *mL 0.020 M
HCl/mL water sample* should agree within 10% for
the two titrations. If it does, move on to the titration
of bottled water; if not, titrate a third sample.

Repeat the entire procedure for at least two samples of bottled water, measuring the sample volume
to an accuracy of ±0.1 mL.

**4. Semi-Quantitative Testing for Sulfate and
Chloride Ions** *Overview of the procedure:* You will
compare the amount of precipitate produced when
$BaCl_2$ solution is added to equal volumes of tap
water, bottled water, and a 30-mg/L solution of sulfate ion to precipitate sulfate ion as $BaSO_4(s)$. (Centrifugation is used to settle the precipitate to the bottom of a test tube.) The same procedure is used for
chloride ion, comparing the amounts of AgCl precipitate produced when $AgNO_3$ solution is added to
equal volumes of tap and bottled water and a 30-mg/L solution of chloride ion.

Procedure: You will need eleven 13 × 100 mm test
tubes (six tubes for testing and five tubes for holding
the three necessary reagents and 30-mg/L sulfate and
chloride solutions), a test tube rack for holding the
tubes, three 1-mL graduated polyethylene transfer
pipets for measuring the added reagents, and a
10-mL graduated cylinder for measuring the water
samples or sulfate or chloride solutions.

Using a waterproof marking pen, number six
tubes so that you can tell them apart later and label
the five tubes that will hold about 5 mL of 1 M HNO_3,
2 mL of 0.05 M $BaCl_2$, 2 mL of 0.1 M $AgNO_3$, 5 mL of
30 mg/L sulfate ion, and 5 mL 30 mg/L chloride ion
solution.

To each of the six numbered test tubes, add the following solutions. (The water samples and the sulfate and chloride solutions may be measured using a 10-mL graduated cylinder, thoroughly rinsing the graduated cylinder with deionized water between samples. Graduated 1-mL polyethylene transfer pipets may be used to add the reagents to the water samples or to the sulfate and chloride solutions.)

Tube 1: 4 mL of tap water + 0.5 mL 1 M HNO_3 + 0.5 mL of 0.05 M $BaCl_2$

Tube 2: 4 mL of bottled water + 0.5 mL 1 M HNO_3 + 0.5 mL of 0.05 M $BaCl_2$

Tube 3: 4 mL of 30 mg/L SO_4^{2-} solution + 0.5 mL 1 M HNO_3 + 0.5 mL of 0.05 M $BaCl_2$

Tube 4: 4 mL of tap water + 0.5 mL 1 M HNO_3 + 0.5 mL 0.1 M $AgNO_3$

Tube 5: 4 mL of bottled water + 0.5 mL 1 M HNO_3 + 0.5 mL 0.1 M $AgNO_3$

Tube 6: 4 mL of 30 mg/L Cl^- solution + 0.5 mL 1 M HNO_3 + 0.5 mL 0.1 M $AgNO_3$

Thoroughly mix the solutions in each test tube with a clean stirring rod, rinsing the stirring rod with deionized water as you move from one test tube to the next.

Allow the solutions to stand for at least 5 min. Then place all six test tubes into a centrifuge for ~3 min or until all the precipitates have been settled to the bottom of the tubes.

By a visual comparison of the amount of precipitate in the following pairs of tubes, decide whether the tap water and bottled water samples contain concentrations of sulfate and chloride ion that are >30 mg/L, approximately equal to 30 mg/L, or <30 mg/L. Enter the appropriate symbol (>30, =30, or <30) in the table in the Report form.

Compare tubes 1 (tap water) and 3 (30 mg/L sulfate)

Compare tubes 2 (bottled water) and 3 (30 mg/L sulfate)

Compare tubes 4 (tap water) and 6 (30 mg/L chloride)

Compare tubes 5 (bottled water) and 6 (30 mg/L chloride)

When you have finished the comparison of the tubes, dispose of their contents as directed in the Waste Collection instructions.

CONSIDER THIS

Suppose you were in a situation where the only fresh water you had access to appeared muddy and was likely to contain pathogenic bacteria, as is common after flooding, hurricanes, tsunamis, and other natural disasters. In some areas of the world, polluted water is a daily fact of life with attendant high rates of infant mortality from waterborne disease. In Bangladesh, the chronic disease effects of drinking water from wells drilled to avoid using polluted surface water were found to be caused by arsenic in the well water.

There are several Web sites that describe water quality interventions that employ simple, reliable, and inexpensive technologies seeking to make water safe through disinfection and safe storage at the point of use. One of these is the "Clearinghouse: Low-cost household treatment technologies for developing countries" (http://www.jalmandir.com), which describes many different approaches. Another is the Center for Disease Control (CDC) of the U.S. Department of Health and Human Services Web site that describes a "Safe Water Systems (SWS)" approach (http://www.cdc.gov/safewater/index.htm). Visit these Web sites and describe one or two approaches that you think would most nearly meet the following criteria: efficacy in producing safe water using locally available materials, simplicity, low cost, energy efficiency, and least dependence on technology (such as pumps requiring a reliable source of electricity).

Bibliography

Belcher, R.; Close, R. A.; West, T. S. "The Complexometric Titration of Calcium in the Presence of Magnesium. A Critical Study," *Talanta* **1958,** *1,* 238–244.

Hildebrand, G. P.; Reilley, C. N. "New Indicator for Complexometric Titration of Calcium in Presence of Magnesium," *Anal. Chem.* **1957,** *29*(2), 258–264.

Lindstrom, F.; Diehl, H. "Indicator for the Titration of Calcium Plus Magnesium with (Ethylenedinitrilo)tetraacetate," *Anal. Chem.* **1960,** *32,* 1123–1127.

McCormick, P. G. "Titration of Calcium and Magnesium in Milk with EDTA," *J. Chem. Educ.* **1973,** *50,* 136–137.

Internet Web sites:

Environmental Working Group (EWG). National Tap Water Quality Database: http://www.ewg.org/tapwater/

The Magnesium Web site. Magnesium, Drinking Water, and Health: http://www.mgwater.com

Mineral Waters of the World. Taste ratings and summaries of the composition of mineral waters from many different countries: http://mineralwaters.org

Natural Resources Defense Council (NRDC). A discussion of water issues and testing results of over 100 brands of bottled water: http://www.nrdc.org/water/drinking/bwinx.asp/

Pacific Institute. A fact sheet on bottled water and energy: http://www.pacinst.org/topics/water_and_sustainability/bottled_water/bottled_water_and_energy.html

United Nations (UN)/World Health Organization (WHO). Web site for Water, Sanitation, and Health (WSH) also contains the "Rolling Revision of the WHO Guidelines for Drinking-Water Quality": http://www.who.int/water_sanitation_health/en/

U.S. Department of Agriculture (USDA) Community Nutrition Mapping Project (CNMap). This Web site takes you to a map of the United States, where you can find a state's estimated nutrient levels, health indicators, and demographics: http://www.ars.usda.gov/Services/docs.htm?docid=15656.

U.S. Environmental Protection Agency (EPA) regulates municipal water supplies. See the documents describing National Primary Drinking Water Regulations at: http://www.epa.gov/OGWDW/contaminants/index.html

U.S. Food and Drug Administration (FDA) regulates bottled water. See the document "Bottled Water Regulation and the FDA" at: http://www.cfsan.fda.gov/~dms/botwatr.html

What's in Your Drinking Water?

Comparing Tap Water and Bottled Water

Name _____

Date _____ Section _____

Locker _____ Instructor_____

Data and Calculations

Brand name and original source of bottled water: _____

1. pH and conductivity of tap water and bottled water

Specify the conductivity units (usually microSiemens/cm or TDS, mg/L).

Sample solution	pH	Conductivity (specify units)	Temperature
Conductivity standard (0.144 g/L NaCl or other) Specify:	—		°C
Tap water			°C
Bottled water			°C

Estimate the sample volumes to be used for the EDTA titration in part 2a for tap water and bottled water samples, using the following equation. (If your meter measures TDS in mg/L, double the TDS reading to get the approximate conductivity in microSiemens/cm.)

$$V_{sample} \cong 5 \text{ mL} \times \frac{208 \text{ microSiemens/cm}}{\text{Sample conductivity (microSiemens/cm)}}$$

Estimated tap water sample volume: _____ mL (rounded to the nearest mL)

Estimated bottled water sample volume: _____ mL (rounded to the nearest mL)

2a. EDTA titration of Ca^{2+} + Mg^{2+} ions at pH 10 using calmagite indicator

Data	Trial 1	Trial 2	Trial 3
Volume of tap water sample (accurate to ±0.1 mL)	mL	mL	mL
Volume of 0.01 M EDTA titrant Initial buret reading	mL	mL	mL
Final buret reading	mL	mL	
Net volume of 0.01 M EDTA titrant	mL	mL	mL
Volume of bottled water sample (accurate to ±0.1 mL)	mL	mL	mL
Volume of 0.01 M EDTA titrant Initial buret reading	mL	mL	mL
Final buret reading	mL	mL	
Net volume of 0.01 M EDTA titrant	mL	mL	mL

Calculate the concentration of Ca^{2+} + Mg^{2+} in the tap water and bottled water samples in mmol/L. Show a sample calculation in the space below:

Calculations	Trial 1	Trial 2	Trial 3
Concentration of Ca^{2+} + Mg^{2+} in tap water samples	mmol/L	mmol/L	mmol/L
Concentration of Ca^{2+} + Mg^{2+} in bottled water samples	mmol/L	mmol/L	mmol/L

Tap water average for Ca^{2+} + Mg^{2+}: _____ mmol/L

Bottled water average for Ca^{2+} + Mg^{2+}: _____ mmol/L

Using the average value of the Ca^{2+} + Mg^{2+} concentrations in millimoles per liter, calculate using Equation (7) the water hardness of the tap water and bottled water in units of mg/L of $CaCO_3$. Show a sample calculation in the space below:

Tap water hardness: _____ mg/L $CaCO_3$ (ppm)

Bottled water hardness: _____ mg/L $CaCO_3$ (ppm)

Name _____ Date _____

2b. EDTA titration of Ca^{2+} at pH 12.5 using calcon indicator

Data	Trial 1	Trial 2	Trial 3
Volume of tap water sample (accurate to ± 0.1 mL)	mL	mL	mL
Volume of 0.01 M EDTA titrant Initial buret reading	mL	mL	mL
Final buret reading	mL	mL	
Net volume of 0.01 M EDTA titrant	mL	mL	mL
Volume of bottled water sample (accurate to ± 0.1 mL)	mL	mL	mL
Volume of 0.01 M EDTA titrant Initial buret reading	mL	mL	mL
Final buret reading	mL	mL	
Net volume of 0.01 M EDTA titrant	mL	mL	mL

Calculate the concentration of Ca^{2+} in the tap water and bottled water samples in mmol/L. Show a sample calculation in the space below.

Calculations	Trial 1	Trial 2	Trial 3
Concentration of Ca^{2+} in tap water samples	mmol/L	mmol/L	mmol/L
Concentration of Ca^{2+} in bottled water samples	mmol/L	mmol/L	mmol/L

Tap water average for Ca^{2+} : _____ mmol/L

Bottled water average for Ca^{2+}: _____ mmol/L

Calculating the Mg^{2+} concentration. By subtracting the average values for Ca^{2+} in part 2b from those for Ca^{2+} + Mg^{2+} in part 2a, calculate the average concentration of Mg^{2+} in the the tap water and bottled water samples in units of mmol/L. Show a sample calculation in the space below:

Tap water average for Mg^{2+}: _____ mmol/L

Bottled water average for Mg^{2+}: _____ mmol/L

3. Determining alkalinity (HCO_3^-) by titration with HCl using bromocresol green indicator

Data	Trial 1	Trial 2	Trial 3
Volume of tap water sample (accurate to ±0.1 mL)	mL	mL	mL
Volume of 0.02 M HCl titrant Initial buret reading	mL	mL	mL
Final buret reading	mL	mL	
Net volume of 0.02 M HCl titrant	mL	mL	mL
Volume of bottled water sample (accurate to ±0.1 mL)	mL	mL	mL
Volume of 0.02 M HCl titrant Initial buret reading	mL	mL	mL
Final buret reading	mL	mL	mL
Net volume of 0.02 M HCl titrant	mL	mL	mL

Calculate the alkalinity (HCO_3^- concentration) in the tap water and bottled water samples in units of mmol/L. Show a sample calculation in the space below, including how you would convert from units of mmol/L to units of mg/L HCO_3^- (taking the molar mass of HCO_3^- as 61 g/mol). Assume that the titration reaction is $H^+ + HCO_3^- \rightarrow H_2O + CO_2$.

Calculations	Trial 1	Trial 2	Trial 3
Alkalinity (HCO_3^- concentration) in tap water samples	mmol/L	mmol/L	mmol/L
Alkalinity (HCO_3^- concentration) in bottled water samples	mmol/L	mmol/L	mmol/L

Average the calculated values for the samples of tap water and bottled water that you titrated in units of mmol/L, and calculate from this value the average value of the HCO_3^- concentration in units of mg/L.

Tap water average for HCO_3^-: _____ mmol/L; _____ mg/L

Bottled water average for HCO_3^-: _____ mmol/L; _____ mg/L

Name _____   Date _____

Summarize in the table below the average values for the ionic constitutents determined in parts 2 and 3 in units of mmol/L and mg/L. (Use the appropriate molar mass of each ionic constituent to convert units of mmol/L to units of mg/L.)

Ionic constituents	Tap water		Bottled water	
$Ca^{2+} + Mg^{2+}$	mmol/L	—	mmol/L	—
Ca^{2+}	mmol/L	mg/L	mmol/L	mg/L
Mg^{2+}	mmol/L	mg/L	mmol/L	mg/L
HCO_3^-	mmol/L	mg/L	mmol/L	mg/L

4. Semi-quantitative testing for sulfate and chloride ions

By visual comparison of the amounts of precipitate described in the part 4 testing procedure, decide whether the concentrations of sulfate and chloride ion in the tap water and bottled water samples are greater than, approximately equal to, or less than 30 mg/L. Enter the appropriate symbol (>30, =30, or <30) in the table below.

Ionic constituents	Tap water	Bottled water
Sulfate ion, SO_4^{2-}	mg/L	mg/L
Chloride ion, Cl^-	mg/L	mg/L

Questions

1. As a matter of preference or habit, do you drink tap water or bottled water? Explain in a few words two or three reasons for your preference (taste, convenience, cost, environmental reasons, health and safety, etc.).

What influence, if any, has this experiment had on your thinking about your preference for tap water or bottled water?

2. Suppose you wanted to prepare drinking water containing 100 mg/L of magnesium ion (Mg^{2+}) by adding magnesium sulfate ($MgSO_4$) to previously purified water. Calculate the amount of $MgSO_4$ that you would have to add per liter to obtain 100 mg/L of Mg^{2+}. If you purchased a kilogram of reagent grade (99%) $MgSO_4$ at the current price of about \$100/kg, how many liters of water having the calculated concentration of $MgSO_4$ could you prepare from a kilogram of $MgSO_4$? What would be the cost per liter of the added $MgSO_4$?

3. Why must you use molar concentration units (mmol/L) when you obtain the concentration of Mg^{2+} by taking the difference using Equation (5) rather than mass concentration units (mg/L)?

4. In the determination of Ca^{2+} by EDTA titration using calcon indicator, the choice of pH is an important factor. We want to choose a pH where Mg^{2+} ion will be essentially completely precipitated as $Mg(OH)_2(s)$ while Ca^{2+} will not precipitate as $Ca(OH)_2(s)$.

Calculate the hydroxide ion concentration at pH 12.5. From this value and the K_{sp} values for $Ca(OH)_2$ and $Mg(OH)_2$ found in Table 9 of the Appendix, calculate the concentrations of Ca^{2+} and Mg^{2+} that would be in equilibrium with their respective solid hydroxides in a pH 12.5 solution. From the calculated values, what conclusions can you draw? Would Mg^{2+} be completely precipitated? If the concentration of Ca^{2+} in the original water sample is typically less than 3 mmol/L, would you expect $Ca(OH)_2(s)$ to precipitate at pH 12.5? Explain.

Name _____ Date _____

████████ **CONSIDER THIS** ████████

Reread the Consider This section at the end of this experiment, visit the Web sites listed (or others that you discover on your own), and describe one or two approaches that you think would most nearly meet the following criteria for producing safe drinking water from contaminated water: efficacy in producing safe water using locally available materials, simplicity, low cost, energy efficiency, and least dependence on technology (such as pumps requiring a reliable source of electricity).

The Solubility Product Constant of Calcium Iodate, $Ca(IO_3)_2$

Purpose

• Understand the relation between the molar solubility and the solubility product constant of a sparingly soluble salt.

• Measure the molar solubility of calcium iodate in pure water and determine the solubility product constant.

• Investigate the common ion effect by measuring the molar solubility of calcium iodate in a solution containing added potassium iodate.

Pre-Lab Preparation

When soluble ionic compounds are dissolved in water, the solution usually contains just the ions that were present in the solid salt. For example, in a saturated solution of KNO_3, the solution contains only K^+ ions and NO_3^- ions, and the dissolved salt is completely dissociated.[1]

$$KNO_3(s) \rightleftharpoons K^+(aq) + NO_3^-(aq) \qquad (1)$$

When a salt is sparingly soluble, it suggests that the attractive forces between the ions in the solid are comparable to the attractive forces between the ions and the solvent (usually water). Many sparingly soluble salts do not completely dissociate into their constituent ions; instead, the ions interact in solution with one another to form soluble aggregates, called *complexes,* or react with water to form new species.

For example, in a saturated solution of calcium sulfate ($CaSO_4$), in addition to $Ca^{2+}(aq)$ and $SO_4^{2-}(aq)$ ions, a significant fraction of the dissolved calcium sulfate exists as molecular $CaSO_4(aq)$. This soluble

form of $CaSO_4$ may be a molecule with polar covalent bonding or a soluble ion pair, $Ca^{2+}SO_4^{2-}(aq)$. Thus, in this system, it takes two equilibria (and two equilibrium constant expressions) to adequately describe the system:

$$CaSO_4(s) \rightleftharpoons Ca^{2+}(aq) + SO_4^{2-}(aq) \qquad (2)$$

$$CaSO_4(s) \rightleftharpoons CaSO_4(aq) \qquad (3)$$

Here's a second example. A saturated solution of $CaCO_3$ contains appreciable concentrations of HCO_3^- and OH^-, as well as Ca^{2+} and CO_3^{2-} ions. In this system, carbonate ion (CO_3^{2-}) reacts with water to produce bicarbonate ion and hydroxide ion. As a result, it takes two equilibria (and two equilibrium constant expressions) to describe the system:

$$CaCO_3(s) \rightleftharpoons Ca^{2+}(aq) + CO_3^{2-}(aq) \qquad (4)$$

$$CO_3^{2-}(aq) + H_2O \rightleftharpoons HCO_3^-(aq) + OH^-(aq) \qquad (5)$$

The behavior shown in Reaction (5) is typical of all salts (such as carbonates, sulfides, and phosphates) in which the anion can react with H_2O to form a weak acid that is only partially ionized (or dissociated).

The two examples whose solubility equilibria are described by Reactions (2) through (5) show that it is not always safe to assume that only the ions produced by complete dissociation of the salt are present. The solubility behavior of every salt must be determined by an experimental investigation.

To summarize, we emphasize two important points about solubility equilibria: (a) *The identity of all of the solution species in equilibrium with the solid must be determined by experiment.* It is not safe to assume that all salts dissociate completely into their constituent ions; (b) *The solubility product constant can be calculated directly from the molar solubility only if the dissolved salt completely dissociates into its constituent ions and if these ions neither react with one another to form complexes nor react with water to form weak acids or bases.*

For the two example salts we described ($CaSO_4$, $CaCO_3$), two equilibria are needed to describe each system. A measurement of just the molar solubility

---

[1] In solid KNO_3, each ion is surrounded by ions of the opposite charge. This is still largely true when the KNO_3 is dissolved in water, but the ions are farther apart, and each ion is surrounded by a cluster of water molecules that are bound to the ion. The ions in solution are therefore said to be "hydrated" or "aquated." (The generic term is *solvated.*)

does not provide enough information to calculate two equilibrium constants. More-detailed discussions of solubility equilibria can be found in the references listed in the bibliography.

The relation between molar solubility and the solubility product constant

Calcium iodate, a salt whose solubility you will study in this experiment, has been shown to completely dissociate in water[2] into calcium ions, Ca^{2+}, and iodate ions, IO_3^-:

$$Ca(IO_3)_2(s) \rightleftharpoons Ca^{2+}(aq) + 2\,IO_3^-(aq) \qquad (6)$$

So a single equilibrium describes the system, and the value of the equilibrium constant can be calculated from the molar solubility.

Reaction (6) shows the dissociation of solid $Ca(IO_3)_2$ in a saturated aqueous solution. The equilibrium constant expression for this reaction (often called the *solubility product* expression) is written as shown in Equation (7), observing the convention that the activity of the pure solid is taken to be equal to 1.00 so that it does not appear in the equilibrium constant expression[3]

$$K_{sp} = [Ca^{2+}][IO_3^-]^2 \qquad (7)$$

For any salt that dissolves to give just the ions originally present in the salt, there is a simple relation between the molar solubility of the salt and the solubility product constant of the salt. The concentration of each ion will be equal to the *molar solubility* or to some multiple of it.

What do we mean by the molar solubility of the salt? This quantity is just the concentration of the dissolved salt, expressed in units of moles per liter. In the calcium iodate example, the calcium ion concentration is equal to the molar solubility; the iodate ion concentration is equal to *two* times the molar solubility. This is true because each mole of calcium iodate that dissolves gives one mole of calcium ions and two moles of iodate ions.

If we let the symbol s represent the molar solubility of calcium iodate (in units of moles per liter), the concentrations of Ca^{2+} and IO_3^- ions will be related to the molar solubility, s, by the equations

$$[Ca^{2+}] = s \qquad (8)$$

$$[IO_3^-] = 2s \qquad (9)$$

If we substitute for the concentration of each ion its equivalent in units of molar solubility, s, by substituting Equations (8) and (9) into Equation (7), we obtain an equation that shows a simple relation between the K_{sp} and the molar solubility of the salt:

$$K_{sp} = [Ca^{2+}][IO_3^-]^2 = s(2s)^2 = 4s^3 \qquad (10)$$

A frequently asked question about this expression is, "Why are you doubling the concentration of IO_3^- and then squaring it?" The short answer is that we aren't doubling the concentration of IO_3^-; its concentration is equal to $2s$.

The solubility of $Ca(IO_3)_2$ in KIO_3 solution: The common ion effect

From Le Châtelier's principle we would predict that the molar solubility of calcium iodate would be smaller in a solution of potassium iodate, KIO_3, which is a strong electrolyte that completely dissociates in water. The hypothesis is that the addition of KIO_3 would shift the equilibrium shown in Reaction (6) toward the left, decreasing the amount of calcium iodate that dissolves. Such a decrease in solubility, which occurs when a salt is dissolving in a solution that already contains one of the salt's ions, is called the *common ion effect*.

To test this hypothesis, you will measure the solubility of calcium iodate in 0.01 M KIO_3. Under these conditions, the concentrations of the ions will be related to the molar solubility of $Ca(IO_3)_2$ in the following way [compare Equations (11) and (12) with Equations (8) and (9)]:

$$[Ca^{2+}] = s \qquad (11)$$

$$[IO_3^-] = 0.01 + 2s \qquad (12)$$

Note that all of the calcium ion must come from dissolved calcium iodate. However, the iodate ion comes from both KIO_3 and dissolved $Ca(IO_3)_2$. As before, we get *two* moles of iodate ions and *one* mole of calcium ions for every mole of calcium iodate that goes into solution.

The relation between the solubility product constant and the molar solubility will also be changed [compare Equation (13) with Equation (10)].

$$K_{sp} = [Ca^{2+}][IO_3^-]^2 = s(0.01 + 2s)^2 \qquad (13)$$

If we know the concentration of potassium iodate, a single measurement of the total iodate concentration

---

[2]The actual composition of the solid that is in equilibrium with the ions in solution is the solid hexahydrate, $Ca(IO_3)_2 \cdot 6\,H_2O$.

[3]The effective concentration, or "activity," of a pure solid that participates in a chemical reaction is constant. (Adding more solid salt to the system does not change the concentration of ions in the solid.) By thermodynamic convention, the pure solid is chosen as the thermodynamic reference state, which is equivalent to assigning the value 1.00 to the activity of the solid.

allows us to calculate the molar solubility of calcium iodate and the solubility product constant.

Determining the molar solubility of calcium iodate

When calcium iodate dissolves in solution, both the solid and the solution remain electrically neutral.[4] This is called the *electroneutrality* principle, meaning that in the case of calcium iodate, two IO_3^- ions will go into solution for each Ca^{2+} ion that goes into solution. So molar solubility can be determined by measuring either the concentration of calcium ion or the concentration of iodate ion. We will use a procedure for measuring the IO_3^- concentration that makes use of the fact that IO_3^- oxidizes iodide ion:

$$IO_3^- + 5\,I^- + 6\,H^+ \rightarrow 3\,I_2 + 3\,H_2O \qquad (14)$$

The I_2 produced is in turn titrated with sodium thiosulfate:

$$I_2 + 2\,S_2O_3^{2-} \rightarrow 2\,I^- + S_4O_6^{2-} \qquad (15)$$

Note from the overall stoichiometry that each mole of IO_3^- will produce enough I_2 to consume 6 mol of $S_2O_3^{2-}$. Therefore, the concentration of IO_3^- ion will be given by

$$[IO_3^-]\left(\frac{mol}{L}\right) = \frac{mL\ S_2O_3^{2-}}{mL\ IO_3^-} \times$$

$$\frac{mol\ S_2O_3^{2-}}{L} \times \frac{1\ mol\ IO_3^-}{6\ mol\ S_2O_3^{2-}} \qquad (16)$$

Experimental Procedure

Special Supplies: *Microscale:* 1-mL microburet (see Introduction at the front of the lab manual); 1-mL serological pipet (also a 2-mL pipet, if students are to standardize the sodium thiosulfate titrant); pipet-filling devices; fine-tipped (50 drops/mL) polyethylene transfer pipets; polyethylene transfer pipets with 1-mL calibration marks. *Macroscale:* 25- or 50-mL buret; two 10-mL pipets or graduated cylinders.

Chemicals: *Microscale:* Standardized 0.15 M sodium thiosulfate (Na$_2$S$_2$O$_3$) solution. *Macroscale:* Standardized 0.05 M sodium thiosulfate (Na$_2$S$_2$O$_3$) solution. KI(s); 1 M HCl; 0.1% starch indicator solution; saturated solution of Ca(IO$_3$)$_2$ • 6 H$_2$O in pure water; saturated solution of Ca(IO$_3$)$_2$ • 6 H$_2$O in 0.0100 M KIO$_3$. The saturated solutions should be prepared a week in advance; see Notes to Instructor.

---

[4]Separation of positive and negative ions on a large scale would produce such a huge force of attraction between the positive and negative charges that any separation is prevented, except on a microscopic scale, where the distance of separation is on the order of atomic dimensions.

Instructions are given for both microscale titrations, using a 1-mL microburet, and macroscale titrations, using a 25- or 50-mL buret. (Direct students to the Introduction in the front of the lab manual for instructions in the use of burets and pipets.)

Microscale: The iodate-containing sample volumes may be measured using 1- and 2-mL serological pipets or by weighing the samples on a balance with milligram accuracy (1.00 g = 1.00 mL). Use 0.15 M sodium thiosulfate (Na$_2$S$_2$O$_3$) as the titrant.

Macroscale: If you prefer to use a 25- or 50-mL buret, increase the volume of iodate-containing samples to 10 mL in parts 1, 2, and 3, and use 0.05 M sodium thiosulfate (Na$_2$S$_2$O$_3$) as the titrant.

Sodium thiosulfate pentahydrate (Na$_2$S$_2$O$_3$ • 5 H$_2$O) that has been kept in a well-sealed bottle is normally pure enough so that a solution made up by accurately weighing the salt will be within 1% of the target concentration, so standardization of the solution is not absolutely necessary. As an option, students may standardize the nominal 0.15 M (or 0.05 M for macroscale) Na$_2$S$_2$O$_3$ if they are supplied with a solution of 0.0100 M KIO$_3$. The titration is performed as described to determine the molar solubility of calcium iodate.

The solid in equilibrium with the saturated solution of calcium iodate is the hexahydrate. We have found, however, that using anhydrous calcium iodate to prepare the saturated solutions gives the same results. Although anhydrous calcium iodate is available from a commercial supplier of laboratory chemicals (G. Frederick Smith Chemical Co.), a quantity of the hexahydrate sufficient for 100 students is easily prepared. Dissolve 0.5 mol (112 g) of KIO$_3$ in 600 mL of hot water in a liter beaker. To the hot solution, slowly add with stirring 0.25 mol of Ca(NO$_3$)$_2$ (or CaCl$_2$) dissolved in 200 mL water. Allow to cool. Filter off the solid on a large Büchner funnel and wash the solid with three 50-mL portions of cold water. Allow the solid to air-dry, or use the moist cake to prepare the saturated solutions.

The saturated solution in pure water is prepared by adding 8 g Ca(IO$_3$)$_2$ • 6 H$_2$O per liter of water. The saturated solution in 0.0100 M KIO$_3$ is prepared by adding 2.14 g KIO$_3$ + 4 g Ca(IO$_3$)$_2$ • 6 H$_2$O per liter of water. Stir the solutions for at least 24 h; allow to stand for 48 h before use. Ideally, the solutions should be thermostatted at 25 °C, but if this is not possible, a temperature correction can be made to estimate the molar solubility at 25 °C, using data at a different temperature (see Question 3 in the Consider This section of the report form). Use a siphon, or centrifuge individual 5-mL samples so that students receive samples free of particles of solid calcium iodate.

1. Standardization of the Sodium Thiosulfate Solution (Optional)
Microscale: Follow the microscale procedure described in part 2, substituting 2.0-mL samples of 0.0100 M KIO_3 for the 1.0-mL samples of saturated $Ca(IO_3)_2$ solution. (The volume of 0.15 M sodium thiosulfate required is between 0.7 and 0.9 mL.)

Macroscale: Follow the macroscale procedure described in part 2, using 10-mL samples of 0.0100 M KIO_3. (The volume of 0.05 M sodium thiosulfate required is between 10 and 14 mL.)

Titrate at least two samples. Calculate the exact molar concentration of sodium thiosulfate, keeping in mind that one mole of KIO_3 produces enough I_2 to consume six moles of sodium thiosulfate. The calculated molarities for the two samples should agree to within 5 to 10%.

2. The Molar Solubility of $Ca(IO_3)_2$ in Pure Water
If you did not perform the standardization procedure in part 1, record the exact concentration of the sodium thiosulfate ($Na_2S_2O_3$) titrant. Also record the temperature of the saturated calcium iodate solution. Read the pages of the Introduction describing how to use burets and pipets.

Microscale: Fill a 1-mL microburet with the 0.15 M sodium thiosulfate solution, taking care to expel all bubbles from the buret. Record the initial buret reading. Put 5 mL of deionized water in a 10- or 25-mL Erlenmeyer flask. Add 0.2 g of KI and swirl to dissolve.

Obtain a 5-mL sample of saturated calcium iodate solution in a 13 × 100 mm test tube. (If the sample shows any visible solid particles, it should be centrifuged to settle them to the bottom of the tube.) Then, using a pipet-filling device to fill the pipet, transfer 1.00 mL of the saturated solution into the Erlenmeyer flask. [A 1.00-mL serological pipet may be used; if not available, weigh out 1.00 ± 0.02 g (1 g = 1 mL) of sample directly into the Erlenmeyer flask, using a fine-tipped polyethylene pipet to dispense the sample solution.] Then add 1 mL of 1 M HCl, using a calibrated polyethylene transfer pipet. The solution should turn brown. Without delay, titrate the solution with sodium thiosulfate until the solution is yellow. Then add 1 mL of 0.1% starch indicator. The solution should turn blue-black, the color of the starch–iodine complex. Continue the titration until you get a sharp change from blue to a colorless solution. Record the final buret reading.

Repeat the titration procedure with a second sample. The volumes should agree to within 5 to 10%. (A possible source of error is the air oxidation of I^- to give I_2. If the solution is allowed to stand for too long before it is titrated, the oxidation will produce values that are too high.)

Macroscale: Fill a 25- or 50-mL buret with 0.050 M sodium thiosulfate solution, taking care to rinse the buret and to expel all bubbles from the tip of the buret. Obtain about 30 mL of saturated calcium iodate solution in a clean, dry flask. (The sample must not show any visible solid particles of undissolved calcium iodate.) Put 50 mL of deionized water and 2 g of KI into a 250-mL Erlenmeyer flask and swirl to dissolve the KI. Then, using a pipet-filling device to fill the pipet, transfer 10 mL of the saturated calcium iodate solution into the 250-mL flask and add 10 mL of 1.0 M HCl. (10-mL graduated cylinders may be used to measure the saturated calcium iodate and 1.0 M HCl solutions.) When the HCl is added, the solution should turn brown. Without delay, titrate the solution with 0.05 M sodium thiosulfate until the solution is yellow. Then add 5 mL of 0.1% starch indicator. The solution should turn blue-black, the color of the starch–iodine complex. Continue the titration until you get a sharp change from blue to a colorless solution. Record the final buret reading.

Repeat the titration procedure with a second sample. The volumes should agree to within 5%. (A possible source of error is the air oxidation of I^- to give I_2. If the solution is allowed to stand for too long before it is titrated, the oxidation will produce values that are too high.)

Calculations: Calculate the concentration of iodate ion in the sample of saturated solution [see Equation (16)]. The molar solubility of calcium iodate will be equal to one half the iodate concentration [see Equations (8) and (9)]. Calculate the solubility product constant for calcium iodate [see Equation (10)].

3. The Molar Solubility of Calcium Iodate in 0.0100 M Potassium Iodate
Record the temperature of the saturated solution. Record the initial and final buret readings for each titration.

Microscale: Using the microscale procedure described in part 2, titrate two 1.00-mL samples of the saturated solution of $Ca(IO_3)_2$ in 0.0100 M KIO_3.

Macroscale: Using the macroscale procedure described in part 2, titrate two 10-mL samples of the saturated solution of $Ca(IO_3)_2$ in 0.0100 M KIO_3.

Calculations: Calculate the total iodate concentration as in part 2. Subtract 0.0100 M from the total concentration to obtain the concentration of iodate ion that comes from dissolved calcium iodate. Divide this result by 2 to obtain the molar solubility of calcium iodate. Is it smaller, as we predicted from Le Châtelier's principle? Calculate the solubility product constant for calcium iodate [see Equation (13)]. Does the calculated value of K_{sp} agree with the value for K_{sp} calculated in part 2? Note that a 10% error in measuring the molar solubility, s, will lead to a larger error in the calculation of K_{sp}, because K_{sp} depends on s^3—cubing the number magnifies the error.

CONSIDER THIS

Using Le Châtelier's principle, we predicted that the solubility of calcium iodate would *decrease* in a solution containing a common ion. Now let's think of how we might *increase* the solubility of calcium iodate, producing a solution that contains more total calcium and total iodate dissolved in solution.

Let's think first about the anion, iodate ion. If the anion is the conjugate base of a weak acid, we might try increasing the solubility of calcium iodate by adding a strong acid such as HCl or HNO₃ to protonate the IO_3^- ion. This would only work if an appreciable amount of undissociated HIO₃ is formed. What about HIO₃, called hydrogen iodate or iodic acid? Is it a strong acid (one that completely dissociates) or a weak acid that has a measurable ionization constant?

Now let's think about the cation, calcium ion. Could we find something that forms a soluble complex with calcium ion or that otherwise interacts with it, for example, forming an ion pair with calcium ion? 1,2-diols, such as ethylene glycol (1,2-ethanediol) have been reported to form soluble complexes with calcium ion. The chelating agent EDTA (see Experiment 29) is known to form soluble complexes in basic solution with nearly any metal ion having +2 or +3 charge.

Finally, there is another rather unexpected possibility. The addition of *any* soluble "inert" salt (one that completely dissociates) to a solution of *any* slightly soluble salt will increase its solubility. This is an *ion–ion interaction effect* or *interionic attraction effect*. In this context, the added "inert" salt is one in which there is no specific chemical interaction between the ions of the inert salt and the ions of the slightly soluble salt—that is, no complexation or ion-pair formation, etc.

The calcium iodate system is ideal for study of these ion–ion interactions because calcium ion has a +2 charge; such interaction effects turn out to depend on the square of the charge. As a result, the interaction effect of a calcium ion is four times that of an ion with +1 charge.

How does this effect operate? Adding inert ions to the solution increases the total concentration of ions, thereby increasing the *ionic strength* of the solution. Because ions are charged, they all interact through the attractive and repulsive forces caused

by their charges, causing each ion to become surrounded by a cluster of ions of the opposite charge. The energy of this interaction acts to change the equilibrium constant by decreasing the *effective concentration* of the ions, called the ion *activity*. Taking account of these effects quantitatively is usually done by introducing an activity coefficient, γ, which is a measure of the departure from ideal behavior:

$$\text{activity of } M^+ = a_{M+} = \gamma[M^+]$$

Interionic attraction effects and activity coefficients are not usually discussed in general chemistry textbooks. How significant is the effect of ionic strength on solubility? Ramette (see the Bibliography) cites measurements showing that the molar solubility of calcium iodate at 25 °C increases by a factor of nearly two in 0.5 M KCl compared to its solubility in pure water—certainly a significant effect.

Bibliography

Butler, J. N.; Cogley, D. R. *Ionic Equilibrium: Solubility and pH Calculations*, John Wiley & Sons, New York, 1998. Everything you need or want to know about ionic equilibria.

Cavaleiro, A. M. V. S. V. "The Teaching of Precipitation Equilibrium: A New Approach," *J. Chem. Educ.* **1996,** *73,* 423–425. Case studies of systems where the anion of the salt reacts with water to form a weak acid and hydroxide ion.

Clark, R. W.; Bonicamp, J. M. "The K_{sp}-Solubility Conundrum," *J. Chem. Educ.* **1998,** *75,* 1182–1185. A critique of solubility and K_{sp} data presented in general chemistry textbooks.

Hawkes, S. J. "Salts Are Mostly *NOT* Ionized," *J. Chem. Educ.* **1996,** *73,* 421–423.

Hawkes, S. J. "What Should We Teach Beginners about Solubility and Solubility Products?" *J. Chem. Educ.* **1998,** *75,* 1179–1181.

Meites, L.; Pode, J. S. F.; Thomas, H. C. "Are Solubilities and Solubility Products Related?" *J. Chem. Educ.* **1966,** *43,* 667–672. A classic study of the solubility of calcium sulfate, including the effects of ion pairing and interionic attraction (ionic activity effects).

Ramette, R. W. *Chemical Equilibrium and Analysis,* Addison-Wesley, Reading, MA, 1981, pp. 109–112. A case study of calcium iodate solubility and ionic strength (interionic attraction) effects.

**The Solubility Product Constant
of Calcium Iodate, Ca(IO₃)₂**

Name _____

Date _____ Section _____

Locker _____ Instructor_____

Data and Calculations

1. Standardization of the sodium thiosulfate solution (optional)

Volume of 0.0100 M KIO_3 samples _____ mL

Data	Trial 1	Trial 2	Trial 3
Volume of $Na_2S_2O_3$ titrant			
Final buret reading	mL	mL	mL
Initial buret reading	mL	mL	mL
Net volume of $Na_2S_2O_3$	mL	mL	mL
Calculated concentration of $Na_2S_2O_3$	M	M	M

For each trial, calculate the concentration of $Na_2S_2O_3$; show a sample calculation in the space below.

Average $Na_2S_2O_3$ concentration _____ M

(If you did not perform the standardization procedure in part 1, record here the concentration of the sodium thiosulfate titrant, $Na_2S_2O_3$.)

2. The molar solubility of Ca(IO₃)₂ in pure water

Temperature of the saturated solution of calcium iodate: _____ °C

Volume (or mass) of saturated calcium iodate solution titrated: _____ mL (or g)

Data	Trial 1	Trial 2	Trial 3
Volume of $Na_2S_2O_3$ titrant			
Final buret reading	mL	mL	mL
Initial buret reading	mL	mL	mL
Net volume of $Na_2S_2O_3$	mL	mL	mL
Calculated concentration of IO_3^-	M	M	M

For each trial, calculate the concentration of IO_3^-; show a sample calculation in the space below.

Average IO_3^- concentration _____ M

Calculate the molar solubility of $Ca(IO_3)_2$ in pure water from the average value of the IO_3^- concentration.

Molar solubility: _____ mol/L

Calculate the solubility product constant, K_{sp}, for a saturated solution of $Ca(IO_3)_2$ in water [see Equation (10)].

$K_{sp} =$ _____

3. The molar solubility of calcium iodate in 0.0100 M potassium iodate

Temperature of the saturated solution of calcium iodate: _____ °C

Volume (or mass) of saturated calcium iodate solution titrated: _____ mL (or g)

Data	Trial 1	Trial 2	Trial 3
Volume of $Na_2S_2O_3$ titrant			
Final buret reading	mL	mL	mL
Initial buret reading	mL	mL	mL
Net volume of $Na_2S_2O_3$	mL	mL	mL
Calculated concentration of IO_3^-	M	M	M

For each trial, calculate the concentration of IO_3^-; show a sample calculation in the space below.

Average IO_3^- concentration _____ M

Subtract the concentration of IO_3^- ion that came from the KIO_3 from the average value of the total IO_3^- concentration to get the iodate ion concentration that came from dissolved $Ca(IO_3)_2$.

Total IO_3^- concentration _____ M

IO_3^- concentration from KIO_3 _____ M

IO_3^- concentration from dissolved $Ca(IO_3)_2$ _____ M

Calculate the molar solubility of $Ca(IO_3)_2$ in 0.0100 M KIO_3 solution.

Molar solubility _____ mol/L

Name _____ Date _____

Compare the molar solubility of $Ca(IO_3)_2$ in 0.0100 M KIO_3 solution with the molar solubility in pure water, determined in part 2. Is the change consistent with Le Châtelier's principle?

Calculate the solubility product constant, K_{sp}, for a saturated solution of $Ca(IO_3)_2$ in 0.0100 M KIO_3 [see Equation (13)].

Considering the magnitude of possible errors in the determination of molar solubility (of the order of 10% maximum error in careful work), is there reasonable agreement between the K_{sp}'s calculated in parts 1 and 2?

Calculate the molar solubility, s, of calcium iodate in 0.020 M $Ca(NO_3)_2$, a completely dissociated strong electrolyte. (NO_3^- ion does not chemically interact with either Ca^{2+} or IO_3^-.) Assume that K_{sp} for $Ca(IO_3)_2 = 2.0 \times 10^{-6}$.

To set up the problem, we can write the following equations:

Material balance for calcium: $[Ca^{2+}] = s + 0.020$

Material balance for iodate: $[IO_3^-] = 2s$

$$K_{sp} = 2.0 \times 10^{-6} = [Ca^{2+}][IO_3^-]^2 = (s + 0.020)(2s)^2$$

The last equation contains only one unknown, the value of s, which we would like to calculate; but s is not negligible compared to 0.020, which leaves us with a nasty equation that is cubic in s. An approach that often works in a situation like this is to rearrange the equation into a more useful form and obtain s by iteration. Thus, we can write

$$4s^2 = \frac{K_{sp}}{s + 0.020} \qquad \text{or} \qquad s = \frac{1}{2}\left(\frac{K_{sp}}{s + 0.020}\right)^{1/2}$$

We must find the value of s that makes both sides of the equation equal. Start first by inserting a trial value of s (say, 0.006 M) in the right-hand side of the latter equation and calculating a value of s. You will get $s = 0.00438$. For your next trial value, take something about halfway between 0.00438 and 0.006. Continue the iteration process until you calculate a value of s that is nearly the same (within 5%) as the value you inserted on the right-hand side of the equation.

Name _____   Date _____

If your data for solubility were obtained at a temperature different from 25 °C, estimate the solubility at 25 °C from your data. Around 25 °C, the molar solubility of calcium iodate in water increases with temperature about 4.5% per degree. Calculate the molar solubility at 25 °C, s_{25}, by correcting your value determined at t °C, which we will call s_t, using the equation

$$s_{25} = s_t \left[\frac{2 - 0.045(t - 25)}{2 + 0.045(t - 25)} \right]$$

where t is the temperature (°C) of the saturated solution that you analyzed. Ramette (see the Bibliography) quotes the molar solubility of $Ca(IO_3)_2$ in pure water at 25 °C as 0.00798 mol/L. Compare your result with the literature value by calculating the percentage relative difference:

$$\text{percentage relative difference} = \frac{s_{25} - 0.00798}{0.00798} \times 100\%$$

▬▬▬▬ **CONSIDER THIS** ▬▬▬▬

Propose the composition of a solution in which you would expect to find that the molar solubility of $Ca(IO_3)_2$ is greater than in pure water. Briefly explain why you would expect to find a greater solubility in the solution you propose.

Identification of the Alkaline Earth and Alkali Metal Ions

Analysis of Cations in Seawater

Purpose

• Study laboratory techniques for qualitative analysis of the most common cations found in seawater.

• Use these techniques for separating and identifying the main Group II alkaline earth cations: Mg^{2+}, Ca^{2+}, Sr^{2+}, and Ba^{2+} ions.

• Learn how to test for the main Group I alkali metal cations, Na^+ and K^+, and for ammonium ion, NH_4^+, which is grouped with the alkali metals because its salts have solubilities that resemble the alkali metal salts.

Introduction

Qualitative analysis is a branch of analytical chemistry that identifies particular substances in a given sample of material. In the analysis of inorganic substances, this branch involves the analysis of both metallic constituents as cations and mostly nonmetallic constituents as anions. This experiment and the next constitute an introduction to qualitative analysis. Experiment 31 deals with the identification of seven common cations: Mg^{2+}, Ca^{2+}, Sr^{2+}, Ba^{2+}, Na^+, K^+, and NH_4^+. Experiment 32 explores the identification of nine anions: S^{2-}, SO_4^{2-}, SO_3^{2-}, CO_3^{2-}, Cl^-, Br^-, I^-, PO_4^{3-}, and NO_3^-. Altogether, six of these sixteen ions are present in seawater at concentrations in the range of 0.5 to 0.01 M and are detectable using the methods described in Experiments 31 and 32.

Qualitative analysis has remained an important part of the laboratory experience in general chemistry for a number of years, even though these analytical methods have been replaced by sophisticated instrumental methods in practical analysis. We believe that qualitative analysis serves three useful purposes: (1) It provides an ideal context for the illustration and application of the principles of ionic equilibria, such as acid–base, solubility, and complex ion equilibria; (2) it provides a systematic framework for the discussion of the descriptive chemistry of the elements; and (3) it helps develop logical and deductive reasoning skills.

Its utility aside, qualitative analysis is fun. It's hands on—not abstract. It's about mixing ions in test tubes and seeing the colors and textures of solutions and precipitates, learning the solubilities of their salts, and then putting all these clues together to identify what's in the solution.

Properties of the alkaline earth and alkali metal cations

The elements making up the alkaline earth and alkali groups are similar and have many properties in common. Their compounds exhibit only one stable oxidation state ($+2$ for the alkaline earth cations and $+1$ for the alkali metal cations), and they are not oxidized or reduced by common chemical reagents. They do not form amphoteric hydroxides (hydroxides that are soluble in excess OH^- by formation of a complex ion with OH^-), and they do not tend to form complex ions with NH_3, as do several transition metal cations including Cu^{2+}, Ag^+, Zn^{2+}, and Cd^{2+}.

Thus, the separation and identification of the main Group II alkaline earth elements depend almost entirely on differences in the solubilities of their salts, which show a regular gradation through the periodic table. The salts of the alkali metals and NH_4^+ are almost all readily soluble and require special tests. Table 31-1 will be useful in understanding the solubility differences that allow separation of the alkaline earth ions.

TABLE 31-1 Solubilities of Alkaline Earth Salts (g/100 g H_2O, at Room Temperature)

	Mg^{2+}	Ca^{2+}	Sr^{2+}	Ba^{2+}
OH^-	0.001	0.16	0.4	3.89
CO_3^{2-}	0.09	0.0015	0.001	0.0022
SO_4^{2-}	35.5	0.2	0.01	0.00024
CrO_4^{2-}	138.	18.6	0.12	0.00044
$C_2O_4^{2-}$	0.015	0.0007	0.005	0.01

Ideally, we would like to have a unique reagent for each cation that gives a very specific reaction (such as a precipitate or color reaction) with that cation and no others. Things are seldom ideal, so we are usually forced to separate the cations into several groups and then work with each group, making use of the different solubilities of the salts of the cations or the tendencies of cations to form complex ions.

This experiment includes four alkaline earth ions: Mg^{2+}, Ca^{2+}, Sr^{2+}, and Ba^{2+}. (In the procedures used in this experiment, Sr^{2+} and Ba^{2+} will be determined together, but they can be separated using hexafluorosilicate ion, SiF_6^{2-}, by means of a procedure described in the reference listed in the Bibliography.) Special tests are used for Na^+, K^+, and NH_4^+ because nearly all of their salts are soluble. We must test for NH_4^+ with a separate portion of the original sample, because we use ammonium carbonate, $(NH_4)_2CO_3$, as a reagent for the alkaline earth ions.

Pre-Lab Preparation

Your instructor will inform you if you will be analyzing seawater and/or unknowns containing alkaline earth and alkali metal cations.

If you are going to analyze seawater, your first task in preparing for this experiment will be to look up the concentrations of the ten most abundant elements in seawater (excluding hydrogen and oxygen found in water) and do the calculations necessary to express their concentrations in units of moles per liter. Summarize the results in a table. If you know how to use a spreadsheet such as Microsoft Excel, do your calculations and tabulate the results in a spreadsheet. Otherwise, construct your table by hand. We will show later some example calculations for converting mass units of concentration into units of moles per liter.

Seawater contains about 35 g of dissolved salts per liter (3.5 mass %) and more than 70 elements with a wide range of concentration levels: *major* (0.5–0.01 M), *minor* (0.01–10^{-5} M), and *trace* level ($<10^{-5}$ M). In this experiment, you will analyze seawater (or an unknown provided by the instructor) for the major cations. In Experiment 32, you will analyze for the major anions in seawater.

Information about seawater composition can be found in most editions of the *CRC Handbook of Physics and Chemistry*, as well as on the Internet. If you use the Google search engine (www.google.com) and the search words *seawater composition*, you will find sites that provide tables of seawater composition. This information is usually provided as the elemental composition by mass, often expressed in

parts per million (1 ppm = 1 mg/kg of seawater). The density of seawater is about 1.02 g/mL. For our purposes, we may assume that 1 kg of seawater has a volume of about 1 L.

▮ NOTES TO INSTRUCTOR ▮

Experiments 31 and 32, taken together, allow students to analyze for the main cations and anions in seawater that comprise >99% of the dissolved salts in seawater. If authentic seawater samples are not available, it is easy to prepare synthetic seawater and/or to analyze unknown samples containing some of the sixteen cations and anions in Experiments 31 and 32. (A recipe for synthetic seawater suitable for these experiments is given in the Instructor's manual and in the online supplemental material that accompanies the article by Selco et al., listed in the Bibliography.)

It is suggested that three lab periods (9 hours) be allowed for doing both cations and anions. This gives sufficient time for discussion of the results, testing the limits of detection for some ions, and repeating work that is problematic.

Many nonmetal elements are not present in seawater in the form of the element but as an oxyanion. For example, the sulfur in seawater is present not as dissolved sulfur or sulfide ion, but as sulfate ion, SO_4^{2-}. Nitrogen is mainly present as the nitrate ion, NO_3^-, and carbon as the bicarbonate ion, HCO_3^-. So you must be careful to note whether the composition given is based on the element, that is, expressed as the mass of elemental S, N, C, or as the mass of the oxyanion, which includes the mass of the oxygen in the anion. Also note that when an element exists as an oxyanion containing just one atom of that element, the molar concentration of that element will be the same as the molar concentration of the oxyanion containing the element.

Example Calculations One source says that seawater contains about 900 ppm (mg/L) of sulfur. Convert to units of moles per liter of sulfate ion, SO_4^{2-}.

$$\frac{900\text{ mg S}}{1\text{ L}} \times \frac{1\text{ g}}{1000\text{ mg}} \times \frac{1\text{ mol S}}{32\text{ g S}} \times \frac{1\text{ mol SO}_4^{2-}}{1\text{ mol S}} =$$

$$0.028\ \frac{\text{mol SO}_4^{2-}}{\text{L}}$$

A second source says that seawater contains 3.5% by weight of dissolved salts, equivalent to 35 g of dissolved salts per liter. The salt contains 7.68 wt %

of sulfate. What is the concentration of sulfate ion, SO_4^{2-}, in units of moles/liter?

$$\frac{35 \text{ g salt}}{1 \text{ L}} \times \frac{7.68 \text{ g SO}_4^{2-}}{100 \text{ g salt}} \times \frac{1 \text{ mol SO}_4^{2-}}{96 \text{ g SO}_4^{2-}} =$$

$$0.028 \frac{\text{mol SO}_4^{2-}}{\text{L}}$$

Laboratory techniques for qualitative analysis

The oft-repeated admonition to *keep your lab bench neat and orderly* will pay big dividends in time saved and in fewer mistakes. Study Figure 31-1. Keep the bench top and the area around the sink clear of unnecessary equipment. Make a convenient arrangement of the most frequently used items, such as clean test tubes, your wash bottle, stirring rods, polyethylene transfer pipets, and indicator test paper, laid out on a clean towel. Keep used or dirty test tubes and other articles in one place. Clean and rinse these with deionized water at the first opportunity so that a stock of clean equipment is always ready. Test tubes don't need to be dried before being reused; just shake out the excess water.

Label any solutions that are to be kept for more than a minute or two. As you prepare the labels, also write in your laboratory notebook the symbols you put on the label along with a brief description of what is in the sample. As you work, *think about what*

you are doing and write in your laboratory notebook what you do and what you observe as you work. Rushing through experimental work without thinking about it, with the idea of trying to understand or recall what you did later, will inevitably lead to mistakes requiring you to repeat your work.

Estimate the Volume of Solutions Use only the small volumes specified. You can estimate most volumes with sufficient precision by counting drops (20 to 25 drops per milliliter) or by estimating the height of the solution in the 100-mm test tube (capacity 8 mL). Some polyethylene transfer pipets have volume calibrations on the stem that you can use to estimate the volume dispensed. Use the 10-mL graduated cylinder only when more accurately known volumes are required.

The Handling of Solutions Review the sections on basic laboratory equipment and procedures in the Introduction. When mixing solutions, use a stirring rod or agitate as shown in Figure 31-2; do not invert or shake the test tube with your unprotected, and possibly contaminated, thumb as the stopper.

When heating solutions, avoid loss by bumping (see Figure 31-3). Do not try to boil more than 2 mL of solution in a 10-cm test tube; transfer larger quantities to a 15-cm test tube. The safest way to heat a solution in a test tube is to immerse the tube in a beaker of water near the boiling point.

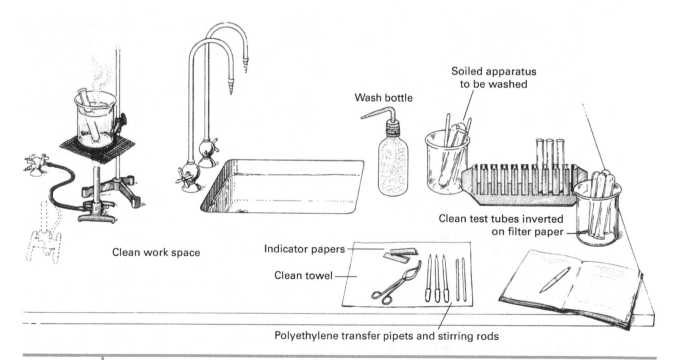

FIGURE 31-1 **Keep your laboratory workbench neat and in order, with the tools of your trade conveniently arranged.**

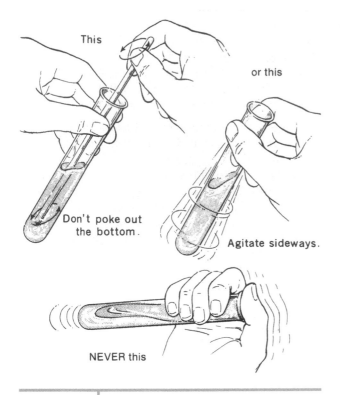

This

or this

Don't poke out the bottom.

Agitate sideways.

NEVER this

FIGURE 31-2 Learn to use good technique in mixing solutions.

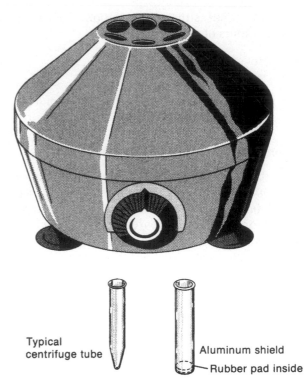

Typical centrifuge tube

Aluminum shield
Rubber pad inside

Opposite pairs of tubes should be filled with equal amounts of liquid to prevent excessive vibration.

FIGURE 31-4 The components of the centrifuge.

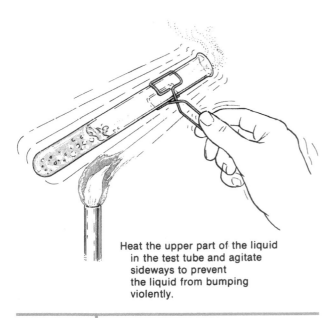

Heat the upper part of the liquid in the test tube and agitate sideways to prevent the liquid from bumping violently.

FIGURE 31-3 Use this technique when you heat a solution in a test tube.

Washing Precipitates When it is desirable to isolate a precipitate and free it from interfering ions in the supernatant solution (the solution above the precipitate) in order to carry out further tests on it, you

must first wash the precipitate. This is done by using a centrifuge, shown in Figure 31-4, to settle the precipitate to the bottom of the test tube. (The test tube holders of the centrifuge should have rubber cushions in the bottom to prevent tube breakage. You can use ordinary test tubes as centrifuge tubes as well as the tapered centrifuge tubes shown in Figure 31-4.)

After the solids have been centrifuged to the bottom of the tube, remove the supernatant solution by carefully pouring it off or by drawing it off with a polyethylene transfer pipet, as shown in Figure 31-5. Then add a few milliliters of deionized water, mixing as shown in Figure 31-2, and recentrifuge the precipitate. The washing can be repeated, if necessary, to remove all interfering ions.

Precipitation with ammonium carbonate

The reagent usually used to precipitate alkaline earth (main Group II) ions is ammonium carbonate, $(NH_4)_2CO_3$. Since this is the salt of both a weak base and a weak acid, it reacts in solution to a substantial extent (nearly 80% reacted):

$$NH_4^+ + CO_3^{2-} \rightleftharpoons NH_3 + HCO_3^- \quad (K \cong 12)$$

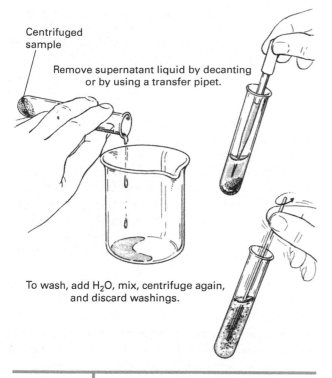

Centrifuged
sample

Remove supernatant liquid by decanting
or by using a transfer pipet.

To wash, add H₂O, mix, centrifuge again,
and discard washings.

FIGURE 31-5 | **The procedure for washing a precipitate that has been centrifuged.**

For this reason, we add excess NH_3 to shift the equilibrium to the left and increase the concentration of CO_3^{2-} to attain more complete precipitation.

The separation of strontium and barium from magnesium and calcium

This separation scheme takes advantage of the solubility differences of the salts listed in Table 31-1. After precipitation of the alkaline earth metals as the carbonates, they are dissolved with acetic acid. Then strontium and barium ions are precipitated as the yellow chromate salts, $SrCrO_4$ and $BaCrO_4$, leaving behind in solution the magnesium and calcium ions. The chromates are then transformed to the still less soluble sulfates, $SrSO_4$ and $BaSO_4$, confirming the presence of strontium and/or barium. The combined strontium and barium chromates dissolve in HCl because the chromate ion can be protonated to form the moderately weak acid $HCrO_4^-$, which in turn forms the orange-colored dichromate ion, $Cr_2O_7^{2-}$, in acid solution, whose salts are more soluble.

$$2\ HCrO_4^- \rightleftharpoons Cr_2O_7^{2-} + H_2O$$

In analyzing seawater samples or unknowns, we will make no attempt to separate strontium and barium, but it is possible to selectively precipitate barium ion with hexafluorosilicate to give the insoluble

$BaSiF_6$ (see the Bibliography). With strontium and barium ions removed, you will be asked to determine which ion in Table 31-1 might be used to selectively precipitate magnesium ion in the presence of calcium. The remaining solution should then contain only calcium ion, if it is present; the presence of calcium can be confirmed by precipitation with oxalate ion in basic solution as calcium oxalate, CaC_2O_4.

Tests for sodium, potassium, and ammonium ions

We will use a sensitive flame test to detect the presence of sodium; a combination of flame test and precipitation of potassium ion by tetraphenylborate ion, $B(C_6H_5)_4^-$, will be used to test for the presence of potassium ion. Finally, we test for ammonium ion by detecting the evolution of gaseous NH_3 when the solution is made very basic by addition of NaOH.

Experimental Procedure

Special Supplies: Centrifuges; nichrome wire; cobalt glass plates; glass stirring rods; red litmus paper; 65-mm and 75-mm diameter Pyrex watch glasses (or 10-mL beaker and plastic wrap).

Chemicals: 0.1 M magnesium chloride, $MgCl_2$; 0.1 M calcium chloride, $CaCl_2$; 0.1 M strontium chloride, $SrCl_2$; 0.1 M barium chloride, $BaCl_2$; 0.1 M sodium chloride, NaCl; 0.1 M potassium chloride, KCl; 0.1 M ammonium chloride, NH_4Cl; 6 M ammonia, NH_3; 3 M ammonium carbonate, $(NH_4)_2CO_3$; 6 M acetic acid, CH_3COOH; 3 M ammonium acetate, NH_4CH_3COO; 1 M potassium chromate, K_2CrO_4; 6 M and 12 M hydrochloric acid, HCl; 0.1 M sodium sulfate, Na_2SO_4; 6 M and 1 M sodium hydroxide, NaOH; 1 M potassium oxalate, $K_2C_2O_4$; 1 M sodium acetate, $NaCH_3COO$; 3% sodium tetraphenylborate, $NaB(C_6H_5)_4$; 0.1% thymol blue indicator.

> **⚠ SAFETY PRECAUTIONS:**
> **The laboratory reagent solutions are corrosive and irritating to the skin and especially the eyes. WEAR YOUR GOGGLES AT ALL TIMES.**
> **Barium is toxic; chromates are toxic and carcinogenic when ingested. Try not to spill any Ba^{2+} or CrO_4^{2-} on your skin. If you do, wash it off immediately. Observe the usual precautions in working with acids such as acetic acid and HCl and bases such as NH_3 and NaOH. Neutralize any spills with $NaHCO_3$ and clean up the residue with a sponge. Wash your hands thoroughly before leaving the laboratory.**

WASTE COLLECTION:
As you work, collect all solutions and precipitates containing barium and chromates into a beaker and later transfer them into a labeled waste container. The remaining solutions may be collected in a beaker, neutralized with $NaHCO_3$, and disposed of down the drain.

A. Typical reactions of the ions

Instead of performing a series of separate tests on each ion by itself to learn the properties of the ions, proceed at once to the study of the following procedure for the analysis; at the same time, complete the flow chart in the Report form by filling in the proper formulas for reagents, precipitates, ions left in solution, or gases evolved, and so on. Note how the procedures for precipitating Mg^{2+}, Ca^{2+}, Sr^{2+}, and Ba^{2+} make use of the solubilities of the alkaline earth salts shown in Table 31-1. Also note that Sr^{2+} and Ba^{2+} are carried along together so that the tests can only confirm that the unknown contains either Sr^{2+} or Ba^{2+} or both ions (which we abbreviate as Sr^{2+} and/or Ba^{2+}).

On your own initiative, you may wish to try the procedures on various combinations of ions and explore the lower limits of detection for a particular ion or for the flame test for K^+ in the presence of Na^+.

B. Analysis of a known solution and seawater sample (or unknown) for Mg^{2+}, Ca^{2+}, Sr^{2+} and/or Ba^{2+}, Na^+, K^+, and NH_4^+

Prepare a known solution by placing 1 mL each of 0.1 M $MgCl_2$, 0.1 M $CaCl_2$, 0.1 M $SrCl_2$, 0.1 M $BaCl_2$, 0.1 M NaCl, 0.1 M KCl, and 0.1 M NH_4Cl into a 13 × 100 mm test tube. Mix the solution using a stirring rod. *Pour half the mixture into a separate test tube to save for separate analysis for NH_4^+ ion in step 8.* Analyze side-by-side the remaining half (approximately 3.5 mL) of the known mixture and an equal volume of the seawater sample (or unknown) as follows.

1. *Precipitation of Mg^{2+}, Ca^{2+}, Sr^{2+}, and Ba^{2+}*
Start with about 3.5 mL of the known and, in a separate test tube, an equal volume of seawater (or unknown). If your solutions are not already basic (test with red litmus paper), add 6 M NH_3 until they are. Then add 10 drops (0.5 mL) 3 M $(NH_4)_2CO_3$ and

mix thoroughly. Let the mixtures stand for 5 min to complete the precipitation of the $MgCO_3$, $CaCO_3$, $SrCO_3$, and $BaCO_3$ salts, then centrifuge them. Add one or two drops of 3 M $(NH_4)_2CO_3$ to test for complete precipitation. Centrifuge again if necessary. *Decant each supernatant solution into separate labeled 13 × 100 mm test tubes and save them for further tests for K^+ and Na^+ described later in step 6. Also save the precipitates for step 2.*

2. *Test for Sr^{2+} and/or Ba^{2+}*
Wash and centrifuge the carbonate precipitates from step 1 and discard the wash water. Add 5 drops (0.25 mL) of 6 M acetic acid (CH_3COOH) and warm as needed to dissolve each precipitate. Then add 1 mL of water and buffer the solutions by adding 5 drops (0.25 mL) of 3 M ammonium acetate (NH_4CH_3COO). Add 3 drops of 1 M potassium chromate (K_2CrO_4), mix, then centrifuge. If each supernatant solution is not yellow, add more 1 M K_2CrO_4, a drop at a time, until the supernatant solutions appear yellow, but avoid adding more than 1 drop past the point where a definite yellow color persists. Mix with a stirring rod. Centrifuge and wash any precipitate. A yellow precipitate indicates the presence of $SrCrO_4$ and/or $BaCrO_4$.

(Pour the supernatant solutions into two clean labeled 13 × 100 mm test tubes and save them for further tests for Mg^{2+} and Ca^{2+}, described in steps 4 and 5.) After pouring off the yellow solutions, confirm the presence of Sr^{2+} and/or Ba^{2+} by adding 3 drops of 6 M HCl to dissolve the yellow precipitates of $SrCrO_4$ and/or $BaCrO_4$. Then add 20 drops (1 mL) of 0.1 M Na_2SO_4. Centrifuge, if necessary, to see whether the precipitates are white in the yellow solution. A white precipitate of $SrSO_4$ and/or $BaSO_4$ confirms the presence of Sr^{2+} and/or Ba^{2+}.

3. *Developing a Test for Mg^{2+}*
Take a look at Table 31-1 (Solubilities of Alkaline Earth Salts). Is there an anion that you might expect to precipitate Mg^{2+} while leaving Ca^{2+}, Sr^{2+}, and Ba^{2+} in solution? (*Hint:* Look at the solubility of the hydroxides.) To test this idea, prepare four solutions in labeled test tubes:

1. The first solution contains 1 mL of 0.1 M $MgCl_2$ + 3 mL of H_2O.

2. The second contains 1 mL of 0.1 M $CaCl_2$ + 3 mL of H_2O.

3. The third contains 1 mL each of 0.1 M $SrCl_2$ and 0.1 M $BaCl_2$ + 2 mL of H_2O.

4. The fourth solution is a fresh 4-mL sample of seawater (or your unknown).

Add 1 drop of 1 M NaOH to each of the solutions (1) to (4), stir each with a clean stirring rod, and test with litmus to see that each is basic. Then add to each solution about 8 drops (0.4 mL) of 1 M NaOH and again stir. (We want the final concentration of OH^- in these solutions to be about 0.05 M.) Allow the solutions to stand a few minutes to allow time for complete precipitation. What do you observe? Centrifuge any solutions in which a precipitate has formed and test for complete precipitation by adding one more drop of 1 M NaOH. *Label and save the test tubes containing the four solutions (and any precipitates) for the test in step 5.*

What conclusions do you draw from your observations? Do your results confirm that only Mg^{2+} ion was precipitated by hydroxide ion?

4. Test for Mg^{2+} Next we apply the test of step 3 to the two yellow supernatant solutions from step 2. Recall that Sr^{2+} and Ba^{2+} have been removed from these solutions by precipitation as chromates and that they are buffered with a mixture of acetic acid and ammonium acetate. One of the solutions came from the known mixture, the other was from a sample of seawater (or unknown).

We must first add enough NaOH to neutralize the buffer in these solutions. From the amounts of 6 M acetic acid and 3 M ammonium acetate used in step 2, we see that the solutions may contain as much as 1.5 mmol of acetic acid and 1.5 mmol of ammonium ion, which would require 3 mmol of 1 M NaOH to neutralize (about 0.5 mL of 6 M NaOH or 3 mL of 1 M NaOH).

We want to avoid a big excess of hydroxide ion, so add 3 drops of an acid–base indicator, 0.1% thymol blue, and titrate the solution carefully by adding the NaOH dropwise with stirring until the indicator just turns blue (which may appear as a slightly blue-green color in the presence of yellow CrO_4^{2-} ion). Then add 4 more drops (0.2 mL) of 1 M NaOH with mixing (which allows for the OH^- necessary to react with Mg^{2+} and gives the desired excess of OH^-). Allow the solutions to stand for a few minutes to give time for any precipitate to form. Then centrifuge to spin down any precipitate. A flocculent white precipitate indicates the presence of magnesium ion in the form of $Mg(OH)_2$. *Label and save the two test tubes containing the solutions (and any precipitates) for the next test in step 5.*

5. Test for Ca^{2+} We next test the four solutions saved from step 3 and the two solutions saved from step 4. Pour off the six solutions into six clean test-tubes which must be labeled so that you know their origin.

The four solutions from step 3 all had Mg^{2+} removed by precipitation as $Mg(OH)_2$. The two solutions from step 4 presumably contain only Ca^{2+}, because Sr^{2+} and Ba^{2+} were removed in step 2 by precipitation as chromates (later converted to the sulfates), and Mg^{2+} was later removed in step 4 by precipitation as $Mg(OH)_2$.

Looking at Table 31-1 again, we see that the oxalates of all the alkaline earth metals are not very soluble. Therefore, we might expect to get a precipitate if we add oxalate ion to any solution containing one of the four alkaline earth metal ions. The test must be performed in a solution that is basic because protonation of the oxalate ion in acid solution reduces the oxalate ion concentration to such a low value that no oxalate precipitate will form. As you can see by reviewing the procedures in steps 3 and 4, all of the solutions should be basic because NaOH has been added. This can be confirmed by testing with litmus paper (red litmus turns blue in basic solution).

Add 10 drops of 1 M $K_2C_2O_4$ (potassium oxalate) to each of the six tubes and stir. Let stand for a few minutes to give time for any precipitate to form.

What do you observe? Did the solution that originally contained only Ca^{2+} give a precipitate with oxalate? Did the solution that contained only Sr^{2+} and Ba^{2+} give a precipitate with oxalate? Did the solution that originally contained only Mg^{2+} show any signs of precipitate? If so, that would mean that the precipitation of Mg^{2+} with hydroxide ion was not complete.

Devise a procedure that would enable you to establish that Sr^{2+} and Ba^{2+} are removed by precipitation with chromate (or sulfate), then test it and record the results of your test.

Note that precipitation with oxalate is not unique to Ca^{2+}. It can only confirm the presence of Ca^{2+} if Mg^{2+}, Sr^{2+}, and Ba^{2+} have been previously removed.

6. Flame Tests for K^+ and Na^+ Test a portion of the solutions saved from step 1 (the solutions left after precipitation of Mg^{2+}, Ca^{2+}, Sr^{2+}, and Ba^{2+} as the carbonates). To test for the presence of Na^+ and K^+, flame tests will be used. In a flame test, a wire loop is repeatedly dipped into concentrated (12 M) HCl and heated to clean it, then dipped into a solution containing sodium or potassium ion. The loop is heated in the flame of a Bunsen burner. Heating the salts in the flame dissociates the salts into neutral atoms. When heated to a high enough temperature, the electrons of the atoms are put into an excited state (a higher electronic energy level). When the electrons drop back down to the normal (ground) state, each kind of metal atom emits light of a characteristic color—an intense yellow color indicates the presence of sodium whereas a fainter violet color indicates potassium. When both are present together, the sodium emission obscures the color of potassium emission, so a filter of cobalt glass is used to absorb or filter out the sodium light so that the violet light of the potassium can be seen. Figure 31-6 illustrates the flame test for potassium (K^+).

You might wonder why there is no mention of light observed from the nonmetal atoms in the anions. The nonmetal atoms also emit light but at shorter wavelengths in the ultraviolet that our eyes cannot see.

7. Tetraphenylborate Test for K⁺ A second way to test for the presence of K^+ is to take 15 drops of the solutions set aside after the precipitation of Mg^{2+}, Ca^{2+}, Sr^{2+}, and Ba^{2+} described in step 1. These solutions contain added NH_4^+ ion that must be removed because it interferes with the test for K^+.[1] We accomplish this by adding 6 drops (0.3 mL) of 1 M NaOH to make the solutions basic. Gently boil the solutions to expel any NH_3 (this prevents later interference by NH_4^+ ion. Then acidify with 2 drops of 6 M HCl and add 15 drops of 1 M sodium acetate ($NaCH_3COO$). (The acetate ions react with excess H_3O^+ and buffer the solution at about pH 5.) Then add 3 drops of 3% sodium tetraphenylborate, $NaB(C_6H_5)_4$. A white precipitate of potassium tetraphenylborate, $KB(C_6H_5)_4$, indicates the presence of potassium.

8. Test for NH₄⁺ This test must always be carried out on the *original solution or sample* because ammonium salts are used as reagents throughout the procedures for the other cations. You need for this test a smaller (65 mm) watch glass, a strip of red litmus paper, and a larger (75 mm) watch glass that can be inverted to cover the smaller one.

Wet a strip of red litmus paper with deionized water and stick it to the concave side of the larger watch glass. Put 20 drops (1 mL) of 0.1 M NH_4Cl into a second, smaller watch glass. Warm this solution on a hot plate to near boiling, then remove it; quickly add 6 drops (0.3 mL) of 6 M NaOH and cover it with the inverted larger watch glass so that the litmus paper stuck to its surface is exposed to any NH_3 vapors coming from the sample.

An alternative procedure is to put the sample plus NaOH into a small (10-mL) beaker, laying the wet litmus paper across the top of the beaker, and cover-

---

[1]Tetraphenylborate ion, $B(C_6H_5)_4^-$, forms a very insoluble precipitate with K^+. However, it also forms precipitates with ions having similar size and charge, including NH_4^+, Ag^+, Rb^+, Cs^+, and Tl^+, so that these ions must be absent when testing for K^+.

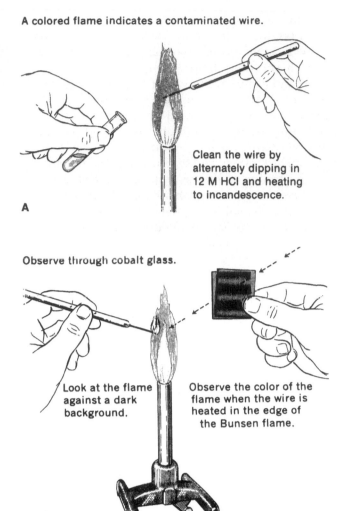

A colored flame indicates a contaminated wire.

A

Clean the wire by alternately dipping in 12 M HCl and heating to incandescence.

Observe through cobalt glass.

Look at the flame against a dark background.

Observe the color of the flame when the wire is heated in the edge of the Bunsen flame.

Dip the clean wire into the substance which has been moistened with 12 M HCl.

B

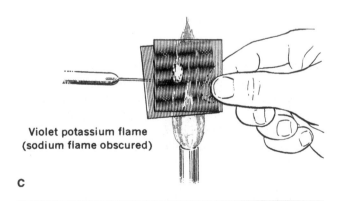

Violet potassium flame (sodium flame obscured)

C

FIGURE 31-6 | **The flame test for sodium and potassium. (A) shows the procedure for cleaning the wire; (B) and (C) illustrate the steps for observing the flame coloration.**

ing the beaker with clear plastic wrap. Then very carefully warm the beaker on a hotplate (low heat setting), taking care not to boil the solution so that NaOH in the solution does not spatter onto the litmus paper, turning it blue.

Allow a few minutes for any released NH_3 to diffuse to the red litmus paper, turning it blue. An even, unspotted blue color caused by the release of NH_3 gas proves the presence of ammonium ion. (Also try using your nose as a sensitive detector of NH_3.) Repeat the test using 20 drops (1 mL) of seawater (or unknown) in place of the 0.1 M NH_4Cl.

C. Analysis of an unknown solution for Mg^{2+}, Ca^{2+}, Sr^{2+} and/or Ba^{2+}, Na^+, K^+, and NH_4^+

If you analyzed a seawater sample along with the known solution, and if time permits, your instructor may direct you to analyze another unknown using the preceding procedures. With the practice you have had, this analysis should go much faster. As you proceed, keep a record of each step in a laboratory notebook or by completing the analysis summary in part C of the Report form, describing your actual observations, negative (no precipitate) as well as positive (precipitate forms), for each unknown.

CONSIDER THIS

The variety of colors seen in fireworks is a result of small amounts of metal salts that are added to the explosive mixtures. Make up ~0.1 M solutions of the chloride salts (nitrate salts work less well) of sodium, potassium, calcium, strontium, barium, copper, lithium, rubidium, and cesium. Use the procedure outlined in Figure 31-6 to perform flame tests on these solutions. How many different colors can you make? Can you make new flame colors by combining solutions? For a brightly colored flame like sodium,

what are the limits of detection of the salt? (That is, how dilute can you make the solution and still detect sodium?) Make up a series of progressively more dilute solutions of a couple of the salts. Can your eyes detect the differences in concentration in a flame test?

Your nose contains a very sensitive ammonia detector. What is the minimum level of NH_3 you can smell in a solution? (See the Introduction, Figure I-1, for the proper technique for a smell test.) For example, can you smell NH_3 from a solution of 0.1 M NH_4Cl? Do your classmates have the same limits as you? (Given that NH_4^+ is a weak monoprotic acid whose $K_a = 5.7 \times 10^{-10}$, show that the equilibrium concentration of NH_3 in a solution of 0.1 M NH_4Cl is about 10^{-5} M.)

From the data contained in Table 31-1, can you propose a statement about the trends in solubilities of the alkaline earth salts? If you check the *CRC Handbook of Chemistry and Physics* for the solubilities of similar beryllium salts, does your statement hold true? Does it work for the nitrate and chloride salts of the alkaline earth elements?

Bibliography

Chandra, R. "Detection of Barium, Strontium, and Calcium with Sodium Rhodizonate," *J. Chem. Educ.* **1962,** *39,* 397.

Pariza, R. "Qualitative Detection of Ba, Sr, and Ca Ions," *J. Chem. Educ.* **1963,** *40,* 417. The use of $(NH_4)_2SiF_6$ to precipitate Ba^{2+} as $BaSiF_6$ is described.

Selco, J. I.; Roberts, Jr., J. L.; Wacks, D. B. "The Analysis of Seawater: A Laboratory-Centered Learning Project in General Chemistry," *J. Chem. Educ.* **2003,** *80,* 54–57.

Identification of Alkaline Earth and Alkali Metal Ions

Analysis of Cations in Seawater

Name _____

Date _____ Section _____

Locker _____ Instructor_____

Pre-Lab Preparation

If you will be analyzing seawater, attach to the report form your table showing the calculations of the concentrations in units of mol/L of the ten most abundant elements in seawater (excluding hydrogen and oxygen).

Data

A. Typical reactions of the ions

NOTE: Indicate any experimental tests you have performed personally in order to be certain of the results.

1. Write net ionic equations for the reactions, *if any,* occurring when dilute solutions of the following are mixed. Write N.R. (for no reaction) if there is none.

(a) NH_4Cl, KCl, NaOH, heated _____

(b) $CaCl_2$, $BaCl_2$, 1 M NH_3 _____

(c) $MgCl_2$, $BaCl_2$, 1 M NaOH _____

(d) $CaCl_2$, K_2CrO_4 _____

(e) Mixture (d) + $BaCl_2$ _____

(f) Mixture (e) + HCl _____

(g) Mixture (f) + Na_2SO_4 _____

(h) KCl, $BaCl_2$, $(NH_4)_2CO_3$ _____

(i) $CaCO_3(s)$, NH_4Cl, 6 M CH_3COOH _____

(j) Mixture (i) + $K_2C_2O_4$, NH_3 _____

2. On the basis of periodic table relationships, what would you predict about the general solubility (indicate soluble, slightly soluble, or insoluble) of each of the following?

(a) $Be(OH)_2$ _____   **(c)** $Ra(OH)_2$ _____

(b) $BeSO_4$ _____   **(d)** $RaSO_4$ _____

3. Why does the addition of HCl dissolve insoluble salts such as $BaCrO_4$ and CaC_2O_4?

B. Analysis of a known solution for Mg^{2+}, Ca^{2+}, Sr^{2+}, Ba^{2+}, Na^+, K^+, and NH_4^+

1. The flow chart summarizes the step procedures, omitting step 3.

Key: *Step numbers* are in bold numerals, *reagents* are on broken-line arrows, *precipitates* or *residues* are in solid-line boxes, and *solutions* are in broken-line boxes.

Flowchart for a known solution containing Mg^{2+}, Ca^{2+}, Sr^{2+}, Ba^{2+}, Na^+, K^+, and NH_4^+

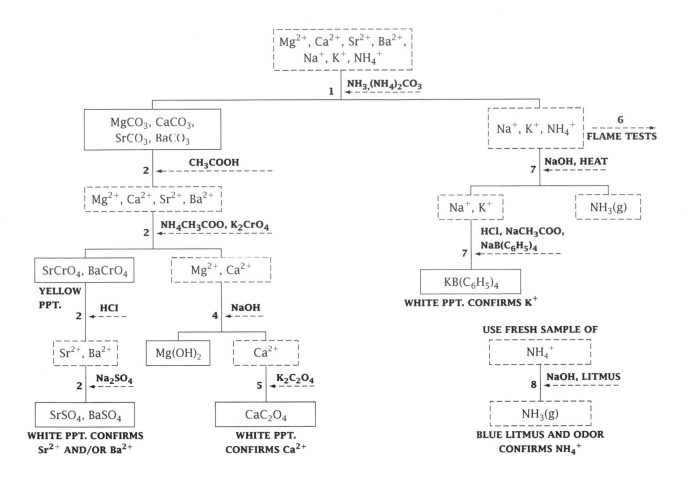

Name _____   Date _____

2. Complete the analysis summary for your known sample (include the unknown in step 3d).

Appearance _____   Ions found _____

Step	Sample description	Reagent(s)	Observations	Conclusions
1				
2				
3	(a) Mg^{2+} (b) Ca^{2+} (c) $Sr^{2+} + Ba^{2+}$ (d) Unknown			
4				
5				
6				
7				
8				
Summary conclusions				

C. Analysis of seawater or unknown solutions for Mg^{2+}, Ca^{2+}, Sr^{2+}, Ba^{2+}, Na^+, K^+, and NH_4^+.

Unknown No. _____ Appearance _____ Ions found _____

Unknown No. _____ Appearance _____ Ions found _____

Analysis summary: For your seawater or unknown samples, report below your actual observations, whether they are negative (no apparent reaction) or positive, for each step of the analysis (omitting step 3) and record the ions found.

Step	Sample description	Reagent(s)	Observations	Conclusions

Name _____   Date _____

▇▇▇▇ CONSIDER THIS ▇▇▇▇

The variety of colors seen in fireworks is a result of small amounts of metal salts that are added to the explosive mixtures. Make up ~0.1 M solutions of the chloride salts (nitrate salts work less well) of sodium, potassium, calcium, strontium, barium, copper, lithium, rubidium, and cesium. Use the procedure outlined in Figure 31-6 to perform flame tests on these solutions. How many different colors can you make? Can you make new flame colors by combining solutions? For a brightly colored flame like sodium, what are the limits of detection of the salt? (That is, how dilute can you make the solution and still detect sodium?) Make up a series of progressively more dilute solutions of a couple of the salts. Can your eyes detect the differences in concentration in a flame test?

Your nose contains a very sensitive ammonia detector. What is the minimum level of NH_3 you can smell in a solution? (See the Introduction, Figure I-1, for the proper technique for a smell test.) For example can you smell NH_3 from a solution of 0.1 M NH_4Cl? Do your classmates have the same limits as you? (Given that NH_4^+ is a weak monoprotic acid whose $K_a = 5.7 \times 10^{-10}$, show that the equilibrium concentration of NH_3 in a solution of 0.1 M NH_4Cl is about 10^{-5} M.)

From the data contained in Table 31-1, can you propose a statement about the trends in solubilities of the alkaline earth salts? If you check the *CRC Handbook of Chemistry and Physics* for the solubilities of similar beryllium salts, does your statement hold true? Does it work for the nitrate and chloride salts of the alkaline earth elements?

The Qualitative Analysis of Some Common Anions

Analysis of Anions in Seawater

Purpose

• Study specific tests for detecting in seawater (or an unknown sample) the presence of nine common anions: sulfide, sulfate, sulfite, carbonate, chloride, bromide, iodide, phosphate, and nitrate ions.

Introduction

Qualitative analysis is a branch of analytical chemistry that identifies particular substances in a given sample of material. In the analysis of inorganic substances, this branch involves the analysis of both metallic constituents as cations and mostly nonmetallic constituents as anions. Experiment 31 deals with the identification of seven common cations and provides an introduction to the techniques of qualitative analysis. This experiment complements Experiment 31 and explores the identification of nine anions: S^{2-}, SO_4^{2-}, SO_3^{2-}, CO_3^{2-}, Cl^-, Br^-, I^-, PO_4^{3-}, and NO_3^-. Altogether, six of these sixteen cations and anions are present in seawater at concentrations in the range of 0.5 to 0.01 M and are detectable using the methods described in Experiments 31 and 32.

Qualitative analysis has remained an important part of the laboratory experience in general chemistry for a number of years, even though these analytical methods have been replaced by sophisticated instrumental methods in practical analysis. We believe that qualitative analysis serves three useful purposes: (1) It provides an ideal context for the illustration and application of the principles of ionic equilibria, such as acid–base, solubility, and complex ion equilibria; (2) it provides a systematic framework for the discussion of the descriptive chemistry of the elements; and (3) it helps develop logical and deductive reasoning skills.

Its utility aside, qualitative analysis is fun. It's hands on—not abstract. It's about mixing ions in test tubes and seeing the colors and textures of solutions and precipitates, learning the solubilities of their salts, and then putting all these clues together to identify what's in the solution.

▓ NOTES TO INSTRUCTOR ▓

Experiments 31 and 32, taken together, allow students to analyze for the main cations and anions in seawater that comprise >99% of the dissolved salts in seawater. If authentic seawater samples are not available, it is easy to prepare synthetic seawater and/or to analyze unknown samples containing some of the sixteen cations and anions in Experiments 31 and 32. (A recipe for synthetic seawater suitable for these experiments is given in the Instructor's Manual and in the online supplemental material that accompanies the article by Selco et al., listed in the Bibliography.)

It is suggested that three lab periods (9 hours) be allowed for doing both cations and anions. This gives sufficient time for discussion of the results, testing the limits of detection for some ions, and repeating work that is problematic.

Pre-Lab Preparation

Your instructor will inform you if you will be analyzing seawater and/or unknowns containing anions.

If you are going to analyze seawater, your first task in preparing for this experiment will be to look up concentrations of the ten most abundant elements in seawater and do the calculations necessary to express their concentrations in units of moles per liter. Summarize the results in a table. If you know how to use a spreadsheet such as Microsoft Excel, do your calculations and tabulate the results in a spreadsheet. Otherwise, construct your table by hand. We will show later some example calculations for converting mass units of concentration into units of moles per liter.

Seawater contains about 35 g of dissolved salts per liter (3.5 mass %) and more than 70 elements with a wide range of concentration levels: *major* (0.5–0.01 M), *minor* (0.01–10^{-5} M), and *trace* level (<10^{-5} M). In Experiment 31, you will analyze seawater (or an unknown provided by the instructor) for the major cations. In this experiment, you will analyze for the major anions in seawater.

Information about seawater composition can be found in most editions of the *CRC Handbook of Physics and Chemistry,* as well as on the Internet. If you use the Google search engine (www.google.com) and the search words *seawater composition*, you will find sites that provide tables of seawater composition. This information is usually provided as the elemental composition by mass, often expressed in parts per million (1 ppm = 1 mg/kg of seawater). The density of seawater is about 1.02 g/mL. For our purposes, we may assume that 1 kg of seawater has a volume of about 1 L.

Many nonmetal elements are not present in seawater in the form of the element but as an oxyanion. For example, the sulfur in seawater is present not as dissolved sulfur or sulfide ion, but as sulfate ion, SO_4^{2-}. Nitrogen is mainly present as the nitrate ion, NO_3^-, and carbon as the bicarbonate ion, HCO_3^-. So you must be careful to note whether the composition given is based on the element, that is, expressed as the mass of elemental S, N, C, or as the mass of the oxyanion, which includes the mass of the oxygen in the anion. Also note that when an element exists as an oxyanion containing just one atom of that element, the molar concentration of that element will be the same as the molar concentration of the oxyanion containing the element.

Example Calculations One source says that seawater contains about 900 ppm (mg/L) of sulfur. Convert to units of moles per liter of sulfate ion, SO_4^{2-}.

$$\frac{900\ \text{mg S}}{1\ \text{L}} \times \frac{1\ \text{g}}{1000\ \text{mg}} \times \frac{1\ \text{mol S}}{32\ \text{g S}} \times \frac{1\ \text{mol } SO_4^{2-}}{1\ \text{mol S}} =$$

$$0.028\ \frac{\text{mol } SO_4^{2-}}{\text{L}}$$

A second source says that seawater contains 3.5% by weight of dissolved salts, equivalent to 35 g of dissolved salts per liter. The salt contains 7.68 wt % of sulfate. What is the concentration of sulfate ion, SO_4^{2-}, in units of moles/liter?

$$\frac{35\ \text{g salt}}{1\ \text{L}} \times \frac{7.68\ \text{g } SO_4^{2-}}{100\ \text{g salt}} \times \frac{1\ \text{mol } SO_4^{2-}}{96\ \text{g } SO_4^{2-}} =$$

$$0.028\ \frac{\text{mol } SO_4^{2-}}{\text{L}}$$

Some general properties of anions

The principles that are employed in the identification of cations can also be applied to the analysis of anions. The qualitative detection of anions in a sample depends on the distinctive solubility properties of particular salts of the ions and specific chemical reactions that are (ideally) unique to a particular ion. In this experiment, we will explore ways to detect the presence of S^{2-}, SO_4^{2-}, SO_3^{2-}, CO_3^{2-}, Cl^-, Br^-, I^-, PO_4^{3-}, and NO_3^-.

Carefully read the following criteria on which the identification of an anion generally depends. (The properties on which these criteria depend are summarized in the tables in the Appendix.)

1. Does the anion form a *precipitate* with certain cations, such as Ag^+ or Ba^{2+}, in neutral or in acidic solutions? (See Table 8 in the Appendix for solubility rules.)

2. Is the ion the *anion of a weak or a strong acid?* (See Table 6 in the Appendix for information on the relative strength of acids.)

3. Is the ion the anion of a *volatile acid?*

4. Can the anion act as an *oxidizing agent* or *reducing agent?* What is its relative strength as an oxidizing or reducing agent? (See the Appendix, Table 10, for redox potentials.)

Review the descriptive chemistry of the anions we will study in this experiment as you apply the preceding criteria to the typical reactions and tests for the following anions.

Typical Reactions of Common Anions

Sulfate ion

Sulfuric acid is a diprotic acid and generally is considered to be a *strong acid;* however, if the concentration of H_3O^+ is 1 M or higher, the sulfate ion is substantially protonated to form HSO_4^-. That is, a 1 M solution of H_2SO_4 has an H_3O^+ concentration of about 1 M, not 2 M as popularly believed. Sulfate ion forms characteristic precipitates with certain metal ions, as with barium ion:

$$Ba^{2+} + SO_4^{2-} \rightleftharpoons BaSO_4(s)$$

Other anions also form insoluble precipitates with barium ion:

$$Ba^{2+} + CO_3^{2-} \rightleftharpoons BaCO_3(s)$$
$$Ba^{2+} + SO_3^{2-} \rightleftharpoons BaSO_3(s)$$
$$3\ Ba^{2+} + 2\ PO_4^{3-} \rightleftharpoons Ba_3(PO_4)_2(s)$$

However, these last three anions are salts of *weak acids* and would all dissolve, or fail to precipitate, in an acid solution. In the presence of excess hydronium ion, the concentration of the free anion is reduced to such a low value that the solubility equilibrium is reversed, and the precipitate dissolves or fails to form in the first place. We may therefore use barium ion in an acid solution ($[H_3O^+] \cong 0.1$ M) as a test reagent for the presence of sulfate ion.

Sulfite ion, sulfide ion, and carbonate ion

Each of these anions of *weak, volatile acids* forms the free acid when a strong acid is added:

$$SO_3^{2-} + 2 H_3O^+ \rightleftharpoons H_2SO_3(aq) + 2 H_2O \rightleftharpoons$$
$$SO_2(g) + 3 H_2O$$

$$S^{2-} + 2 H_3O^+ \rightleftharpoons H_2S(g) + 2 H_2O$$

$$CO_3^{2-} + 2 H_3O^+ \rightleftharpoons H_2CO_3(aq) + 2 H_2O \rightleftharpoons$$
$$CO_2(g) + 3 H_2O$$

The sharp odor of sulfur dioxide (SO_2) and the rotten egg smell of hydrogen sulfide (H_2S) are usually sufficient identification. Sulfite ion (SO_3^{2-}) can be oxidized to sulfate ion by bromine water (Br_2) or hydrogen peroxide (H_2O_2), then tested as the sulfate ion is tested. Sulfide ion (S^{2-}) can be further confirmed by allowing the $H_2S(g)$ formed to come into contact with a lead salt solution on filter paper. The result will be the formation of a black deposit of lead sulfide (PbS).

The sulfite ion can be reduced to free sulfur by sulfide ion, which in turn is oxidized also to sulfur, so that a yellow precipitate of sulfur would be the main observable product. The equation is

$$SO_3^{2-} + 2 S^{2-} + 6 H_3O^+ \rightarrow 3 S(s) + 9 H_2O$$

These ions, therefore, are not likely to be found together in the same solution, particularly if it is acidic. A solution of sulfite is somewhat unstable because of its ease of oxidation by atmospheric oxygen:

$$2 SO_3^{2-} + O_2(g) \rightarrow 2 SO_4^{2-}$$

The conventional oxidation number of sulfur in sulfite ion is +4, which increases to +6 in sulfate ion.

In the test for carbonate, the addition of acid causes the evolution of a colorless gas having a nose-tickling odor like that of a carbonated beverage. This is an indication, but not proof, that the gas is carbon dioxide. Confirmation that the sample contains carbonate is evidenced by precipitation of the ion as an insoluble carbonate salt, such as strontium carbonate ($SrCO_3$). We do this in two steps. First we trap the evolved CO_2 in a dilute solution of sodium hydroxide, where it forms the carbonate ion:

$$CO_2(g) + 2 Na^+(aq) + 2 OH^-(aq) \rightleftharpoons$$
$$2 Na^+(aq) + CO_3^{2-}(aq) + H_2O$$

In the second step, we add strontium chloride solution to the basic solution containing carbonate ion to precipitate strontium carbonate:

$$Sr^{2+}(aq) + CO_3^{2-}(aq) \rightleftharpoons SrCO_3(s)$$

The overall net ionic reaction is

$$CO_2(g) + 2 OH^-(aq) + Sr^{2+} \rightleftharpoons SrCO_3(s) + H_2O$$

It's important that the trapping solution remain basic in order to capture the evolved CO_2, but we use a relatively low concentration (0.1 M) of hydroxide ion in order to avoid precipitating in the second step the moderately soluble $Sr(OH)_2$, whose solubility is about 0.14 M.

The trapping solution must remain basic because a neutral aqueous solution does not capture CO_2 efficiently, and the CO_2 that is captured does not form enough carbonate ion to precipitate $SrCO_3$ because CO_2 is such a weak acid:

$$CO_2(g) + H_2O \rightleftharpoons H_2CO_3(aq) \rightleftharpoons$$
$$2 H^+ + CO_3^{2-} \quad (K_{eq} = K_{a1}K_{a2} \approx 10^{-17})$$

Sulfite ion, if also present in the unknown, will interfere with the carbonate ion test, since when the sample is acidified, it will liberate sulfur dioxide gas (SO_2) along with the carbon dioxide gas and will form a strontium sulfite ($SrSO_3$) precipitate when the gases are absorbed in the sodium hydroxide and added $SrCl_2$. (Sulfide ion, S^{2-}, would not interfere, since SrS is soluble.) To overcome this interference, it is necessary first to oxidize any sulfite ion to sulfate ion, thus preventing the formation of any sulfur dioxide gas.

Chloride ion, bromide ion, and iodide ion

Like the sulfate ion, the chloride, bromide, and iodide ions are the anions of *strong acids*. Their salts with silver ion are insoluble, whereas the silver salts of most other anions, which are also insoluble in neutral solution, will dissolve in acid solution. Silver sulfide (Ag_2S), however, is so very insoluble that the presence of hydrogen ion does not reduce the sulfide ion concentration sufficiently to dissolve it.

We can test for the presence of *chloride* ion in the mixed precipitate of AgCl, AgBr, and AgI. We dissolve the AgCl from the still more insoluble AgBr and AgI by adding ammonia solution, thus forming the complex ion $Ag(NH_3)_2^+$ and Cl^-. When the centrifugate (or filtrate) is again acidified, the AgCl is reprecipitated:

$$AgCl(s) + 2\ NH_3(aq) \rightleftharpoons Ag(NH_3)_2^+ + Cl^-$$

$$Ag(NH_3)_2^+ + Cl^- + 2\ H_3O^+ \rightleftharpoons$$
$$AgCl(s) + 2\ NH_4^+ + 2\ H_2O$$

It is therefore possible, by carefully controlling the NH_3 and Ag^+ concentrations, to redissolve the AgCl without appreciably redissolving the more insoluble AgBr and AgI. A further distinguishing characteristic is the color of the precipitates: AgCl is white, AgBr is cream-colored, and AgI is light yellow.

The distinctions that enable us to separate and test for *bromide ion* and *iodide ion* in the presence of chloride ion depend on differences in their ease of oxidation. Iodide ion is easily oxidized to iodine (I_2) by adding ferric ion (Fe^{3+}), but ferric ion does not affect the others. After adding hexane (C_6H_{14}) to identify (purple color in hexane) and remove the iodine, we can next oxidize the bromide ion with potassium permanganate or chlorine water solution, extract the bromine (Br_2) that forms into hexane, and identify it by the brown color produced in the hexane layer.

Phosphate ion

Phosphate ion is the anion of a moderately weak, nonvolatile acid—phosphoric acid (H_3PO_4). It is tested for by the typical method of forming a characteristic precipitate. The reagent used is ammonium heptamolybdate solution, to which an excess of ammonium ion is added in order to shift the equilibrium farther to the right. An acid solution is necessary. The net ionic reaction is

$$21\ NH_4^+ + 12\ [Mo_7O_{24}^{6-}] + 7\ H_3PO_4 + 51\ H^+ \rightleftharpoons$$
$$7\ (NH_4)_3PO_4 \cdot 12\ MoO_3(s) + 36\ H_2O$$

The yellow precipitate is a mixed salt, ammonium phosphomolybdate. Sulfide ion interferes with this test but can be removed first by acidifying the solution with HCl and boiling it.

Nitrate ion

Since all nitrates are soluble, we cannot use a precipitation method to test for nitrate ion. Instead, we use a two-step process. First, nitrate ion is an oxidizing agent that is reduced to nitrite ion in slightly basic

$NaHCO_3$ solution by copper-coated zinc metal, a reducing agent:

$$NO_3^- + H_2O + Zn(s) \rightarrow NO_2^- + Zn(OH)_2$$

Second, when acid and excess iodide ion are added, nitrite ion is protonated to form nitrous acid, which oxidizes the iodide ion to form iodine (or triiodide ion):

$$2\ HNO_2 + 2\ H_3O^+ + 3\ I^- \rightarrow 2\ NO + 4\ H_2O + I_3^-$$

In the presence of starch indicator, the deep blue color of the starch–triiodide ion complex is formed. Substances such as chlorine, chlorate, sulfite, and sulfide ions interfere, so they must be removed before testing for nitrate.

Summary

Prepare a summary of your results by writing balanced chemical equations for the reactions that were used to test each anion and the observations that you made to confirm the presence or absence of each anion. Table 32-1 and the Pre-Lab Preparation discussion should serve as valuable resources in identifying the important chemical species. This table should also help you to answer the questions in the Report form.

Experimental Procedure

Special Supplies: 50- and 10-mL beakers, as shown in Figure 32-1; polyethylene transfer pipets with 1-mL calibration marks; red litmus paper; clear plastic wrap.

Chemicals: 6 M HCl; 6 M HNO_3; 6 M NH_3; 0.1 M sodium sulfide, Na_2S (freshly made); 0.1 M sodium sulfate, Na_2SO_4; 0.1 M sodium sulfite, Na_2SO_3 (freshly made); 0.1 M sodium carbonate, Na_2CO_3; 0.1 M sodium chloride, NaCl; 0.1 M potassium bromide, KBr; 0.1 M potassium iodide, KI; 1.0 M potassium iodide, KI (used in nitrate ion test); 0.1 M sodium phosphate, Na_3PO_4; 0.1 M potassium nitrate, KNO_3; 0.01 M sodium nitrite, $NaNO_2$; lead acetate paper, $Pb(C_2H_3O_2)_2$; 0.1 M barium chloride, $BaCl_2$; saturated bromine water, $Br_2(aq)$ (see Bibliography); 3% hydrogen peroxide, H_2O_2; 0.1 M silver nitrate, $AgNO_3$; 0.1 M iron(III) chloride, $FeCl_3$; hexane, C_6H_{14}; saturated chlorine water, Cl_2 (see Bibliography); 0.5 M ammonium heptamolybdate (9.8 g/100 mL $[NH_4]_6Mo_7O_{24} \cdot 4\ H_2O$); 30-mesh zinc metal, $Zn(s)$; 0.03 M $CuSO_4$ in 1 M acetic acid, CH_3COOH (0.75 g $CuSO_4 \cdot 5\ H_2O$ and 6 g acetic acid in 100 mL of water); 1 M sodium bicarbonate, $NaHCO_3$; 0.5% starch (freshly prepared by boiling for 20 min 5 g potato starch/liter of water); 5 M (55% by weight)

TABLE 32-1 The Behavior of Some Negative Ions with the Metallic Ions Ag^+, Ba^{2+}, Ca^{2+}, and Pb^{2+}

Negative Ions	Metallic Ions			
	Ag^+	Ba^{2+}	Ca^{2+}	Pb^{2+}
NO_3^-	Soluble	Soluble	Soluble	Soluble
Cl^-, Br^-, I^-	AgCl, white; AgBr, cream; AgI, yellow. All insoluble in HNO_3. AgCl soluble, AgBr slightly soluble, and AgI insoluble, in NH_3	Soluble	Soluble	$PbCl_2$ and $PbBr_2$, white, soluble in hot water. PbI_2, yellow, slightly soluble in hot water
SO_4^{2-}	Moderately soluble	$BaSO_4$, white, insoluble in HNO_3	$CaSO_4$, white, slightly soluble	$PbSO_4$, white, insoluble in HNO_3
SO_3^{2-}	Ag_2SO_3, white, soluble in NH_3 and HNO_3	$BaSO_3$, white, soluble in HNO_3	$CaSO_3$, white, soluble in HNO_3	$PbSO_3$, white, soluble in HNO_3
S^{2-}	Ag_2S, black, soluble in hot, conc. HNO_3	Soluble	Soluble	PbS, black, soluble in HNO_3
PO_4^{3-}	Ag_3PO_4, yellow, soluble in NH_3 and in HNO_3	$Ba_3(PO_4)_2$, white, soluble in HNO_3	$Ca_3(PO_4)_2$, white, soluble in HNO_3	$Pb_3(PO_4)_2$, white, soluble in HNO_3
CO_3^{2-}	Ag_2CO_3, white, soluble in NH_3 and in HNO_3	$BaCO_3$, white, soluble in HNO_3	$CaCO_3$, white, soluble in HNO_3	$PbCO_3$, white, soluble in HNO_3

tartaric acid (it's OK if the tartaric acid solution is yellow or light-brown); 0.1 M sodium hydroxide; NaOH; pH paper or universal indicator; 0.1 M strontium chloride, $SrCl_2$.

FIGURE 32-1 | Setup for the carbonate test. The inner beaker contains the test sample, 30-mesh zinc metal, and HCl. Any CO_2 evolved is swept out by H_2 gas produced by the reaction of zinc and HCl. The evolved CO_2 is trapped in the 0.1 M NaOH contained in the outer beaker. The plastic wrap hinders escape of the evolved CO_2.

⚠ SAFETY PRECAUTIONS:

H_2S gas is very toxic and has a characteristic odor of rotten eggs. You should avoid inhaling H_2S in part 1 (Test for Sulfide Ion). The SO_2 gas produced in part 3 (Test for Sulfite Ion) is irritating and toxic and has a strong, suffocating odor. Avoid inhaling SO_2 gas. In part 5 (Test for Chloride Ion), boiling to remove H_2S should be done in a fume hood, or you should use an inverted funnel connected to an aspirator (as shown in Figure 5-1) to remove H_2S.

Note that 6 M HCl, 6 M HNO_3, and 6 M NH_3 are corrosive. If they contact your skin, wash them off immediately.

☠ WASTE COLLECTION:

While most of the solutions and solids formed in this experiment are nontoxic, some should not get into the environment in the form resulting from this lab. Rather than present an extensive list of hazards, we emphasize that all waste solutions and solids should be collected in an appropriately labeled container. Your instructor may wish to have separate containers for lead waste, silver compounds, and barium products.

First, familiarize yourself with the following test procedures, using 2-mL samples of 0.1 M solutions of each ion to be tested. Any tests that yield unsatisfactory results should be repeated to improve your technique.

Second, answer the review questions and statements in the experiment report before analyzing the unknowns.

Third, obtain one or more unknowns and perform analyses for each of the ions, using a fresh 2-mL sample for each test.

1. Test for Sulfide Ion To about 2 mL of the test solution, add a slight excess of 6 M HCl. You may notice an odor of H_2S, which is easily detectable in the low parts-per-million range in air. As a second sensitive test, place a piece of moistened lead acetate paper over the mouth of the test tube.[1] Heat the test tube gently in the fume hood. A darkening of the paper indicates S^{2-} in the original solution. (If S^{2-} is found to be present, SO_3^{2-} is not likely to be present—why not?—and the SO_3^{2-} test may then be omitted. Note also the modification of the PO_4^{3-} test when S^{2-} is present.)

2. Test for Sulfate Ion To 2 mL of the test solution, add 6 M HCl drop by drop until the solution is slightly acidic. Then add 3 mL of 0.1 M BaCl2 solution, or more as needed to complete the precipitation. A white precipitate of $BaSO_4$ proves the presence of SO_4^{2-}. (Save the solution for the sulfite test.)

3. Test for Sulfite Ion If you noticed a sharp odor of SO_2 when the solution from the previous sulfate ion test was made acidic with HCl, then SO_3^{2-} is present. If you are in doubt, filter or centrifuge the solution to obtain a clear filtrate, add a drop or more of 0.1 M $BaCl_2$ to be sure all SO_4^{2-} is precipitated, and if necessary add more $BaCl_2$; then refilter or recentrifuge. To the clear solution add 1 to 2 mL of bromine water to oxidize any SO_3^{2-} to SO_4^{2-}. Formation of a second white precipitate of $BaSO_4$ now proves the presence of SO_3^{2-}.

4. Test for Carbonate Ion Besides the test sample, you will need 50-mL and 10-mL beakers; solutions of 0.1 M NaOH, 3% H_2O_2, 6 M HCl, and 0.1 M $SrCl_2$; 30-mesh zinc metal; red litmus paper; and a piece of clear plastic wrap.

First, place 4 mL of 0.1 M NaOH in the large (50-mL) beaker, cover it with plastic wrap, and set it aside for the moment.

Then place 3 mL of the unknown or seawater sample in the small (10-mL) beaker. (As a known test solution, use 1 mL of 0.1 M Na_2CO_3 + 2 mL of H_2O.) If sulfite ion is present in the unknown, add 1 mL of 3% H_2O_2 to oxidize it to sulfate ion. Then add 0.20–0.25 g of 30-mesh zinc metal.

Now remove the plastic wrap cover from the 50-mL beaker containing 0.1 M NaOH and carefully lower the 10-mL beaker into the 50-mL beaker, as shown in Figure 32-1. When it is ready, use a polyethylene transfer pipet to add 1 mL of 6 M HCl to the 10-mL beaker and immediately cover the top of the 50-mL beaker again with the plastic wrap. The purpose of the plastic wrap is to keep CO_2 from escaping from the 50-mL beaker. (When adding the 6 M HCl to the 10-mL beaker inside the larger beaker, you must take care not to spill any of it outside the 10-mL beaker, where even one drop would neutralize the NaOH in the larger beaker. If this happens, rinse the beakers and start over.)

Any carbonate or hydrogen carbonate ion in the test solution will be protonated by the HCl to form carbonic acid (H_2CO_3), which rapidly dehydrates to form $CO_2(g)$. The HCl also reacts with the zinc metal to form hydrogen gas, which helps to sweep the CO_2 out of the 10-mL beaker so that it can diffuse to the surface of the 0.1 M NaOH solution contained in the 50-mL beaker, where it reacts to form carbonate ion.

Allow the HCl to react with the zinc metal in the 10-mL beaker for 5 min, then carefully remove the 10-mL beaker, taking care not to spill any of the solution contained in it.

Test the solution remaining in the 50-mL beaker with red litmus paper. The litmus should turn blue, indicating the solution is still basic. (If the litmus remains red, you probably spilled HCl into the solution and you must start over.) Finish the test for carbonate by adding 1 mL of 0.1 M $SrCl_2$ to the basic solution in the 50-mL beaker. A cloudy white precipitate indicates the formation of $SrCO_3$ and the presence of CO_3^{2-} or HCO_3^- in the original test sample.

5. Test for Chloride Ion To a 2-mL portion of the test solution, add a few drops of 6 M HNO_3, as needed, to make the solution slightly acid. (Test with litmus paper.) Any sulfide ion present can be removed by boiling the solution for a moment in a fume hood. The free sulfur formed in the oxidation of S^{2-} by HNO_3 does not interfere. Add 3 mL of 0.1 M $AgNO_3$. (Here, no precipitate proves the absence of Cl^-, Br^-, or I^-.) Centrifuge the mixture. Test the clear filtrate with 1 drop of 0.1 M $AgNO_3$ for complete precipitation. If necessary, centrifuge again. Discard the filtrate. Wash the precipitate with distilled water to remove excess acid and silver ion. To this precipi-

---

[1]If lead acetate paper is not available, a suitable replacement can be made by taking a small strip of filter paper and moistening it with a few drops of 0.1 M lead nitrate solution.

tate, add 2 mL of distilled water, 1.5 mL of 6 M NH_3, and 0.5 mL of 0.1 M $AgNO_3$. (The proportions are important, since we want to dissolve only the AgCl from any mixture of AgCl, AgBr, and AgI.) $Ag(NH_3)_2^+$ and Cl^- will form. Shake the mixture well, and centrifuge. Transfer the clear solution to a clean test tube and acidify with 6 M HNO_3. A white precipitate of AgCl confirms Cl^-.

6. Test for Iodide Ion To 2 mL of the test solution, add 6 M HCl to make the solution acid. If S^{2-} or SO_3^{2-} is present, boil the solution to remove the ion. Add 3 mL of 0.1 M $FeCl_3$ to oxidize any I^- to I_2. (Br^- is not oxidized by Fe^{3+}.) Add 1 mL of hexane (C_6H_{14}) and agitate the mixture. A purple color of I_2 in the hexane layer indicates that I^- was present in the original sample. (Save the mixture for the Br^- test.)

7. Test for Bromide Ion If no I^- was present in the *preceding* mixture, add 2 mL of chlorine water and agitate it. A brown color in the hexane layer indicates Br^-. If I^- was present, separate, by means of a medicine dropper, as much as possible of the preceding iodide test solution below the hexane layer that contains the I_2, and place it in a clean test tube. Again extract any remaining I_2 by adding 1 mL of hexane, agitating the mixture, and separating the solution. Repeat this process as many times as necessary to eliminate the violet I_2 color from the hexane layer. The solution may be boiled for a moment to remove any remaining trace of I_2. Then add 2 mL of chlorine water and 1 mL of hexane and agitate the mixture. A brown color indicates Br^-.

8. Test for Phosphate Ion First mix about 1 mL of 0.5 M $(NH_4)_6Mo_7O_{24}$ reagent with 1 mL of 6 M HNO_3. (If a white precipitate forms, dissolve it by making the solution basic with 6 M NH_3, then reacidify with 6 M HNO_3.) If S^{2-} has been found in the unknown, first make a 2-mL portion of the test sample distinctly acidic with 6 M HCl, boil it a moment to remove all H_2S, then add this (or, if no S^{2-} is present, a 2-mL sample of the original unknown) to the clear molybdate solution. If iodide ion, I^-, is found to be present, add 1 mL of 6 M HNO_3 to the test solution and place the test tube in a boiling water bath for 5 min to remove the iodide. Then add 1 mL of the ammonium molybdate reagent. A yellow precipitate of ammonium phosphomolybdate, $(NH_4)_3PO_4 \cdot 12\ MoO_3$, appearing at once or after the mixture has been warmed a few minutes to about 40 °C, indicates the presence of PO_4^{3-}.

9. Test for Nitrate Ion Read again the synopsis of the method in the Pre-Lab section. A blank containing unreduced nitrate ion and a control using 1.0×10^{-4} M sodium nitrite solution are run, along with the test solution containing nitrate ion.

Preparing Copper-Coated Zinc Metal. Weigh 1.0 g of 30-mesh zinc metal and place it in a 13 × 100 mm test tube. Add to the zinc metal 2 mL of 0.03 M $CuSO_4$/1 M acetic acid solution. Shake the mixture or use a clean polyethylene transfer pipet to thoroughly mix the solution. (Place the tip of the pipet about 1 cm above the zinc metal in the bottom of the test tube, draw up the solution into the pipet, and rapidly expel it.) Initially, the solution turns cloudy from the evolution of small bubbles of hydrogen gas. After vigorous mixing for about 2 min, the solution will become clear and the blue color of Cu(II) ion will no longer be visible.

Next, draw off and discard the solution with a transfer pipet, leaving in the test tube the zinc metal, now coated with copper metal. Add 5 mL of deionized water. Mix thoroughly for a few seconds, then draw off and discard the solution. Repeat the washing procedure two more times with 5-mL portions of deionized water, drawing off and discarding the wash water.

Completing the Test. **Note:** The test for nitrate ion is very sensitive. Solutions containing concentrations of nitrate greater than 0.001 M should be diluted with deionized water until the nitrate ion concentration is about 0.001 M. When working with unknown samples containing mixtures of anions, use pH paper or universal indicator to check the pH of a sample of the diluted solution suspected of containing nitrate ion. The pH should be in the pH range of 6–8. If not, adjust the pH by adding very small amounts of NaOH or HCl until the solution is approximately neutral.

Preparing known and unknown test solutions and the blank and control solutions

Preparing the unknown test solution. Add 2 mL of seawater or an approximately neutral unknown solution (tested with pH paper) containing ~0.001 M nitrate ion, NO_3^-, to the test tube containing the washed copper-coated zinc metal. Then add 2 drops of 1 M $NaHCO_3$ to make the solution slightly basic. Using the same technique used in preparing the copper-coated zinc metal, mix the solution and the copper-coated zinc metal vigorously for about 2 min. During the mixing, some of the nitrate ion is reduced to nitrite ion, NO_2^-. After mixing for 2 min, draw off 1 mL of the solution and place it in a clean 13 × 100 mm test tube. Label the test tube as the unknown. Wash the copper-coated zinc metal with several portions of deionized water to rinse out the remaining unknown solution.

Preparing the known test solution. Prepare 0.001 M KNO_3 solution by diluting 1 mL of 0.1 M KNO_3 with 99 mL of deionized water. Mix this solution thoroughly and save a portion of it for preparing the blank solution, described later. Add 2 mL of the 0.001 M KNO_3 solution to the test tube containing the washed copper-coated zinc metal. Add 2 drops of 1 M $NaHCO_3$ and mix the sample vigorously with the copper-coated zinc. After mixing for 2 min, draw off 1 mL of the solution and place it in a clean 13 × 100 mm test tube. Label the test tube as the known. (Wash the copper-coated zinc metal with several portions of deionized water before using it for another sample.)

Preparing the blank solution. Place 1 mL of 0.001 M KNO_3 in a clean test tube. Add 1 drop of 1 M $NaHCO_3$ and mix. Label the test tube as the blank.

Preparing the control solution. Prepare a 1 × 10^{-4} M sodium nitrite, $NaNO_2$, solution by diluting 1 mL of 0.01 M $NaNO_2$ solution with 99 mL of deionized water. Mix the diluted solution thoroughly. Place 1 mL of the 1 × 10^{-4} M nitrite solution in a clean test tube, add 1 drop of 1 M $NaHCO_3$, and mix thoroughly. Label the test tube as the control.

Testing the Prepared Solutions for Nitrite Ion To each of the four solutions that you have prepared, contained in the four labeled test tubes, add the following reagents in the order specified: 2 drops of 1 M KI, 2 drops of 0.5% starch indicator, and 5 drops of 5 M tartaric acid, then mix thoroughly. If nitrite ion is present, the solutions should, within a few seconds, form the intense blue color of the starch–triiodide ion complex. (*Note:* Chlorine or the ions chlorate, sulfite, or sulfide must be absent because they interfere by reacting with nitrite or iodine produced in the test.)

Blank solution. No blue color should form in the blank solution, indicating the absence of iodine and nitrite ion. If a faint blue color does form, it may indicate that there has been some air oxidation of the KI solution or interfering substances previously mentioned may be present.

Control solution. An intense blue color should immediately appear. This confirms that nitrite ion is present and that the starch indicator is functioning properly. (The starch indicator has a finite shelf life and should have been recently prepared.)

Usually, the intensity of the blue color in the 1 × 10^{-4} M nitrite ion control is greater than that observed with the sample of 0.001 M KNO_3 solution, even though the nitrite control has a tenfold smaller concentration. This suggests that less than 10% of

the nitrate originally present is reduced to form nitrite ion. Even though nitrate ion is only partially reduced, the test is still very sensitive, easily detecting the presence of 0.001 M nitrate ion.

CONSIDER THIS

In the test for Cl^- and SO_4^{2-}, explain fully why the addition of acid will dissolve other insoluble silver salts and barium salts, respectively, which would interfere with the tests if the solution were neutral (see Table 32-1).

Suppose a white precipitate is obtained when adding $BaCl_2$ reagent to a neutral unknown solution. What might the precipitate be? Write formulas for as many possible substances as you can, considering both positive and negative ions as given in Table 32-1.

A positive test for SO_4^{2-} is to be expected whenever you have SO_3^{2-} in an unknown solution, even when no SO_4^{2-} has been placed in the solution. What could explain this?

Why cannot both S^{2-} and SO_3^{2-} be present in the same solution? Explain, and write the equation.

Both CO_3^{2-} and HCO_3^- respond to the usual carbonate ion test. How could you distinguish between solutions of 1 M $NaHCO_3$ and 1 M Na_2CO_3? (*Hint:* One way would be to consider the pH of the solution.)

Bibliography

Baker, Jr., R. C. "Checking Trace Nitrate in Water and Soil Using an Amateur Scientist's Measurement Guide," *J. Chem. Educ.,* **1995,** *72,* 57–59.

Feigl, F. *Spot Tests in Inorganic Analysis,* Elsevier Publishing, New York, 1958, p. 332.

Hiegel, G. A.; Abdala, M. H.; Burke, S. V.; Beard, D. P. "Methods for Preparing Aqueous Solutions of Chlorine and Bromine for Halogen Displacement Reactions," *J. Chem. Educ.,* **1987,** *64,* 156.

Holmes, Jr., L. H., "An Easy Way to Make Chlorine Water," *J. Chem. Educ.,* **1997,** *74,* 1326.

Selco, J. I.; Roberts, Jr., J. L.; Wacks, D. B. "The Analysis of Seawater: A Laboratory-Centered Project in General Chemistry," *J. Chem. Educ.* **2003,** *80,* 54–57.

The Qualitative Analysis of Some Common Anions

Analysis of Anions in Seawater

Name _____

Date _____ Section _____

Locker _____ Instructor_____

Attach your table of seawater composition if you are analyzing seawater.

Summary of the Experiment

Write balanced chemical equations to describe the reactions that you used to identify the anions in this experiment. For this summary, ignore the complications that may be caused by the presence of other ions. Also describe the observations (i.e., what you actually saw) to confirm the presence or absence of each ion. The entries for phosphate have been filled in as an example (and because it is one of the more complicated reactions).

Species	Equations	Observations
1. Sulfide		
2. Sulfate		
3. Sulfite		
4. Carbonate		
5. Chloride		
6. Iodide		
7. Bromide		
8. Phosphate	$21\ NH_4^+ + 12\ [Mo_7O_{24}^{6-}] + 7\ H_3PO_4 + 51\ H^+ \rightleftharpoons$ $7\ (NH_4)_3PO_4 \cdot 12\ MoO_3(s) + 36\ H_2O$	A yellow precipitate confirms phosphate, PO_4^{3-}
9. Nitrate		

Some More Brainteasers

1. Give the formulas of all new molecules or ions formed when solutions of the following are mixed. Write N.R. (for no reaction) if there is none.

   (a) Cl^-, Br^-, I^-, Fe^{3+} _____

   (b) Fe^{2+}, Br_2 water _____

   (c) Cl^-, Br^-, I^-, Cl_2 water _____

   (d) $CO_2(g)$, NaOH, $SrCl_2$ _____

   (e) $SO_2(g)$, $CaCl_2$, HCl _____

   (f) SO_3^{2-}, H_2O_2 (acid) _____

   (g) S^{2-}, SO_3^{2-} (acid) _____

   (h) $Pb(C_2H_3O_2)_2$, $H_2S(g)$ _____

   (i) Ag^+, Cl^-, NH_3 (excess) _____

   (j) Ag^+, I^-, NH_3 (excess) _____

2. An unknown solution (I) is to be tested for the anions, or their derivatives, listed below right. (Derivatives are, e.g., PO_4^{3-} or H_3PO_4.) The following tests are performed on *separate* portions of the unknown.

   (a) When a test portion is acidified with HCl, and lead acetate is held in the mouth of the test tube, the paper turns black.

   (b) A test portion is acidified with HCl, boiled, and cooled; then Cl_2 water and hexane are added. After the mixture is shaken, the hexane layer is a clear, light reddish brown color.

   (c) Addition of an excess of HNO_3 produces a colorless gas that forms a white precipitate when bubbled through sodium hydroxide followed by addition of $SrCl_2$.

On the basis of these tests considered as a whole, mark, at the right, a plus sign (+) for each ion or its derivative that is definitely present in the original solution, a minus sign (−) if the ion is definitely absent, or a question mark (?) if there is no evidence that the ion is present or absent.

Cl^-	_____
$Br-$	_____
I^-	_____
SO_4^{2-}	_____
SO_3^{2-}	_____
S^{2-}	_____
NO_3^-	_____
CO_3^{2-}	_____
PO_4^{3-}	_____

Name _____ Date _____

Analysis of Seawater or Unknown Solutions for S^{2-}, SO_4^{2-}, SO_3^{2-}, CO_3^{2-}, Cl^-, I^-, Br^-, PO_4^{3-}, and NO_3^-

Unknown No. _____ Appearance _____ Ions found _____

Unknown No. _____ Appearance _____ Ions found _____

Analysis summary: For your seawater or unknown samples, report below your actual observations, whether they are negative (no apparent reaction) or positive, for each test you perform and record the ions found.

Test No.	Sample description	Reagent(s)	Observations	Conclusions

In the tests for Cl^- and SO_4^{2-}, explain fully why the addition of acid will dissolve other insoluble silver salts and barium salts, respectively, which would interfere with the tests if the solution were neutral (see Table 32-1).

Suppose a white precipitate is obtained on adding $BaCl_2$ reagent to a neutral unknown solution. What might the precipitate be? Write formulas for as many possible substances as you can, considering both positive and negative ions as given in Table 32-1.

A positive test for SO_4^{2-} is to be expected whenever you have SO_3^{2-} in an unknown solution, even when no SO_4^{2-} has been placed in the solution. What could explain this?

Why cannot both S^{2-} and SO_3^{2-} be present in the same solution? Explain, and write the equation.

Both CO_3^{2-} and HCO_3^- respond to the usual carbonate ion test. How could you distinguish between solutions of 1 M $NaHCO_3$ and 1 M Na_2CO_3? (*Hint:* One way would be to consider the pH of the solution.)

An Introduction to Green Chemistry

Making Biodiesel and Soap from Vegetable Oils and Fats

Purpose

- Learn the principles and goals of green chemistry.

- Make two products, biodiesel and soap, that illustrate green chemistry in action.

▮ NOTES TO INSTRUCTOR ▮

Working efficiently, it is possible to prepare biodiesel and soap in one three-hour laboratory period, but soap requires some aging for the saponification to be complete. So the comparison of the properties of prepared soap and commercial soaps will benefit from having two lab periods available for the experiment. This also gives more time for students to look into the issues of sustainability and green chemistry, the Love Canal story, the role of biofuels in our energy future, and the interesting story of how soap has been prepared and used throughout history.

Introduction

Green chemistry is part of the broader concept of sustainability. A sustainable global society has been defined as one in which people today meet their needs without compromising the ability of future generations to live equally well. Green chemistry is an important step toward creating a sustainable global society, one that does not squander natural resources and degrade the environment.

Green chemistry embodies a set of principles whose goal is to reduce or eliminate the use or generation of hazardous substances in the design, manufacture, and use of chemical products. You will find a commonly quoted statement of "Twelve Principles of Green Chemistry" in Table 12 of the Appendix. Reading these principles, you will see that most of them are aimed at the chemical industry. However,

several of them can be applied in the general chemistry laboratory; the goal of this experiment is to illustrate some of the principles of green chemistry in action. But before we do that, let's look at some of the history of green chemistry and how it has come to be an active area of research and practice in contemporary chemistry.

Human progress is sometimes a response to a crisis or disaster. In the 1970s and 1980s, a number of environmental disasters in the United States were followed by a public outcry that spurred legislation embodying a regulatory approach focusing on cleanup and remediation, with fines and penalties for violators. Here are some notable examples of actions responding to public concern.

Beginning as early as 1868, the Cuyahoga River that flows through Cleveland, Ohio, into Lake Erie has suffered about ten river fires ignited in floating oil slicks and debris. A large river fire in 1952 caused over $1 million in damage to boats and a riverfront office building, and the last fire in June 1969 finally awakened public concern about the sad condition of American rivers, lakes, and estuaries and spurred legislation resulting in the Clean Water Act of 1972.

The publicity surrounding the Cuyahoga River fire in 1969 was quickly followed in December 1970 by creation of the U.S. Environmental Protection Agency (EPA) by congressional approval of a presidential executive order drawing together in one independent agency several smaller units of different federal agencies. The EPA is charged with protecting human health and with safeguarding our air, water, and land environment; it is chiefly responsible for formulating the environmental policy of the United States.

In 1980, the Comprehensive Environmental Response, Compensation, and Liability Act (CERCLA)—commonly known as the Superfund law—was enacted in response to the Love Canal disaster in Niagara Falls, New York, in which an old canal bed had been used as a chemical waste dump from the 1930s to the early 1950s, then covered over to build homes on the site. This law is designed to protect people and communities from heavily contaminated toxic waste sites that have been abandoned. (See the reference in the Bibliography to the article by Erika Engelhaupt on the thirtieth anniversary of the Love Canal disaster.)

The Emergency Planning and Community Right-to-Know Act of 1986 modified the Superfund law; it was a response to the Bhopal, India, tragedy in December 1984, in which an accidental release of 40 tons of toxic methyl isocyanate gas killed more than 3000 people within a few days and caused an estimated 20,000 premature deaths in the following years.

In 1990, the U.S. Congress passed the Pollution Prevention Act (PPA), which seeks to prevent pollution at its source, and marked a shift to a policy direction emphasizing prevention. At about the same time, several significant events marked the birth and development of the green chemistry movement, beginning with the influential 1987 report of the World Commission on Environment and Development that defined sustainable development as "meeting the needs of the present without compromising the ability of future generations to meet their own needs."

In 1989, Paul Anastas, now a leading advocate and spokesman for green chemistry, accepted a position in the EPA's Office of Pollution Prevention and Toxics (OPPT), which administers the Pollution Prevention Act of 1990. By 1991, the term *green chemistry* had been coined by Anastas and his colleagues at the EPA.

The Presidential Green Chemistry Challenge Awards Program was inaugurated in 1995 as an effort to recognize individuals and businesses for innovations in green chemistry. There are five awards each year, one in each of five categories.

The Green Chemistry Institute (GCI) was incorporated in 1997 after a year of planning by individuals from industry, government, and academia. In 2001, the GCI came under the umbrella of the American Chemical Society (ACS) as the ACS Green Chemistry Institute (ACS GCI).

In 1998, Paul Anastas and John C. Warner published their introductory text, *Green Chemistry: Theory and Practice*, which enunciated the widely quoted twelve principles of green chemistry (see Appendix Table 12). In the same year, the (British) Royal Society of Chemistry (RSC) launched a new journal, *Green Chemistry*, celebrating in 2008 its tenth year as a leading journal in the field.

Since 2000, the green chemistry movement has spread to the point that there are now in the United States and around the world several green chemistry institutes dedicated to the promotion and advancement of green chemistry principles. A few U.S. universities are beginning to offer courses in green chemistry as well as programs leading to graduate degrees in green chemistry. However, it remains to be seen whether the green chemistry movement will have a major impact on the chemical industry, which traditionally only adopts new technologies that offer a significant competitive economic advantage and has been reluctant to include social benefits in its balance sheets.

Green Chemistry in the General Chemistry Laboratory In this experiment, we will see how a renewable resource or feedstock—vegetable oils from a variety of plant sources—can be used to make biodiesel and soap. Glycerol is a by-product of the industrial scale operations for making these products. As biodiesel production increases, a glut of by-product glycerol results. So a question arises: What can we do with glycerol, which is now being produced in quantities that far exceed the need for its traditional uses as an emollient ingredient in skin lotions and cosmetics?

Here's a possible new use. In 2006, Professor Galen Suppes from the University of Missouri–Columbia received the Presidential Green Chemistry Challenge Award in the academic category for his system of converting waste glycerol (from biodiesel production) to propylene glycol. Propylene glycol produced this way could replace the more toxic ethylene glycol that is the primary ingredient in automobile antifreeze.

The Chemical Composition of Fats and Oils Animal and vegetable fats and oils are mainly triglycerides (more properly known as *triacylglycerols*). Those that are solids at room temperature or are of animal origin are commonly called *fats*, while those that are liquid or are of vegetable origin are called *oils*—but this is not a hard and fast distinction. The general chemical structure of a triglyceride is shown in Figure 33-1, enclosed in the brackets on the left. Triglycerides are formed from one molecule of glycerol and three fatty acid molecules. Glycerol is a trifunctional alcohol (containing three —OH hydroxyl groups) whose structure is shown in Figure 33-1, enclosed in the brackets on the right.

Glycerol can combine with one to three fatty acid molecules to form a monoglyceride, a diglyceride, or a triglyceride. Compounds formed by reaction of an alcohol R′—OH with a carboxylic acid HO—C(O)—R are

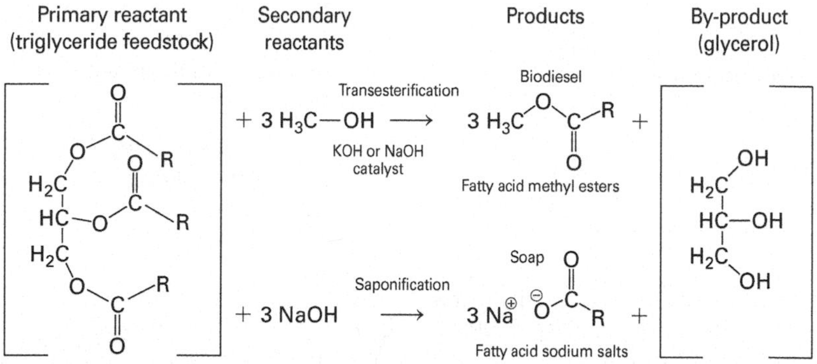

FIGURE 33-1 The overall reaction scheme shows the synthesis of biodiesel (upper reaction) and soap (lower reaction). In the reaction scheme, R represents the different hydrocarbon chains of the fatty acids, which typically have a length of 8 to 20 carbon atoms (including the acyl [OC(O)] carbon atom.) Glycerol is a by-product of both reactions.

known as *esters*. When glycerol—which we represent as R'(OH)$_3$—reacts with three fatty acid molecules, HO—C(O)—R, three *ester* linkages are formed with elimination of three molecules of water:

$$R'(OH)_3 + 3\ HO\text{—}C(O)\text{—}R \rightarrow R'[O\text{—}C(O)\text{—}R]_3 + 3\ H_2O$$
glycerol   fatty acid   triglyceride ester   water

(In this reaction, R' represents the hydrocarbon backbone of glycerol and R represents the hydrocarbon chain, or "tail," of a fatty acid.)

The [—O—C(O)—] group of the ester is called the *acyl* group—leading to the more systematic nomenclature of triglycerides as triacylglycerols.

The fatty acids produced by splitting (hydrolyzing) the three ester linkages in the triglyceride have linear (unbranched) chains of different lengths, typically containing 8 to 20 carbon atoms linked together and having a weakly acidic carboxyl HOC(O)— group on one end and terminating with a methyl —CH$_3$ group on the other end. (Figure 33-2 shows the structures of the most common fatty acids found in fats and oils, while Table 33-1 shows the fatty acid composition of some common fats and oils.)

We might guess that there would be a more or less random distribution of fatty acids in the triglyceride molecules. In any single triglyceride molecule, all three fatty acids might be different, or two of them could be the same, or all three could be the same. Also, there can be positional isomers, the central ester linkage to glycerol having a chemical environment different from the ester linkages on each end of the glycerol molecule. To summarize, natural triglycerides are mixtures of dozens of different molecules containing fatty acids of different chain lengths. Consequently, triglycerides do not have sharp melting points and their molar masses can be defined only as a weighted average of their fatty acid composition.

Saturated fatty acids are chains of —CH$_2$— groups terminated by a methyl group, —CH$_3$. The C—C single bonds of saturated fatty acids allow the chain to be flexible since there is nearly free rotation about a C—C single bond. Unsaturated fatty acids have at various positions along the chain one or more —CH=CH— groups, in which there is a C=C double bond. In this context, the term *unsaturated* means that the fatty acids contain less than the maximum number of C—H bonds. With the aid of a catalyst, a molecule of hydrogen, H$_2$, can be added to each —CH=CH— group, converting the unsaturated fatty acid into a saturated fatty acid having only —CH$_2$—CH$_2$— groups in the chain.

A C=C double bond is rigid in the sense that there is no free rotation about the C=C double bond. Thus, there are two possible configurations around a —CH=CH— double bond: *cis* or *trans*. In the *cis* configuration, both hydrogen atoms bonded to the carbon atoms are on the same side of the double bond; in the *trans* configuration, they are on opposite sides. The double bonds in natural fats and oils typically have the *cis* configuration. A *cis* configuration produces a change of direction, or kink, in the fatty acid chain, and gives a fishhook shape to linolenic acid, which has three C=C double bonds with *cis* configuration, as shown in Figure 33-2.

There are several different ways of naming unsaturated fatty acids. They have traditional common names as well as more systematic names with numbering schemes that specify the number of carbon atoms and the position and configuration of the double bonds in unsaturated fatty acids. In the scheme that we use in Figure 33-2 and Table 33-1, the position

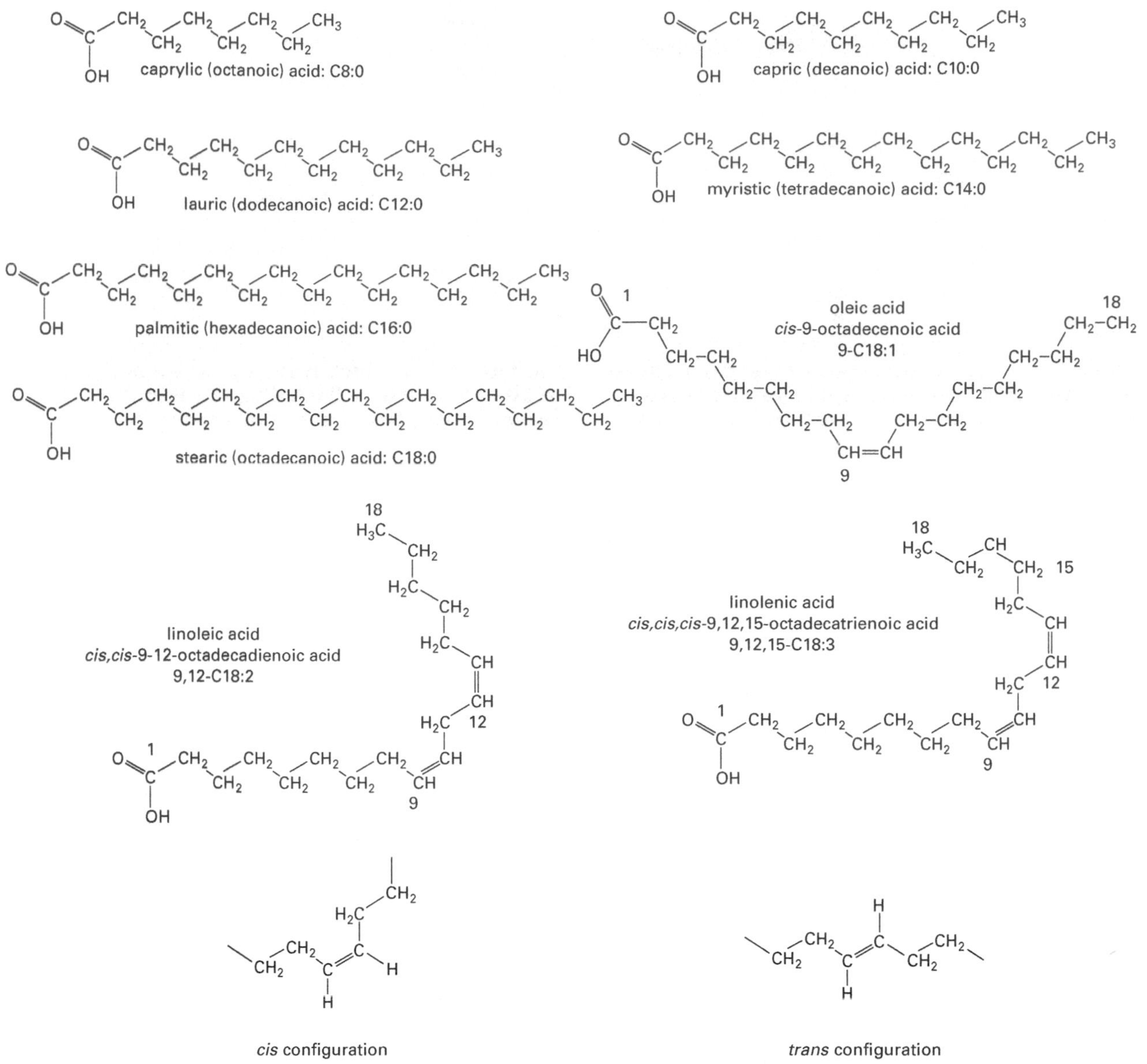

O
‖
C—CH₂—CH₂—CH₂—CH₃
| CH₂ CH₂ CH₂
OH   caprylic (octanoic) acid: C8:0

O
‖
C—CH₂—CH₂—CH₂—CH₂—CH₃
| CH₂ CH₂ CH₂ CH₂
OH   capric (decanoic) acid: C10:0

O
‖
C—CH₂—CH₂—CH₂—CH₂—CH₂—CH₃
| CH₂ CH₂ CH₂ CH₂ CH₂
OH   lauric (dodecanoic) acid: C12:0

O
‖
C—CH₂—CH₂—CH₂—CH₂—CH₂—CH₂—CH₃
| CH₂ CH₂ CH₂ CH₂ CH₂ CH₂
OH   myristic (tetradecanoic) acid: C14:0

O
‖
C—CH₂—CH₂—CH₂—CH₂—CH₂—CH₂—CH₂—CH₃
| CH₂ CH₂ CH₂ CH₂ CH₂ CH₂ CH₂
OH   palmitic (hexadecanoic) acid: C16:0

O
‖
C—CH₂—CH₂—CH₂—CH₂—CH₂—CH₂—CH₂—CH₂—CH₃
| CH₂ CH₂ CH₂ CH₂ CH₂ CH₂ CH₂ CH₂
OH   stearic (octadecanoic) acid: C18:0

oleic acid
cis-9-octadecenoic acid
9-C18:1

linoleic acid
cis,cis-9-12-octadecadienoic acid
9,12-C18:2

linolenic acid
cis,cis,cis-9,12,15-octadecatrienoic acid
9,12,15-C18:3

cis configuration

trans configuration

FIGURE 33-2 **The molecular structures of the most common fatty acids found in triglycerides. The labels show the common name, the systematic name, and a shorthand notation that specifies the position of the C=C double bonds, the number of carbon atoms, and the number of double bonds in the fatty acid (see text). The two** possible configurations about the C=C double bonds are shown in greater detail at the bottom of the figure: In the *cis* configuration, the two H atoms are on the same side of the double bond; in the *trans* configuration, the two H atoms are on opposite sides.

of the double bonds is specified by counting from the carboxyl end (the number 1 carbon atom). For example, linolenic acid, represented by the notation *cis,cis,cis*-9,12,15-C18:3, is a fatty acid with 18 carbon atoms and 3 double bonds (hence C18:3). All three double bonds have the *cis* configuration, and the locations of the double bonds in the fatty acid chain are specified by counting from the carboxyl end.

Green Chemistry in Action: Turning Fats and Vegetable Oils into Biodiesel Fuel and Soap Starting with vegetable oils, we will study two processes. The first is a catalytic reaction, leading to biodiesel; the second is a stoichiometric reaction leading to soap. Figure 33-1 shows in a concise way the chemical reactions involved in making biodiesel and soap from triglycerides.

TABLE 33-1 Fatty Acid Composition of Some Common Fats and Oils

Name of Oil or Fat	Triglyceride Weighted Average (molar mass, g/mol)	Fatty Acid Methyl Ester Weighted Average (molar mass, g/mol)	Total Fatty Acid (mass %)	Fatty Acid Composition (mass %)								
				C8:0 Caprylic	C10:0 Capric	C12:0 Lauric	C14:0 Myristic	C16:0 Palmitic	C18:0 Stearic	C18:1 Oleic	C18:2 Linoleic	C18:3 Linoleic
Canola	880	295	100					4	2	62	22	10
Olive	875	293	98					13	3	71	10	1
Soybean	873	292	100					11	4	24	54	7
Palm	848	284	100				1	45	4	40	10	
Coconut	678	227	99	8	6	47	18	9	3	6	2	
Lard (pork fat)*	860	288	99				2	29	14	44	10	
Beef tallow*	858	287	97				3	29	14	47	3	1

*Lard and beef tallow contain about 3% C16:1 palmitoleic acid, which is included with C16:0 palmitic acid in this table.

TABLE 33-1 | The typical fatty acid composition of selected vegetable and animal triglycerides are shown, together with their weighted average molar masses and the weighted average molar masses of the methyl esters derived from them. The fatty acid composition may vary with genetic, soil, and climate factors.

Biodiesel. We begin with biodiesel, whose history parallels the development of the internal combustion engine, beginning at about the end of the nineteenth century. At the 1900 World Exposition in Paris, an early diesel engine supplied by the French Otto company was demonstrated that ran on peanut oil. Although vegetable oils can be directly used as a fuel in modified diesel engines, they are not used much today because of their high viscosity and tendency to thicken and congeal at low temperatures, making them less suitable in cold climates.

A way to improve the fuel properties of vegetable oils is to react them with methanol using a small amount of a basic catalyst that promotes the reaction. In this process—called *transesterification*—the ester linkage of the fatty acids to glycerol is split and a new ester linkage of the fatty acid to methanol is formed, converting the triglycerides to the methyl esters of the fatty acids. The methyl esters have about one-third the molecular weight of the original vegetable oil and a much lower viscosity. This product is commonly called *biodiesel* because it can be used as fuel in unmodified diesel engines or mixed in various proportions with petroleum-derived diesel fuel. Such mixtures are designated with the symbols B-10, B-20, etc. that specify the percentage of biodiesel in the mixture. Besides methanol, other alcohols such as ethanol or butanol can be used in the transesterification process to make biodiesel.

Modern diesel engines depend on the lubricating properties of the fuel to minimize wear on critical parts such as the fuel injectors. When low-sulfur diesel fuel derived from petroleum was recently introduced, one of the concerns was the loss of the sulfur compounds in the fuel that served as lubricants. Biodiesel has good lubricating properties, making it a desirable additive to low-sulfur petroleum diesel.

In making biodiesel from vegetable oil and methanol, concentrated aqueous solutions of sodium or potassium hydroxide can be used as a catalyst. [However, the presence of too much water gives an undesirable side reaction leading to the production of fatty acid salts (soaps) that lead to emulsions and foaming. This problem can be avoided by using as the catalyst a solution of sodium methylate (NaOCH$_3$) in methanol.]

Soap. Soap has a long history that can be traced back thousands of years, but until the mid-1800s it was not widely used for personal hygiene. Soap is made by reacting the triglycerides in fats and oils with an alkaline base such as sodium hydroxide. The overall reaction, shown in Figure 33-1, is called *saponification* (after *sapo*, the Latin word for "soap"). In the saponification process, hydroxide ion (OH$^-$) reacts with the ester groups of the triglyceride to split (hydrolyze) them, forming in the process the fatty acid salts that we call soap and by-product glycerol.

Until the early 1800s, alkaline bases were available only from hardwood ashes (rich in potassium) or the ashes of salt-tolerant marine plants (rich in sodium). This limited production, and in England soap was treated as a valuable commodity and taxed until around 1853.

Two developments contributed to the manufacture of good quality soap on an industrial scale beginning in the nineteenth century. The first and probably most important was the invention of processes for making soda ash (sodium carbonate) on a large scale—the Leblanc process first patented in France in 1791—later displaced by the Solvay process developed into its modern form in the 1860s by Ernest Solvay and still used today. The second was the rapid development in the early 1800s of chemistry as a science and the 1823 publication of the work of French chemist Michel Chevreul on the chemical composition of fats and oils and of several fatty acids that could be derived from them by hydrolysis. This work led to a better understanding of the chemistry of soapmaking, which up to that time had been performed by soapmakers using closely guarded recipes worked out by trial and error over the years.

In industrial soap production, the by-product glycerol is usually separated from the soap, purified, and used as an emollient (skin-softening) ingredient for preparing fine toilet soaps, hand lotions, and other cosmetic products and as a food additive where its humectant (moisture-conserving) properties help to keep baked goods moist.

The soap we will prepare retains all of the glycerol by-product, enhancing the richness of its lather and its emollient qualities. In making our soap, we will aim to add an amount of NaOH that is slightly less than the theoretical amount required to completely react with the fat or oil. Adding an excess of NaOH can make the soap harsh and drying to the skin. On the other hand, too much fat or oil may leave the soap with a greasy feel.

Experimental Procedure

Special Supplies: *Part 1 (Biodiesel):* 13 × 100 mm culture tubes (rimless test tubes); centrifuge to hold 13 × 100 mm test tubes; size 00 rubber stoppers (bottom diameter 11 mm); plastic wrap; graduated 1-mL polyethylene transfer pipets (Samco Scientific 222 or equivalent); 10-mL graduated cylinder (glass or transparent-plastic); $\frac{1}{16}$-in.-diameter Nylon or acetal (Delrin) balls [Nylon balls were obtained from www.product-components.com/balls.html; acetal (Delrin) balls were obtained from www.smallparts.com]; stopwatch or timer; forceps.

Part 2 (Soap): hot plate (or improvised water bath); thermometer (digital thermometer with stainless steel probe, or non-mercury thermometer, or mercury thermometer with Teflon sheath); vegetable shortening (Crisco); stainless steel scoop (half-round, tapered at one end) or wooden stirrers (popsicle or craft sticks); graduated 1-mL polyethylene transfer pipets (Samco Scientific 222 or equivalent); 10-mL graduated cylinder; 8-oz Styrofoam cups; 1–14 pH indicator paper such as pHydrion® paper (www.microessentiallab.com); 13 × 100 mm culture tubes (or test tubes).

Chemicals: *Part 1 (Biodiesel):* Vegetable oil (canola oil); methanol; 10 M KOH (40 wt% KOH).
Part 2 (Soap): Vegetable shortening (Crisco); 10 M NaOH (30 wt% NaOH); 15-g/L (~0.05 M) solutions of Kirk's Original Coco-Castile soap and Ivory soap; 0.5 M CaCl$_2$; 0.5 M Na$_3$PO$_4$.

1. Making Biodiesel from Vegetable Oil

Obtain two clean and dry 13 × 100 mm culture tubes (rimless test tubes), two size 00 rubber stoppers to fit the mouth of the test tubes, and a 150-mL beaker. (Using a marking pen, mark them as tubes 1 and 2. Tube 1 will be the reaction tube, and tube 2 will be used to temporarily store the final biodiesel product.)

Overview of the procedure: We will place all the reactants in a test tube (tube 1), stopper and shake the test tube for 15–20 min, allow it to stand for 45–60 min, then centrifuge it to separate the reaction products into two distinct layers.

To allow an accurate mass balance on the reactants and products, we need a mass accuracy of at least ±0.01 g. Weigh and record the mass of the 150-mL beaker (or tare the beaker to zero on an electronic balance). Then place a clean, dry 13 × 100 mm culture tube (tube 1) in the beaker and weigh and record the mass of the empty tube to the nearest 0.01 g. Then place 5 mL (~4.6 g) of canola oil in tube 1 and record the mass of *tube 1 + oil*. Next, using a 1-mL graduated polyethylene transfer pipet, add 1 mL (~0.8 g) of methanol and weigh and record the mass of *tube 1 + oil + methanol*. **Observe and record:** Does the added methanol mix with the vegetable oil, float on top of the oil, or sink to the bottom? Finally, add to the tube 2 drops (~0.07 g) of 10 M KOH (40 wt% KOH) and weigh and record the mass of *tube 1 + oil + methanol + KOH solution*.

⚠️ **SAFETY PRECAUTION:**
Before you proceed to mix the reactants, make sure that your eyes and those of your neighbors are protected with safety goggles. A lot of shaking comes next—stopper the tube with a rubber stopper or cork, cover the stopper and top of the tube with plastic film as an extra safety precaution, and vigorously shake the tube for a total of 15–20 min. A rotation of the wrist works well—that way your arm won't get so tired. Keep your thumb or finger firmly on the stopper during the entire shaking process so the stopper can't pop out of place. It's OK to pause every few minutes to give your wrist or arm a rest. Go by the clock so you don't shortchange the shaking process. *Observe and record*: What is the appearance of the reaction mixture after shaking? What is its color? Is it clear or cloudy?

Next, we allow the tube of reaction mixture to stand for at least 30–60 min while the reaction continues in the emulsified state. (Move on to part 2, making soap, while waiting for the reaction to be completed.) When the waiting period is over, we separate the components of the emulsified products by centrifuging them for 3–4 min. The glycerol reaction product is not soluble in biodiesel—composed of the methyl esters of the fatty acids—and appears as a layer having a much smaller volume than the biodiesel layer. Why does it appear below the biodiesel layer while methanol floats on top of the original vegetable oil?

Using a clean polyethylene transfer pipet, carefully remove the upper layer of biodiesel and transfer it to tube 2. After transferring, weigh and record the final mass of *tube 1* from which the biodiesel has been removed, standing up the tube in a tared 150-mL beaker as before. Stopper and save the tube of biodiesel for further tests and measurements.

2. Making Soap from Vegetable Shortening Weigh 30 ± 0.2 g of Crisco vegetable shortening into a tared 250-mL beaker. Record the mass of Crisco. Put 10 ± 0.1 mL of 10 M (30 mass%) NaOH into a 10-mL graduated cylinder using a 1-mL graduated polyethylene transfer pipet to adjust the volume to the 10-mL mark. Leave the empty transfer pipet standing in the graduated cylinder and place the cylinder containing the NaOH in a 150-mL beaker so that it can't tip over.

Place the beaker containing the Crisco on a hot plate (low to medium heat setting). (If a hot plate is not available, you can use a water bath improvised by placing the 250-mL beaker in a larger 400-mL beaker containing about 100 mL of water, supporting the water bath over a Bunsen burner on wire gauze resting on an iron ring supported by a ringstand. Use a second iron ring surrounding the 400-mL beaker to keep the water bath centered and prevent it from falling off the supporting iron ring and wire gauze.)

Heat the Crisco until it melts and reaches a temperature of around 50 °C, while stirring the molten oil. If the temperature rises above 50 °C, remove the beaker from the hot plate (continue stirring) and lower the temperature setting of the hot plate. (A digital thermometer with stainless steel probe is ideal for monitoring the temperature—you can use the probe as a stirrer. An ordinary non-mercury thermometer or a mercury thermometer protected with a Teflon sheath is also suitable, but if the thermometer bulb is unprotected, it's best to use a spatula or wooden popsicle stick to stir so that you don't risk breaking the thermometer bulb.)

When the Crisco has melted, and the temperature is around 50 °C, begin adding the 10 M NaOH in 1-mL increments, stirring vigorously for 25–30 s between each addition. Continue until you have added the total volume contained in the 10-mL graduated cylinder. It's important to stir vigorously during the addition process so that the oil and aqueous NaOH will not separate into two layers. As the NaOH begins to react, it forms soap—the sodium salt of the fatty acids in the oil—which acts to help emulsify the mixture. The process in which fats or oils are reacted with a base to form soap is called *saponification*—derived from *sapo*, the Latin word for "soap."

Keep stirring the mixture for 15–20 min while keeping the mixture in the 40–50 °C temperature range. (It's OK to pause now and then to give your arm a rest.) When the mixture thickens to the point where it leaves a "trace"—a raised trail that does not immediately flatten out when some of the mixture is dripped from the stirrer over the surface—transfer the mixture to an 8-oz Styrofoam cup. Using a marking pen, write on the cup your name or initials and the type of fat or oil used.

At this point, the hydrolysis of the triglyceride by NaOH to form soap is normally sufficiently complete that the resulting emulsion will not separate into two layers on standing, but there still remains some unreacted NaOH. On standing for a few hours or overnight, the mixture will harden as the reaction continues. Keep the cup in a warm place or put the cup in a closed polyethylene storage container and place it in a spot where it gets the direct rays of the sun. This will hasten the reaction so that it will be complete in 3–4 days or by the next lab period a week later. If not kept warm, the soap might take as long as two weeks for the saponification reaction to be complete.

You can check the progress of the reaction by monitoring the pH using universal pH indicator paper—put a drop of deionized water on the soap and touch a piece of 1–14 pH indicator paper to the water. A soap solution is normally alkaline, around pH 8.5, because it is the salt of a weak carboxylic acid; but a pH above 9 indicates the reaction is not complete. On standing, some of the water in the soap evaporates so that the soap will continue to harden as it ages.

▨ NOTES TO INSTRUCTOR ▨

For this part of the experiment, the student will need about 15 mL of biodiesel, a larger quantity than made in part 1. This may be obtained by pooling the biodiesel samples made by several students, or by providing them with biodiesel made ahead of time by scaling up the procedure described in part 1. Allow 24–48 hours for the emulsion to separate into two phases without centrifugation.

---

3. Comparing the Properties of Vegetable Oil and Biodiesel With the instructor's permission, work together in pairs. One student will determine the density and estimate the viscosity of vegetable oil, the other the density and viscosity of biodiesel. They will then exchange data so that each student has the data in the lab notebook for both vegetable oil and biodiesel. The procedures require a 10-mL graduated cylinder, 15 mL of vegetable oil or biodiesel, a timer, and five or six $\frac{1}{16}$-in.-diameter Nylon or acetal (Delrin) balls to make a semiquantitative comparison of viscosity. [Nylon balls have a density of about 1.15 g/mL; acetal (Delrin) balls have a density of about 1.36 g/mL and will fall faster than Nylon balls.]

Density procedure: Weigh a clean, dry 10-mL graduated cylinder to an accuracy of ±0.01 g. Obtain about 15 mL of vegetable oil or biodiesel in a 50-mL beaker. Fill the graduated cylinder with vegetable oil (or biodiesel) to precisely the 10-mL mark using a polyethylene pipet to adjust the volume. (The bottom of the rounded meniscus should be just at the 10-mL mark.) Then reweigh the *graduated cylinder + oil* to obtain the mass of 10 mL of oil. Calculate the density in units of g/mL.

Viscosity by the falling ball method: A spherical ball falling in a fluid under the influence of gravity reaches a terminal velocity related to the viscosity of the fluid and the difference in densities of the falling ball and the fluid. (See http://en.wikipedia/wiki/Viscometer for an explanation of the falling sphere viscometer.)

To carry out the procedure, obtain a timer and five or six $\frac{1}{16}$-in.-diameter Nylon or acetal (Delrin) balls. Using the same 10-mL graduated cylinder used for the density procedure, fill the cylinder to the 10-mL mark with vegetable oil or biodiesel. Holding a ball in forceps (take care—they're very small), position the ball at the center of the column of oil, just above the surface of the oil. Ready the timer, then drop the ball, starting the timer after the ball has fallen about 1 cm. In other words, start the timer when the ball passes the 9-mL mark on the 10-mL graduated cylinder, and stop the timer when the ball hits the bottom. The time will be short—from a second or two to 20 s or more; it might be helpful to work in pairs, one person observing the ball and operating the timer while the other concentrates on dropping the ball.

Repeat the measurement with another ball until you have three reasonably consistent trials, recording the time for each trial. If it is necessary to make more trials, you can pour the oil back into the 50-mL beaker, recover the balls, blot them with a tissue to remove the oil, and refill the graduated cylinder to the 10-mL mark. Finally, measure and record the height of the column of oil in the graduated cylinder (the distance from the 9-mL mark to the bottom).

Take care not to lose the balls. They can be reused and are moderately expensive, so when you are through with the viscosity measurements, recover the balls, blot them with tissue, and put them in a container reserved for the purpose.

The viscosity is inversely proportional to the velocity of the ball. Since *velocity = distance/time*, the viscosity is directly proportional to the time that it takes for a ball to fall a fixed distance.

4. Comparing the Properties of Soaps For this part, you will need five 13 × 100 mm test tubes with stoppers, a 1-mL graduated transfer pipet, universal indicator pH paper (such as pH 1–14 pHydrion® paper), instructor-prepared solutions of Coco-Castile and Ivory soaps containing 15 g/L (~0.05 M) of soap, and 0.5 M solutions of $CaCl_2$ and Na_3PO_4.

We will be comparing the pH and lathering properties of each of the three soaps in deionized water, hard water produced by addition of $CaCl_2$, and soft water in which the Ca^{2+} ion has been precipitated by addition of phosphate ion (as Na_3PO_4). Soap made from Crisco has a fatty acid composition similar to soybean oil, except the soybean oil has been partially hydrogenated, converting some unsaturated fatty acids to saturated and partially unsaturated fatty acids. Coco-Castile soap is made mainly from coconut oil, which contains shorter fatty acid chains than does Ivory soap made from sodium tallowate (beef tallow) and/or sodium palmate. (See Table 33-1 for the fatty acid composition of soybean oil, coconut oil, and beef tallow or palm oil.) Our aim is to discover whether the fatty acid composition (length of the fatty acid chains) has any influence on the lathering properties of the soap.

Begin by freeing the soap from the Styrofoam cup that you used as a mold. Weigh and record the mass of the soap and compare its mass to the sum of the masses of the materials used to prepare it—Crisco (or other vegetable oil you used) and 10 M NaOH solution. (The mass of the NaOH solution may be calculated from its density, 1.33 g/mL.) Explain any difference in the mass of the soap and the total mass of the materials used to prepare it.

pH and lathering properties: Place a piece of weighing paper on a balance pan and, using a knife blade or metal spatula, shave off about 0.3 g of the soap you prepared. Transfer the soap shavings to a 30- or 50-mL beaker, add 20 mL of deionized water, and use a glass stirring rod to disperse and dissolve the soap.

Obtain five 13 × 100 mm test tubes and stoppers and three 3-cm lengths of 1–14 pH indicator paper. Number or otherwise label the test tubes so they can be told apart. Three of the tubes will receive 1-mL portions of three different soap solutions— 1 mL of your prepared soap solution in the first tube, 1 mL of Ivory soap solution in the second tube, and 1 mL of Coco-Castile soap solution in the third tube. (Each of the three soap solutions has a concentration of 15 g/L of soap in deionized water.) In the fourth tube, place a 1-mL portion of 0.5 M $CaCl_2$ and in the last tube, place 1 mL of 0.5 M Na_3PO_4.

pH measurement: Estimate the pH of each of the three soap solutions by dipping a clean glass stirring rod into the solution and touching the tip of the rod to a length of pH 1–14 indicator paper. For each solution, immediately compare the color with the color chart and record the color and estimated pH in your lab notebook.

Lathering in deionized water: Next, stopper each of the three test tubes holding the soap solutions and shake them vigorously for a few seconds. Observe and describe in your lab notebook the lathering properties of the three soaps in pure deionized water—qualities such as the size and stability of the soap bubbles, how long they last, etc.

Lathering in hard water: Next add 0.2 mL (about 4 drops) of 0.5 M $CaCl_2$ solution to each of the three soap solutions to simulate hard water. Stopper and shake as before, recording your observations about the lathering properties of the three soaps in hard water. Look for the formation of any precipitates or soap scum formed by precipitation of fatty acid anions $RC(O)O^-$ by Ca^{2+} ions. We can represent the general formula for this insoluble calcium salt as $Ca[RC(O)O]_2$, where R represents the hydrocarbon tail of the fatty acid.

Lathering in soft water: Finally, add to each soap solution 0.2 mL (about 4 drops) of 0.5 M Na_3PO_4 solution. Stopper and shake vigorously. This simulates the addition of a water-softening agent that removes from the solution ions such as Ca^{2+} or Mg^{2+}. [Ca^{2+} ion forms with PO_4^{3-} ion a very insoluble precipitate of $Ca_3(PO_4)_2$ that reverses the formation of the more soluble $Ca[RC(O)O]_2$, freeing up the fatty acid anions to cleanse away oils and dirt.] Again, observe and record the lathering properties of the three soaps in the presence of phosphate ion.

If bars of Coco-Castile and Ivory soap are available, try using them and your prepared soap to wash your hands in ordinary tap water, comparing the feel and lathering qualities of the three soaps.

As a result of your observations, is it possible to draw any conclusions about differences in the lathering ability of soaps in hard and soft water related to their fatty acid composition? Which soap lathered best in hard water—soaps such as Ivory with longer hydrocarbon "tails" in their fatty acids or coconut oil–based soap with shorter hydrocarbon "tails" in their fatty acids?

Atom Economy Calculations

Atom economy is one aspect of synthetic efficiency, and maximizing it is a key goal of green chemistry. In simplest terms, atom economy is merely a quantitative measure of how much of the reactants end up in the products. The goal would be that all the reactants end up in the products with no waste by-products. This would correspond to an atom economy of 100%. Let's calculate now the maximum atom economy we could expect for the biodiesel synthesis of part 1 if we mixed with the canola oil an exactly stoichiometric amount of methanol: 1 mol canola + 3 mol methanol, as shown in Figure 33-1. We will call this the *ideal atom economy*.

1 mol canola oil (880 g) + 3 mol methanol
  (3 × 32 g/mol = 96 g) →
                    3 mol biodiesel methyl ester
  (3 × 295 g/mol = 885 g) + 1 mol glycerol (92 g)

Note first that the total mass of the products, 997 g, is equal (within round-off error) to the total mass of the reactants, 996 g, illustrating the conservation of mass in chemical reactions. The (ideal) % atom economy for this reaction is given by

$$\% \text{ atom economy} = \frac{\text{mass of desired product}}{\text{total mass of reactants}} \times 100\%$$
$$= \frac{885 \text{ g}}{976 \text{ g}} \times 100\% = 90.7\% \quad (1)$$

Now what has been called the *experimental atom economy* is less than 90.7% because we did not include in the total mass of reactants the mass of the catalyst (KOH solution). Neither did we take account of the fact that in the experimental reaction, we used an excess of methanol beyond the stoichiometric amount required to form the methyl ester. So in computing the % atom economy for the reaction experiment that we carried out in part 1, we would include the total mass of vegetable oil, methanol, and catalyst:

experimental % atom economy =
$$\frac{\text{mass of biodiesel}}{\text{total mass of vegetable oil+methanol+KOH sol'n}}$$
$$\times 100\% \quad (2)$$

We will leave it to you to calculate in the Report form the experimental % atom economy for the real reaction using your data from part 1 of the experiment.

% Yield. A common conventional measure of reaction efficiency is the % yield. In the biodiesel synthesis, we are assuming that the reaction goes completely to the right with no side-reactions. The amount of desired biodiesel product under these conditions would be 885 g, which is the maximum possible amount of product that could be obtained from the stoichiometric reaction of the reactants.

$$\% \text{ yield} = \frac{\text{mass of product obtained}}{\text{maximum possible mass of product}} \times 100\%$$
$$= \frac{885 \text{ g}}{885 \text{ g}} \times 100\% = 100\% \quad (3)$$

Notice that even for the experimental (or actual) reaction, the % yield calculation takes no account of the fact that we added KOH solution as catalyst, nor that we used excess methanol so that the vegetable oil becomes the limiting reagent, nor that by-product glycerol is produced. In other words, the % yield calculation takes no account of the fact that there is a waste stream of by-product glycerol, catalyst, and excess methanol. Put another way, the experimental % atom economy is a better (or more realistic) measure of the reaction efficiency, or "greenness," of a synthesis than is the % yield.

CONSIDER THIS

A 42-gal barrel of crude oil can generate about 44 gal of petroleum products. (The difference reflects the lower density of the products compared to the crude oil.) About 20 gal of gasoline and 7 gal of diesel fuel can be produced from each barrel of crude oil.

In 2004, the United States consumed about 140 billion gallons of gasoline, more than any other country. An additional 40 billion gallons of diesel are consumed in a year for on-road transportation. The 75 million gallons of biodiesel produced in 2005 represents only a tiny fraction (0.06%) of the total diesel fuel consumption.

About 80% of the biodiesel now produced in the United States comes from soybean oil. Most of the rest comes from other vegetable oils, waste vegetable oils and fats from restaurants, and various animal fats that are by-products of meat processing.

An interesting question to ask is this: How many acres of soybeans would we have to harvest to provide enough biodiesel to match the present consumption of 40 billion gallons? To answer this question, we need some information about the productivity of soybeans and the amount of oil that can be derived from the harvested soybeans. In the Consider This section of the Report form, you will find sufficient information, obtained from various state and federal government Internet sources, to answer this question, along with some questions for you to think about and respond to.

Bibliography

Anastas, Paul T.; Warner, John C. *Green Chemistry: Theory and Practice*, Oxford University Press Inc., New York, 1998 (first published new as paperback 2000).

Biodiesel: See the www.wikipedia.org article titled "Biodiesel."

Chemical and Engineering News (C&EN) is the news magazine of the American Chemical Society (ACS) and a valuable resource. Individual articles may be accessed at http://www.cen-online.org/cen/. Beginning at the *C&EN* home page, you may select

Archive, or search for the particular article you want to access. (Some articles may be accessed only by ACS members or institutional subscribers.) On the *C&EN* home page, you will also find at the bottom a list of topics under *Features*, about green chemistry and sustainability. Selecting a feature produces a list of recent *C&EN* articles on the topic that can be accessed, except for the articles marked by the ACS logo.

Editorial staff, JCE Classroom Activity #14: Soapmaking, *J. Chem. Educ.* **1999,** *76*(2), 192A–192B.

Engelhaupt, Erika. "Happy Birthday, Love Canal," *Chemical and Engineering News* **2008,** *86*(46), 46–53 (Nov. 17, 2008 issue). Reproduced from *Environ. Sci. Technol.* **2008,** *42*(22), 8179–8186. Article and audio slideshow available online at www.cen-online.org.

Green Chemistry: http://pubs.acs.org/cen/ greenchemistry.html.

Hoag, Hannah, "The Greening of Chemistry," *Chemical Heritage* **2008,** *26*(2), 26–30. (A publication of the Chemical Heritage Foundation: www.chemheritage. org/pubs/ch-v26n2-articles/feature_greening_p1.html.)

Kostka, Kimberly L.; McKay, David D. "Chemists Clean Up: A History and Exploration of the Craft of Soapmaking," *J. Chem. Educ.* **2002,** *79*(10), 1172–1175.

Mabrouk, Suzanne T. "Making Usable, Quality Opaque or Transparent Soap," *J. Chem. Educ.* **2005,** *82*(10), 1534–1537.

Ritter, Stephen K. "Calling All Chemists," *Chemical and Engineering News* **2008,** *86*(33), 59–68 (Aug. 18, 2008 issue). This special issue contains several articles on sustainability on pages 42–77. The article cited describes green chemistry and how it can make contributions to sustainability.

Ryan, Mary Ann; Tinnesand, Michael (eds.). *Introduction to Green Chemistry: Instructional Activities for Introductory Chemistry*, American Chemical Society, 2002.

Soapmaking and the history of soap: See the www.wikipedia.org articles titled "Soap"; "Procter & Gamble"; "Colgate-Palmolive."

Tundo, Pietro; Aricò, Fabio. "Green Chemistry on the Rise," *Chemistry International* **2007,** *29*(5), 4–7. (A publication of the International Union of Pure and Applied Chemistry available online at www.iupac.org/publications/ci/2007/2905/ 1_tundo.html)

Introduction to Green Chemistry

Making Biodiesel and Soap from Vegetable Oils and Fats

Name _____

Date _____ Section _____

Locker _____ Instructor_____

Data, Calculations, and Observations

1. Making biodiesel from vegetable oil

Name of vegetable oil used: _____

TABLE (A)

Test tube 1 measurements (If you are using an electronic balance with a tare feature, place an empty 150-mL beaker on the balance and tare to zero before weighing the test tube and contents. Then you will not have to use the same beaker for each weighing. If this is not possible, you must use the same 150-mL beaker for all weighings.)	Mass
Empty 150-mL beaker (if you tare beaker mass to zero, enter zero for mass)	g
Beaker + empty test tube	g
Beaker + test tube + 5 mL vegetable oil	g
Beaker + test tube + vegetable oil + 1 mL methanol	g
Beaker + test tube + vegetable oil + methanol + 2 drops 10 M KOH	g
Beaker + test tube + by-products after removal of biodiesel	g

From the mass measurements on test tube 1 in Table (A), calculate the net mass and the number of moles of each reactant and product and enter their values in Table (B). To calculate the moles of vegetable oil reactant and the biodiesel product, use the weighted average molar masses from Table 33-1 for the oil and the biodiesel (fatty acid methyl esters). Assume that the 10 M KOH is 40 mass % KOH (40 g KOH/100 g solution).

Reactants	Net mass	Molar mass	Number of moles
Vegetable oil	g	g/mol	mol
Methanol	g	g/mol	mol
KOH (as 40 mass% aqueous solution)	g	56.1 g/mol	mol KOH = g KOH sol'n $\times \dfrac{40 \text{ g KOH}}{100 \text{ g sol'n}} \times \dfrac{1 \text{ mol KOH}}{56.1 \text{ g KOH}}$ =
Products	**Net mass**	**Molar mass**	**Number of moles**
Biodiesel (fatty acid methyl esters)	g	g/mol	mol
By-product (after removing biodiesel)	g	—	The by-product is a mixture of glycerol, water, KOH, and excess methanol; we won't attempt to calculate here the moles of each component.

In Table (C), use the calculated mole values in Table (B) to calculate the mole ratio, defined as (*mol of reactant or product*) / (*mol of vegetable oil*). Simply divide the number of moles of each reactant and product by the number of moles of vegetable oil initially used to make the biodiesel. The mole ratio for the vegetable oil is 1.00 because we are dividing the number of moles of vegetable oil by itself. Compare these mole ratios with the reaction coefficients shown in Figure 33-1 for the synthesis of biodiesel.

TABLE (C)

Reactants	Number of moles Copy from Table (B)	mol ratio (mol/mol of vegetable oil)
Vegetable oil	mol	1.00
Methanol	mol	
KOH	mol	
Product	**Number of moles** Copy from Table (B)	**mol ratio** (mol/mol of vegetable oil)
Biodiesel	mol	

2. Making soap from vegetable shortening

Name of shortening: _____

Record below the information from the label about the composition of the shortening:

Making soap	Mass or volume
Mass of vegetable shortening taken	g
Total volume of 10 M (30 mass%) NaOH added	mL

Record the initial and final temperature of the soap mixture at the beginning and after the addition of NaOH in 1-mL increments.

Initial temperature: _____ Final temperature: _____

Describe the changes in the texture of the soap mixture during the stirring following the addition of the NaOH solution.

3. Comparing the properties of vegetable oil and biodiesel

Density procedure:

Vegetable oil density measurement	
Mass of empty 10-mL graduated cylinder	g
Mass of cylinder filled with vegetable oil	g
Calculated density	g/mL
Biodiesel density measurement	
Mass of empty 10-mL graduated cylinder	g
Mass of cylinder filled with biodiesel	g
Calculated density	g/mL

Viscosity procedure:

Vegetable oil viscosity measurement	Trial 1	Trial 2	Trial 3
Distance traveled by the falling ball	cm	cm	cm
Time for the ball to fall	s	s	s
Calculated velocity of the falling ball	cm/s	cm/s	cm/s
Biodiesel viscosity measurement	**Trial 1**	**Trial 2**	**Trial 3**
Distance traveled by the falling ball	cm	cm	cm
Time for the ball to fall	s	s	s
Calculated velocity of the falling ball:	cm/s	cm/s	cm/s

Average falling ball velocity:

Vegetable oil: _____ cm/s;   Biodiesel: _____ cm/s

4. Comparing the properties of soaps

After a few days or a week of aging in a warm place, free the soap you prepared from the mold and weigh it.

Mass of prepared soap: _____ g

Mass of Crisco used to make the soap: _____ g

Mass of added 10 M NaOH solution (assume a density of 1.33 g/mL): _____ g

Sum of Crisco and NaOH solution masses: _____ g

Compare the mass of the soap you prepared with the sum of the masses of the reactants (Crisco + NaOH solution) used to prepare the soap. Explain any difference.

pH test:

Soap	Color of pHydrion paper	pH value	Classification: acidic, basic, or neutral
Prepared soap			
Ivory soap			
Coco-Castile soap			
Deionized water			

Lathering (foaming) in deionized, hard, and softened water: Prepared soap, Ivory soap, and Coco-Castile soap

Soap	Amount of foaming in deionized water	Amount of foaming after adding $CaCl_2$ (hard water)	Amount of foaming after adding Na_3PO_4 (softened water)
Prepared soap			
Ivory soap			
Coco-Castile soap			

If you washed your hands in tap water using bars of Coco-Castile, Ivory, and your prepared soap, compare the feel and lathering qualities of the three soaps:

From the mass of Crisco (or other vegetable oil) taken, calculate the number of moles. [You may assume without serious error that Crisco has the same molar mass as soybean oil (see Table 33-1).]

Moles of Crisco: _____

From the volume and molar concentration of the added NaOH, calculate the number of moles of NaOH

added: _____

Calculate the ratio: mol NaOH/mol Crisco: _____

Divide this ratio by the stoichiometric ratio 3.00 shown in Figure 33-1 for the synthesis of soap and express as a percentage by multiplying by 100%: _____

Which reactant is in excess, Crisco or NaOH—i.e., is the amount of NaOH added less than or greater than 100% of the stoichiometric amount?

5. Atom economy for the biodiesel synthesis

From Equations (2) and (3), calculate for the biodiesel synthesis the following quantities using the data of part 1 and assuming the stoichiometry shown in Figure 33-1 for the biodiesel-forming reaction.

Experimental % atom economy calculation:

Mass of biodiesel formed:_____

Total mass of vegetable oil + methanol + KOH solution:_____

Calculated experimental % atom economy: _____

% yield calculation:

Mass of biodiesel formed: _____

Maximum possible mass of biodiesel that could be formed based on the actual starting mass of vegetable oil used in the experiment. (Use the molar mass values found in Table 33-1):

_____

Calculated % yield: _____

Questions

1. In making biodiesel by reacting vegetable oil with methanol, we did not analyze the reaction mixture to confirm that the reaction shown in Figure 33-1 took place. (Superficially, the reactants and products look much alike.) An analysis might be accomplished by one of several chromatographic techniques—gas chromatography (GC), high-performance liquid chromatography (HPLC), or thin layer chromatography (TLC)—that allow you to separate the compounds in the biodiesel product. Perhaps your instructor has access to one of these techniques and could provide an analysis of a representative sample.

The question might be raised: How do we know that a reaction took place? Let's look at the evidence we have from the experiment itself:

- The visible separation of reactants and products into two layers
- Density measurements
- Viscosity measurements

(a) *Density measurements.* We measured the densities of vegetable oil and the biodiesel reaction product. Compare their densities. Is there a significant difference?

Biodiesel density: _____ g/mL

Vegetable oil density: _____ g/mL

Look up the densities of methanol and glycerol in the *Handbook of Chemistry and Physics* or some other source and write them below. Are the densities of methanol, vegetable oil, biodiesel, and glycerol consistent with the methanol layer floating atop the vegetable oil and the glycerol layer resting beneath the biodiesel layer after centrifugation? Explain.

Methanol density: _____ g/mL

Glycerol density: _____ g/mL

(b) *Viscosity measurements.* Compare the times it took for the ball to fall a fixed distance in vegetable oil and in the biodiesel. Are the differences in viscosity sufficiently great to suggest that a reaction took place? Explain.

Vegetable oil fall time: _____ s

Biodiesel fall time: _____ s

2. Read the twelve principles of green chemistry, found in Appendix Table 12. Identify by number two or three of the twelve principles that you think best apply to making (a) biodiesel and (b) soap. Briefly explain the reasons for your choices.

(a) Biodiesel:

(b) Soap:

3. Some research suggests that varieties of algae that produce fatty acids could be used to produce biodiesel. What advantages might there be in growing algae to make biodiesel as opposed to using crops such as soybeans, canola (rapeseed) and palm oil? (*Hint:* Use an Internet google.com search on "algae biodiesel" to get current information about the prospects of growing algae to produce biodiesel.)

CONSIDER THIS

A 42-gal barrel of crude oil can generate about 44 gal of petroleum products (the difference reflects the lower density of the products compared to the crude oil.) About 20 gal of gasoline and 7 gal of diesel fuel can be produced from each barrel of crude oil.

In 2004, the United States consumed about 140 billion gallons of gasoline, more than any other country. About 40 billion gallons of diesel are consumed in a year for on-road transportation. The 75 million gallons of biodiesel produced in 2005 represents only a tiny fraction (0.06%) of the total diesel fuel consumption.

About 80% of the biodiesel now produced in the United States comes from soybean oil. Most of the rest comes from other vegetable oils, waste vegetable oils and fats from restaurants, and various animal fats that are by-products of meat processing.

An interesting question to ask is this: How many acres of soybeans would we have to harvest to provide enough biodiesel to match the present consumption of 40 billion gallons of diesel fuel? To answer this question, we need some information about the productivity of soybeans and the amount of biodiesel that can be derived from the harvested soybeans. Here is some information that was obtained from various state and federal government Internet sources.

- The United States has about 1 billion acres of farmland, about half of which (500 million acres) is planted to crops; the rest is mainly for pasture, grazing, woodlands, unfarmed surplus, etc. This crop land is now mostly used to produce food for people or feed for animals (cattle, pigs, chickens, pets, etc.). Some crops, particularly soybeans, are now exported to other countries.

- About 40 billion gallons of biodiesel are required for on-road transportation.

- It takes about 40 lb of soybeans to make 1 gal of biodiesel.

- 1 bushel of soybeans weighs about 60 lb.

- 1 acre of soybeans will produce about 40 bushels of soybeans.

(a) Use the information above to calculate the number of acres of soybeans required to produce 40 billion gallons of biodiesel each year. How does this number compare with the farmland now planted to all crops? What does this comparison suggest about the plausibility of the notion that we can meet our present diesel fuel needs for transportation by turning crops into biofuels? How would adding in the 140 billion gallons of gasoline consumed each year affect your conclusions?

(b) Discuss some of the disadvantages of growing seed crops (soybeans, canola, sunflower, etc.) planted on a large scale to produce biodiesel. For example, what do you think would be the effect on the food supply and the price of food?

(c) Explain why using a plant source as a source of biofuel is approximately carbon-neutral, that is, would not materially increase the amount of CO_2 in the atmosphere, while burning fossil fuels such as petroleum and coal increases CO_2 in the atmosphere.

Oxidation–Reduction

Electron Transfer Reactions

Purpose

• Study a number of chemical reactions that involve the transfer of electrons from a reducing agent to an oxidizing agent.

• Establish a qualitative chemical redox couple table for seven redox couples by mixing the oxidized form of one redox couple with the reduced form of another couple and observing whether a reaction takes place.

• Place the following redox couples into our table in order of increasing oxidizing or reducing strength: Cu(II)/Cu; Zn(II)/Zn; Fe(II)/Fe; H^+/H_2; Fe(III)/Fe(II); Br_2/Br^-; and I_2/I^-.

Pre-Lab Preparation

A definition of oxidation and reduction

There is a problem in the traditional definition of an oxidation–reduction, or redox, reaction as "a chemical reaction that involves the transfer of electrons from the substance oxidized to the substance reduced" because, strictly speaking, every chemical reaction involves changes in the electron density on atoms and therefore involves charge transfer from one atom to another. As a result, the traditional definition could encompass every chemical reaction.

For this reason, some have suggested (see the Bibliography) that there is no consistent basis for distinguishing so-called oxidation–reduction reactions from other chemical reactions other than by a change in oxidation numbers; and as we have pointed out, oxidation numbers are assigned using arbitrary rules, so these numbers do not accurately reflect the actual distribution of charge in a molecule or polyatomic ion.

This somewhat unsatisfactory state of affairs forces us to conclude that the classification of a reaction as an oxidation reaction is somewhat artificial

and arbitrary, based on the purely arbitrary notion of oxidation number. To be logically consistent, *we must define oxidation as an increase in oxidation number and reduction as a decrease in oxidation number.*

Only for simple reactions involving elements and their ions will there be a direct connection between the changes in oxidation number of the elements and the transfer of electrons. For example, when you put a piece of zinc metal in an aqueous solution of copper(II) sulfate, the Zn reacts with Cu^{2+} to form Zn^{2+} and Cu metal. Thus, each Cu^{2+} ion has gained two electrons, and each Zn atom has lost two electrons. In other words, there has been a net transfer of two electrons from a Zn atom to a Cu^{2+} ion.

Likewise, Zn goes from oxidation state 0 to +2, and each Cu(II) ion is reduced from oxidation state +2 to 0. The oxidation number of Zn has increased, so we say it has been oxidized. The oxidation number of Cu has decreased, so we say it has been reduced.

In the reaction of Zn with Cu^{2+}, represented schematically in Figure 34-1, Zn acts as a *reducing*

Solution containing copper ions

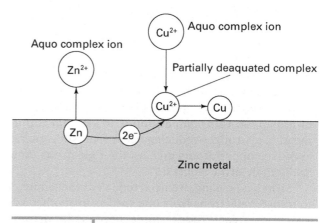

FIGURE 34-1 | **Direct chemical reaction of Zn with Cu^{2+}. As the Zn dissolves, Cu is plated out on the surface of the Zn. A Cu^{2+} ion is reduced by transfer of two electrons from a Zn atom.**

Cu^{2+}/Cu

Zn^{2+}/Zn

agent, or *reductant,* and Cu^{2+} acts as an *oxidizing agent,* or *oxidant.*

Writing the net stoichiometric reaction as the sum of two half-reactions shows explicitly that the transfer of electrons is involved:

Oxidation half-reaction:

$$Zn(s) \rightarrow Zn^{2+}(aq) + 2\ e^- \qquad (1)$$

Reduction half-reaction:

$$Cu^{2+}(aq) + 2\ e^- \rightarrow Cu(s) \qquad (2)$$

Net reaction:

$$Zn(s) + Cu^{2+}(aq) \rightarrow Zn^{2+}(aq) + Cu(s) \qquad (3)$$

Note that *no electrons appear in the overall reaction.* This is why it may sometimes be difficult to look at a chemical reaction and to decide whether it can be classified as an oxidation–reduction reaction. Keep in mind the fact that this decision is arbitrary. Certain conventional rules have been adopted about the definition of oxidation number. We will classify a reaction as an oxidation–reduction reaction *only* if there is a change in oxidation state (or oxidation number) of atoms involved in the reaction according to our previously defined rules.

The relative strength of oxidizing and reducing agents

Oxidation–reduction processes involve a *relative competition* between substances for electrons. The stronger oxidizing agents are substances with greater affinity for additional electrons; the stronger reducing agents are substances with the least attraction for the electrons that they already possess. Thus, silver(I) ion is a stronger oxidizing agent than is copper(II) ion because the reaction

$$2\ Ag^+ + Cu(s) \rightarrow Cu^{2+} + 2\ Ag(s) \qquad (4)$$

takes place in the forward direction but not appreciably in the reverse direction. From this fact we may deduce that Ag^+ is a stronger oxidizing agent than Cu^{2+}, meaning that silver(I) ion has a greater affinity for electrons than copper(II) ion does. Another way of looking at this competition is to say that copper metal is a stronger reducing agent than silver metal, because copper atoms give up their electrons more easily than silver atoms do.

Strictly speaking, we need to look at the tendency to transfer electrons from one species to another as a property of each *redox couple,* because (a) the tendency of an Ag(I) ion to accept an electron and become a silver atom depends on the difference in the Gibbs energy between an Ag(I) ion and a silver atom and because (b) the tendency of the copper atom to give up two electrons and become a Cu(II)

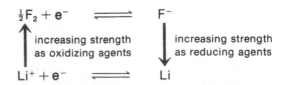

FIGURE 34-2 Schematic diagram illustrating the strongest and the weakest oxidizing and reducing agents in the table of standard reduction potentials. The strongest oxidizing agent (in the upper left-hand corner) has the most positive standard reduction potential. The strongest reducing agent (in the lower right-hand corner) has the most negative standard reduction potential.

ion depends on the difference in the Gibbs energy of the copper atom and the Cu(II) ion. In other words, we are really comparing the tendency of redox couples to exchange electrons in an oxidation–reduction reaction. As a result, it is only possible to arrange the various oxidizing and reducing agents as *redox couples* in a table, listing them according to their relative tendencies to transfer electrons. In this experiment, we shall explore such relative tendencies for a limited number of reactions.

Fluorine is one of the strongest oxidizing agents known, and lithium is one of the strongest reducing agents; therefore, these two redox couples would be at opposite ends of the table, with most other redox couples falling somewhere in between (see Figure 34-2).

A qualitative chemical redox scale

In this experiment, we establish qualitatively the relative position of a limited number of oxidation–reduction couples in a redox couple table. We will start with only three metals and their ions, then expand our study until we have considered seven couples in all.

NOTE **Before beginning this experiment, complete the preliminary exercise in your experiment report form. As directed there, indicate the substances oxidized and reduced in the several equations given and the change in oxidation state, if any.**

Experimental Procedure

Chemicals: Small pieces of copper, zinc, and iron (in the form of wire or 5×15 mm strips); the iron may be in the form of small, bright nails or clean steel wool (to minimize rusting, nails can be cleaned in 6 M H_2SO_4, then rinsed and dried in an oven); copper, Cu (turnings); 6 M hydrochloric acid, HCl; 6 M ammonia, NH_3; 0.1 M copper(II) sulfate, $CuSO_4$; 0.1 M zinc sulfate, $ZnSO_4$; iron(II) sulfate, $FeSO_4 \cdot 7H_2O(s)$; bromine water, Br_2 (saturated solution); 0.05 M iodine, I_2, in methanol; hexane, $C_6H_{14}(l)$; 0.1 M iron(III) chloride, $FeCl_3$; 0.1 M potassium bromide, KBr; 0.1 M potassium iodide, KI; 0.1 M potassium ferricyanide, $K_3Fe(CN)_6$; 0.1 M silver nitrate, $AgNO_3$; 6 M sulfuric acid, H_2SO_4.

> **⚠ SAFETY PRECAUTIONS:**
> **Hexane is volatile and flammable. Carry out the tests requiring hexane in a fume hood away from all open flames.**

> **☠ WASTE COLLECTION**
> **The metal wire or strips should be collected; they can be reused if they are cleaned off. A waste container should be provided, in the hood if possible, for the hexane- and methanol-containing mixtures. Also collect the metal ion solutions in a beaker, neutralize with $NaHCO_3(s)$, and put the solid residue in a waste container.**

1. A Qualitative Redox Series for Copper, Zinc, Iron, and Their Ions Explore the behavior of small pieces of Cu, Zn, and Fe metals (if the iron is rusty, it should first be cleaned in 6 M H_2SO_4, then rinsed in water), putting each metal in 3 mL of 0.1 M solutions of the ions of the other metals—that is, Cu with Zn^{2+} and with Fe^{2+}, Zn with Cu^{2+} and with Fe^{2+}, and Fe with Cu^{2+} and with Zn^{2+}—to determine the order in which the redox couples should be placed in a table like that shown in Figures 34-2 and 34-3. The *metal*

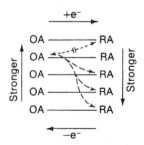

FIGURE 34-3 | **Tendencies for reaction between oxidizing and reducing agents. The dashed arrows indicate possible redox reactions. An oxidizing agent can react with any reducing agent that lies below it in the table.**

ion/metal couple that appears lowest in the table will correspond to the metal that visibly reacts with all other metal ions and is therefore the strongest reducing agent.

Sometimes the evidence for reaction is subtle. Keep in mind that when most metal ions are reduced chemically, they do not produce a shiny metal surface. More often the metal deposited is in the form of finely divided crystals that appear black or brown. Write equations for the reactions that occur.

Solutions containing Fe^{2+} should be freshly prepared by dissolving 0.1 g of $FeSO_4 \cdot 7H_2O$ in 3 mL of water.

Note that the oxidized form of one redox couple is always mixed with the reduced form of the other couple, or vice versa. (The oxidized form of the couple is always the one with the higher charge or higher oxidation number.) Under part 1 of the Experimental Data, prepare a table as directed in the report form.

2. The Hydrogen Ion/Hydrogen Couple Test the ability of each metal to reduce H^+ to $H_2(g)$ by placing a small piece of copper, zinc, and iron metal in three separate test tubes; add enough 6 M HCl to cover the metal. (Nitric acid, HNO_3, cannot be used in place of HCl in this test because nitrate ion in acid solution is a stronger oxidizing agent than H^+; therefore, the HNO_3 would react first and confuse the results.) If there is no immediate reaction visible, warm the solution gently (be careful not to confuse bubbles of water vapor for hydrogen gas). If H^+ reacts with a particular metal, the H^+/H_2 couple should be placed above the *metal ion/metal* couple, as shown in Figure 34-3. Your table should now contain four redox couples.

3. The Oxidizing Power of the Halogens First, if you are not familiar with the colors of the free halogens in hexane, add a little Br_2 water (water saturated with bromine) to 3 mL of H_2O and 1 mL of hexane. (**Caution:** Hexane is flammable.) Mix the liquids together by shaking, and observe the color of the hexane layer floating on top. Dispose of the hexane solution mixture in the appropriate waste container. Repeat, using 0.05 M I_2 in methanol instead of Br_2 water. Dispose of the mixture in the appropriate waste container.

Now, explore the behavior of 3-mL samples of 0.1 M solutions of KBr and KI (with 1 mL of hexane added to each) when each is treated with a little of the other free halogen—that is, mix KBr with a few drops of I_2 in methanol, and KI with a few drops of Br_2 water—to determine the relative order of the Br_2/Br^- and I_2/I^- redox couples. Dispose of these mixtures in the appropriate containers. Write equations for the reactions that take place. Then arrange the two halogens in a separate redox couple table, as directed in the Report form.

4. The Iron(III) Ion/Iron(II) Ion Couple Determine whether the Fe^{3+}/Fe^{2+} couple is a stronger or weaker oxidizing couple than the I_2/I^- or Br_2/Br^- couple by adding 1 mL of 0.1 M $FeCl_3$ to 2 mL each of 0.1 M KBr and 0.1 M KI in separate test tubes. [Note that this mixes the oxidized form of the Fe(III)/F(II) couple with the reduced forms of the halogens, I^- and Br^-.] Add 1 mL of hexane to each test tube, shake to mix, and from the color of the hexane layer note any evidence for the formation of free halogen. Dispose of these mixtures appropriately. To conduct a separate test for any reduction of Fe^{3+} to Fe^{2+}, prepare two fresh reaction mixtures, omitting the addition of hexane, and add to each a little potassium ferricyanide solution, $K_3Fe(CN)_6$. If Fe^{2+} is present, the deep blue precipitate of $Fe_3[Fe(CN)_6]_2$ will form.

Write equations for any reactions in which iron(III) is reduced. Note that Fe(III) does not go to metallic iron. Place the Fe^{3+}/Fe^{2+} couple in its proper place in your table of redox couples already containing the halogens.

5. The Reaction of the Halogens with Metals Now we want to join the table produced in parts 1 and 2 with the table produced in parts 3 and 4. We will test the ability of the halogens Br_2 and I_2 to oxidize copper metal, as described by the reactions

$$Cu(s) + Br_2(aq) \rightarrow Cu^{2+}(aq) + 2\,Br^-(aq) \quad (5)$$

and

$$Cu(s) + I_2(methanol) \rightarrow$$
$$Cu^{2+}(methanol) + 2\,I^-(methanol) \quad (6)$$

The tests work like this. Some copper turnings are shaken with Br_2 water. In a separate test tube, copper metal is shaken with I_2 dissolved in methanol. (Methanol is used as the solvent because I_2 is not very soluble in water.) After allowing any precipitate that forms to settle, we will divide the reaction solution in each test tube into two separate portions. One portion of the reaction solution is tested for the presence of Cu^{2+} (formed by oxidation of Cu metal), and the second portion is tested for halide ion (Br^- or I^-), formed by reduction of Br_2 or I_2.

Begin by placing 2 g of copper metal turnings into each of two 15 × 125 mm test tubes. Add 10 mL of Br_2 water to the first tube and 10 mL of 0.05 M I_2 in methanol to the second tube. Stopper the tubes and shake each tube vigorously for 30 s or until the color of the halogen has faded.

NOTE If a precipitate forms, allow it to settle for a few minutes. The precipitates correspond to formation of Cu(I) bromide (CuBr) or Cu(I) iodide (CuI) by partial oxidation of Cu to Cu^+, as shown in Reactions (7) and (8):

$$2\,Cu(s) + Br_2 \rightarrow 2\,CuBr(s) \quad (7)$$
$$2\,Cu(s) + I_2 \rightarrow 2\,CuI(s) \quad (8)$$

This intermediate (+1) oxidation state of copper is stabilized by the formation of the very insoluble salts CuBr and CuI. If Cu^+ is formed as an intermediate oxidation product of Cu metal, it may be immediately precipitated by bromide or by iodide ions. It is also interesting to note that the existence of this intermediate (+1) oxidation state of copper makes it possible for Cu^{2+} and I^- to react with one another to form CuI and I_2, as described by the net reaction

$$2\,Cu^{2+} + 4\,I^- \rightarrow 2\,CuI(s) + I_2 \quad (9)$$

Note that this reaction of Cu^{2+} and I^- is not the reverse of Reaction (6), because no copper metal is produced in the reaction. Rather it is a partial reduction of Cu^{2+} that produces a (+1) oxidation state of copper, Cu(I); Cu(I) then is stabilized by the formation of the very insoluble CuI(s).

Pour off 2–3 mL of the Cu/Br_2 reaction solution into each of two 13 × 100 mm test tubes. Add to the first tube 5–6 drops of 0.1 M $AgNO_3$. The formation of a precipitate indicates the presence of Br^-, which forms the very insoluble precipitate AgBr(s).

Add to the second portion of the Cu/Br_2 reaction solution 1 mL of 6 M NH_3. Formation of a blue $Cu(NH_3)_4^{2+}$ complex ion indicates the presence of Cu^{2+} ion. If both tests are positive, it is reasonable to conclude that the copper metal reacted with the bromine.

Now divide the Cu/I_2 reaction solution into two portions in 13 × 100 mm test tubes. Test the solution for the presence of I^- using 0.1 M $AgNO_3$ and for the presence of Cu^{2+} ion with 6 M NH_3 in exactly the same way as you tested the Cu/Br_2 reaction mixture. Were both tests positive? If so, it's reasonable to conclude that the copper metal reacted with the I_2.

Further Exploration. Try adding a few drops of 0.1 M Cu^{2+} to 2–3 mL of 0.1 M KI. Do you observe any evidence of reaction like that shown in Reaction (9)?

Considering the results of the tests with copper metal and looking at the table you established for the three *metal ion/metal* redox couples, what would you predict about the ability of other metals, such as Fe and Zn, to be oxidized by bromine or iodine?

6. Summary of Data Your experimental observations, together with the preceding data, will enable you to combine your two separate redox couple tables into one general redox couple table. The final table will contain all seven redox couples and will show the relative tendencies of the various elements and ions to lose or gain electrons. Note that each redox couple is written so that the change from left to right represents gain of electrons (reduction). In the table in your Report form, designate clearly (1) the reducing agents, (2) the oxidizing agents, (3) the end of each column that is the stronger (S), and (4) the end that is the weaker (W).

Such a table can be expanded to include many more oxidation–reduction couples and is useful in predicting the course of many reactions. Refer to Table 10 in the Appendix, noting that *any oxidizing agent (in the left column) has the possibility of reacting with any reducing agent (in the right column) that lies below it in the table* (see Figure 34-3). The table makes no prediction, however, of the *rate* of a given reaction—some reactions are too slow to be practical. Also, you should note that the *concentration* of the ions in a solution has a definite effect on the tendency for reaction, in accordance with the Le Châtelier principle. This factor is considered in Experiments 23 and 36.

CONSIDER THIS

Before you come to the laboratory to do this experiment, answer the questions posed on the first page of the Report form. After you have finished the experiment, answer the questions posed on the last page of the Report form.

Bibliography

Sisler, H. H.; VanderWerf, C. A. "Oxidation–Reduction: An Example of Chemical Sophistry," *J. Chem. Educ.* **1980,** *57,* 42–44.

Redox Titrations

The Oxidizing Capacity of a Household Cleanser or Liquid Bleach

Purpose

- Learn the techniques of a redox titration, including the identification of the end point of the titration.

- Show how a redox titration (the oxidation of iodide ion to iodine with thiosulfate ion) can be used for the quantitative determination of the total amount of oxidizing agent in household cleansers and liquid hypochlorite bleaches.

Pre-Lab Preparation

Oxidation–reduction reactions, like acid–base reactions, are widely used as the basis for the analytical determination of substances by titration.[1] In a volumetric titration, a known volume of the *titrant*, usually contained in a buret, is added to the substance being determined, called the *titrand*. The conditions needed for redox titrations are the same as for any titration. The reaction between the titrant and the titrand must be rapid, stoichiometric, and quantitative; that is, the kinetics of the reaction as well as the equilibrium must greatly favor the products. In addition to these basic requirements, the titrant solution must be stable, and there must be some means of determining its concentration accurately. Finally, there must be some means of detecting the end point of the titration reaction.

Oxidizing and reducing agents

Oxidizing agents available in pure form, such as $K_2Cr_2O_7$, may be weighed out directly to form a titrant solution of known concentration. Potassium permanganate and cerium(IV) salts are often used as oxidizing agents, but they are not pure enough to be

weighed out directly; ordinarily, we must titrate them against reducing substances of known high purity (such as As_2O_3) to determine their concentrations accurately. This process is called *standardization*.

The redox potentials of several commonly used oxidizing agents are

$$E° \text{ (V)}$$

$Ce^{4+} + e^- \rightleftharpoons Ce^{3+}$	$+1.61$
$MnO_4^- + 8 H^+ + 5 e^- \rightleftharpoons Mn^{2+} + 4 H_2O$	$+1.51$
$Cr_2O_7^{2-} + 14 H^+ + 6 e^- \rightleftharpoons 2 Cr^{3+} + 7 H_2O$	$+1.33$
$I_2(aq) + 2 e^- \rightleftharpoons 2 I^-$	$+0.54$

Cerium(IV) is one of the strongest oxidizing agents available for use as a titrant, and its solutions are very stable. However, it is more expensive than either potassium permanganate, $KMnO_4$, or potassium dichromate, $K_2Cr_2O_7$. Iodine is a rather weak oxidizing agent and is most often used in indirect procedures in which iodide ion is first oxidized to iodine and then the iodine is titrated with sodium thiosulfate solution.

Most strong reducing agents are easily oxidized by the oxygen in the air, so they are not often employed as titrants. However, they are commonly employed in analysis to reduce a substance to a lower oxidation state before it is titrated with a standard solution of an oxidizing agent. Reducing agents of moderate strength, whose solutions are stable, also exist. One of these is sodium thiosulfate, $Na_2S_2O_3$, which is most often used to titrate I_2.

End-point detection in redox titrations

The most general methods of detecting the end points in redox titrations are like those used in acid–base titrations. When both the oxidized and reduced species are soluble in solution (as MnO_4^-/Mn^{2+}, for example), an inert platinum or gold electrode will respond to the redox potential of the system, provided that the electron transfer processes at the electrode surface are

---

[1] Techniques of volumetric analysis are described in the Volumetric Measurements of Liquids section of the Introduction.

reasonably fast so that the equilibrium potential is established. Such an inert electrode may be used in conjunction with a suitable reference electrode in much the same way that the glass electrode is used in acid–base titrations. At the end point of the titration, the redox potential of the system changes abruptly by a few hundred millivolts—a voltmeter can be used to detect the change. Such a titration is called a *potentiometric redox titration*.

Redox indicators are available that function like acid–base indicators except that they respond to changes in the redox potential of the system rather than to changes in the hydrogen ion concentration. For example, the redox indicator *ferroin* is red in its reduced form, at potentials of less than $+1.12$ V. In titrations in which strong oxidizing agents, such as cerium(IV), are used, the potential jumps abruptly at the end point to potentials of greater than $+1.12$ V, and the indicator changes to its oxidized form, which is blue in color.

In some titrations, a titrant or the substance titrated acts as its own indicator. For example, the permanganate ion, MnO_4^-, is a very deep reddish-purple color, whereas the reduced product of its reaction with reducing agents in acid solution is the colorless manganese(II) ion, Mn^{2+}. Solutions titrated with standard MnO_4^- are usually colorless until the end point, then turn pink when no reducing species are left to react with the added MnO_4^-. Conversely, in the titration of iodine solutions with standard thiosulfate, $S_2O_3^{2-}$, the solutions are brown or yellow until the end point is reached, then become colorless when all of the iodine is titrated. Also, in these titrations, starch is added just before the end point. The starch forms a blue-black colored complex with I_2 and I^- that turns colorless when the I_2 is used up, making the end point sharper and easier to detect.

Redox titration of the oxidizing agents in a household cleanser or liquid bleach

In this experiment, we will make use of two oxidation–reduction reactions to determine the oxidizing capacity of a household cleanser. To a solution containing the sample of cleanser will be added an unmeasured excess of KI in acid solution. Oxidizing agents contained in the cleanser, such as sodium hypochlorite, oxidize the iodide ion to iodine according to the reaction

	$E°$ (V)
$HClO(aq) + H^+ + 2\,e^- \rightleftharpoons Cl^- + H_2O$	$+1.49$
$2\,I^- \rightleftharpoons I_2 + 2\,e^-$	-0.54
$HClO(aq) + H^+ + 2\,I^- \rightleftharpoons I_2 + Cl^- + H_2O$	$+0.95$

The iodine produced in the solution will then be determined by titration with a standardized thiosulfate solution, which reduces iodine stoichiometrically according to the reaction

	$E°$ (V)
$2\,S_2O_3^{2-} \rightleftharpoons S_4O_6^{2-} + 2\,e^-$	-0.09
$I_2 + 2\,e^- \rightleftharpoons 2\,I^-$	$+0.54$
$2\,S_2O_3^{2-} + I_2 \rightleftharpoons 2\,I^- + S_4O_6^{2-}$	$+0.45$

Even though the method is indirect, only one standard solution is required.

Establishing the end point in the iodine–thiosulfate titration

Starch (amylose) forms an intensely colored blue complex with traces of iodine (I_2) in the presence of iodide ion. Iodine reacts with I^- to form a complex with the formula I_3^-. This complex has a linear structure ($I—I—I^-$) that fits nicely into the helically coiled starch molecule, as shown in Figure 35-1. The formation of the starch I_3^- complex is reversible (as is the formation of the deep blue color) and may be used as a sensitive indicator for traces of iodine in solution:

$$\text{starch} + I_2 + I^- \rightleftharpoons (\text{starch } I_3^- \text{ complex})$$
$$\textbf{BLUE}$$

Iodine in concentrations as low as 10^{-6} M can be detected, provided that the concentration of iodide ion is 10^{-3} M or greater. Most iodine-containing solutions are titrated with standard thiosulfate to the disappearance of the blue color at the end point. In such titrations, the starch indicator should not be added until just before the end point, when the iodine concentration is low. If it is added too early in the titration, the formation of the blue is not as easily reversible. In practice, therefore, withhold the starch indicator until the last tinge of yellow due to excess iodine has almost disappeared; then add the starch and quickly complete the titration.

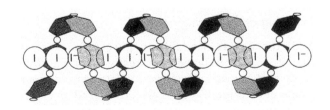

FIGURE 35-1 | **Starch–triiodide ion complex. The starch (amylose) molecule forms a helix with a cavity large enough to contain the triiodide ions.**

Iodide ion is oxidizable by oxygen in air according to the reaction

$$4\,I^- + O_2 + 4\,H^+ \rightleftharpoons 2\,I_2 + 2\,H_2O$$

Although this reaction is slow in neutral solution, the rate increases with acid concentration. If an iodimetric titration must be carried out in acid solution, exclude the air as much as possible and perform the titration quickly.

Experimental Procedure

Special Supplies: Several brands of household cleansers (such as Ajax and Comet) or liquid bleach containing approximately 5% sodium hypochlorite, NaOCl; 1-mL microburet (described in the Introduction); 2-mL pipet; two fine-tipped polyethylene transfer pipets.

Chemicals: Potassium iodide, KI(s); 1 M sulfuric acid, H_2SO_4; 0.0100 M potassium iodate, KIO_3; 3% ammonium heptamolybdate, $(NH_4)_6Mo_7O_{24} \cdot 4\,H_2O$, catalyst; 0.15 M sodium thiosulfate, $Na_2S_2O_3$; 0.1% starch indicator freshly made (prepare 1 L by making a paste of 1 g of soluble starch in about 30 mL of water; pour into 1 L of boiling water, and heat about 30 min, or until the solution is clear; cool and store in a bottle).

> ⚠ **SAFETY PRECAUTIONS:**
> **Liquid bleaches containing 5% sodium hypochlorite are corrosive to the skin and eyes. Your eyes must be protected, as always, by safety goggles. When handling the liquid bleach, take care not to get the solution on your skin. Note that bleach or other hypochlorite-containing substances, such as household cleanser, should never be mixed with ammonia, NH_3; doing so could produce chloramine compounds, such as H_2NCl and $HNCl_2$, which are toxic and volatile.**

� NOTES TO INSTRUCTOR ▰

The procedures described employ reduced-scale titrations and a 1-mL microburet. If you prefer to use a 50-mL buret instead of a 1-mL microburet, use 0.05 M $Na_2S_2O_3$ as the titrant and scale up the solution volumes in the standardization procedure of part 1, using a 25-mL sample of 0.01 M KIO_3 and adding 25 mL of water, 2 g of KI(s), and 10 mL of 1 M H_2SO_4. In part 2, use a 2-g sample of cleanser in a 250-mL beaker, adding 100 mL of water, 30 mL of 1 M

H_2SO_4 added drop by drop, 5 drops of 3% ammonium molybdate catalyst, and 2 g of KI. For liquid bleach, use a 1-g sample in a 250-mL Erlenmeyer flask, adding 50 mL of water, 2 g of KI, 5 drops of 3% ammonium molybdate catalyst, and 10 mL of 1 M H_2SO_4.

---

1. Standardizing the Sodium Thiosulfate Solution

If the approximately 0.15 M $Na_2S_2O_3$ solution has not been standardized, it may be titrated against a solution made by dissolving pure dry potassium iodate, KIO_3, to a concentration of 0.0100 M. KIO_3 reacts with excess KI in acid solution according to the reaction

$$IO_3^- + 5\,I^- + 6\,H^+ \rightleftharpoons 3\,I_2 + 3\,H_2O$$

Fill a 1-mL microburet with 0.15 M $Na_2S_2O_3$, taking care to eliminate all bubbles. Record the initial volume reading of the buret. To perform the standardization, dispense 2.00 mL of 0.0100 M KIO_3 solution, accurately measured by pipet, into a 25-mL Erlenmeyer flask. Add 3 mL of distilled water and 0.2 g of solid KI. Swirl the contents of the flask until the KI dissolves; then add 1 mL of 1 M H_2SO_4 and mix by swirling. A deep brown color should appear, indicating the presence of iodine. Titrate immediately with 0.15 M $Na_2S_2O_3$ solution contained in a 1-mL microburet until the brown fades to a pale yellow. Then add 1 mL of starch indicator solution and continue the titration until the deep blue color of the starch indicator disappears. The change from blue to a colorless solution is very sharp. Record the volume of sodium thiosulfate solution used in the titration. Repeat the procedure in a duplicate titration. In your report form, calculate the accurate concentration of the $Na_2S_2O_3$ solution.

2. Determining the Oxidizing Capacity of a Household Cleanser or Liquid Bleach

Most formulations of household cleanser contain a mild abrasive and cleaning agents such as calcium carbonate and sodium carbonate. They comprise more than 98% of the mass of the cleanser. In addition, the cleanser usually contains small amounts of a bleaching (oxidizing) agent, anionic detergent, fragrance, and coloring agent. In determining the oxidizing capacity of the cleanser, enough acid must be added to neutralize the carbonates present and provide an acid medium for oxidation of iodide ion by the oxidizing agent. The addition of acid to a carbonate produces a considerable quantity of carbon dioxide gas, causing the solution to foam, so the neutralization step must be carried out slowly, adding the 1 M H_2SO_4 with vigorous stirring to keep the sample from foaming out of its container.

The reduced-scale determination of the oxidizing agent in a powdered household cleanser is carried out by weighing 1.0 g (weighed to nearest milligram) of the cleanser into a 150-mL beaker, adding 50 mL of deionized water and neutralizing the carbonates in the sample through a drop-by-drop addition of 15 mL of 1 M H_2SO_4, accompanied by vigorous stirring to disperse the foam. When all of the 1 M H_2SO_4 has been added and the foaming has subsided, add 1 g of KI(s) and 1–2 drops of 3% ammonium heptamolybdate catalyst. Stir to dissolve the KI. The solution should turn brown, indicating the formation of iodine. Immediately titrate the solution with 0.15 M $Na_2S_2O_3$ contained in a 1-mL microburet until the brown color fades to yellow. Then add 2 mL of 0.1% starch indicator and continue the titration until the blue color of the starch indicator disappears. Record the volume of thiosulfate solution used in the titration. If time permits, make a duplicate titration. Then, if time permits, repeat the procedure, using a second brand of cleanser.

When titrating samples of cleansers that contain powdered scouring agents (and possibly dyes), the end point will be harder to see. It may be helpful to make up a sample of cleanser dispersed in the same volume of water for you to compare with the sample you are titrating, so that the changes can be more easily seen.

A liquid bleach containing about 5% sodium hypochlorite can be titrated by a procedure that closely resembles the standardization titration in part 1. Using a fine-tipped polyethylene transfer pipet, dispense 0.080–0.090 g of liquid bleach (about 4 drops) directly into a 25-mL Erlenmeyer flask on the pan of a balance. The mass of the sample should be determined to the nearest milligram. (**Caution:** Liquid bleaches are corrosive to the skin and eyes and to balances; wipe up any spills immediately.) Add 5 mL of deionized water and 0.2 g of KI(s). Swirl to dissolve the KI; then add 1–2 drops of 3% ammonium heptamolybdate catalyst and 1 mL of 1 M H_2SO_4. The solution should turn brown, indicating the formation of iodine. Immediately titrate the solution with 0.15 M $Na_2S_2O_3$ contained in a 1-mL microburet until the brown color fades to yellow. Then add 1 mL of 0.1% starch indicator and titrate until the blue color of the starch–iodine complex turns colorless. The end point is very sharp. Record the volume of sodium thiosulfate titrant used.

From the concentration and volume of the added sodium thiosulfate solution used to titrate the different brands of cleanser, calculate the weight of oxidizing agent present, assuming it to be sodium hypochlorite, NaOCl. Express the final result as the percentage by weight of sodium hypochlorite: (g NaOCl/g cleanser) × 100%.

If you have the time to compare two cleansers, remember that the effectiveness of a cleanser is influenced by several factors. First, the ability of a cleanser to remove food stains will be related to its oxidizing (bleaching) power. Second, its ability to remove stains from smooth surfaces is also assisted by the abrasive action of its polishing agent, which is usually calcium carbonate. Finally, detergents are also added to cleansers to provide foaming action and to emulsify greases and dirt. In your comparison, you must consider all three of these factors, as well as the cost per unit weight of the cleanser.

3. Titration of the Oxidizing Capacity of an Unknown In this optional determination, your instructor will provide a sample containing an unknown amount of an oxidizing agent. Using the sample mass specified by the instructor, carry out the titration for the unknown in the same manner as is described in part 2 for the liquid bleach. Calculate the weight of oxidizing agent present, assuming it to be sodium hypochlorite, NaOCl. Express the final result as the percentage by weight of sodium hypochlorite: (g NaOCl/g sample) × 100%.

CONSIDER THIS

The oxidizing power of household bleach solutions can be lost when they are exposed to heat or sunlight, or due to something as simple as leaving the cap off the bottle. Design experiments to test these storage variables for bleach. Be sure to follow laboratory safety rules as you handle bleach solutions. Test the same variables for a cleanser.

The techniques of this experiment should be easily adaptable to measure the concentration of many common oxidizing agents. Develop a method for testing hydrogen peroxide solutions available at the supermarket or drugstore. Devise strategies for testing the stability of these solutions in various storage conditions: temperature, sunlight, etc.

REPORT 35

Redox Titrations

The Oxidizing Capacity of a Household Cleanser or Liquid Bleach

Name _____

Date _____ Section _____

Locker _____ Instructor_____

Data and Calculations

In all of the following calculations, be sure to show the units of each quantity and the proper number of significant figures.

1. Standardizing the sodium thiosulfate solution

	Trial 1	Trial 2	Trial 3
Volume of KIO_3 solution			
Final buret reading (if buret is used)	mL	mL	mL
Initial buret reading	mL	mL	mL
Net volume of KIO_3	mL	mL	mL
Volume of $Na_2S_2O_3$ solution			
Final buret reading (if buret is used)	mL	mL	mL
Initial buret reading	mL	mL	mL
Net volume of $Na_2S_2O_3$	mL	mL	mL
Molarity of the $Na_2S_2O_3$ solution	M	M	M

Average _____ M

Show one of your molarity calculations

2. Determining the oxidizing capacity of a household cleanser or liquid bleach

(a) Record the following data for the titration of a household cleanser (or bleach).

Brand name of cleanser or bleach _____

Data	Trial 1	Trial 2	Trial 3
Mass of sample	g	g	g
Volume of $Na_2S_2O_3$ titrant			
Final buret reading	mL	mL	mL
Initial buret reading	mL	mL	mL
Net volume of $Na_2S_2O_3$	mL	mL	mL
Calculation of percentage NaOCl			
Moles of $Na_2S_2O_3$ used	mol	mol	mol
Moles of NaOCl reacted	mol	mol	mol
Mass of NaOCl in sample	g	g	g
Percentage by weight NaOCl (g NaOCl/g sample) $\times$ 100%	%	%	%
		Average _____ %	

(b) If you have titrated a sample of another cleanser or bleach, record your data below.

Brand name of cleanser or bleach _____

Data	Trial 1	Trial 2	Trial 3
Mass of sample	g	g	g
Volume of $Na_2S_2O_3$ titrant			
Final buret reading	mL	mL	mL
Initial buret reading	mL	mL	mL
Net volume of $Na_2S_2O_3$	mL	mL	mL
Calculation of percentage NaOCl			
Moles of $Na_2S_2O_3$ used	mol	mol	mol
Moles of NaOCl reacted	mol	mol	mol
Mass of NaOCl in sample	g	g	g
Percentage by weight NaOCl (g NaOCl/g sample) $\times$ 100%	%	%	%
		Average _____ %	

Name _____   Date _____

3. Titration of the oxidizing capacity of an unknown

Record the following data for the titration of the unknown.

Data For Unknown No. _____	Trial 1	Trial 2	Trial 3
Mass of sample	g	g	g
Volume of $Na_2S_2O_3$ titrant			
Final buret reading	mL	mL	mL
Initial buret reading	mL	mL	mL
Net volume of $Na_2S_2O_3$	mL	mL	mL
Calculation of percentage NaOCl			
Moles of $Na_2S_2O_3$ used	mol	mol	mol
Moles of NaOCl reacted	mol	mol	mol
Mass of NaOCl in sample	g	g	g
Percentage by weight NaOCl (g NaOCl/g sample) $\times$ 100%	%	%	%
		Average _____ %	

Exercises

1. Some cleansers may contain bromate salts as oxidizing agents. These salts will react with iodide ion under the conditions we are using according to the reaction

$$BrO_3^- + 6\ H^+ + 6\ I^- \rightarrow 3\ I_2 + Br^- + 3\ H_2O$$

What percentage by weight of $KBrO_3$ would a cleanser have to contain in order to produce an amount of iodine equivalent to that produced by an equal weight of cleanser containing 0.50% NaOCl by weight? (*Hint*: Start with the amount of NaOCl in a 100-g sample of cleanser and convert it to an equivalent mass of $KBrO_3$.)

2. Many household bleaching solutions are essentially solutions of sodium hypochlorite, NaOCl. Calculate the sample weight of liquid bleach containing 5.0% by weight of NaOCl that would oxidize enough I^- to consume 0.80 mL of 0.15 M sodium thiosulfate.

The oxidizing power of household bleach solutions can be lost when they are exposed to heat or sunlight, or due to something as simple as leaving the cap off the bottle. Design experiments to test these storage variables for bleach. Be sure to follow laboratory safety rules as you handle bleach solutions. Test the same variables for a cleanser.

The techniques of this experiment should be easily adaptable to measure the concentration of many common oxidizing agents. Develop a method for testing hydrogen peroxide solutions available at the supermarket or drugstore. Devise strategies for testing the stability of these solutions in various storage conditions: temperature, sunlight, etc.

Electrochemical Cells

Purpose

• Show how an electrochemical cell can be constructed based on two half-reactions that are physically separated so that electrons are transferred through an external circuit.

• Illustrate how cell voltages are related to the concentrations of the components of the electrochemical cell.

• Demonstrate how measurements of a cell voltage can be used to calculate the value of an equilibrium constant.

Pre-Lab Preparation

Electrochemistry and everyday life

As you read these words, nerve impulses are traveling along the optic nerves from your eyes to your brain, propagated by electrochemical discharges across the cell walls of the nerves. So important is this electrical activity that one definition of death is the cessation of electrical activity in the brain.

In our everyday lives, we use electrochemical cells or their products without thinking about them. When you turn the ignition key of your car, its starter motor is powered by the current from a lead–acid battery. The chromium plate on the car's trim was deposited electrochemically. The aluminum in the engine parts was produced in a cell by electrolysis. Electrochemical reactions are involved in all of these processes, which either use electrical energy to produce chemical substances or vice versa.

The nature of electrochemical cells

Whenever an oxidation–reduction reaction occurs involving two elements and their ions, electrons are transferred from the substance oxidized to the substance reduced. Thus, when zinc is oxidized by copper(II) ion, the zinc atom loses two electrons and the copper(II) ion gains two electrons. We can express this event as two separate half-reactions:

$$Zn(s) \rightarrow Zn^{2+} + 2\ e^- \qquad \text{(oxidation)}$$

$$Cu^{2+} + 2\ e^- \rightarrow Cu(s) \qquad \text{(reduction)}$$

The sum of these two half-reactions gives the net chemical reaction

$$Zn(s) + Cu^{2+} \rightarrow Zn^{2+} + Cu(s)$$

The net (or total) reaction does not contain any electrons because all of the electrons lost by the zinc are gained by copper(II) ion. An *electrochemical cell* is simply a device used to physically separate a chemical reaction into two component half-reactions in such a way that the electrons are transferred through an external circuit rather than by direct mixing of the reactants. If the chemical reaction proceeds spontaneously, creating a current flow in the external circuit, we call the cell *self-driven.*

It's interesting that we can construct electrochemical cells based on chemical reactions that we do not ordinarily regard as oxidation–reduction reactions. For example, consider the electrochemical cell based on the following half-reactions:

$Ag^+(aq) + e^- \rightarrow Ag(s)$	(reduction)
$Ag(s) + Cl^-(aq) \rightarrow AgCl(s) + e^-$	(oxidation)
$Ag^+(aq) + Cl^-(aq) \rightarrow AgCl(s)$	(net cell reaction)

In this cell, the net reaction is the combination of Ag^+ with Cl^- to form AgCl, which we ordinarily think of as a precipitation reaction, not an oxidation–reduction reaction.

Or consider the cell based on these two half-reactions:

$Cu^{2+}(1\ M) + 2\ e^- \rightarrow Cu(s)$	(reduction)
$Cu(s) \rightarrow Cu^{2+}(0.1\ M) + 2\ e^-$	(oxidation)
$Cu^{2+}(1\ M) \rightarrow Cu^{2+}(0.1\ M)$	(net cell reaction)

This is called a *concentration cell,* and the net result of its reactions is the transfer of Cu^{2+} from the more concentrated solution to the more dilute solution, a process that we might not even call a chemical reaction.

The message of these two examples is clear: An electrochemical cell can be based on any reaction or process that can be separated into two half-reactions involving the transfer of electrons to or from an external circuit. The net reaction does not have to be a conventionally defined oxidation–reduction reaction. (Keep in mind that the definition of an oxidation–reduction reaction is somewhat arbitrary, as we discussed in Experiment 34, and every chemical reaction involves a redistribution of electronic charge.) In other words, chemical reactions generally involve at least partial *charge transfer* from one atom to another.

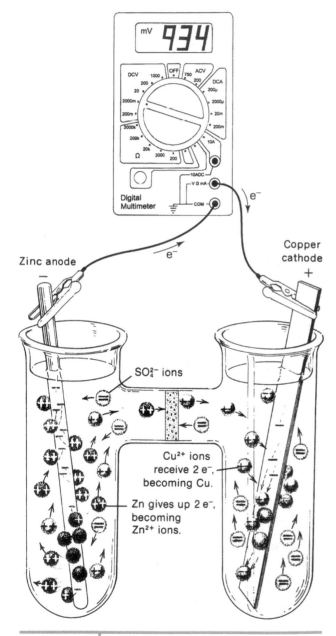

FIGURE 36-1 | **The Daniell cell, a simple electrochemical cell that transforms the energy liberated by a chemical reaction into electrical energy. The electric current in the solution consists of sulfate ions moving toward the left, and of zinc(II) and copper(II) ions moving toward the right.**

An apparatus for carrying out the reaction of Zn with Cu^{2+} in an electrochemical cell is shown in Figure 36-1. In the cell, oxidation takes place at the zinc electrode (the anode),[1] liberating electrons to the external circuit. Reduction takes place at the copper electrode (the cathode),[2] consuming electrons coming from the external circuit. By isolating each half-reaction in its own compartment, we have arranged things so that electron transfer must take place through the external circuit made of metallic wire. It is not possible for the electrons to travel long distances through the solution because they are much too reactive, reacting rapidly with water to reduce it $(e^- + H_2O \rightarrow \frac{1}{2}H_2 + OH^-)$.

The current flow in solution consists of positive ions moving in one direction and negative ions moving in the opposite direction. This ionic current flow in solution is a direct result of electron transfer that takes place at the surface of the electrodes. As current flows in the cell, there is a tendency for excess positive charge (in the form of Zn^{2+} ions) to accumulate in solution around the zinc anode as zinc atoms are oxidized. Likewise, excess negative charge (in the form of SO_4^{2-} ions) accumulates around the copper cathode as Cu^{2+} ions are removed from solution by reduction to copper metal. These excess charges create an electric field that causes the ions to *migrate*, positive ions (cations) migrating toward the cathode and negative ions (anions) migrating toward the anode. This migration of ions between the two compartments of the cell constitutes the cell current in the solution. To summarize, *ions are the charge carriers in solution, and electrons are the charge carriers in the external circuit.*[3]

In Figure 36-1, a porous barrier (for example, a glass frit) is shown separating the two compartments of the cell. This prevents gross mixing of the solutions in the two compartments by slowing down simple diffusion, but ions can still pass through the porous barrier. If it became plugged or blocked so that no ions could pass through, current flow in the cell would cease, just as if you had cut the wire in the

---

[1]This statement may be regarded as a definition of *anode:* The anode is the electrode at which oxidation takes place.

[2]This statement may be regarded as a definition of *cathode:* The cathode is the electrode at which reduction takes place.

[3]In the wires, electrons are the charge carriers, and the flow of current in a wire consists entirely of a flow of negative charge. Since current is conventionally defined as a flow of positive charge, the conventional current flow is opposite to the electron flow. The movement of an electron in one direction in the wire is equivalent to the movement of a hypothetical positive charge in the opposite direction.

external circuit. This could happen, for example, if the reaction of ions in the two compartments produced a precipitate, because the precipitate would form right in the porous plug where the two solutions come in contact. Another point to note is that all of the ions in solution migrate under the influence of the electric field, even ions from an added inert electrolyte such as K_2SO_4, which are not involved in the redox reactions. Each ion moves at an average speed determined by the strength of the field and the size of the ion. Big, bulky ions move more slowly (and carry a smaller fraction of the current) than small, fast-moving ions.

It is important to realize that the net chemical result of the operation of the cell shown in Figure 36-1 is exactly the same as the net stoichiometric reaction: One Cu^{2+} ion is reduced for each zinc atom that is oxidized. But there are some important nonchemical differences between the direct reaction and the reaction carried out in the cell. By using the cell, we are able to convert a fraction of the chemical energy of the reaction directly into electrical energy that can be used to do useful work (like running an electric motor). The direct chemical redox reaction wastes all of the chemical energy as heat—the random thermal motion of the metallic atoms and the ions in solution.

There is another practical difference in the two reactions. In the direct reaction, copper is plated out on the zinc metal as the reaction proceeds, so that after a period of time the zinc atoms get coated with a layer of copper until the reaction slows down and practically stops. In the cell, the reaction can proceed until the cell reaches equilibrium (where the cell voltage is zero). For this reaction, where equilibrium lies far to the right, current would flow either until the Zn anode is practically consumed or until practically all of the Cu^{2+} ions are plated out on the copper cathode.

The Schematic Representation of a Cell

We can always construct, at least in principle, a half-cell that corresponds to a particular redox half-reaction. For example, the redox half-reactions that take place in the Daniell cell are

$$Cu^{2+} + 2\ e^- \rightarrow Cu(s)$$

$$Zn(s) \rightarrow Zn^{2+} + 2\ e^-$$

The sum of these two half-reactions gives the net chemical reaction

$$Zn(s) + Cu^{2+} \rightarrow Zn^{2+} + Cu(s)$$

We will represent such a cell by the diagram

$$Zn \mid ZnSO_4 \vdots CuSO_4 \mid Cu$$

in which a vertical bar ($\mid$) represents a phase boundary and the dashed vertical bar ($\vdots$) represents the boundary between two miscible ionic solutions (liquid junction). A double dashed vertical bar ($\vdots\vdots$) will be used to represent a double liquid junction through an intermediate ionic solution called a *salt bridge*. Salt bridges are used to minimize the liquid junction potential and to prevent mixing of the components of two half-cells.

Cell Voltage[4]

The *volt* (V) is the unit of electrical potential, or driving force. The product of the voltage times the charge of the electron, e, is a measure of the work done when this unit electric charge is transferred from one substance to another. The *voltage of a cell*—sometimes called its *electromotive force* or *potential*—is thus a *quantitative value expressing the tendency of the chemical reaction occurring in the cell to take place.* The magnitude of this voltage depends on the relative strengths of the oxidizing and reducing agents. If the oxidizing agent has an affinity for electrons that is stronger than the tendency of the reducing agent to hold electrons, the electrical potential, or voltage, is correspondingly large.

Standard Electrode Potentials

Although we cannot measure a single half-cell potential, we can construct a scale of half-cell potentials by choosing a single *reference* half-cell and measuring the potential of all other half-cells with respect to it. The reference half-cell chosen is based on the half-reaction

$$2\ H^+ + 2\ e^- \rightleftharpoons H_2(g)$$

This couple is arbitrarily assigned a potential of zero, so that the total cell voltage is ascribed to the other couple. For example, in a cell composed of the Zn^{2+}/Zn half-reaction and the H^+/H_2 half-reaction, where all species are at unit activity,[5] the potential of the cell is found to be -0.763 V, with the zinc electrode being more negative than the hydrogen electrode. This value, -0.763 V, is called the *standard electrode potential* for the Zn^{2+}/Zn couple (see Figure 36-2). All standard electrode potentials are the values of the voltage obtained when all substances in solution are present at unit

---

[4]The voltmeter used to measure the cell voltage must draw only a small current from the cell, in order not to load the cell and change the concentrations at the electrode surface. The cell voltage under load will be smaller than the open-circuit voltage.
[5]The activity of an ion in solution is usually less than the molar concentration because of ionic interactions. It may be crudely thought of as the "effective concentration."

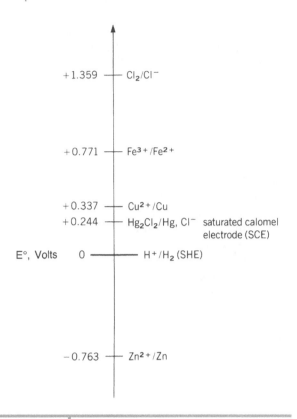

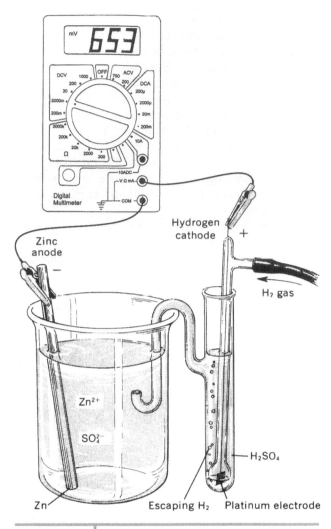

FIGURE 36-2 On the potential scale, the H^+/H_2 half-reaction is arbitrarily assigned the value zero. The zinc electrode has a voltage of -0.76 V measured against this standard hydrogen electrode (SHE). This value is assigned as the standard electrode potential of the Zn^{2+}/Zn couple. This procedure is analogous to measuring elevation from sea level (rather than from the center of the earth), with sea level being assigned zero in the scale of elevation. The saturated calomel electrode (SCE) is more often used as a practical reference electrode.

FIGURE 36-3 Hydrogen gas, adsorbed on the platinum electrode and in contact with 1 M H^+, forms the reference half-cell. When this half-cell is coupled with a Zn electrode in contact with 1 M Zn^{2+} to form the cell (Pt) | H_2 | H^+ ⦙ Zn^{2+} | Zn, the meter indicates that the zinc electrode is more negative than the hydrogen electrode. The cell voltage will not be the same as the theoretical $E°$ (-0.76 V versus SHE for the Zn^{2+}/Zn half-reaction) because of activity and junction potential effects.

activity (approximately 1 M), all gases are at unit fugacity (approximately 1 atm pressure), and the temperature is at a fixed, convenient value, usually 25 °C (see Table 10 of the Appendix).

The Standard Hydrogen Electrode (SHE) and Practical Reference Electrodes A practical hydrogen reference electrode is shown in Figure 36-3, but it is not possible to construct a standard hydrogen electrode (SHE) whose composition corresponds to the arbitrarily chosen reference state.[6] This is because the choice of the standard state for an elec-

trolyte is a solution with a concentration that is 1 *m* and in which the ionic activity coefficient is also equal to unity. The standard state thus corresponds to a *hypothetical* solution (one that cannot be made in the laboratory), because the activity coefficients of real solutions are usually less than 1 for 1 *m* solutions. In order to obtain a standard electrode potential ($E°$), we extrapolate measured cell voltages to conditions that correspond to the standard reference state. This procedure is too cumbersome for everyday measurements, so in practice we measure cell

[6]See T. Biegler and R. Woods. "The Standard Hydrogen Electrode—A Misrepresented Concept," *J. Chem. Educ.* **1973,** *50,* 604; and O. Robbins, Jr. "The Proper Definition of Standard EMF," *J. Chem. Educ.* **1971,** *48,* 737.

potentials versus reference half-cells whose potentials have been very accurately determined with respect to the SHE. The saturated calomel electrode (SCE) is a popular reference electrode based on the half-reaction

$$Hg_2Cl_2(s) + 2\,e^- \rightarrow 2\,Hg(l) + 2\,Cl^-\ (sat.\ KCl(aq))$$

$$Hg \mid Hg_2Cl_2(s) \mid sat.\ KCl(aq) \vdots$$

$$E = 0.244\ V\ versus\ SHE\ at\ 25\ °C$$

The saturated KCl electrolyte reduces liquid-junction potentials to a small and reproducible value (a few millivolts or less). Half-cell potentials measured versus the saturated calomel electrode are easily converted to the standard hydrogen electrode scale by adding +0.244 V (see Figure 36-2).

Calculating the Standard Potential of an Electrochemical Cell

When two half-cells are combined to make an electrochemical cell, the standard $E°$ of the cell will equal the difference of the standard half-cell potentials of the two redox couples. For example, the $E°$ for the electrochemical cell formed from the half-cell couples Zn^{2+}/Zn and Cl_2/Cl^- is +2.122 V. Note that this is just the difference between the two standard half-cell potentials on the scale of redox potentials shown in Figure 36-2. The net reaction that occurs when the cell operates spontaneously is

$$Zn(s) + Cl_2(g) \rightarrow Zn^{2+} + 2\,Cl^-$$

Electrode potentials and the principle of Le Châtelier

The Effect of Concentration

We have observed that the tendency for an oxidation–reduction reaction to take place is measured by the voltage created when the reaction takes place in an electrochemical cell. Thus, the electromotive force of 2.122 V created by the cell in the preceding paragraph, for the reaction

$$Zn(s) + Cl_2(g) \rightleftharpoons Zn^{2+} + 2\,Cl^-$$

will be attained when the reactants and products are in their standard states.

According to Le Châtelier's principle, and in accord with observed fact, an increase in the concentration (or pressure) of chlorine gas, $Cl_2(g)$, will increase the voltage of the cell. Conversely, an increase in the concentration of zinc ion, or of chloride ion, will favor the reverse process and therefore decrease the voltage.

The Nernst Equation

In 1889, Walther Nernst (1864–1941) introduced the well-known equation that expresses the quantitative relationship between the voltage of a cell and the concentrations of the reactants and products,

$$E = E° - \frac{2.3\,RT}{nF}\log Q$$

where E is the measured voltage of the cell, $E°$ is the standard potential as calculated from the standard electrode potentials of the half-reactions, R is the gas constant ($8.314\ J\cdot mol^{-1}\cdot K^{-1}$), T is the temperature (in kelvins), n is the number of moles of electrons transferred per mole of net chemical reaction, F (called the *Faraday constant*) is the number of coulombs per mole of electrons (96,485), and Q is the product of the activities of the reaction products divided by the product of the activities of the reactants. In determining the value of Q, the activity of each substance is raised to a power equal to its coefficient in the net chemical reaction. (The expression for Q is formulated exactly like the equilibrium constant, K, for the net chemical reaction, but the activities of reactants and products can assume any arbitrary value so that Q is not a constant.)

In all of our applications of the Nernst equation, we will make the approximation that the activity coefficients are all equal to 1, replacing *activities* by molar concentrations and the *fugacities* of gases by their *pressures* in atmospheres. It is also convenient to remember that

$$\frac{2.3RT}{F} = 0.059 \quad at\ 25\ °C\ (298\ K)$$

To see the ramifications of the Nernst equation, consider the following. In the preceding reaction, we found that

$$Zn(s) + Cl_2(g,\ 1\ atm) \rightarrow Zn^{2+}(1\ M) + 2\,Cl^-(1\ M)$$

and $E° = 2.122$ V. To calculate the corresponding voltage if the $Cl_2(g)$ were at 4.0 atm, the Zn^{2+} were 0.01 M, and the Cl^- remains at 1 M, we would write

$$E = E° - \frac{0.059}{2}\log\frac{(0.010)(1)^2}{(1)(4.0)} =$$

$$2.122 + 0.077 = 2.199\ V$$

Both the increase in the $Cl_2(g)$ pressure and the decrease in the Zn^{2+} concentration have a modest effect in increasing the voltage of the cell.

Calculating an equilibrium constant by measuring the voltage of an electrochemical cell

Electrochemical cells can be used to determine the equilibrium constant of chemical reactions. As an example of this, let's consider a cell composed of the half-reactions

$$CuCO_3(s) + 2\,e^- \rightarrow Cu(s) + CO_3^{2-}\,(1\text{ M})$$

$$Cu(s) \rightarrow Cu^{2+}\,(1\text{ M}) + 2\,e^-$$

$$\overline{CuCO_3(s) \rightarrow Cu^{2+}\,(1\text{ M}) + CO_3^{2-}\,(1\text{ M})}$$

(NET CELL REACTION) (1)

We can represent the cell by the schematic diagram

$$Cu \mid Cu^{2+} \vdots CO_3^{2+}\,(1\text{ M}) \mid CuCO_3(s) \mid Cu \qquad (2)$$

The voltage of the cell is given by the Nernst equation:

$$E_{cell} = E° - \frac{2.3RT}{nF}\log Q \qquad (3)$$

Because $Q = 1$ for the cell represented by Reaction (1) (neglecting activity coefficients), Equation (3) will reduce to $E_{cell} = E°$:

$$Q = [Cu^{2+}][CO_3^{2-}] = (1)(1) = 1 \qquad (4)$$

If

$$Q = 1, \qquad \text{then} \qquad \log Q = 0 \qquad (5)$$

and

$$E_{cell} = E° \qquad (6)$$

If the cell is allowed to operate spontaneously, E_{cell} will gradually decrease until it reaches zero. At this point, the chemical reaction has reached equilibrium and $Q \rightarrow K_{eq}$, the equilibrium constant for the chemical reaction. Therefore, at equilibrium we can rewrite Equation (3) as

$$E_{cell} = 0 = E° - \frac{2.3RT}{nF}\log K_{eq} \qquad (7)$$

which we can rearrange to

$$\log K_{eq} = \frac{nE°}{2.3RT/nF} = \frac{nE°}{0.059} \qquad (8)$$

where we have introduced the constant $2.3RT/F = 0.059$ at 25°C. This gives the interesting result that the equilibrium constant for the net chemical reaction of the cell can be calculated if we know $E°$ for the cell.

For the chemical reaction of the cell we have described, the equilibrium constant is the solubility product constant, K_{sp}, for the dissociation of $CuCO_3(s)$ in aqueous solution.

NOTES TO INSTRUCTOR

For the voltage measurements in this experiment, a voltmeter with internal resistance of at least 20,000 ohms/V is recommended. The measured voltage will then approach the open-circuit voltage since only a small current (5 μA or less) will be drawn from the cell. A digital voltmeter is even better. The mV scale of a pH meter also works well for these measurements. The voltmeter should have leads ending in alligator clips convenient for attaching the leads to the electrodes of the cells.

For the construction of the cells in this experiment, the use of glass tubes closed with cellulose dialysis tubing is preferred. These are constructed as shown in Figure 36-4 and described in part 1; their use is illustrated in Figure 36-5. These cells are inexpensive, easily renewed, and require only a small volume of solution.

The simple classic arrangement employing two beakers or test tubes to hold the half-cells and an inverted U-tube salt bridge is also satisfactory. The salt bridge is conveniently filled with 1 M KNO_3 or NH_4NO_3 in a 1% agar gel. The resistance of a 5-mm-ID bridge made in this way is about 1000 ohms (Ω), satisfactory for use with a voltmeter with an internal resistance of at least 20,000 Ω/V. If a digital voltmeter is used, a 50-fold higher salt-bridge resistance could be tolerated, but it is best to keep cell resistance as low as possible.

An alternative arrangement, recommended in earlier editions of this manual, employs a porous porcelain cup (3 in. tall and 1 in. in diameter) containing one half-cell. This is placed in a 150-mL beaker containing the other half-cell. Although easy to use, these cups are expensive and are difficult to clean when inadvertently clogged with precipitates.

Experimental Procedure

Special Supplies: Voltmeter (see following note); 9-cm lengths of 13-mm-OD glass tubing (7 per student), lightly fire polished on both ends (see Figure 36-4); cellulose dialysis tubing (cut in 3 × 3 cm pieces and stored in water 24 h before use); 12–15-mm lengths of ½-in.-ID by 1/16-in.-wall polyvinyl chloride (Tygon) tubing; wire or 0.5 × 10 cm strips of Cu, Zn, and Fe metal (or a large iron nail) for use as metal electrodes; 10 cm of 20- to 22-gauge platinum or gold wire (or 0.6 × 10 cm graphite rods) for use as inert electrodes.

Chemicals: 1.0 M copper(II) nitrate, $Cu(NO_3)_2$; 0.10 M copper(II) nitrate, $Cu(NO_3)_2$; 0.10 M zinc nitrate, $Zn(NO_3)_2$; 0.10 M iron(II) sulfate, $FeSO_4$, prepared just at the beginning of the experiment); 0.10 M iron(III) chloride, $FeCl_3$; 1 M potassium nitrate, KNO_3; bromine water, Br_2 (saturated solution); 0.10 M potassium bromide, KBr; 0.05 M iodine, I_2, dissolved in methanol; 0.10 M potassium iodide, KI; 6 M ammonia, NH_3; 1 M sodium sulfide, Na_2S; 1 M sodium carbonate, Na_2CO_3.

Half-cell tube

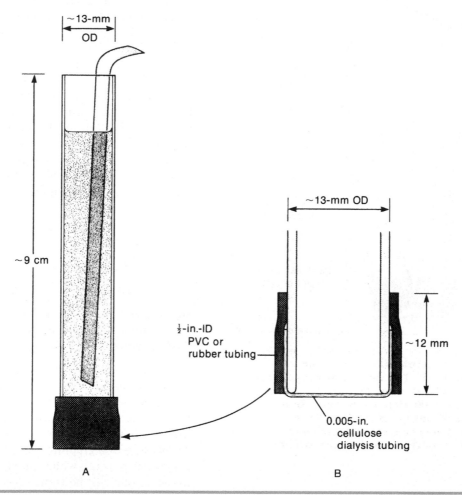

~13-mm OD

~9 cm

½-in.-ID
PVC or
rubber tubing

~13-mm OD

~12 mm

0.005-in.
cellulose
dialysis tubing

A

B

FIGURE 36-4 | **(A) A method of making simple metal/metal ion half-cells. (B) Detail of the construction of the ion-permeable junction.**

In part 1 of this experiment, you will measure the potentials of several cells composed of redox couples that you studied in Experiment 34. In part 2, you will relate Le Châtelier's principle to the effect of concentration changes on the voltage of a cell. In part 3, you will calculate an approximate value for the K_{sp} of $CuCO_3$ from the measured voltage of the cell represented by cell diagram (2).

Before beginning the experiments, it is necessary to put cellulose film on the end of seven half-cell tubes, if this has not already been done (see Figure 36-4). Take a 3 × 3 cm piece of dialysis tubing that has been stored in water to keep it flexible and center it over the end of the tube. Soften a 12–15 mm length of ½-in.-ID Tygon tubing in hot water, and work it between your fingers until it is pliable. Then force the tubing over the cellulose film until the tubing is about two-thirds of the way on. Now use a

razor blade to trim away excess cellulose film around the top of the tubing and slide the tubing the rest of the way on. At the end of this operation, the film should be stretched tightly and wrinkle free across the end of the tube. Fill the tubes with water to check for leaks (no water should drip through, but the membrane should be moist, since water can diffuse through it). Store the tubes in water until you are ready to use them. When you are through with them at the end of the laboratory period, rinse them out and store them in water.

1. Standard Cell Potentials In your Report, rewrite the redox potential series for the couples studied in Experiment 34 (it is part 6—Summary of the Data—of your Report for that experiment). By referring to the Appendix, Table 10, also enter the standard electrode potential ($E°$) of each couple you have listed.

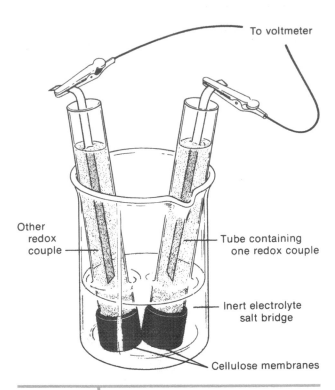

To voltmeter

Other
redox
couple

Tube containing
one redox couple

Inert electrolyte
salt bridge

Cellulose membranes

FIGURE 36-5 | **A simple way to make an electrochemical cell. The two redox couples are contained in half-cell tubes that are closed with cellulose membranes that are permeable to water and ions but prevent gross mixing of the solutions. Electrical contact in solution is provided by the inert electrolyte salt bridge.**

(a) M^{2+}/M *Half-Cells.* For the experimental part of this section, prepare half-cells for Cu, Zn, and Fe by filling three half-cell tubes about three-quarters full of 0.1 M $Cu(NO_3)_2$, 0.1 M $Zn(NO_3)_2$, and 0.1 M $FeSO_4$ (freshly prepared), and placing electrodes of the appropriate metal in each of the three half-cells. (Clean the electrodes with sandpaper or steel wool, then with detergent solution, and rinse with deionized water. A large iron nail, cleaned by immersion in a test tube of 6 M HCl, then rinsed, may be used for the iron electrode.)

(b) *A Simple Daniell Cell.* To make a simple Daniell cell, Zn | $Zn(NO_3)_2$ ⫶ $Cu(NO_3)_2$ | Cu, place the Zn^{2+}/Zn and Cu^{2+}/Cu half-cells you have made in a 150-mL beaker containing 25 mL of 1 M KNO_3, which functions as a salt bridge (see Figure 36-5). Which of these two metals will give up its electrons more readily? To determine the answer, connect the metal electrodes by means of alligator clips and leads to the voltmeter, first with both half-cells in the beaker, then with one half-cell removed from the beaker. Read the meter

carefully to the nearest 0.01 V, noting which electrode is the negative terminal (the electron source or anode). (As further proof of this polarity, recall the comparative behavior of Zn metal in Cu^{2+} solution, and of Cu metal in Zn^{2+} solution, from Experiment 34, part 1 of the Experimental Procedure.) Explain fully all aspects of the operation of the Daniell cell and complete a diagram of this cell in your Report form. Be sure you understand the following: (1) What constitutes the electric current in the wire? (2) What constitutes the electric current in the solution? (3) Why must there be actual contact of the two solutions? (4) What are the chemical reactions at each electrode?

In the same manner, put the appropriate pairs of half-cells in the beaker to form the cells

$$Fe \mid FeSO_4 ⫶ Cu(NO_3)_2 \mid Cu$$

$$Zn \mid Zn(NO_3)_2 ⫶ FeSO_4 \mid Fe$$

Measure and record the voltage of each cell, noting carefully which electrode is positive. Save the Zn^{2+}/Zn and Cu^{2+}/Cu half-cells for later use.

(c) *Nonmetal Half-Cells.* When the oxidized and reduced form of a redox couple are water soluble, electrical contact is made by placing an inert electrode in the solution. It conducts electrons to and from the external circuit. Most often, a platinum or gold wire or a graphite rod is used. (Graphite is somewhat porous, which presents a problem if the electrode will be transferred from one solution to another; it is hard to rinse off solution that has entered the pores.) Prepare in three small beakers or test tubes the following solutions, which contain equimolar amounts of the oxidized and reduced forms of the redox couple:

5 mL Br_2 water (sat. soln.) + 5 mL 0.1 M KBr

5 mL 0.1 M $FeCl_3$ + 5 mL 0.1 M $FeSO_4$

5 mL 0.05 M I_2 in methanol + 5 mL 0.1 M KI

(Because I_2 reacts almost quantitatively with an excess of I^- to form I_3^-, the last mixture is essentially 0.025 M in I_3^- and 0.025 M in I^-.)

Now pour about 7 mL of each mixture into separate half-cell tubes. A half-cell is completed by inserting an inert electrode, preferably a platinum or gold wire. In the same way as before, a complete cell is formed by placing two half-cells in a 150-mL beaker containing 25 mL of 1 M KNO_3.

Working in this fashion, measure and record the voltages of the following cells, being sure to note which electrode is positive:

$$Zn \mid Zn(NO_3)_2 \;\vdots\; Br_2, KBr \mid Pt$$

$$Zn \mid Zn(NO_3)_2 \;\vdots\; FeCl_3, FeSO_4 \mid Pt$$

$$Zn \mid Zn(NO_3)_2 \;\vdots\; KI_3, KI \mid Pt$$

Use the same platinum or gold wire for each half-cell in turn, rinsing it in water between measurements.

For all six cells for which you made measurements, compare the voltage of each cell (including the sign) with that calculated by combining the $E°$s of the two half-cells that constitute the cell. In each case, the $E°$ of a cell is the algebraic difference of the $E°$ values for the two half-cells. (Why is it necessary to subtract one $E°$ from the other?) You will probably notice that the measured cell voltage is less than the calculated $E°$ for the cell. The discrepancies may be attributed to differences in the activities of the ions from the hypothetical state of unit activity of the ions that was discussed earlier. However, note that for the seven half-cells studied, the measured values are still sufficient to establish a table of redox couples that is consistent with the order shown in Table 10 of the Appendix.

2. The Effect of Concentration Place 50 mL of 0.1 M $Cu(NO_3)_2$ (or $CuSO_4$) in a 150-mL beaker, along with a copper electrode. Also place the Zn^{2+}/Zn half-cell you saved from part 1 into the beaker and connect the electrodes to a voltmeter. Read and record the voltage; then add, stirring continuously, about 5 mL of 6 M NH_3 to the Cu^{2+} solution, until the deep blue $Cu(NH_3)_4^{2+}$ complex ion is obtained, thus reducing the concentration of $Cu(H_2O)_6^{2+}$. Read and record the voltage. Further reduce the concentration of $Cu(H_2O)_6^{2+}$ by adding, while stirring, an excess (10 mL) of 1 M Na_2S. Again, read and record the voltage. Interpret the voltage changes you observe in terms of the Le Châtelier principle.

3. Calculating an Equilibrium Constant from a Cell Voltage Measurement Place 50 mL of 1.0 M Na_2CO_3 and a clean copper strip in a 100-mL beaker. Add 5 drops of 1.0 M $Cu(NO_3)_2$ to form a precipitate of $CuCO_3$ and stir the solution. Fill a half-cell tube (like that shown in Figure 36-4) about two-thirds full of 1.0 M $Cu(NO_3)_2$ and place a clean copper electrode in the tube. Place the tube in the beaker containing the precipitate of $CuCO_3$, connect the voltmeter to the copper strips, and record the voltage. Which electrode is the positive electrode?

Because the concentrations of all the reactants and products in the cell are 1 M, the cell voltage will be equal to $E°_{cell}$, neglecting activity coefficients. (Remember that the activity of a pure solid is defined to be 1.)

From Equation (8) you can calculate K_{sp}, the equilibrium constant for the reaction

$$CuCO_3(s) \rightarrow Cu^{2+}(aq) + CO_3^{2-}(aq)$$

In using Equation (8) you must know the value of n and the magnitude and sign of $E°_{cell}$. What is the value of n that appears in the half-reactions that are summed to give Reaction (1)? The cell diagram is shown in cell diagram (2). By convention, the cell voltage, E_{cell}, is defined as the potential of the right-hand electrode (the Cu strip dipping in 1 M CO_3^{2-}) measured with respect to the left-hand electrode. If the right-hand electrode is the negative electrode, E_{cell} is negative. As defined by this convention, is the $E°_{cell}$ you measured positive or negative?

CONSIDER THIS

Choose one of the electrochemical cells in this experiment and study the effect of concentration on the cell voltage. Does it follow the Nernst equation?

Prepare a copper or zinc concentration cell and explore the values of E obtained at a variety of concentrations.

Devise an electrochemical cell to measure the equilibrium constant for the reaction

$$Ag^+(aq) + 2\,NH_3(aq) \rightleftharpoons Ag(NH_3)_2^+(aq)$$

Bibliography

Ramette, R. W. "Silver Equilibria via Cell Measurements," *J. Chem. Educ.* **1972**, *49*, 423–424.

Electrochemical Cells

Name _____

Date _____ Section _____

Locker _____ Instructor_____

1. Standard cell potentials

Complete the diagram at the right by filling in the formulas for the composition of the electrodes and of the ions in the solutions. Place these formulas near the arrows leading into the solutions. Also indicate at the top which electrode is the anode and which electrode is the cathode.

What constitutes an electric current in a wire?

Indicate the direction of these particles by drawing an arrow over the wires connected to the voltmeter.

What constitutes an electric current in a solution?

Indicate the direction of movement of these particles in the solution by writing their formulas on the proper arrow below the sketch.

Why must the two solutions come into actual contact?

Write the half-reaction taking place at the cathode.

Write the half-reaction taking place at the anode.

Write the total cell reaction.

Give the experimental voltage for this cell.

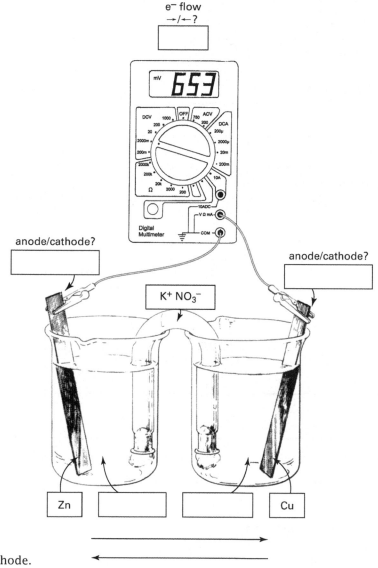

In the right-hand column, enter the standard electrode potential for each half-cell reaction that you studied in part 1 of the Experimental Procedure, as obtained from Table 10 in the Appendix. In the space below, calculate the $E°$ for each cell whose voltage you measured in part 1. Write the calculated values in the table below. Also supply the equation for each cell reaction measured in part 1. Indicate which element is the positive electrode and which the negative, and give the experimental and calculated voltages in each case.

Redox couples	
Couple	**$E°$, volts**
Zn^{2+}/Zn	
Fe^{2+}/Fe	
Cu^{2+}/Cu	
I_3^-/I^-	
Fe^{3+}/Fe^{2+}	
Br_2/Br^-	

Cell reaction	Positive electrode	Negative electrode	Experimental voltage	Calculated voltage

Name _____   Date _____

2. The effect of concentration

Record your observed voltages and write net ionic equations for the reactions, first of NH_3, and then of Na_2S with Cu^{2+}.

Initial voltage _____   _____

Voltage with
NH_3 added _____   _____

Voltage with
Na_2S added _____   _____

Explain the reasons for the effects on the cell voltage you observed when NH_3 and Na_2S were added to the Cu^{2+}/Cu half-cell.

How would you adjust the concentrations of Cu^{2+} and Zn^{2+} in the cell in order to obtain the maximum voltage possible for this cell?

3. Calculating an equilibrium constant from a cell voltage measurement

Give the magnitude of the cell voltage and indicate which electrode is positive for the following cell.

Cell voltage _____ $Cu \mid Cu^{2+} \vdots CO_3^{2-}$ (1 M) $\mid CuCO_3(s) \mid Cu$

Calculate $\log K_{sp}$ (and K_{sp}) using Equation (8) and your cell voltage data.

$$\log K_{sp} =$$

$$K_{sp} =$$

Compare your calculated value with the value for K_{sp} of $CuCO_3$ given in the Appendix, Table 9.

Suggest reasons why the value of K_{sp} that you calculated might be considerably larger than the equilibrium value given in Table 9.

Choose one of the electrochemical cells in this experiment and study the effect of concentration on the cell voltage. Does it follow the Nernst equation?

Prepare a copper or zinc concentration cell, and explore the values of E obtained at a variety of concentrations.

Devise an electrochemical cell to measure the equilibrium constant for the reaction

$$Ag^+(aq) + 2\,NH_3(aq) \rightleftharpoons Ag(NH_3)_2{}^+(aq)$$

(*Hint:* Find two half-reactions in Table 10 of the Appendix that can be combined to give the net chemical reaction.)

Electrochemical Puzzles

Metal Corrosion and Anodic Protection

The Golden Penny Experiment

Purpose

• Observe how iron may be corroded or protected from corrosion depending on whether it is connected to copper or to zinc metal.

• Observe the deposition of zinc metal on a penny that is in contact with metallic zinc.

• Develop a hypothesis to explain the observed deposition of zinc.

Pre-Lab Preparation

Electrochemistry is everywhere

In general, whenever two condensed phases (solid or liquid) are brought into contact, a potential (or voltage) difference develops across the interface. Because the interface region is very thin, transfer of even a small amount of charge across the interface can create a very large electric field. For example, transferring about one picomole (10^{-12} mole) of electron charge per square centimeter of area will typically create a potential difference of approximately 1 V across an interface layer about 1 nanometer (nm) thick. The electric field in this interface region would be about 10^9 V/m. Electric fields this large can cause the transfer of electrons across an interface layer or the transfer of ions between the inside and outside of cells in living organisms. Because contacts between condensed phases are very common in nature, electrochemical phenomena are very common, even though we are often unaware of them. At the cellular level, electrochemical phenomena are crucial to the propagation of nerve impulses, the timing of muscular contractions of the heart, and activity in your brain cells.

Most of the electrochemical technology created by humans involves the simplest kind of chemical change: electron transfer across an interface. Often, the interface is between a good electron conductor, called an *electrode,* and a solution containing molecules or ions. The electrode might be a solid (like platinum or copper metal or graphite), or it could be liquid (like mercury metal). When electrons are transferred from the electrode to a molecule, we say the molecule has been *reduced.* Electron transfer in the opposite sense (from molecule to electrode) is called *oxidation.*

What is an electrochemical cell?

As we saw earlier, an electrochemical cell is a device used to separate the reactants in a chemical reaction in such a way that the electrons are transferred from one reactant to another through an external circuit rather than by direct mixing of the reactants. We will call the cell a *galvanic,* or *self-driven,* cell if the chemical reaction proceeds spontaneously, creating a current flow in the external circuit. A self-driven electrochemical cell converts the energy of a chemical reaction into electrical energy.

This process can be reversed in an *electrolytic,* or *driven,* electrochemical cell. That is, current can be forced to flow through a cell by using an external power source, such as a battery or direct current (DC) power supply. This process, called *electrolysis,* can create new chemical substances that have greater chemical energy than the reactants, thereby "storing" energy. Upon demand, this chemical energy can be converted back into electrical energy.

The electrolysis of water by means of the cell shown in Figure 37-1 is a familiar example of a driven cell. Less well known is the fact that chemical energy stored in the electrolysis products (hydrogen and oxygen gases) can be directly converted into

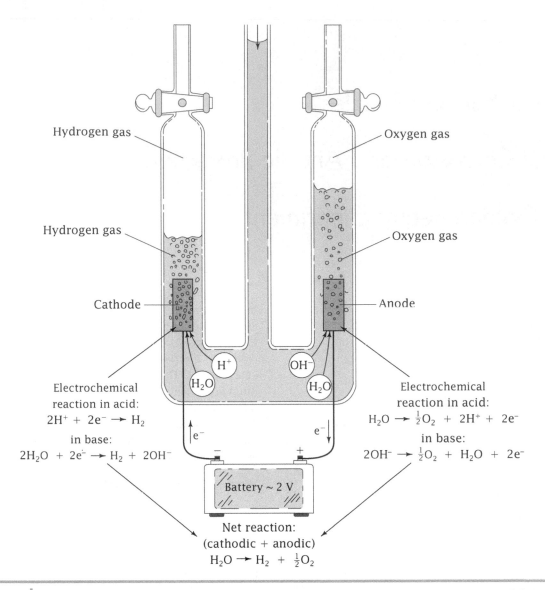

Hydrogen gas

Hydrogen gas

Oxygen gas

Oxygen gas

Cathode

Anode

Electrochemical
reaction in acid:
$$2H^+ + 2e^- \rightarrow H_2$$
in base:
$$2H_2O + 2e^- \rightarrow H_2 + 2OH^-$$

Electrochemical
reaction in acid:
$$H_2O \rightarrow \tfrac{1}{2}O_2 + 2H^+ + 2e^-$$
in base:
$$2OH^- \rightarrow \tfrac{1}{2}O_2 + H_2O + 2e^-$$

H^+

OH^-

H_2O

H_2O

e^-

e^-

Battery ~ 2 V

Net reaction:
(cathodic + anodic)
$$H_2O \rightarrow H_2 + \tfrac{1}{2}O_2$$

FIGURE 37-1 | **Production of hydrogen and oxygen by electrolysis of water in an electrolytic (driven) cell.**

electrical energy by means of the self-driven fuel cell[1] shown in Figure 37-2.

Several important industrial chemicals are produced by electrochemical processes. Chlorine is produced by the electrolysis of brine (NaCl) solution. Aluminum is produced by high-temperature electrolysis of alumina (Al_2O_3) dissolved in molten cryolite (Na_3AlF_6). In these examples, driven cells are used as "substance-producing" devices. Driven cells are also used in electroplating, in which a thin metal layer is plated out on a conducting substrate, and in the passivation of metals (like aluminum) to produce a thin layer of oxide that provides a protective (and decorative) coating.

Faraday's laws and electrolysis

Michael Faraday (1791–1867) studied the effects of passing current through cells and published his findings in 1833 and 1834. His experimental conclusions have come to be known as Faraday's laws. Paraphrased in modern language, they may be stated as follows:

1. The amount of chemical change produced is proportional to the total amount of charge passed through the cell.

2. The passage of a mole of electric charge produces an equivalent of chemical change.

An equivalent of chemical change is equal to a mole of chemical change divided by the moles of electrons transferred per mole of chemical change.

---

[1]A *fuel cell* is any electrochemical cell in which a net reaction is supported by a supply of reactants external to the cell. Often the oxidant is oxygen from the air.

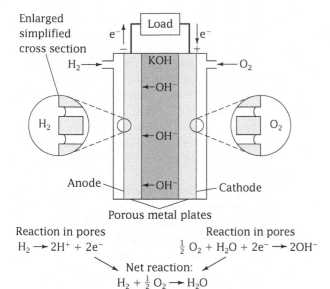

Reaction in pores
$H_2 \rightarrow 2H^+ + 2e^-$

Reaction in pores
$\frac{1}{2} O_2 + H_2O + 2e^- \rightarrow 2OH^-$

Net reaction:
$H_2 + \frac{1}{2} O_2 \rightarrow H_2O$

FIGURE 37-2 | **A scheme for obtaining electrical energy from a galvanic (self-driven) hydrogen–oxygen fuel cell. The net chemical reaction in the cell is equivalent to the combustion of hydrogen.**

The amount of electric charge is expressed in coulombs (C) and can be determined experimentally by measuring the electric current (flow of charge). The current is measured in units of amperes (A). One ampere is equal to a charge flow of one coulomb per second. If the current is constant, the total charge transferred is equal to the current multiplied by the time in seconds:

$$\text{coulombs} = \text{amperes} \left(\text{or} \frac{\text{coulombs}}{\text{second}} \right) \times \text{seconds} \quad (1)$$

When a mole of silver (107.87 g) is deposited at a cathode, the half-cell equation

$$Ag^+ + e^- \rightarrow Ag(s) \quad (2)$$

indicates that an Avogadro's number (6.022×10^{23}) of individual silver ions must be reduced by the same number of electrons—"a mole of electrons." The charge on a mole of electrons is

$$6.0221 \times 10^{23} \text{ electrons} \times 1.6022 \times$$
$$10^{-19} \frac{\text{coulombs}}{\text{electron}} = 96,485 \text{ coulombs} \quad (3)$$

which is called a *faraday* and given the symbol *F*. When copper is plated out at the cathode, the half-cell equation

$$Cu^{2+} + 2\,e^- \rightarrow Cu(s) \quad (4)$$

shows that 1 mol of electrons (1 faraday) will produce only ½ mol of copper [$(63.54/2) = 31.77$ g].

To summarize, when 1 *F* (96,485 C of charge) is passed through a cell, chemical change equivalent to transfer of 1 mol of electrons (oxidation at the anode and reduction at the cathode) occurs at each electrode. This quantity of material is called one *electrochemical equivalent.*

Only a few electrochemical reactions are known that give a simple, well-defined reaction conforming exactly to Faraday's laws. More often, an electrochemical reaction is complex, yielding several products. This may occur when more than one electron transfer reaction takes place on the electrode or when the primary products are very reactive in some side reaction.

Electrochemists use the terms *current efficiency* or *yield* and define these terms as the percentage of total charge that is effective in producing the desired substance. It was just these effects that made it difficult to discover these laws, which may seem self-evident to students already familiar with atomic theory. But in 1834, modern theories of atomic structure, chemical bonding, or ionization did not exist, nor was anything known about the existence and properties of electrons. These laws, which eluded many who were studying electricity, should be regarded as a lasting tribute to Faraday's ingenuity and careful experimental work.

Electrochemistry and corrosion

When it is not desired, the dissolution of a metal is called *corrosion,* and much money and effort are spent in preventing or combating it. A simple example of corrosion is the dissolution of a metal in acid, where the oxidation–reduction processes are those shown in Figure 37-3.

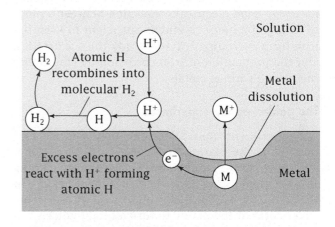

FIGURE 37-3 | **In corrosion, dissolution of the metal liberates electrons. If these electrons can be taken up by some other electrochemical reaction (like H₂ evolution), the metal continues to dissolve.**

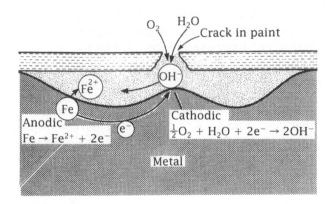

FIGURE 37-4 At a crack in the protective paint, corrosion often develops under the paint. It is the (invisible) part of the steel under the paint that becomes anodic and dissolves. Electron flow in the metal and ionic flow in the corroded area couple the two half-reactions.

Corrosion is a common affliction of automobiles in regions with cold climates, where corrosive salts are spread on the roads to melt ice. When a hole or crack in the paint exposes the steel of the auto body, the steel often begins to corrode. This suggests that the corrosion process is primarily electrochemical in nature, two separated half-reactions being mediated by flow of electrons in the steel (see Figure 37-4). Unless the corroded area is cleaned and repainted, corrosion will continue until the metal is rusted through over a large area.

Anodic Protection By coating steel (which is mostly iron metal) with a more active metal like zinc, a process called *galvanizing,* the steel's corrosion can be retarded or entirely prevented. Often, simply making a good electrical connection between a piece of iron and a piece of zinc is sufficient to keep the iron from corroding. We will study the acceleration and the prevention of iron corrosion by connecting the iron to various metals.

The golden penny experiment

An experiment that simultaneously charms and puzzles is the plating of a penny with zinc metal. First, the penny is immersed in a solution containing 1 M NaOH and granular zinc. Subsequent heating of the penny for a few seconds on a hot plate causes the silver color of the penny to turn a bright golden yellow. Explaining the details of the process presents a challenging chemical puzzle. Some questions that arise may only be answered by doing further experiments, so this experiment can be somewhat open-ended.

Experimental Procedure

Special Supplies: *Part 1:* Nominal 100×15 mm disposable polystyrene Petri dishes (three per student group); fine steel wool; approximately one soldering kit for every six students, consisting of a 140-W soldering iron, rosin-core solder, and one 6×6 in. ceramic fiber square (available from Flinn Scientific Inc.); digital multimeters with alligator clip leads; and short lengths (3 cm) of Pt or nichrome wire to use as voltage probes.

Part 2: Digital multimeters with alligator clip leads; pennies (preferably clean and bright); 20-gauge copper wire; hot plates; polypropylene or stainless steel forceps; approximately one soldering kit (see the description in part 1) for every six students.

Chemicals: *Part 1:* Agar (powder), 1% phenolphthalein indicator, 0.1 M potassium ferricyanide [hexacyanoferrate(III)], $K_3Fe(CN)_6$; two zinc metal strips, 6×40 mm, cut from 0.01-in.-thick zinc foil; two copper metal strips, 6×40 mm, cut from 0.01- (or 0.005-) in.-thick copper foil; ungalvanized finishing nails, 4-penny size, 40 mm long (before use, clean by soaking briefly in 3 M H_2SO_4 acid, rinsing with deionized water, and drying in an oven).

Part 2: 30-mesh zinc metal; zinc metal powder; 6×100 mm strips of zinc metal, cut from 0.01-in.-thick zinc foil; 20-gauge copper wire; 1 M sodium hydroxide, NaOH.

■■■ **NOTES TO INSTRUCTOR** ■■■

We have profitably spent a whole lab period or two on just part 2 of this experiment, the golden penny experiment. It presents an interesting challenge to explain the observations and many opportunities for open-ended experimentation. Besides, it's fun. Part 1 involves observations over a period of a week or so; as a result, parts 1 and 2, done together, make a good 2-week lab, with time in between for discussion and planning of other experiments to try.

 SAFETY PRECAUTIONS:
WEAR EYE PROTECTION AT ALL TIMES. Sodium hydroxide is corrosive. You may want to provide latex rubber gloves for handling pennies that have been in contact with 1 M NaOH.

1. Metal Corrosion and Anodic Protection Obtain two 6×40 mm strips of zinc foil, two 6×40 mm strips of copper foil, and two 4-penny (40 mm long) ungalvanized iron finishing nails (which should have been previously cleaned by immersion in 3 M H_2SO_4, then rinsed with deionized water, and dried in an oven). Clean the zinc and copper strips with steel wool to produce a clean shiny surface. Use detergent to remove the film of oil on the strips, rinse them with deionized water, and dry them with a tissue.

General Soldering Instructions. Go to a soldering station where you will find a ceramic fiber square, soldering iron, and rosin-core solder. When soldering, place the zinc strip on a ceramic fiber square and position the copper strip so the ends of the two strips overlap by about 4–5 mm. Ask a partner to apply pressure to the copper strip, holding it in place while you are soldering the joint. (Your partner may use almost any tool for this—except a bare finger, because the strip will get very hot.) Pull the trigger of the soldering iron to let it heat up. Then tin the tip of the iron with solder, wiping off any excess with a damp sponge.

Place the freshly tinned tip on the zinc strip, next to the copper strip, angling the iron to get good thermal contact. Feed the solder into the area between the copper strip and the tip of the soldering iron. When the zinc strip is hot enough, the solder will flow into the joint. Don't dab at the joint with the tip of the iron while soldering. The tip must be kept in continuous good thermal contact with the joint so that the zinc strip heats up. Using steel wool, remove any rosin remaining from the soldering operation; then rinse the soldered joints, and dry the metal strips with a tissue.

Now you have a Zn/Cu piece. Next, make Cu/Fe and Zn/Fe pieces. (Once these soldered bimetal pieces have been made, they can be cleaned and reused several times, so don't throw them away at the end of the experiment unless your instructor directs you to.)

Add 3 g of agar to 225 mL of boiling deionized water in a 600-mL beaker. At this point, it's best to turn down the heat and stir until the agar is dissolved. Be careful not to heat the agar so much that it scorches or boils over. While the agar suspension is hot, add 2 mL of 1% phenolphthalein indicator. Then mix with a stirring rod.

Get three nominal 100×15 mm polystyrene Petri dishes. Put each of the three soldered bimetal pieces in the bottom (taller) half of a Petri dish. Protecting your hands from the beaker containing the hot agar solution, pour the agar over the metal pieces in the Petri dishes. Cover the metal pieces, but do not fill the dish beyond half its depth. There should be no voids or bubbles underneath the metal strips.

After the agar cools and gels, use a fine-tipped polyethylene transfer pipet to place one small (0.02-mL) drop of 0.1 M potassium ferricyanide [$K_3Fe(III)(CN)_6$] about 1 cm away from the middle of both iron nails. The ferricyanide salt will diffuse radially outward. If the ferricyanide ions encounter any Fe^{2+} formed by oxidation of Fe, they will react with the Fe(II) ions to form a dark blue compound formulated as $KFe(III)Fe(II)(CN)_6$. [It's reported that you get the same product by mixing Fe^{3+} with $K_4Fe(II)(CN)_6$.] Evidently, in the final product, the iron atoms have exchanged an electron so that the Fe^{2+} ion is oxidized to Fe(III), and the Fe(III) originally in the ferricyanide ion is reduced to Fe(II).

During the course of the lab sessions, make periodic observations of the Petri dishes. Look for evidence of formation of hydroxide ion, which will turn the phenolphthalein pink; the formation of insoluble metal ion-hydroxide salts, which will appear as a cloudy band; or the formation of a blue compound in those dishes containing iron nails [with an added drop of 0.1 M potassium ferricyanide, $K_3Fe(CN)_6$], indicating oxidation of Fe to form Fe^{2+}.

Optional. Obtain two short lengths (about 3 cm long) of platinum or nichrome wire and a digital multimeter with leads connected to alligator clips to hold short lengths of wire that will be used as voltage probes. Adjust the voltmeter to its most sensitive voltage range (200 mV). First, immerse the probes in the agar about 1 mm from the edge of the two metals soldered together about midway along the length of the metals. (Keep the probes upright, perpendicular to the Petri dish, and make sure the probes do not touch the metal strips.) Note whether there is any voltage difference. Note the polarity. Which metal is nearest the positive (+) end? Any voltage difference indicates an electric field between the two points in the agar, created by the formation of positive and negative ions in the two regions. Considering the polarity of the measured field, what ions do you think might be responsible for the presence of the electric field? Write plausible reactions for the formation of positive ions (metal atom oxidation) and negative ions (reduction of water or oxygen).

Next, touch the probes directly to the two metals at their midpoints and note any voltage reading. (The voltage reading is expected to be zero volts because metals are such good electronic conductors that only a tiny electric field can exist in the two metals connected together.) The metals soldered together are said to form an *equipotential surface* (a surface where the potential is constant, so that the voltage difference between any two points on the metal is zero.)

Put the top cover on your Petri dish and tape it in place with two or three short strips of tape. Write your initials or other identifying marks on the tape.

Continue visual observations over a period of several days, looking for evidence of formation of any pink color or any visible precipitates. Make sketches and write descriptions of the changes you observe.

Reactions That Might Form OH⁻ Ion When a pink color develops around a metal in a gel containing phenolphthalein indicator, it means that the solution next to the metal is basic. In an aqueous gel, the pink color means that some hydroxide ions have been formed.

Although any electrons given up when a reactive metal is oxidized might react at the spot where the oxidation occurs, they can also readily travel to any other spot on the surface of the two joined pieces of metal. That means it is possible that the point where metal atoms are oxidized could be some distance from the point where hydroxide ions are produced.

Now let's think about what might be most likely to accept these available electrons. Metal atoms typically don't accept electrons to form negatively charged metal ions. Rather, metal ions tend to give up electrons to form positive ions. Things that are easy to reduce have the most positive standard reduction potentials, like the halogens, but we don't have any halogens in our system. The gel surrounding the metal consists mainly of water with about 1% of agar. Although water is not easy to reduce, because water has a negative standard reduction potential in basic solution, this substance can be reduced when the reaction is coupled to the oxidation of Zn metal in basic solution, as shown by the following standard reduction potentials:

$$Zn(OH)_4^{2-} + 2\ e^- \rightarrow Zn(s) + 4\ OH^-$$
$$(E° \text{ in } 1\text{ M OH}^- = -1.28\text{ V})$$

$$2\ H_2O + 2\ e^- \rightarrow H_2(g) + 2\ OH^-$$
$$(E° \text{ in } 1\text{ M OH}^- = -0.80\text{ V})$$

Agar is a polysaccharide (like starch), and polysaccharides are not easy to reduce. Finally, we must not forget that the Petri dishes are open to the air, so the agar gel also contains dissolved oxygen, a good acceptor of electrons. At least two reactions involving oxygen deserve serious consideration:

$$O_2(g) + H_2O + 2\ e^- \rightarrow HO_2^- + OH^-$$
$$(E° \text{ in } 1\text{ M OH}^- = -0.065\text{ V})$$

$$O_2(g) + 2\ H_2O + 4\ e^- \rightarrow 4\ OH^-$$
$$(E° \text{ in } 1\text{ M OH}^- = +0.40\text{ V})$$

The $E°$ for reduction of oxygen in basic solution is considerably more positive than for the reduction of water. So we definitely must consider the possibility that oxygen might be the species that could most easily be reduced, with OH⁻ (and possibly hydrogen peroxide) being the reduction product.

The reduction of either water or oxygen produces hydroxide ions, but the formation of a pink color with phenolphthalein does not tell us which reaction might be responsible. Thermodynamics (as measured by the standard reduction potentials) favors reduction of oxygen over reduction of water. However, the reduction of oxygen on many metals is known to have a large activation energy which usually causes the reaction to be slow. Thus, kinetics may favor the reduction of water, particularly because the concentration of water is much greater than the concentration of oxygen in the agar gel.

Can you think of an experiment that might allow you to distinguish whether water or oxygen is the major species being reduced?

2. The Golden Penny Experiment Work in groups of two or three students. Put 8 g of 30-mesh zinc in the bottom of a 400-mL beaker. Tilt and tap the beaker on the benchtop in order to get the granular zinc to cover about half the bottom of the beaker (one-half of the beaker should be covered, one-half not; see Figure 37-5). Carefully pour 200 mL of 1 M NaOH down the side of the beaker, being careful not to disturb the distribution of zinc. Use a stirring rod or spatula to clear any remaining granules so that half of the beaker bottom is completely free of zinc granules. Place the beaker on a hot plate in the fume hood and turn the hot plate to medium heat. The solution should be heated to about 80–90 °C; if it is heated to boiling, the distribution of zinc granules will be disturbed.

While waiting for the solution to heat, buff six copper pennies with steel wool until they are shiny. Solder 10-cm lengths of 20-gauge copper wire to two of the pennies, overlapping the wire and penny about 2–3 mm from the edge. Solder the free end of one of the copper wires to a 5 × 100 mm strip of zinc metal, as shown in Figure 37-5. Clean any rosin off the soldered joints with steel wool, and rinse with water.

When soldering, place the penny on a ceramic fiber square, with the end of a copper wire overlapping the penny 2–3 mm. Ask a partner to apply pressure to the wire to hold it in place while you solder the joint. (Almost anything will do, but don't use a bare finger because the wire will get very hot.) Review the General Soldering Instructions in part 1. Place the freshly

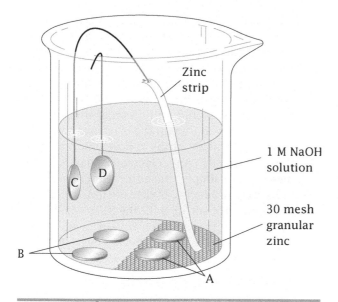

FIGURE 37-5 | **Four arrangements of copper pennies in the golden penny experiment. The bottom of the beaker is half-covered with 30-mesh zinc metal. (A) Two pennies are lying on top of the 30-mesh zinc. (B) Two pennies are lying on the bottom of the beaker but not in contact with 30-mesh zinc. (C) A penny soldered to copper wire is immersed in solution, the other end of the copper wire being soldered to a strip of zinc metal in contact with 30-mesh zinc on the bottom of the beaker. (D) A penny soldered to copper wire is immersed in solution. The solution in the beaker is 1 M NaOH.**

tinned tip on the penny, next to the wire, angling the iron to get good thermal contact. Don't dab at the joint with the tip of the iron while soldering.

When the solution has warmed, use forceps to place two copper pennies on top of the granular zinc metal and two pennies in the area that is free of granular zinc (make sure that the pennies on the uncovered side do not contact even one grain of zinc). Bend a small "foot" on the free end of the zinc strip and place the strip and attached penny in the beaker, as shown in Figure 37-5. The foot of the zinc strip should rest on the granular zinc so that both are in direct contact. The penny should be completely immersed in solution but should not contact any granular zinc metal on the bottom of the beaker. Finally, hang the last penny (the one with only copper wire soldered to it) over the edge of the beaker so that the penny is completely immersed in solution (see Figure 37-5).

Leave the pennies in the beaker until some of them turn a silvery color. This may take anywhere from 5 to 30 min, depending on the temperature of the solution. (Some of the pennies will never turn silver even after waiting an hour or more.)

Which pennies turn a silvery color? Is it the three pennies that are in contact with the solution and with zinc, either directly or through the copper wire? Or is it the three pennies in contact with the solution but not in direct or indirect contact with zinc metal?

Using a pair of forceps, or wearing latex gloves for protection, remove the pennies that have turned a uniform silvery color, rinse them with water, and put the pennies on a hot plate for a few seconds. Watch what happens to the silver-colored pennies as they heat on the hot plate. Keep the solution warm in the beaker in case you need to repeat some part of the experiment or try some new experiment, as described below.

Further Experiments and Clues to the Puzzle. Solder another length of copper wire to a shiny clean penny. Connect one lead of a voltmeter to a zinc strip and the other lead to the copper wire soldered to the penny. Using the same solution you prepared earlier, immerse the zinc strip and penny in the solution. Is there a voltage difference between the zinc strip and the copper wire soldered to the copper penny? Which metal is the electron source (the negative terminal of this electrochemical cell)?

Do you think a current flows in the copper wire connecting the zinc strip and copper penny when both are immersed in the solution? If so, in which direction will electrons flow, and what are the anode and cathode reactions? Put the digital voltmeter into its current measuring mode on its most sensitive (microampere) scale. Then see if any current is flowing when you connect the meter in series between the copper wire soldered to the penny and the zinc strip—both the penny and the zinc strip should be immersed in the hot 1 M NaOH solution. The series connections should look like this: Penny/copper wire/(+)ammeter(−)/zinc strip. How large a current flows? Is the current (charge flow) from penny to zinc strip or vice versa?

In wires, the charge carriers are electrons. Current (defined as a flow of positive charge) is opposite to the flow of electrons. In which direction are electrons flowing: from penny to zinc strip or vice versa?

Construct a Hypothesis. Describe your observations, and develop a hypothesis to explain what's going on. Why do the copper pennies turn silver? What causes the silver color? Do copper pennies not in contact with zinc turn a silver color? If not, why not? Is this consistent with the observations in Experiment 34, in which you learned that Cu metal does not reduce Zn^{2+} ion? Why do the silver colored pennies turn gold when heated on the hot plate? Is this the realization of the alchemists' dreams of turning base metals into gold? Or is this just a brazen trick?

▰▰▰ CONSIDER THIS ▰▰▰

Our goal is to try to use the observations we made to construct a mental image of what the atoms are doing and why they are doing it. First, let's consider the observations and how they provide clues about what's happening at the atomic level.

1. Metal Corrosion and Anodic Protection

(a) Phenolphthalein is an acid–base indicator that turns pink (or red) in basic solution. When it turns pink in solution, that means some hydroxide ions have been formed. The only plausible source of hydroxide ions in the system is the reduction of water and/or oxygen (which could come from the air). Where do you think the electrons required for the reduction came from?

(b) Potassium ferricyanide [$K_3Fe(III)(CN)_6$] forms a blue color with Fe^{2+}. If a blue color forms at the spot near the iron metal, it's plausible that the iron is being oxidized to form Fe^{2+}.

(c) When freshly made, a dilute $Fe(II)SO_4$ solution is light green, nearly colorless. On standing in contact with air, the solution turns yellow, indicating the formation of Fe(III), which must have resulted from oxidation of Fe(II) by oxygen in the air.

(d) If a band of precipitate forms, it seems likely that it would be formed from metal cations reacting with anions (most likely hydroxide ion). The identity of the metals present and the color of the precipitate give us clues about what metal ion might be involved.

(e) Since many metals are easier to oxidize than water, it seems a good bet to assume that if water or oxygen is being reduced to form hydroxide ion, some metal must be getting oxidized. When two different metals are connected together, it's reasonable to assume that the most active metal will get oxidized. The most active metal has the most negative M^{+n}/M standard reduction potential.

(f) One way of looking at the bimetal strips in the Petri dishes is to think of them as an electrochemical cell formed by putting some agar gel that contains phenolphthalein into a beaker and embedding strips of two different metals in the gel. A voltmeter connected externally to the strips will measure some voltage difference between the two strips. An ammeter connected between the two metal electrodes will measure a small current. (You can easily try this to see if it is true.)

Now imagine shorting the strips together in the external circuit by soldering them together.

Electrons are able to easily flow from one metal to the other. The electrochemical processes will still go on at the surface of the metal strips, just as when they were connected through the ammeter. Now imagine adding more gel until the electrodes are completely covered. The same electrochemical processes would still go on at the surface of the metals, even though the electrical connection between the metal strips is now "internal" rather than "external." Any electrons that leave the metal would simply cause the reduction of water or oxygen, producing hydroxide ion.

Using this set of hypotheses to assist your thinking, try to construct in your own words an explanation at the atomic level of everything happening in the three Petri dishes.

2. The Golden Penny Experiment

(a) Assume that the silver coating on the copper penny is zinc metal. From previous experience, we know that zinc metal dissolves in acids. Try putting a few drops of 1 M HCl on a silver-coated penny. What happens?

(b) If the penny is coated with zinc, how is the zinc getting to the penny connected to the copper wire soldered to the zinc strip? Can zinc atoms travel over the wire? Or do they travel through the solution? Do you think they travel from the zinc to the penny as zinc atoms or zinc ions?

(c) If the zinc gets to the copper penny by traveling through solution as zinc ions, where would the zinc ions come from, since the 1 M NaOH solution that we started with did not contain any added zinc salts? Some possible reactions to consider that form zinc ions include

$$Zn + 2 H_2O + 2 OH^- \rightarrow H_2(g) + Zn(OH)_4^{2-}$$

$$2 Zn + O_2(g) + 2 H_2O + 4 OH^- \rightarrow 2 Zn(OH)_4^{2-}$$

NOTE Zinc is known to be amphoteric in the sense that $Zn(OH)_2$ is soluble in a solution containing excess hydroxide ion. When a drop of hydroxide ion solution is added to $Zn(NO_3)_2$ solution, a white precipitate of $Zn(OH)_2(s)$ forms. If an excess of hydroxide is added, the precipitate dissolves to give a colorless solution containing the tetrahydroxyzincate complex ion, $Zn(OH)_4^{2-}$.

The first reaction (the reduction of water by zinc in basic solution) would produce hydrogen gas. Did you notice any evidence of evolution of gas that might indicate that hydrogen is being formed? Could the reaction be so slow that the concentration of hydrogen never builds up to the point where bubbles of hydrogen form? To increase the rate of reaction by increasing the surface area of zinc exposed to the 1 M NaOH, try adding powdered zinc to 1 mL of 1 M NaOH in a test tube. Do you see gas evolution?

Is there any way to tell if oxygen is involved in the oxidation of zinc metal? Could both reactions be going on simultaneously?

(d) If the zinc coating on the copper penny is produced by reduction of zinc ion on the surface of the penny, here are two possibilities to consider:

(i) Direct reduction of zinc ion by copper metal:

$$Zn(OH)_4^{2-} + Cu(s) \rightarrow Zn(s) + Cu^{2+}(aq)$$

(ii) Reduction of zinc ion by electrons traveling over the wire. These electrons are produced when zinc atoms are oxidized at the surface of the zinc metal in contact with the solution.

Can copper metal on the surface of the penny directly reduce zinc ions in solution to form zinc atoms on the surface of the copper? If so, would you expect the penny soldered to a copper wire and immersed in solution—but not connected to zinc metal—to get covered with zinc? Did that penny turn a silver color?

A direct test: Immerse a clean shiny strip of copper metal in a solution of 1 M NaOH/0.01 M $Zn(NO_3)_2$ in a test tube. Heat the test tube in a beaker of hot water for 20 min. What happens?

(e) What do the measurements with the digital multimeter, operating as a voltmeter and as an ammeter, tell you about the possibility that zinc ion is reduced to zinc metal on the surface of the copper penny by a flow of electrons in the copper wire connecting the zinc metal to the copper penny? If you did the optional tests, did you measure a current? If so, which direction were electrons flowing? From zinc to copper or vice versa?

If this is the process by which zinc ion is reduced on the surface of the copper penny, how long would you expect it to continue? Putting the question another way, when do you think the reaction would stop? How thick would the zinc coating have to get before the surface of the

penny began to look (from the viewpoint of the solution) like a piece of zinc metal? Would you expect the process to stop after a single layer (a monolayer) of zinc atoms coated the surface? Or perhaps a thickness of several monolayers is required for the surface of the zinc-coated penny to behave like a zinc metal surface.

(f) *Using Faraday's laws.* Estimate what average constant current would have to flow to produce a monolayer of zinc atoms on the surface of the copper penny in a period of 30 min. Make the following assumptions.

Assume that the surface area of a penny is 6.6 cm^2 and that the diameter of a zinc atom is 2.5×10^{-8} cm. Assuming that a monolayer has a thickness equal to the diameter of a zinc atom, calculate the volume of a monolayer. Calculate the volume of a zinc atom assuming that it is like a little cube of edge length 2.5×10^{-8} cm. (Zinc atoms are presumably spherical, and therefore could be more closely packed as spheres than as cubes, but this is just an approximate, back-of-the-envelope calculation.)

The ratio *volume of a monolayer/volume of a zinc atom* gives the number of zinc atoms required to form a monolayer, assuming they are packed together like little cubes on the surface of the penny. From the number of atoms, you can calculate the number of moles of zinc. From the moles of zinc, you can calculate the moles of electrons (two moles of electrons must be transferred per mole of Zn^{2+} reduced). From the moles of electrons, you can calculate the total charge transferred in coulombs. The charge in coulombs divided by the total time in seconds gives the average current.

We estimated an average current of about 2 microamperes (μA), a current that is measurable with most digital multimeters; the initial current would likely be greater than the average current. If more than a single monolayer were deposited, the average current would be even larger.

(g) What happens when you heat the zinc-coated pennies on the hot plate? Heating could allow the zinc and copper atoms to slip between one another, a process called diffusion. This would form a new phase, a mixture of copper and zinc atoms. Brass is an alloy of copper and zinc, so brass could be formed when the copper and zinc atoms interdiffuse. Would this account for the change in color? What is the color of brass? Unscrupulous alchemists apparently employed this trick in the Middle Ages to fool people into thinking they could make gold from base metals.

Bibliography

Bonneau, M. C. "Gold Pennies," *J. Chem. Educ.* **1995,** *72,* 389–390.

Karpenko, V. "Transmutation: The Roots of the Dream," *J. Chem. Educ.* **1995,** *72,* 383–385.

Shakhashiri, B. Z. "Copper to Silver to Gold," *Chemical Demonstrations,* Vol. 4, The University of Wisconsin Press, 1992, Demonstration 11.33, pages 263–268.

Szczepankiewicz, S. H.; Bieron, J. F.; Kozik, M. "The Golden Penny Demonstration: An Explanation of the Old Experiment and the Rational Design of the New and Simpler Demonstration," *J. Chem. Educ.* **1995,** *72,* 386–388. (*Note:* In our hands the new recipe does not work as well as the older one, which we use in the procedure.)

VanderZee, C.; Mosher, M. "Helping Students to Develop an Hypothesis about Electrochemistry," *J. Chem. Educ.* **1992,** *69,* 924–925. Corrosion and anodic protection.

Electrochemical Puzzles

Metal Corrosion and Anodic Protection
The Golden Penny Experiment

Name _____

Date _____ Section _____

Locker _____ Instructor_____

1. Metal corrosion and anodic protection

(a) In the space below, make annotated sketches of what you observed in the three Petri dishes containing bimetal strips in agar gel.

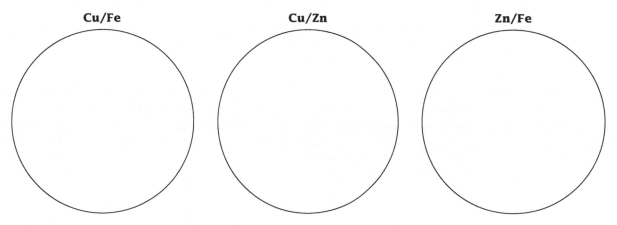

Cu/Fe   Cu/Zn   Zn/Fe

(b) Summarize your observations on the three Petri dishes in the table below:

	Phenolphthalein test: pink color is an indication of OH⁻	$K_3Fe(CN)_6$ test: blue color is an indication of Fe^{2+}	Describe the formation of any bands of precipitate	Other observations
Cu/Fe				
Cu/Zn				
Zn/Fe				

(c) If you used a digital voltmeter to probe the voltages in the three Petri dishes, answer the following questions by filling in the table.

When the probes are inserted 1–2 mm from the metal strips, at about the center of the metal strips, do you measure a voltage difference? If so, what is the magnitude of the voltage and next to which metal is there a region of positive charge? What is the magnitude of the voltage when the probes are firmly contacting the metal strips?

	Magnitude of the voltage difference next to the metals, millivolts	The region of positive charge is next to which metal?	Magnitude of the voltage difference when the probes are contacting the metals
Cu/Fe			
Cu/Zn			
Zn/Fe			

(d) Ions migrate in an electric field. So if positive ions are formed at the surface of one metal and negative ions at the surface of the other metal, the ions of opposite charge will migrate toward one another under the influence of the electric field created by the charge separation. How might this explain the formation of bands of precipitate? Describe the color of any bands of precipitate that form. What would be plausible chemical formulas for the precipitates you observed?

(e) Did the iron nail connected to the zinc metal strip corrode? What's your hypothesis (in terms of what the atoms are doing) to explain how the iron nail is protected from corrosion by being connected to zinc metal?

(f) Did the iron nail connected to the copper strip corrode? What's your hypothesis (in terms of what the atoms are doing) to explain this result? Plumbers use a dielectric union when connecting steel water pipes to a copper pipe plumbing system. The union contains an insulating sleeve to prevent a direct electrical contact between the steel and copper pipes. Based on what you observed in the Petri dish, what do you think would happen if you directly connected the steel and copper pipes?

2. The golden penny experiment

General observations

(a) What happens to pennies resting in contact with the 30-mesh zinc on the bottom of the beaker?

(b) What happens to pennies resting on the bottom of the beaker but not in contact with zinc?

(c) What happens to a copper penny soldered to copper wire and immersed in solution when the copper wire is not in contact with zinc metal?

(d) What happens to a penny soldered to copper wire, suspended in solution, when the other end of the copper wire is soldered to a strip of zinc metal which is in turn touching the 30-mesh zinc?

(e) Do you observe the evolution of any gas from the 30-mesh zinc in 1 M NaOH which might come from the reaction $Zn + 2 H_2O + 2 OH^- \rightarrow H_2(g) + Zn(OH)_4^{2-}$?

Does zinc powder react with 1 M NaOH to form a gas?

(f) Is there a voltage difference between a strip of zinc metal and the free end of a copper wire soldered to a penny when both the zinc strip and penny are immersed in solution?

If there is a voltage difference, which is the negative electrode (the electron source)? The copper wire soldered to the penny? Or the strip of zinc metal?

If there is a voltage difference, does it remain constant? Or does it change with time?

Constructing a hypothesis to explain the whole process of copper to silver to gold color.

(a) What do you think makes up the silver-colored coating on the penny? Does it behave in any ways similar to zinc metal?

(b) If we assume the silver-colored coating is zinc metal, how do you think the atoms of zinc get from the 30-mesh zinc to the copper penny? By diffusing from the zinc metal to the penny or traveling through the wire connecting the zinc metal to the penny? Or through the solution?

(c) If the zinc that coats the copper penny comes from the solution, is the zinc in solution more likely to be in the form of Zn atoms or Zn^{+2} ions? Explain your answer.

(d) If you analyzed the solution at the end of the experiment, would you expect to find zinc ions [or $Zn(OH)_4^{2-}$ complex ions] in solution? If so, by what chemical (or electrochemical) reaction could they have been formed?

(e) When the silver-plated pennies are heated on the hot plate, they turn a golden color, which is the color of brass, a copper–zinc alloy. Explain how heating could promote the formation of an alloy and cause a color change from silver to gold.

(f) Summarize the key observations and your hypothesis for explaining the whole process by which a copper penny acquires a shiny gold color.

The Chemistry of Some Nonmetals: Nitrogen, Sulfur, and Chlorine

Purpose

• Become acquainted with the chemistry of three nonmetals (nitrogen, sulfur, and chlorine), whose chemistry is representative of the elements of Groups 15, 16, and 17.

• Show that the nonmetals can exist in a variety of formal oxidation states in their compounds with oxygen and hydrogen.

• Understand how the chemistry of these compounds can be correlated with the electronic structures of the nonmetal elements and their positions in the periodic table.

Pre-Lab Preparation

The oxidation states of some elements and compounds of the nonmetals

The oxidation states of some representative Group 15, 16, and 17 elements are shown in Figure 38-1.

This chart illustrates three important points that we will discuss in the following paragraphs.

1. *Each of the elements in these groups can have many formal oxidation states.* This leads to a rich variety of compounds whose chemistry is complex in comparison with the chemistry of most metals in Groups 1 and 2.

2. *The electronic structures of the elements in the three groups place definite limits on the formal oxidation states accessible to each element.* Nitrogen, whose atoms have five valence electrons, can share these five electrons with electronegative elements like oxygen to produce a formal oxidation state as high as +5. It can accept a share of as many as three electrons from less electronegative elements like hydrogen to produce a formal oxidation state as low as −3. In both the +5 and −3 oxidation states, the nitrogen atoms have a share of eight valence electrons (like the electron configuration of the inert gases).

Because a neutral sulfur atom has six valence electrons, the formal oxidation state of sulfur ranges

Oxidation state	Group 15	Group 16	Group 17
		Oxidizing strength →	
+7			Cl_2O_7, $HClO_4$, ClO_4^-
+6		SO_3, H_2SO_4, HSO_4^-, SO_4^{2-}	Cl_2O_6
+5	$N_2O_5(g)$, HNO_3, NO_3^-		$HClO_3$, ClO_3^-
+4	$NO_2(g)$, $N_2O_4(g)$	SO_2, H_2SO_3, HSO_3^-, SO_3^{2-}	ClO_2
+3	$N_2O_3(g)$, HNO_2, NO_2^-		$HClO_2$, ClO_2^-
+2	$NO(g)$	$S_2O_3^{2-}$	
+1	$N_2O(g)$		Cl_2O, $HClO$, ClO^-
0	$N_2(g)$	S_8	Cl_2
−1	NH_2OH, NH_3OH^+		Cl^-
−2	N_2H_4, $N_2H_5^+$, $N_2H_6^{2+}$	H_2S, HS^-, S^{2-}	
−3	$NH_3(g)$, NH_4^+		

(left axis: Reduction / Oxidation)

FIGURE 38-1 | Compounds representing a wide variety of oxidation states of three nonmetals.

from +6 in sulfate ion to −2 in sulfide ion. The oxidation states of chlorine range from +7 to −1 because a neutral chlorine atom has seven valence electrons.

We must emphasize, however, that the oxidation states are assigned to atoms by formal rules. *The formal oxidation number is not the actual charge on the atom as it exists in the compound.* The actual charge on the atoms of real compounds is usually a fractional charge in the range of +1 to −1, depending on the electronegativities of the atoms. This is consistent with the electroneutrality principle, which reflects the fact that the distribution of electron density around the nuclei always adjusts itself so that large net charges on adjacent atoms are not formed.

As an example, the formal oxidation state we assign to the oxygen atoms in the perchlorate ion (ClO_4^-) is −2. The chlorine atom is therefore assigned oxidation state +7. The net charge is given by $4(-2) + 7 = -1$. The actual partial charges have been estimated by Sanderson (see the Bibliography) to be −0.211 on each oxygen atom and −0.155 on the chlorine atom. (Verify that these sum to a net charge of −1 for the perchlorate ion.)

The addition of one electron to a neutral chlorine atom produces stable chloride ion, having an electron configuration like that of argon. (Chloride ion and argon have the same number of electrons; therefore, we say they are *isoelectronic*.) In an ion like perchlorate, ClO_4^-, chlorine shares its electrons with the electronegative oxygen atoms, and we can draw Lewis (electron-dot) structures that give each atom a share of eight electrons.

The Lewis structures for some of the compounds mentioned above are shown in Figure 38-2.

3. *As the formal oxidation state of the nonmetal increases in a compound, the compound will tend to be a stronger oxidizing agent.* Figure 38-1 shows this trend. The compounds at the top of the diagram have the largest formal oxidation number assigned to the nonmetal element and are the strongest oxidizing agents. Thus nitric, sulfuric, and perchloric acids are all strong oxidizing agents, particularly if they are hot and concentrated.

There is also a tendency for the compounds with the most negative oxidation states to be good reducing agents. This is particularly true of NH_3 and hydrazine, N_2H_4, as well as sulfide ion, S^{2-}. However, chloride ion is not a good reducing agent, and in fact all of the chlorine compounds except chloride ion are strong oxidizing agents. This can be understood qualitatively as the effect of the increasing nuclear charge when we go from Group 15 to Group 17 elements. The increased nuclear charge makes Group 17 elements hold on to their electrons more tightly. Therefore, Group 17 compounds tend to be stronger oxidizing agents.

The chemistry of representative elements of Groups 15, 16, and 17

It would be impossible to explore the chemistry of all the elements in these three groups in the course of one laboratory session; therefore, we will choose one member of each group as representative of the chemistry of that particular group. First, we will study the chemistry of some of the compounds of nitrogen (a Group 15 element), some compounds of sulfur (Group 16), and of chlorine (Group 17).

Oxygen is so electronegative that it exists primarily in stable compounds in the formal oxidation state −2, with a few compounds (the peroxides) having oxidation state −1. For these reasons, we have chosen sulfur as the representative element of the Group 16 elements (chalcogens).

Similarly, we have chosen chlorine as the representative element of the Group 17 elements (halogens) rather than the first member of the series, fluorine, because fluorine is so electronegative that all stable fluorine compounds contain fluorine with a formal oxidation state of −1.

Nitrogen oxides and sulfur dioxide in the atmosphere

Nitrogen monoxide is produced in large amounts in steelmaking furnaces and by reaction of nitrogen with oxygen at high temperature in automobile engines. The nitrogen monoxide quickly reacts with oxygen to yield nitrogen dioxide, a brownish gas. Under the action of sunlight, nitrogen dioxide can be photolyzed to give NO + O. The free oxygen atom can then react with oxygen to give ozone, O_3. Ozone is a lung irritant and is also very toxic to many species of pine trees. It can also react with unburned hydrocarbons in the atmosphere to yield peroxyacetyl nitrate (PAN), which is very irritating to the eyes and also damages leafy crops such as spinach.

The combustion of coal containing sulfur produces large quantities of sulfur dioxide by the reaction

$$S + O_2 \rightarrow SO_2$$

In regions in which an appreciable amount of ozone has been formed by a photochemical reaction with

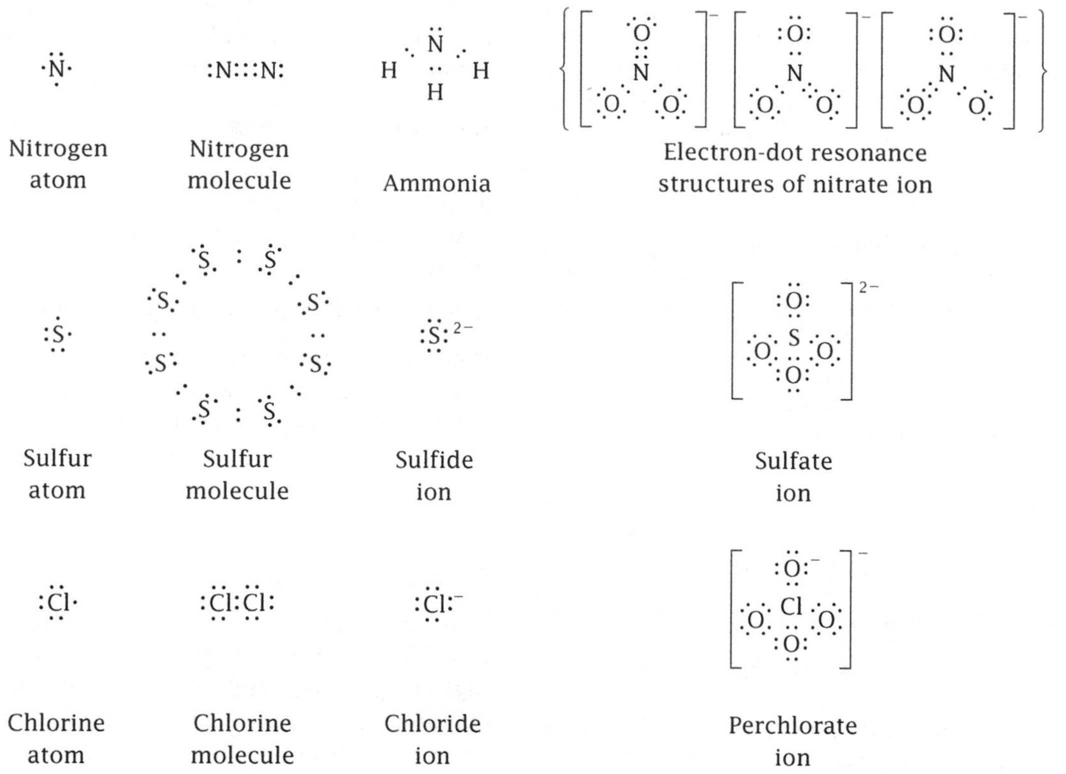

FIGURE 38-2 | **Electron-dot structures of three nonmetallic elements and some of their compounds.**

nitrogen dioxide, the ozone can further oxidize the SO_2 to sulfur trioxide, SO_3, by the reaction

$$SO_2 + O_3 \rightarrow SO_3 + O_2$$

Sulfur trioxide is the anhydride of sulfuric acid. Thus, it can combine with water to produce an aerosol of sulfuric acid mist, or, if ammonia is present (produced in large quantities in agricultural areas by livestock), ammonium sulfate particles can be produced. The haze that is associated with photochemical smog often contains a considerable amount of ammonium sulfate as well as organic matter.

Sulfur dioxide is used in agriculture to kill the wild yeasts that are normally present on grapes used in winemaking. The grapes are then inoculated with a pure strain of yeast that produces a higher quality of wine than the wild strains would. SO_2 is also used in bleaching dried fruit, such as apricots and raisins. One can demonstrate this bleaching action easily by taking a few petals of a red rose, boiling them in 30 mL of ethanol to extract the red pigment, acidifying the extract with a drop of 6 M HCl, and bubbling SO_2 through the solution. The solution will turn pale as

the sulfur dioxide reduces the red pigment to a colorless chemical form of the pigment.

Some Chemistry of Nitrogen Compounds

Nitrogen, the first element in Group 15, is an important nonmetal that forms compounds illustrating all the oxidation states from -3 to $+5$. In the zero oxidation state, nitrogen is particularly stable and has the electron-dot structure :N:::N: The commercially important compounds are those in the -3 and the $+5$ oxidation states. Ammonia and its salts are important as soluble fertilizers. Nitric acid is an important oxidizing agent and is used in making explosives.

The −3 oxidation state

The nitride ion, N^{3-}, forms with a few Group 1 or 2 metals. Once ignited, magnesium metal burns in air with a brilliant white light (dangerous to look at), forming an ash that is predominantly white MgO. However, the reaction temperature is so high that nitrogen also reacts with hot magnesium, producing a small amount of yellow magnesium nitride, Mg_3N_2. Nitrides react with water to form ammonia, NH_3, in which nitrogen has a formal oxidation state of -3 but the bonding is predominantly covalent:

$$Mg_3N_2(s) + 6\ H_2O(l) \rightarrow 2\ NH_3(aq) + 3\ Mg(OH)_2(s)$$

You are probably familiar with ammonia as a household cleaning solution. However, ammonia is actually a gas at room temperature. It is extremely soluble in water, and household ammonia is a solution of about 2 M in NH_3. The volatility of ammonia in this solution makes it easy to detect the characteristic odor of $NH_3(g)$. This odor serves to identify solutions that contain NH_3.

Ammonia is a weak base and forms ammonium ion, its conjugate acid, along with hydroxide ion when it is dissolved in water:

$$NH_3(aq) + H_2O(l) \rightarrow NH_4^+(aq) + OH^-(aq)$$

$$K_b = 1.8 \times 10^{-5}$$

You may still see old texts containing aqueous ammonia labeled ammonium hydroxide (NH_4OH). This formula stems from a notion that ammonia forms a molecule of NH_4OH by combining with a water molecule, and then this weak base, ammonium hydroxide, dissociates slightly to give NH_4^+ and OH^-. There is some controversy over whether the NH_4OH molecule has an independent existence. It is probably more accurate to think of an aqueous ammonia solution as containing mainly NH_3 molecules dissolved in water.

The addition of ammonia to aqueous solutions of certain metal ions causes precipitates to form. These precipitates are usually not nitrogen compounds but hydroxides, formed from the small amounts of hydroxide produced by reaction of ammonia with water. Naturally, these precipitates form only with metal ions that have very insoluble hydroxides ($K_{sp} < 10^{-11}$ for M^{2+} cations). Because of its lone pair of electrons,

$$\ddot{N} \atop H\ H\ H$$

ammonia can act as a Lewis base, forming donor–acceptor complexes with Lewis acids. An example of this behavior is the formation of ammine complexes of the transition metals such as

$$Ag^+(aq) + 2\ NH_3(aq) \rightleftharpoons Ag(NH_3)_2^+(aq)$$

and

$$Cu^{2+}(aq) + 4\ NH_3(aq) \rightleftharpoons Cu(NH_3)_4^{2+}(aq)$$

This copper complex, with its intense violet-blue color, is often used in the laboratory as an indicator for the presence of ammonia.

The -2 oxidation state

The -2 oxidation state of nitrogen is represented by hydrazine, N_2H_4. This compound can be produced by reacting ammonia with hypochlorite ion in basic solution:

$$2\ NH_3(aq) + ClO^-(aq) \rightarrow N_2H_4(aq) + H_2O(l) + Cl^-(aq)$$

However, hydrazine is very toxic, and this mixture also produces toxic and explosive chloramine compounds, such as $ClNH_2$. This is the reason for the statement on laundry bleach bottles (5% aqueous NaOCl) warning that bleach should not be mixed with ammonia solutions.

The oxides of nitrogen

None of the nitrogen oxides form spontaneously at room temperature and pressure. They can form at high temperatures and pressures, with some beneficial and some unpleasant consequences. Nitrogen oxides are formed in lightning discharges, and this serves as a natural form of "fixed" nitrogen fertilizer for plants as it is washed into the soil by rain. Nitrogen monoxide, NO (also called nitric oxide), forms at the high temperatures characteristic of a flame or an internal combustion engine. When this NO(g) reacts with atmospheric oxygen, it forms (as we will observe) nitrogen dioxide, NO_2. This reddish-brown gas is an irritating component of the smog found in many large cities.

We also want to examine an important nitrogen acid, nitric acid (HNO_3). Nitric acid is useful because it is a strong acid and because the nitrate ion does not form precipitates with metal ions, since nitrate salts are soluble. In addition, the nitrate ion in acidic solution is an excellent oxidizing agent. Because of this property, nitric acid can dissolve some metals that do not react with other acids such as HCl. Depending on the concentration of the nitric acid and the strength of the reducing agent, nitric acid can be reduced to $NO_2(g)$, NO(g), or all the way to the -3 oxidation state of ammonia.

Experimental Procedure
Nitrogen Compounds

NOTES TO INSTRUCTOR

The chemicals required for the study of each element (N, S, Cl) are listed separately under the heading for each element.

Special Supplies: pH paper or universal indicator solution or pH meter; No. 4 one-hole and two-hole rubber stoppers; 25 × 250 mm, 15 × 125 mm, 13 × 100 mm test tubes; 6-mm-OD glass tubing bent to the shapes shown in Figures 38-3 and 38-4; 3/16-in.-ID rubber tubing; wood splints.

Chemicals: 1 M ammonia, NH_3; 6 M ammonia, NH_3; 1 M ammonium chloride, NH_4Cl; 0.10 M iron(III) nitrate, $Fe(NO_3)_3$; 1 M sodium hydroxide, NaOH; 0.1 M copper(II) sulfate, $CuSO_4$; 0.1 M sodium chloride, NaCl; 0.1 M silver nitrate, $AgNO_3$; 6 M nitric acid, HNO_3; 16 M nitric acid, HNO_3; 3% hydrogen peroxide, H_2O_2; 3 M iron(III) chloride, $FeCl_3$; copper metal, Cu (turnings).

> **⚠ SAFETY PRECAUTIONS:**
> **Nitrogen oxides (NO and NO_2) will irritate eyes and lungs. Work in a fume hood when generating nitrogen monoxide, NO. This experiment should not be performed if a fume hood is not available.**
> **Nitric acid is very corrosive and can cause severe burns. Handle it carefully. If spills occur, wash with plenty of water. Wear eye protection at all times in the laboratory.**

> **☠ WASTE COLLECTION:**
> **A waste container should be provided for the silver compound formed in part 1c. Acid and base wastes should be collected separately or rinsed down the drain after neutralization with sodium bicarbonate, $NaHCO_3$.**

> **NOTE** In your written report of this experiment, include all observations of the properties of the products formed and write equations for all reactions.

1. Ammonia and Ammonium Ion: The −3 Oxidation State

(a) *Acid–Base Properties.* Determine the pH of 1.0 M NH_3 and 1.0 M NH_4Cl with pH paper, a few drops of universal indicator solution, or a pH meter.

(b) *Precipitation Reactions.* Place 1 mL of 0.1 M $Fe(NO_3)_3$ in each of three 13 × 100 mm test tubes. To the first tube, add 1 mL of 1 M NaOH. Note the precipitate, and describe its characteristics. To the second tube, add 1 mL of 6 M NH_3. To the third tube, add 1 mL of 1 M NH_4Cl. Note the reaction(s) (if any), and write net ionic equations to describe the reaction(s).

(c) *Ammonia as a Complexing Agent.* Put 1 mL of 0.1 M $CuSO_4$ in a test tube, and add 2 drops of 1 M NH_3. Describe the precipitate. Continue adding 1 M NH_3 with mixing (about 20 drops total). Observe the solution. Is it the same color as the original $CuSO_4$ solution? (See the Pre-Lab Preparation discussion of the −3 oxidation state for the formula of the product.) Put 1 mL of 0.1 M $CuSO_4$ in another test tube and add 1 mL (20 drops) of 1.0 M NH_4Cl. Do you observe any changes?

Prepare some solid silver chloride by adding 10 drops of 0.10 M NaCl to 10 drops of 0.10 M $AgNO_3$ in a test tube. Then add 3 mL of 1 M NH_3. Mix thoroughly with a glass stirring rod for 30 s. Does the precipitate dissolve? (Again, see the Pre-Lab Preparation for the formula of the product.) Dispose of the product in the silver waste container provided.

2. The Preparation of Nitrogen Monoxide by Reduction of HNO_3 with Copper: Oxidation States +2, +4, and +5
Skip this section if a fume hood is not available. If you have access to one, assemble the generator shown in Figure 38-3, which uses a 20 × 200 mm test tube, and place about 3 g of copper turnings in it. Prepare to collect three 15-cm test tubes of nitric oxide by displacement of water. Add 10 mL of 6 M HNO_3 to the generator, replace the delivery tube connection, and warm the test tube gently to initiate the reaction. After the air has been displaced from the apparatus and the gas bubbling through is colorless, collect two full test tubes of the gas and a third test tube about half full. Do not allow the delivery tube to remain under water while the heated test tube cools; if you do, water will be drawn into the test tube. What is the reaction for the reduction of dilute nitric acid by copper?

Test the nitric oxide in one test tube with a glowing splint to see if it supports combustion. Note the colored gas produced when the tube is exposed to the air. Write the equation for the reaction that accounts for this change.

Test the second sample of nitrogen monoxide for solubility in water by swirling the test tube with its mouth under the water to allow contact of fresh water with the gas. Note whether the water level in the test tube rises.

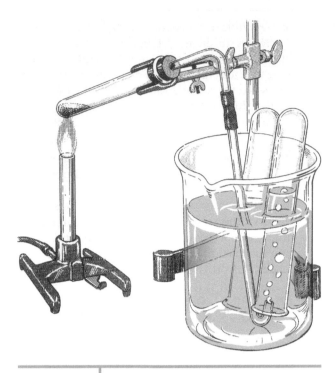

FIGURE 38-3 | **Apparatus for the preparation of nitrogen gas or of the oxides of nitrogen.**

Now take the test tube out of the water for a few seconds and allow the oxygen of the air to react with the gas, as will be evidenced by the formation of a brown gas. Invert the test tube under the water again, and swirl it to note the solubility of the brown gas. Write the equation for the reaction occurring when this gas dissolves in water.

Mark the water level in the third test tube, which is about half full of nitrogen monoxide, with a wax pencil or gummed label. Set up the small oxygen generator shown in Figure 38-4, placing about 10 mL of 3% hydrogen peroxide in the test tube. Draw up a few milliliters of 3 M $FeCl_3$ into the medicine dropper and replace the two-hole stopper in the mouth of the generator tube. When ready to begin generating oxygen, release a few drops of $FeCl_3$. After a brief period, the $FeCl_3$ will catalyze the decomposition of H_2O_2, producing oxygen gas. After the air has been displaced from the generator, place the delivery tube under the marked test tube and allow 8 to 10 bubbles of oxygen to enter. Note whether the level of the water is lowered by the addition of oxygen.

Recall that in the balanced equation for the reaction that is taking place, two volumes of nitrogen monoxide react with one volume of oxygen to produce two volumes of nitrogen dioxide. Now swirl the test tube with its mouth under the water and note what happens to the water level as the NO_2

reacts with the water. Allow more oxygen to bubble into the tube until the gases turn brown. Note the water level, and again allow the gases to dissolve in water. Repeat the process until the water level approaches the top of the tube. Remember that excess oxygen is not soluble in water. What substances are present in the water solution in the test tube? Apply a simple test to verify your answer. What part of one of the commercial processes involved in the production of nitric acid does this experiment illustrate?

Some Chemistry of Sulfur Compounds

In this experiment, we study some chemical properties of some common compounds of sulfur in oxidation states −2, +2, +4, and +6. Sulfur is a Group 16 element with the electron configuration $1s^2 2s^2 2p^6 3s^2 3p^4$. Its electron-dot structure is [∶S̈·]. By sharing two electrons with two hydrogen atoms, it forms the covalent compound H_2S. A saturated aqueous solution of H_2S is weakly acidic and contains small concentrations of H^+ and HS^- ions. In strongly basic solutions, the concentrations of HS^- and S^{2-} ions are greater.

By sharing its electrons with more electronegative elements, such as oxygen, sulfur attains positive

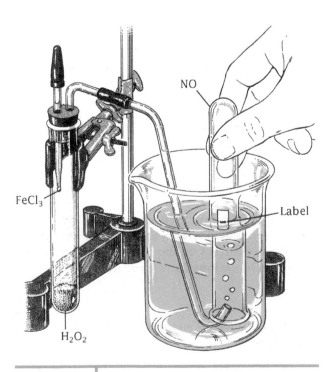

FIGURE 38-4 | **Generator for the preparation of a small amount of oxygen by the reaction of $FeCl_3$ with H_2O_2. This oxygen is then reacted with NO(g).**

oxidation states. In sulfur dioxide, four of the sulfur electrons are involved in bonding with oxygen:

$$:\ddot{O}: \quad \ddot{O}:\ddot{S}: \;+\; H:\ddot{O}: \;\rightleftharpoons\; H:\ddot{O}:\ddot{S}:\ddot{O}:H$$
$$\overset{}{\underset{H}{}} \qquad :\ddot{O}:$$

SO_2 is formally the anhydride of H_2SO_3, a weak acid called sulfurous acid, but there is no evidence for the existence of H_2SO_3 in solution although it is known to exist in the gas phase. So the acid ionization of "sulfurous acid" is best represented as:

$$SO_2 + 2\,H_2O \rightleftharpoons H_3O^+ + HSO_3^-$$

$$HSO_3^- + H_2O \rightleftharpoons H_3O^+ + SO_3^{2-}$$

In basic solution, the equilibria are shifted to the right, forming more hydrogensulfite and sulfite ions.

In sulfur trioxide, all six of the sulfur electrons are involved in bonding:

$$:\ddot{O}: \;\; :\ddot{O}. \qquad\qquad \overset{H}{\underset{}{}}$$
$$\underset{:\ddot{O}:}{\overset{}{S}} \;+\; H:\ddot{O}: \;\rightleftharpoons\; H:\ddot{O}:\ddot{S}:\ddot{O}:$$
$$\qquad\qquad H \qquad\qquad :\ddot{O}:$$

Sulfur trioxide is the anhydride of sulfuric acid, H_2SO_4, one of the most important industrial inorganic chemicals. It is a strong acid whose aqueous solutions contain large concentrations of H^+ and HSO_4^- ions. In basic solutions, the SO_4^{2-} ion is the predominant species.

The electron-dot structures of sulfite, sulfate, and thiosulfate ions are represented as

$$:\ddot{O}:\ddot{S}:\ddot{O}:^{2-} \qquad \underset{:\ddot{O}:}{\overset{:\ddot{O}:^{2-}}{\ddot{S}}} \qquad \underset{:\ddot{O}:}{\overset{:\ddot{S}:^{2-}}{\ddot{S}}}$$
$$:\ddot{O}:$$

Sulfite   Sulfate   Thiosulfate

Note that in the thiosulfate ion, the sulfur atom that replaces the oxygen in the sulfate structure may be assigned a -2 oxidation number and that the central sulfur atom has an oxidation number of $+6$, just as it has in sulfate. The $+2$ oxidation number assigned to sulfur in thiosulfate is obtained by finding the average of $+6$ and -2: $(+6 - 2)/2 = +2$.

The experiments you perform will show that H_2SO_4 and SO_4^{2-} are mild oxidizing agents, whereas H_2S, S^{2-}, H_2SO_3, and SO_3^{2-} are reducing agents. In addition, since SO_2, SO_3^{2-}, and H_2SO_3 represent intermediate oxidation states, they can act as oxidizing agents with a strong reducing agent and as reducing agents with a strong oxidizing agent.

Experimental Procedure
Sulfur Compounds

Special Supplies: Source of H_2S gas.[1]

Chemicals: Crushed roll sulfur, S; iron, Fe (filings); 0.1 M lead(II) nitrate, $Pb(NO_3)_2$; 0.1 M tin(IV) chloride, $SnCl_4$; 0.1 M zinc nitrate, $Zn(NO_3)_2$; 0.05 M iodine, I_2 (0.05 M I_2 in 0.1 M KI); 0.1 M calcium nitrate, $Ca(NO_3)_2$; 0.1 M barium chloride, $BaCl_2$; 0.1 M barium hydroxide, $Ba(OH)_2$; solid sodium sulfite, Na_2SO_3; 0.1 M sodium sulfate, Na_2SO_4; saturated bromine water, Br_2; sugar (sucrose) crystals; pieces of zinc, Zn, and copper, Cu; sodium chloride, NaCl(s); potassium bromide, KBr(s); lead acetate test paper, $Pb(C_2H_3O_2)_2$; source of hydrogen sulfide, H_2S gas; 6 M hydrochloric acid, HCl; concentrated sulfuric acid, H_2SO_4; 3 M sulfuric acid, H_2SO_4.

SAFETY PRECAUTIONS:
This experiment should not be performed if a fume hood is not available. Hydrogen sulfide has a noxious odor and is very toxic. Part 1 should be carried out in a fume hood.

In part 3, the addition of concentrated (18 M) sulfuric acid to NaCl, KBr, and KI should be carried out in a fume hood because the products formed (HCl, HBr, and HI) will irritate eyes and lungs.

When you are asked to note the odor of a substance, do so very cautiously. Do not stick your nose into the mouth of a test tube and inhale. Gently fan the vapor toward your nose with your hand.

WASTE COLLECTION:
A waste container should be provided for the lead sulfide, lead acetate paper, and lead sulfate compounds in parts 1b, 1c, and 3c. A separate container should be available for the barium compounds in parts 2b and 3c.

Note to Students: In your written report, include all observations and equations for all reactions.

---

[1]Consult your instructor about your best source of H_2S gas. A cylinder of the compressed gas (in a fume hood) or a Kipp generator charged with FeS and 6 M HCl (in a fume hood) is often used. A little "Aitch-tu-ess" (a commercial mixture of sulfur, paraffin, and asbestos) heated in a test tube that is fitted with a gas delivery tube is a convenient method.

1. Sulfides and Hydrogen Sulfide: Oxidation State −2

(a) *Preparation of a Sulfide.* Mix approximately 3.5 g of iron filings with 2 g of crushed sulfur in a crucible supported on a triangle. Place a lid on the crucible and heat with a Bunsen burner until the reaction is initiated, removing the burner and lid occasionally to note whether the reaction continues with the evolution of heat. Burn off any excess sulfur and allow the crucible to cool. Place a small piece of the compound in a small test tube. Add a few milliliters of 6 M HCl and note cautiously the products of the reaction.

(b) *Hydrogen Sulfide as a Precipitating Agent.* Metallic sulfides, other than those of the alkali and alkaline earth metals, are sparingly soluble in water. In qualitative analysis, many metal ions are identified by precipitating them as sulfides. Using the source of H_2S gas available in your laboratory (see footnote 1), saturate 3 mL of each of the following solutions with H_2S gas: 0.1 M $Pb(NO_3)_2$, 0.1 M $Zn(NO_3)_2$, and 0.1 M $SnCl_4$. Record the color and formula of each precipitate. Dispose of the lead sulfide in the container provided.

(c) *Hydrogen Sulfide as a Reducing Agent.* Saturate each of the following solutions with H_2S: (1) 5 mL of warm 3 M HNO_3, (2) 5 mL of 0.05 M I_2, and (3) a freshly prepared solution of H_2SO_3 made by adding a few crystals of Na_2SO_3 and a drop of 6 M H_2SO_4 to 5 mL of water. Test the vapors with moistened lead acetate paper. Note the products and write balanced equations for each reaction.

2. Sulfur Dioxide, Sulfurous Acid, Sulfite Ion: Oxidation State +4

(a) *Preparation of Sulfur Dioxide.* Since sulfur is in an intermediate oxidation state in SO_2, this compound can be prepared by oxidation of sulfides (as is done in metallurgical roasting), by oxidation of elemental sulfur, or by the reduction of hot concentrated H_2SO_4 by copper. It can also be conveniently prepared (without a change in oxidation state) from metal sulfites by the addition of an acid.

Place about 2 g of Na_2SO_3 in a large test tube. Add 6 M HCl drop by drop until you can smell (**Caution!**) the odor of the gas (SO_2) given off. Now add about 15 mL of water and stir until all the solid dissolves. Save for part 2b.

(b) *Chemical Properties of Sulfurous Acid.* Divide the solution equally among three smaller test tubes. To one portion, add a few milliliters of 0.1 M $Ba(OH)_2$ until the solution is basic to litmus. What is the precipitate formed? Is it soluble in 6 M HCl added a drop at a time? (Ignore a slight turbidity, which is due to air oxidation of sulfite to sulfate.) Discard the solution in the barium waste container. To another portion, add 5 to 6 mL of saturated bromine water, drop by drop. How do you account for the decolorization that takes place? Now add a few milliliters of 0.1 M $Ba(OH)_2$ to this test tube. What is the precipitate formed? Is it soluble in 6 M HCl? Dispose of the solid in the barium waste container.

Add a little of the third portion, a few drops at a time, to a test tube containing 5 mL of 3 M H_2SO_4 and some mossy zinc. Observe the odor of the gas, and note the precipitate formed.

3. Sulfuric Acid, Sulfates: Oxidation State +6

(a) *Physical and Chemical Properties of Sulfuric Acid.* While stirring, cautiously add a few mL of 18 M H_2SO_4, drop by drop, to 50 mL of tap water in a small beaker. Note the temperature change. To what do you attribute this result?

Place a few drops of 18 M H_2SO_4 on a few crystals of sugar in a small evaporating dish. Repeat the test on a small piece of paper or wood (such as a matchstick). How do you explain the results?

(b) *Effect of Concentrated Sulfuric Acid.* Investigate the oxidizing strength of concentrated (18 M) sulfuric acid by adding in a fume hood 1 mL of concentrated H_2SO_4 to 1 g of each of the following salts in 20 × 150 mm test tubes: NaCl, KBr, and KI. Note the color of any gases evolved. (Place the tube containing KI in a beaker of hot water.) Test for acidity of the gases evolved by holding a piece of moistened blue litmus paper near the mouth of each test tube. Are any halide ions oxidized to the elementary halogens?

(c) *Solubility of Sulfates.* Add 3 mL of 0.1 M Na_2SO_4 to 3 mL each of the following solutions in separate 10-cm test tubes: 0.1 M $Ca(NO_3)_2$, 0.1 M $BaCl_2$, 0.1 M $Pb(NO_3)_2$. Test the solubility of any precipitates in dilute nitric acid by adding 1 mL of 6 M HNO_3 to each precipitate. Dispose of the lead and barium compounds in their respective waste containers.

Some Chemistry of Chlorine Compounds

Figure 38-5 illustrates the variety of compounds and oxidation states that exist for chlorine, which is a prototype of the halogens.

This chart of the oxidation states of chlorine emphasizes that this element, which in varying

		+7	(Cl_2O_7), $HClO_4$, ClO_4^-	Cl_2O_7 is unstable; $HClO_4$ is a strong oxidizing agent. Reduced to Cl^-
		+5	$HClO_3$, ClO_3^-	Strong oxidizing agent. Reduced to Cl^-
		+4	ClO_2	Unstable, explosive
O	R	+3	$HClO_2$, ClO_2^-	Good oxidizing agent. Reduced to Cl^-
		+1	Cl_2O, $HClO$, ClO^-	Good oxidizing agent. Reduced to Cl^-
		0	Cl_2	Good oxidizing agent. Reduced to Cl^-
		−1	Cl^-	

FIGURE 38-5 | **Chlorine compounds.**

degrees is typical of all the Group 17 halogen elements, is *strongly electronegative*. Note that *in all oxidation states except the lowest very stable −1 state, chlorine forms compounds that are strong oxidizing agents.* They have a strong affinity for additional electrons. This tendency is extremely strong for fluorine, which exists only in the oxidation states −1 and 0. The other halogens, in order of their increasing size and atomic mass, are somewhat less electronegative.

All the halogen elements have the same Lewis structures as chlorine. Note the following structures for a chlorine atom, a chlorine molecule, Cl_2, and a chloride ion, Cl^-.

$$:\!\overset{..}{\underset{..}{Cl}}\!\cdot \qquad :\!\overset{..}{\underset{..}{Cl}}\!:\!\overset{..}{\underset{..}{Cl}}\!: \qquad :\!\overset{..}{\underset{..}{Cl}}\!:^-$$

Chlorine atom	Chlorine molecule	Chloride ion

In each halogen, the free element can add one more electron to fill its valence orbitals, thus giving the ion the electron structure of the inert noble gas next in order in the periodic table.

Chlorine and its compounds show a marked tendency to undergo self- or auto-oxidation reduction, in which some molecules or ions of a species are oxidized to a higher state while others are reduced to the stable −1 state. This process is called *disproportionation*.

It is possible to oxidize Cl^- (oxidation state −1) to free chlorine, $Cl_2(g)$ (oxidation state 0), then carry out a series of disproportionation reactions in which the chlorine is successively oxidized to the +1, +5, and finally the +7 oxidation states, as indicated on the flow chart in Figure 38-6. This could be done as follows:

1. $Cl_2(g)$ is passed into a cold basic solution, in which it is auto-oxidized to ClO^- and auto-reduced to Cl^-. (This is the reaction that occurs in the commercial preparation of 5% $NaClO$ bleaching solution, the bleach sold in grocery markets.)

2. When heated, the ClO^- in basic solution is further oxidized to ClO_3^-, and a portion is reduced back to Cl^-. [Suitable crystallization of the salts from this solution produces the commercially important oxidant $KClO_3$ or the weed killers $NaClO_3$ and $Ca(ClO_3)_2$.]

3. Maintaining $KClO_3$ crystals at a temperature just above their melting point results in further auto-oxidation of the ClO_3^- to ClO_4^- and reduces a portion of it to Cl^-. (Perchlorates are important oxidants in solid rocket fuels.)

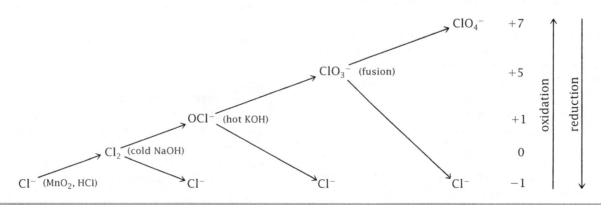

FIGURE 38-6 | **The interconversions of some ions containing chlorine in several different oxidation states.**

The chloride ion and its compounds: The −1 oxidation state

The −1 oxidation state of the halogens, known collectively as the halides, is by far the most common form of the halogens. We are familiar with chloride ions in the form of salt, both solid NaCl and salt solutions such as seawater and blood. The salt in seawater is predominantly sodium chloride, but bromide and iodide ions are also present in smaller amounts.

Hydrochloric acid, HCl, is one of the common laboratory acids. Hydrochloric acid solutions are formed by dissolving gaseous hydrogen chloride molecules, HCl(g), in water. These acid solutions have powerful, choking odors when they are concentrated due to the volatilization of the gaseous hydrogen halides from these solutions.

Each of the halide ions forms precipitates with Ag^+, Pb^{2+}, and Hg_2^{2+} ions. The solids formed in these reactions are useful in the laboratory for identifying these metal ions in solution or for identifying these halide ions in solution. The silver halide solids serve as the basis for photographic materials because of their sensitivity to light. (Light activates the decomposition of these solids into the halogen element and silver metal.)

$$2 \text{ AgBr(s)} \xrightarrow[\text{developer}]{\text{light}} 2 \text{ Ag(s)} + Br_2(l)$$

The halide ions can also serve as ligands in forming complex ions with metal ions. While the halide ions are not the best ligands, we will observe the formation of some complexes by noting color changes and the dissolution of a solid.

Elemental chlorine: The 0 oxidation state

As elements, the halogens are identifiable by their color and state: Cl_2 is a pale, greenish-yellow gas at room temperature; Br_2 is a dark, reddish-brown fuming liquid; and iodine is a purple-black solid that gives off a violet vapor. Because these molecules are nonpolar, they are not highly soluble in water—but they are soluble in nonpolar solvents such as hexane, C_6H_{14}. The distinguishing characteristics of the halogens is their oxidizing ability, more commonly recognized as their bleaching action.

The halogens are not such strong oxidizing agents that they can oxidize water to $O_2(g)$, but solutions of the halogens are unstable because of their tendency to disproportionate. For example,

$$Cl_2(aq) + H_2O(l) \rightarrow Cl^-(aq) + HClO(aq) + H^+(aq)$$

Solutions of chlorine water must be freshly made or they will not contain much chlorine in the form Cl_2.

Hypochlorite and hypochlorous acid: The +1 oxidation state

The solutions that we call chlorine bleach do not contain Cl_2 but are actually solutions of sodium hypochlorite, NaClO, prepared by dissolving chlorine gas into sodium hydroxide solutions:

$$Cl_2(aq) + 2 \text{ NaOH(aq)} \rightarrow$$
$$NaClO(aq) + NaCl(aq) + H_2O(l)$$

The hypochlorite ion itself is a good oxidizing agent, so these solutions possess the same desired bleaching action.

Although it is similar to chlorine in this oxidizing power, the hypochlorite ion is very different from the chloride ion. Because hypochlorous acid, HClO, is a weak acid, the hypochlorite ion is a base of moderate strength. This is in contrast to the neutral character of the chloride ion, the conjugate base of the strong acid, HCl.

Chlorate and perchlorate: Higher oxidation states

If basic solutions containing ClO^- are heated, the hypochlorite disproportionates to give chlorate and chloride ions.

$$3 \text{ ClO}^-(aq) \rightarrow ClO_3^-(aq) + 2 \text{ Cl}^-(aq)$$

Chlorate can be further oxidized to the perchlorate ion, ClO_4^-. We will not be exploring the chemistry of these higher oxidation states because of their instability and hazardous (sometimes explosive) nature.

Experimental Procedure
Chlorine Compounds

Chemicals: 0.1 M sodium chloride, NaCl; 0.1 M silver nitrate, $AgNO_3$; 6 M ammonia, NH_3; 6 M nitric acid, HNO_3; 0.1 M copper(II) sulfate, $CuSO_4$; 12 M hydrochloric acid, HCl; sodium hypochlorite, NaOCl solution (a commercial bleach solution containing 5% NaOCl); red and blue litmus paper; 6 M sodium hydroxide, NaOH; 0.1 M potassium iodide, KI; hexane, C_6H_{14} (or petroleum ether, b.p. 60–90 °C); 0.1 M potassium bromide, KBr.

⚠ SAFETY PRECAUTIONS:
Observe the usual precautions, especially eye protection, with the strong acids and bases used in this section as well as the bleach solution. Do not mix NH_3 and NaOCl solutions.

⚠ **WASTE COLLECTION:**
A waste container should be provided for the silver compound formed in part 2b. A container for the hexane-containing mixtures should be made available, in the hood if possible.

1. The Chloride Ion: Oxidation State −1

(a) *Solubility and Stability of Chloride Salts.* To 1 mL of 0.1 M NaCl, add 1 mL of 0.1 M $AgNO_3$. Mix the test tube well. Add an equal volume of 6 M NH_3 (about 3 M in the mixture) to the test tube. Agitate the test tube thoroughly to break up any lumps of precipitate. Finally, add a slight excess of 6 M HNO_3, and mix. Explain the results on your Report form.

(b) *Transition Metal Complexes with Chloride.* To 2 mL of 0.1 M $CuSO_4$, add 2 mL of 12 M (concentrated) HCl; dilute this with about 5 mL of water. Write equations and interpret the color changes you observed. Remember that the copper sulfate solution actually contains $[Cu(H_2O)_6]^{2+}$ ions. Assume that the new complex formed is $CuCl_4^{2-}$.

To 1 mL of 0.1 M $AgNO_3$, add 3 mL of 12 M (concentrated) HCl; agitate this well for several minutes to redissolve the precipitated AgCl. Now dilute this with about 5 mL of distilled water. Write equations for the reactions and interpret the changes you observed, assuming that the complex initially formed was $AgCl_2^-$.

2. The Hypochlorite Ion
Since solid NaOCl cannot be isolated easily without decomposition, we will test portions of the solution of a commercial bleach obtained at the grocery store. It was prepared by passing chlorine into a solution of NaOH.

(a) *Litmus Reaction.* Put several drops of NaOCl solution on red and blue litmus to note its acidity or basicity. Note any bleaching effect.

(b) *Reacting to $AgNO_3$.* To a 3-mL portion of the NaOCl solution, add 1 mL of 0.1 M $AgNO_3$. What is the precipitate? (Compare with the behavior of a drop of 6 M NaOH on 0.1 M $AgNO_3$.) Is it soluble in 6 M HNO_3, and does any other precipitate remain?

Explain your observations. Dispose of the silver compounds in the silver waste container.

(c) *Oxidizing Strength.* Place 2 mL of 0.1 M KI and 1 mL of hexane (or petroleum ether) in a test tube. (**Caution:** Work in a fume hood. Hexane is flammable. There must be no open flames nearby.) Add 5% NaOCl solution one drop at a time, shaking the test tube after each drop, and note any color change in the hexane layer. Is there any evidence for formation of I_2? An excess of NaOCl must be avoided because it will remove the color, owing to further oxidation of the initial product to the colorless IO_3^- ion. Dispose of the hexane mixture in the proper waste container.

Repeat this test, using 2 mL of 0.1 M KBr in place of KI. Is there any evidence for the formation of Br_2 detected by a color change in the hexane layer? Where would you place the ClO^- ion with respect to Br_2 and I_2 in a scale of oxidizing strength? Now acidify the test solution with 6 M HCl and shake, noting any color formed in the hexane layer. Does the oxidizing strength of ClO^- change when the solution is acidified? Properly dispose of the hexane mixtures.

Record all of your observations and write equations to explain the reactions.

■ CONSIDER THIS ■

From thermodynamic tables find the enthalpies of formation of the nitrogen oxides listed in Figure 38-1. Using Le Châtelier's principle, determine whether these oxides are likely to form at high or low temperatures and at high or low pressures. Use this knowledge to discuss the formation of nitrogen oxides in an internal combustion engine and how you would design a catalytic converter to eliminate them from automobile exhaust.

Are the same issues involved in the formation of sulfur oxides if the gasoline (or coal) contains sulfur as it is burned?

Bibliography

Sanderson, R. T. *Polar Covalence*, Academic Press, New York, 1983, p. 194.

The Chemistry of Some Nonmetals: Nitrogen, Sulfur, and Chlorine

Name _____

Date _____ Section _____

Locker _____ Instructor_____

Data, Observations, and Conclusions: Nitrogen Compounds (Group 15)

1. Ammonia and ammonium ion: The −3 oxidation state

(a) What is the pH of 1 M NH_3? _____

What is the pH of 1 M NH_4Cl? _____

Write net ionic equations to describe the reactions that cause these solutions to be acidic or basic.

$NH_3 + H_2O \rightarrow$ _____

$NH_4^+ + H_2O \rightarrow$ _____

(b) *Precipitation reactions*

What is the precipitate formed when $Fe(NO_3)_3$ and NaOH solutions are mixed?

Does the precipitate formed from $Fe(NO_3)_3$ and NH_3 appear to be the same as the previous one?

Write net ionic equations for the reactions of Fe^{3+} with OH^- and with NH_3 in water.

$Fe^{3+} + OH^- \rightarrow$ _____

$Fe^{3+} + NH_3 + H_2O \rightarrow$ _____

(c) *Ammonia as a complexing agent*

What precipitate forms when a small amount of NH_3(aq) is added to $CuSO_4$(aq)?

Net ionic equation _____

What evidence do you have that a reaction has occurred when excess NH_3 is added?

What is the formula of the copper-containing product? (See the Pre-Lab Preparation on the −3 oxidation state of nitrogen.)

Why doesn't a precipitate form when NH_4Cl is added to Cu^{2+}(aq)?

Describe what happens when you add an excess of 1 M NH_3 to a precipitate of AgCl.

Net ionic equation _____

2. The preparation of *nitrogen monoxide* by reduction of HNO_3 with copper: Oxidation states +2, +4, and +5

Give the electron-dot structures proposed for NO.

Why is nitrogen monoxide called an *odd* molecule?

Write the equation for the reaction of dilute nitric acid with copper.

_____

List the properties of NO: color _____ ; ability to support combustion _____

What is the colored gas formed when NO is exposed to air? _____

Write the equation for this reaction.

_____

To what extent is NO soluble in water? _____

Is NO_2 soluble in water? _____

How do you account for this? _____

Write the equation for the reaction of NO_2 with H_2O assuming the products are HNO_3 and HNO_2.

_____

Write the equation for the production of oxygen from H_2O_2.

_____

Describe the changes taking place in the test tube half full of NO gas as oxygen is added.

    Color of gases _____

    Change in water level _____

Explain these observations by writing chemical equations for the reactions that are most likely to be taking place.

_____

When the test tube is swirled underwater, the water level becomes _____ and the color of the gases becomes _____ . Account for these observations.

What results were obtained when the swirling process was repeated after exposure to the air?

What substances are present in the aqueous solution in the test tube at the end of the process?

_____

Data, Observations, and Conclusions: Sulfur Compounds (Group 16)

1. Sulfides and hydrogen sulfide: Oxidation state −2

    (a) Write the equation for the preparation of a sulfide.

_____

    Is the reaction endothermic or exothermic? _____

Write the reaction of ferrous sulfide with hydrochloric acid.

_____

(b) *Hydrogen sulfide as a precipitating agent*

Note the formula and color of the insoluble sulfides prepared.

Pb^{2+} _____   _____

Zn^{2+} _____   _____

Sn^{4+} _____   _____

(c) Write the equations for oxidation of H_2S by

3 M HNO_3 _____

0.05 M I_2 _____

H_2SO_3 _____

2. **Sulfur dioxide, sulfurous acid, sulfite ion: Oxidation state +4**

   (a) Write the equation for the preparation of SO_2.

   _____

   (b) Write the equation for reactions of H_2SO_3 (or SO_2 + H_2O) with

   (i) $Ba(OH)_2$

   Solubility of precipitate in dilute HCl

   _____

   (ii) Saturated bromine water (Br_2)

   Reaction of product with $Ba(OH)_2$

   _____

   Solubility of precipitate in dilute HCl

   _____

   (iii) Zn and H_2SO_4

   _____

3. **Sulfuric acid, sulfates: Oxidation state +6**

   (a) *Physical and chemical properties of sulfuric acid*

   Temperature effect on dilution _____

   Explanation

   Effect on sugar _____   on paper _____

   Explanation

(b) *Effect of concentrated sulfuric acid*

Note what happens when concentrated sulfuric acid is added to the following.

Substance	Observations	Equations
NaCl		
KBr		
KI		

(c) *Solubility of sulfates*

Formulate a solubility rule for sulfates of the common metals.

Data, Observations, and Conclusions: Chlorine Compounds (Group 17)

1. The chloride ion: Oxidation state −1

(a) *Solubility and stability of chloride salts*

Write net ionic equations to describe the reactions that you observed.

    (i) $NaCl(aq)$ was mixed with $AgNO_3(aq)$.

_____

    (ii) 6 M NH_3 was added to the precipitate that formed in (i).

_____

    (iii) 6 M HNO_3 was added to the solution from (ii).

_____

(b) *Transition metal complexes with chloride*

What evidence do you have that a reaction occurred when 12 M HCl was added to the $CuSO_4$ solution?

Did another reaction occur when the mixture was diluted with water?

Write a chemical equation to describe the processes that you observed in part b.

_____

Write the equation for the reaction that occurred when the concentrated HCl was added to the $AgNO_3$ solution.

_____

Write the equation for the reaction that occurred when distilled water was added to the previous solution.

_____

2. The hypochlorite ion

(a) Is the 5% NaOCl solution acidic or basic?

Does it have bleaching properties?

(b) Explain and write equations for the reaction with $AgNO_3$.

_____

(c) Write the equation for the reaction of OCl^- in basic solution with I^-.

_____

Write the equation for the reaction of OCl^- in acidic solution with Br^-.

_____

Place the ClO^- (base), ClO^- (acid), I_2, and Br_2 in order of increasing oxidizing strength.

████████   **CONSIDER THIS**   ████████

From thermodynamic tables, find the enthalpies of formation of the nitrogen oxides listed in Figure 38-1. Using Le Châtelier's principle, determine whether these oxides are likely to form at high or low temperatures and at high or low pressures. Use this knowledge to discuss the formation of nitrogen oxides in an internal combustion engine and how you would design a catalytic converter to eliminate them from automobile exhaust.

Are the same issues involved in the formation of sulfur oxides if the gasoline (or coal) contains sulfur as it is burned?

Equilibria of Coordination Compounds

Purpose

• Illustrate the tendency of metal ions to form coordination compounds (metal complexes) with ions and neutral polar molecules that act as electron-pair donors toward metal ions.

• Study metal ion complexes with ammonia, chloride ion, and hydroxide ion.

• Determine the equilibrium constant for the dissociation of the diamminesilver(I) complex ion, $Ag(NH_3)_2^+$.

Pre-Lab Preparation

The hydration of ions

We have already studied the most common class of coordination compounds. We have noted that all ions in solution attract polar water molecules to them to form ion–dipole bonds. Thus, the cupric ion, Cu^{2+}, does not exist as a bare ion in water but rather as the hexaaquo ion, $Cu(H_2O)_6^{2+}$. The origin of this attraction lies in the polar nature of the water molecule. Figure 39-1 shows the bent shape of the water molecule, which gives rise to the high dipole moment (a measure of the separation of charge in a molecule) of water.

The bond angle formed by the oxygen atom and the two hydrogen atoms in a water molecule is known to be 105°. This is not far from the 109.5° for ideal tetrahedral coordination, so the oxygen atom in a water molecule can be viewed as having four orbitals arranged in an approximately tetrahedral fashion, two of them forming the two bonds to the hydrogen atoms and the remaining two each contain-

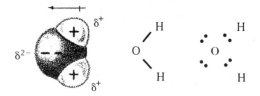

FIGURE 39-1 | **Three representations of the water molecule.**

ing a lone pair of electrons. It is these lone pairs that are centers of excess electron density. In Figure 39-1, the symbol ↔ indicates the negative direction of the dipole moment, and the small delta symbols (δ) indicate where the partial charges exist in the water molecule.

The negative side of the water molecule is strongly attracted to cations. In general, the greater the charge on the cation and the larger the cation, the greater the number of coordinated water molecules. Examples are $H(H_2O)^+$, $Be(H_2O)_4^{2+}$, $Cu(H_2O)_6^{2+}$, $Al(H_2O)_6^{3+}$, $Fe(H_2O)_6^{3+}$.

Anions attract the positive side of the water molecule and are also hydrated. However, the attraction is not as large, the hydrates are less stable, and fewer water molecules are coordinated.

Coordination compounds

These hydrates are one example of the type of substances called *coordination compounds*. Other neutral, but polar, molecules such as ammonia, NH_3, and also a number of anions, such as OH^-, Cl^-, CN^-, S^{2-}, $S_2O_3^{2-}$, and $C_2O_4^{2-}$, likewise can form similar very stable coordination groupings about a central ion. Such coordination compounds result from the replacement of the water molecule from the hydrated ion by other molecules or ions when they are present in a solution at high concentration, forming a still more stable bond. The resulting coordination compound may be a positively or negatively charged ion (a complex ion), or it may be a neutral molecule, depending on the number and kind of coordinating groups attached to the central ion.

What bonding forces hold the atoms of such a complex coordination compound together? As always, it is the mutual attraction of positively charged atom nuclei and negatively charged electrons. Where large differences of electronegativity are present, as in AlF_6^{3-}, the bonding has a large amount of ionic character. In the majority of cases, the bonding is mainly covalent, often with partial ionic character. The type of bond orbitals determines the spatial geometry and affects the stability of a given complex. Study the examples in Figure 39-2.

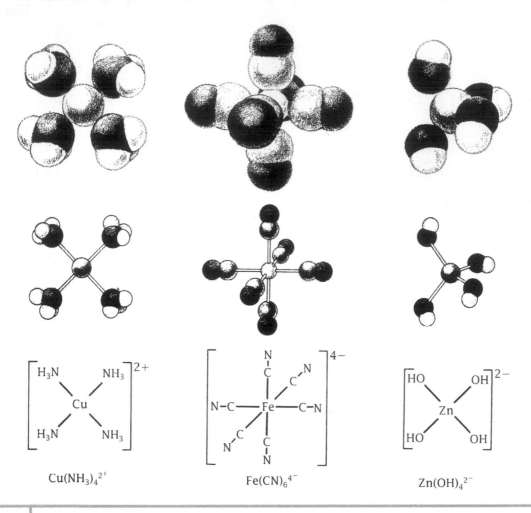

$Cu(NH_3)_4^{2+}$ $\qquad$ $Fe(CN)_6^{4-}$ $\qquad$ $Zn(OH)_4^{2-}$

FIGURE 39-2 **Diagrams showing the spatial arrangement of the coordinating groups about a central ion in the formation of a complex ion. Above each conventional formula, the structural formula represents each covalent electron-pair bond by a single line. The spatial geometry is determined by the type of orbitals represented:** dsp^2 **(square planar) by Cu(NH$_3$)$_4^{2+}$,** d^2sp^3 **(octahedral) by Fe(CN)$_6^{4-}$, and** sp^3 **(tetrahedral) by Zn(OH)$_4^{2-}$. The ball-and-stick models indicate these geometric patterns more clearly, and the space-filling models in the top row portray the atoms according to their accepted ionic diameters and bond lengths.**

Ammonia Complex Ions Some of the important ammonia complexes are

$Ag(NH_3)_2^+$	$Cu(NH_3)_4^{2+}$	$Ni(NH_3)_4^{2+}$
$Au(NH_3)_2^+$	$Cd(NH_3)_4^{2+}$	$Ni(NH_3)_6^{2+}$
$Cu(NH_3)_2^+$	$Zn(NH_3)_4^{2+}$	$Co(NH_3)_6^{2+}$

These complexes are formed by adding ammonia to a solution containing the hydrated cation. Ammonia molecules are bound by the cation one at a time as the concentration of ammonia increases. At low concentrations of ligand, smaller numbers of ammonia molecules may be bound. For instance, the two NH$_3$ molecules bound to the Ag$^+$ bind in successive steps. The equilibrium constant is known for each step of the reaction sequence

$$Ag^+ + NH_3 \rightleftharpoons Ag(NH_3)^+$$

$$K_1 = \frac{[Ag(NH_3)^+]}{[Ag^+][NH_3]} = 1.6 \times 10^3$$

$$Ag(NH_3)^+ + NH_3 \rightleftharpoons Ag(NH_3)_2^+$$

$$K_2 = \frac{[Ag(NH_3)_2^+]}{[Ag(NH_3)^+][NH_3]} = 6.8 \times 10^3$$

The equilibrium constant for the overall reaction can be obtained by the addition of the two reactions and the corresponding multiplication of their equilibrium constants.

TABLE 39-1 Some Important Amphoteric Hydroxides

Simple Ion[a] (acid solution)	Precipitate	Hydroxide Complex Ion[b] (strongly basic solution)
Pb^{2+}	$Pb(OH)_2$	$Pb(OH)_3^-$, trihydroxyplumbate(II) ion
Zn^{2+}	$Zn(OH)_2$	$Zn(OH)_4^{2-}$, tetrahydroxyzincate(II) ion
Al^{3+}	$Al(OH)_3$	$Al(OH)_4^-$, tetrahydroxyaluminate ion
Cr^{3+}	$Cr(OH)_3$	$Cr(OH)_4^-$, tetrahydroxychromate(III) ion
Sn^{2+}	$Sn(OH)_2$	$Sn(OH)_3^-$, trihydroxystannate(II) ion
Sn^{4+}	$Sn(OH)_4$	$Sn(OH)_6^{2-}$, tetrahydroxystannate(IV) ion

[a]So highly charged an ion as Sn^{4+} probably does not exist as such. In strong HCl solution, tin(IV) salts dissolve as the chloride complex, $SnCl_6^{2-}$.
[b]Formerly these ions were written in the anhydrous form: PbO_2^{2-}, ZnO_2^{2-}, AlO_2^-, CrO_2^-, SnO_2^{2-}, and SnO_3^{2-}. These formulas may be derived from the hydroxide complex ion formulas simply by subtracting the appropriate number of H_2O molecules or H_3O^+ ions.

$$Ag^+ + 2\,NH_3 \rightleftharpoons Ag(NH_3)_2^+$$

$$K_{formation} = K_1 K_2 = \frac{[Ag(NH_3)^+]}{[Ag^+][NH_3]}$$
$$\times \frac{[Ag(NH_3)_2^+]}{[Ag(NH_3)^+][NH_3]}$$
$$= \frac{[Ag(NH_3)_2^+]}{[Ag^+][NH_3]^2}$$
$$= 1.1 \times 10^7$$

In part 5 of the Experimental Procedure, you will determine the dissociation constant, K_{diss}, for the $Ag(NH_3)_2^+$ complex ion.[1] Because the $[NH_3]$ is so high under the experimental conditions that are employed, the $Ag(NH_3)^+$ species may be neglected.

Hydroxide Complex Ions, or Amphoteric Hydroxides The hydroxides of most metals are relatively insoluble in water. Thus, when sodium hydroxide is added to a metal ion in solution, such as lead ion, a precipitate is formed:

$$Pb(H_2O)_4^{2+} + 2\,OH^- \rightleftharpoons Pb(H_2O)_2(OH)_2(s) + 2\,H_2O$$

or, using the simple unhydrated metal ion formula,

$$Pb^{2+} + 2\,OH^- \rightleftharpoons Pb(OH)_2$$

By Le Châtelier's principle, excess hydroxide ion would give more complete precipitation. Instead, the precipitate dissolves. This is explained by the tendency of lead ion to form a more stable coordination compound[2] with excess hydroxide ion:

$$Pb(H_2O)_2(OH)_2(s) + OH^- \rightleftharpoons Pb(H_2O)(OH)_3^- + H_2O$$

or

$$Pb(OH)_2 + OH^- \rightleftharpoons Pb(OH)_3^-$$

Other ions react similarly. For example, hydrated aluminum ion, $Al(H_2O)_6^{3+}$, reacts to form the hydroxide precipitate, $Al(H_2O)_3(OH)_3$, or, with excess OH^-, the hydroxide complex ion $Al(H_2O)_2(OH)_4^-$. Traditionally, chemists use the unhydrated formulas in ordinary chemical equations just because they are simpler, except where it is important to emphasize the hydrated structure.

The reactions that form these hydroxide complex ions are entirely reversible. The addition of acid to the above strongly basic $Pb(OH)_3^-$ solution reacts first to reprecipitate the hydroxide,

$$Pb(OH)_3^- + H^+ \rightleftharpoons Pb(OH)_2(s) + H_2O$$

and then, with excess acid,

$$Pb(OH)_2(s) + 2\,H^+ \rightleftharpoons Pb^{2+} + 2\,H_2O$$

Such metal hydroxides, which may be dissolved by an excess of either a strong acid or a strong base, are called *amphoteric hydroxides*.

The more important metal ions whose hydroxides are amphoteric are given in Table 39-1.

Experimental Procedure

Chemicals: 1 M ammonium chloride, NH_4Cl; 1 M ammonia, NH_3; 15 M (concentrated) ammonia, NH_3; 6 M ammonia, NH_3; 6 M sodium hydroxide, NaOH; 6 M nitric acid, HNO_3; 12 M hydrochloric acid, HCl; 6 M hydrochloric acid, HCl; copper(II) sulfate pentahydrate, $CuSO_4\cdot 5H_2O(s)$; 0.1 M copper(II) sulfate, $CuSO_4$; 0.1 M sodium chloride, NaCl; 0.1 M silver nitrate, $AgNO_3$; 0.1 M zinc nitrate, $Zn(NO_3)_2$; 0.1% phenolphthalein, alizarin yellow R, and indigo carmine indicators.

---

[1]The overall reaction may be written as a formation or as a dissociation, with $K_{formation} = 1/K_{diss}$.
[2]There is some uncertainty as to whether the lead hydroxide complex (and also the stannous hydroxide complex) will coordinate further to form $Pb(OH)_4^{2-}$ [and $Sn(OH)_4^{2-}$].

SAFETY PRECAUTIONS:
Concentrated (12 M) HCl and 6 M HCl, 6 M HNO₃, and 6 M NH₃ are lung irritants. These solutions and 6 M NaOH are hazardous to the skin and eyes. If they contact your skin, wash them off immediately. Dispense the solutions in a well-ventilated fume hood. Clean up any spills immediately.

WASTE COLLECTION:
Waste containers should be provided for the copper, zinc, and silver compounds formed in this experiment.

1. The Formation of Complex Ions with Ammonia

To 3 mL of 0.10 M $CuSO_4$, add a drop of 6 M NH_3. Mix this. (Record your observations and write the equation for the reaction.) Continue to add NH_3 a little at a time, with mixing, until a distinct change occurs. Save this solution. Is this result contrary to the law of Le Châtelier? Obviously the OH^- concentration was increasing while the $Cu(OH)_2$ dissolved. How must the Cu^{2+} concentration have changed? Did it increase or decrease?

To learn which of the substances present in an ammonia solution (NH_4^+, OH^-, NH_3, H_2O) is responsible for the change you noted, try the following tests: (a) Add 1 mL of 1 M NH_4Cl to 1 mL of 0.10 M $CuSO_4$. (b) Add 2 drops (an excess)[3] of 6 M NaOH to 2 mL of 0.10 M $CuSO_4$. (c) Add ammonia gas by placing several crystals of $CuSO_4 \cdot 5H_2O(s)$ in a small dry beaker. At one side in the beaker, place a piece of filter paper moistened with concentrated (15 M) NH_3. Cover with a watch glass and observe any changes. From this evidence, write an equation to show the formation of this new substance when excess NH_3 is added to Cu^{2+}.

To 1 mL of this cupric ammonia complex ion solution, add 6 M HNO_3 in excess. Explain the result and write the equation for the reaction. Dispose of the solution in the copper waste container.

2. The Formation of Amphoteric Hydroxides

To 5 mL of 0.10 M $Zn(NO_3)_2$, add 6 M NaOH drop by drop, with mixing, until the precipitate that first forms just redissolves. Avoid undue excess of NaOH. Divide the substance into two portions; test one portion with alizarin yellow R and the other with indigo carmine indicator. Estimate the OH^- concentration (for later comparison in part 3), using the information on the color changes and pH intervals of the indicators given in Table 5 of the Appendix. Now to one portion, add 6 M HCl, drop by drop, until a precipitate forms (what is it?), then redissolves as more HCl is added. Interpret all these changes as related to Le Châtelier's law and to the relative concentration of the various constituents (the zinc in its various forms, H^+, and OH^-), both in words and in net ionic equations.

3. The Reaction of Zinc Ion with Ammonia

When ammonia is added gradually to Zn^{2+}, does the precipitate of zinc hydroxide that first forms redissolve as zincate ion, $Zn(OH)_4^{2-}$, owing to the excess base added, or does it redissolve as $Zn(NH_3)_4^{2+}$, owing to the NH_3 molecules added? To test this point, to 3 mL of 0.10 M $Zn(NO_3)_2$ add 6 M NH_3 drop by drop, with mixing, until the precipitate that first forms just redissolves. Divide this mixture, test one portion with phenolphthalein and the other portion with alizarin yellow R. Estimate the approximate OH^- concentration and compare this with the corresponding situation in part 2, where NaOH was used (see the Appendix, Table 5). What can you conclude about the possibility of forming $Zn(OH)_4^{2-}$ by adding NH_3 to a zinc salt solution? Explain. Write the equation for the equilibrium that you have verified.

4. Chloride Complex Ions

(a) To 2 mL of 0.10 M $CuSO_4$, add 2 mL of 12 M (concentrated) HCl; then dilute the solution with about 5 mL of water. Write equations and interpret the color changes you observed, assuming that the complex formed is $CuCl_4^{2-}$.

(b) To 1 mL of 0.10 M $AgNO_3$, add 3 mL of 12 M (concentrated) HCl; then agitate the solution well for several minutes to redissolve the precipitated AgCl. Now dilute this mixture with about 5 mL of distilled water. Write equations for the reactions and interpret the changes you observed, assuming that the complex formed is $AgCl_2^-$. Dispose of the tube contents in the silver waste container.

---

[3]This provides an excess of OH^-, a much stronger base than NH_3. The strong base OH^- shows some amphoteric effect (see part 2) with copper(II) salts but is far from complete.

5. The Equilibrium Constant of an Ammonia Complex Ion

The dissociation of diamminesilver(I) complex ion is represented by the equilibrium

$$Ag(NH_3)_2^+ \rightleftharpoons Ag^+ + 2\,NH_3 \qquad (1)$$

and the corresponding equilibrium constant expression

$$\frac{[Ag^+][NH_3]^2}{[Ag(NH_3)_2^+]} = K_{diss} \qquad (2)$$

If you add sufficient Cl^- gradually to an equilibrium mixture of Ag^+ and NH_3 [represented by Reaction (1)] so that you can just barely begin precipitation of AgCl(s), a second equilibrium is established simultaneously without appreciably disturbing the first equilibrium. This may be represented by the combined reactions

$$Ag(NH_3)_2^+ \rightleftharpoons Ag^+ + 2\,NH_3$$
$$+$$
$$Cl^-$$
$$\updownarrow$$
$$AgCl(s) \qquad (3)$$

By using a large excess of NH_3, you can shift Reaction (1) far to the left, with reasonable assurance that the Ag^+ is converted almost completely to $Ag(NH_3)_2^+$ rather than to the first step only, $Ag(NH_3)^+$. From the measured volumes of NH_3, Ag^+, and Cl^- solutions used, you can determine the concentrations of the species in Reaction (1) and calculate the value of K_{diss}.

To prepare the solution,[4] place 3.0 mL of 0.10 M AgNO$_3$ (measure it accurately in a 10-mL graduate) in a 15 × 150 mm test tube. Add 3.0 mL (also carefully measured) of 1.0 M NH$_3$. Now prepare some 0.020 M NaCl by diluting 2.0 mL of 0.10 M NaCl to 10.0 mL in your 10-mL graduate. Mix this thoroughly and note the exact volume. Then, from a medicine dropper, add it to the mixture of AgNO$_3$ and NH$_3$, about 1 to 1.5 mL at first, then drop by drop until a very faint, permanent milky precipitate of AgCl remains. Return any excess NaCl from the medicine dropper to the graduate and note the exact volume used. From these data, K_{diss} can be calculated. Dispose of the solution in the silver waste container.

CONSIDER THIS

If a visible spectrophotometer is available (such as a Spectronic 20®), develop a procedure for determining the equilibrium constant of the colored complexes such as Cu(NH$_3$)$_4^{2+}$.

Cobalt(II) chloride is often used as a humidity meter since its color changes from purple to pink as it forms the hexahydrate form. Estimate the equilibrium constant of this reaction by observing the color of the compound in air at various relative humidities.

---

[4]If desired, some improvement in precision can be obtained by using larger volumes—20.0 mL each of 0.10 M AgNO$_3$ and 1.0 M NH$_3$. Then dilute 10.0 mL of 0.10 M NaCl to about 50.0 mL in your 50-mL graduate, mix this well, and note the exact volume. Add first about 15 mL, then very small portions, to the Ag$^+$–NH$_3$ mixture, stirring as you do so, until a very faint permanent milky precipitate of AgCl remains. Note the total volume of 0.020 M NaCl used.

Equilibria of Coordination Compounds

Name _____

Date _____ Section _____

Locker _____ Instructor_____

Observations and Data

1. The formation of complex ions with ammonia

The net ionic equation for the reaction of excess $CuSO_4$ with NH_3 is

_____

The predicted effect on the reaction above of adding excess NH_3 (based on Le Châtelier's principle) is

_____

Describe the observed results when

excess NH_3 is added to $CuSO_4$ solution: _____

NH_4Cl and $CuSO_4$ solutions are mixed: _____

excess NaOH and $CuSO_4$ solutions are mixed: _____

$CuSO_4 \cdot 5H_2O$(s) is exposed to NH_3 gas: _____

Considering the equilibrium equation $NH_3 + H_2O \rightleftarrows NH_4^+ + OH^-$, explain which substance (NH_3, NH_4^+, or OH^-) causes the deep blue color when ammonia is added to a copper(II) solution, and give the equation for the reaction.

The observed effect and the net ionic equation for the reaction of HNO_3 on this deep blue solution is

2. The formation of amphoteric hydroxides

The net ionic equation for the reaction of excess $Zn(NO_3)_2$ with NaOH is

_____

The predicted effect on the reaction above of adding excess NaOH (based on Le Châtelier's principle) is

_____

Explain in your own words why $Zn(OH)_2$(s) dissolves with excess OH^-, and write the net ionic equation for the reaction.

| Color with alizarin yellow R _____ | indigo carmine _____ | OH^- concentration _____ |

Explain the effect of adding a moderate amount of HCl to this strongly basic solution, and give the net ionic equation for the reaction.

What further change occurs when excess HCl is added? (Give the equation and explain.)

3. The reaction of zinc ion with ammonia

Note your observations below on the addition of indicators to the solution formed by adding, one drop at a time, 6 M NH_3 to $Zn(NO_3)_2$(aq) to first form, then just redissolve, the precipitate.

Color with   Alizarin   Estimated OH^-
phenolphthalein _____   yellow R _____   concentration _____

(To estimate the OH^- concentration, use the information on the color changes and pH intervals of the indicators given in Table 5 of the Appendix.)

Which coordination compound, $Zn(OH)_4^{2-}$ or $Zn(NH_3)_4^{2+}$, forms when Zn^{2+} reacts with excess NH_3 solution? Compare with part 2; explain fully.

Therefore, the equation for the formation of this equilibrium complex ion is

_____

4. Chloride complex ions

(a) Explain the successive changes observed when concentrated HCl, then H_2O, is added to a $CuSO_4$ solution; give the equations for the reactions.

(b) Explain the changes observed when concentrated HCl, then H_2O, is added to an $AgNO_3$ solution, and give the equations.

5. The equilibrium constant of an ammonia complex ion

(Indicate your calculations for each step in the spaces provided.)

Volumes of solutions:

0.10 M   1.0 M   0.020 M   Total
AgNO$_3$ _____   NH$_3$ _____   NaCl _____   volume _____

(a) Concentration of $Ag(NH_3)_2^+$

[Assume all the silver to be present as the complex ion, $Ag(NH_3)_2^+$, ignoring the traces of free Ag^+ and $Ag(NH_3)^+$ remaining.]

_____ M

(b) Concentration of Cl^-

[Ignore any trace of Cl^- removed as $AgCl(s)$.]

_____ M

(c) Concentration of Ag^+ (Use the Cl^- concentration above and the solubility product relationship, $[Ag^+][Cl^-] = 1.8 \times 10^{-10}$.)

_____ M

(d) Concentration of free NH_3

[First calculate the NH_3 concentration as if none combined with Ag^+, then subtract twice the concentration of $Ag(NH_3)_2^{2+}$ found above.]

_____ M

(e) Use the values found in 5(a), 5(c), and 5(d) to calculate the value of the equilibrium constant from Equation (2):

_____

(f) Compare your value for the equilibrium constant with the value given in the Appendix, Table 7.

Application of Principles

1. Which reagent, NaOH or NH_3, will enable you to precipitate the first-named ion from a solution containing each of the following pairs of ions, leaving the second ion in solution? Give also the formula of the precipitate and the exact formula of the other ion in solution. You will need to refer to Tables 7 and 9 in the Appendix to determine which ions form hydroxide precipitates or hydroxide complexes, and ammonia complexes.

Pair	Reagent	Precipitate	Ion in solution
(a) Al^{3+}, Zn^{2+}			
(b) Cu^{2+}, Pb^{2+}			
(c) Pb^{2+}, Cu^{2+}			
(d) Fe^{3+}, Al^{3+}			
(e) Ni^{2+}, Sn^{2+}			
(f) Sn^{2+}, Ni^{2+}			
(g) Mg^{2+}, Ag^+			

2. When ammonia is added to $Zn(NO_3)_2$ solution, a white precipitate forms, which dissolves on the addition of excess ammonia. But when ammonia is added to a mixture of $Zn(NO_3)_2$ and NH_4NO_3, no precipitate forms at any time. Suggest an explanation for this difference in behavior.

3. Calculate the OH^- concentration in (a) 1.0 M NH_3 and (b) a solution that is 1.0 M in NH_3 and also 1.0 M in NH_4Cl. ($K_b = 1.8 \times 10^{-5}$ for NH_3; see the Appendix, Table 6.)

4. Suppose you are given the following experimentally observed facts regarding the reactions of silver ion.

(a) Ag^+ reacts with Cl^- to give white $AgCl(s)$.

(b) Ag^+ reacts with ammonia to give a quite stable complex ion, $Ag(NH_3)_2^+$.

(c) A black suspension of solid silver oxide, $Ag_2O(s)$, shaken with NaCl solution, changes to white $AgCl(s)$.

(d) $AgCl(s)$ will dissolve when ammonia solution is added, but $AgI(s)$ does not dissolve under these conditions.

Write equations for any net reactions in the cases above; then, based on these observations, arrange each of the substances $AgCl$, AgI, Ag_2O, and $Ag(NH_3)_2^+$ in such an order that their solutions with water would give a successively decreasing concentration of Ag^+.

(1) _____

(2) _____

(3) _____

(4) _____

CONSIDER THIS

If a visible spectrophotometer is available (such as a Spectronic 20®), develop a procedure for determining the equilibrium constant of the colored complexes such as $Cu(NH_3)_4^{2+}$

Cobalt(II) chloride is often used as a humidity meter since its color changes from purple to pink as it forms the hexahydrate form. Estimate the equilibrium constant of this reaction by observing the color of the compound in air at various relative humidities.

Atmospheric Chemistry of Nitrogen Monoxide

Smog and Acid Rain

Purpose

- Study the reaction between nitrogen monoxide and oxygen in the presence of water.

- See how this reaction can lead to the formation of acid rain.

Pre-Lab Preparation

The chemistry of nitrogen monoxide (NO) is so rich that a study of it could occupy a chemist for a lifetime. It plays a significant role in (1) the formation of photochemical smog and ozone in the troposphere and the destruction of ozone in the stratosphere, (2) the formation of acids in the atmosphere, and (3) the biological chemistry of humans and other animals.

In the troposphere (the layer of the atmosphere closest to the earth's surface), ozone is undesirable because it is a lung irritant; but in the stratosphere, ozone is desirable because it shields us from ultraviolet radiation from the sun. In its atmospheric reactions, nitrogen monoxide cuts two ways, playing a primary role in the formation of "bad" ozone in the part of the atmosphere where we live and breathe, but catalyzing the decomposition of "good" ozone in the stratosphere, where ozone has a beneficial protective effect.

More recently it has been discovered that nitrogen monoxide plays an important role in biological systems, acting as a neurotransmitter (or messenger molecule) and having profound effects on the vascular system, where it enlarges blood vessels by relaxing the smooth muscles in their walls.

In this experiment, we will focus on the role of nitrogen monoxide in the formation of acids in the atmosphere. In rainy weather, these acids can return to earth as acid rain. In dry weather, the acids can react with ammonia from dairies and cattle feed lots to form solid particulates such as ammonium nitrate. The tiny solid particles scatter light like the tiny droplets of water in fog, forming a haze that reduces visibility.

The role of nitrogen monoxide in the formation of photochemical smog and acids

One of the noxious pollutants that comes out of the tailpipes of autos and trucks is nitrogen monoxide, NO. In fact, any high-temperature combustion that takes place in the presence of air produces some NO—for example, cooking over a gas burner or heating with a gas-fired furnace. The higher the temperature of the combustion, the more NO is produced. It results from the reaction at high temperature of nitrogen and oxygen:

$$N_2(g) + O_2(g) \rightarrow 2\ NO(g) \qquad (1)$$

Although the amount of NO produced is small, in the concentration range of a few parts per million, this small amount of NO has significant environmental effects when it is multiplied in a large urban area by the millions of people driving autos or heating homes and cooking with natural gas. At high concentrations, NO rapidly reacts with oxygen to form nitrogen dioxide (NO_2), a brown noxious gas. The net reaction of NO with oxygen is reported to be

$$2\ NO(g) + O_2(g) \rightarrow 2\ NO_2(g) \qquad (2)$$

Kinetically, the reaction is second order in NO, so at the low concentrations found in polluted ambient air the reaction is much slower, taking hours to completely react (Shaaban, 1990).

NO_2 plays a key role in the formation of what is called photochemical (or Los Angeles–type) smog, which contains ozone, a very reactive and irritating gas that is responsible for the painful breathing that you feel if you exercise outdoors during a severe smog alert.

How is the ozone produced? NO_2 has an absorption band centered at 393 nm, near the borderline between the visible and the ultraviolet regions of the spectrum. As the sun shines on a mass of air containing NO_2, some NO_2 is dissociated by absorption of a photon ($h\nu$) of light:

$$NO_2(g) + h\nu \rightarrow NO(g) + O(g) \qquad (3)$$

The oxygen atom produced is very reactive and in the presence of a third molecule, M, which can be a nitrogen or oxygen molecule, it reacts with a dioxygen molecule to form ozone:

$$O(g) + O_2(g) + M(g) \rightarrow O_3(g) + M^*(g) \quad (4)$$

The molecule M* (the asterisk means it is in an excited state) carries away excess energy of reaction that would otherwise keep the O atom and the O_2 molecule from forming a stable chemical bond. The NO produced in the photochemical step, Reaction (3), can recycle to again form NO_2, so ozone can be formed anytime that sunlight shines on air containing nitrogen oxides (Ainscough & Brodie, 1995).

The brown gas NO_2 is very soluble in water. If you attempt to collect it over water, it is completely absorbed, leaving no measurable volume of gas. Depending on reaction conditions, several different reactions have been reported for the reaction of NO_2 (or NO and O_2) with water. According to some sources, the NO_2 reacts with water to form nitrous and nitric acids by the net reaction (Bonner & Steadman, 1996; Cotton et al., 1999)

$$2 NO_2(g) + H_2O(l) \rightarrow HNO_2(aq) + HNO_3(aq) \quad (5)$$

It is also reported that, when heated, nitrous acid decomposes to nitric acid and NO (Bonner & Steadman, 1996):

$$3 HNO_2 \rightarrow HNO_3 + 2 NO + H_2O \quad (6)$$

The NO formed could then keep recycling and reacting with oxygen, so that when the last bit of nitrous acid, HNO_2, has disappeared, the net overall reaction in the presence of water and excess oxygen can be written as (Hornack, 1990)

$$4 NO(g) + 3 O_2(g) + 2 H_2O \rightarrow 4 HNO_3 \quad (7)$$

Finally, it has been reported that when both NO and O_2 are dissolved in aqueous solution, they react to produce only nitrous acid (or nitrite ion in the presence of base), corresponding to the stoichiometry (Bonner & Steadman, 1996; Cotton et al., 1999)

$$4 NO(aq) + O_2(aq) + 2 H_2O \rightarrow 4 HNO_2(aq) \quad (8)$$

Reactions (2) + (5), (7) (if HNO_2 readily decomposes), or (8) all ultimately form acids starting with nitrogen monoxide, NO, and O_2. If the reactions take place in an area of the atmosphere containing water droplets in a cloud, acid rain will be produced.

Some possible reactions of NO and O_2 in the presence of water

Looking at Reactions (2) and (5)–(8), we see that there are three distinguishable net chemical reactions that might take place in the presence of water, depending on the reaction conditions:

A: Reactions (2) + (5): $2 NO(g) + O_2(g) + H_2O \rightarrow HNO_2(aq) + HNO_3(aq)$

B: Reaction (7): $4 NO(g) + 3 O_2(g) + 2 H_2O \rightarrow 4 HNO_3(aq)$

C: Reaction (8): $4 NO(aq) + O_2(aq) + 2 H_2O \rightarrow 4 HNO_2(aq)$

All three reactions predict one mole of total acid produced per mole of NO reacted, so no distinction among the three is possible based solely on the titration of total acid produced. But there are three clearly distinguishable outcomes based on the predicted NO/O_2 reaction stoichiometry and the stoichiometric amounts of HNO_2 and HNO_3 produced, as shown in Table 40-1. In this experiment, we will prepare NO and O_2 and make a qualitative and semi-quantitative study of their reaction products in the presence of water.

Measuring the pH and total amount of acid produced

Nitric acid (HNO_3) and/or nitrous acid (HNO_2) are reported as reaction products when NO_2 reacts with water. Nitric acid is a strong acid; nitrous acid is a

TABLE 40-1 Three Possible Net Reactions of NO and O_2 in the Presence of Water

A: $2 NO(g) + O_2(g) + H_2O \rightarrow HNO_2(aq) + HNO_3(aq)$
B: $4 NO(g) + 3 O_2(g) + 2 H_2O \rightarrow 4 HNO_3(aq)$
C: $4 NO(aq) + O_2(aq) + 2 H_2O \rightarrow 4 HNO_2(aq)$

Reaction	Mols of O_2 per mol of NO	Total mols of acid per mol of NO	Mols of nitrite ion per mol of NO	Mols of nitrate ion per mol of NO
A	0.50	1.0	0.5	0.5
B	0.75	1.0	0.0	1.0
C	0.25	1.0	1.0	0.0

weak acid. By measuring both the pH of the reaction products and the total concentration of acid we can, in principle, tell if the solution contains only a strong acid, only a weak acid, or a mixture of weak and strong acids. A properly standardized pH meter is necessary for making the pH measurements.

We can determine the total amount of acid in an aqueous solution by titration with a base. This amounts to adding base to the acid solution until the solution is neutralized. We will add a color indicator that allows us to know when the end point of the titration is reached. The acid/base neutralization is

$$HNO_2(aq) + OH^-(aq) \rightarrow NO_2^-(aq) + H_2O(l) \quad (9)$$

and/or

$$HNO_3(aq) + OH^-(aq) \rightarrow NO_3^-(aq) + H_2O(l) \quad (10)$$

Semiquantitative testing for the presence of nitrite ion

We will use a very sensitive test for nitrous acid, described more fully in the Experimental Procedure. In acid solution, nitrous acid oxidizes iodide ion to iodine, which in the presence of starch indicator forms a deep blue starch–iodine complex. The test easily detects nitrous acid at a concentration of 1×10^{-4} M.

Experimental Procedure

Special Supplies: pH meter; 10-in. length forceps; each student (or student group) will require the following: disposable Luer tip syringes (two 60-mL size and one 10-mL size); two syringe tip caps (available from Flinn Scientific Inc.); two Titeseal polyethylene shell vial caps (Kimble 1-dram shell vials, 15 mm OD × 45 mm length, with Titeseal caps); 3-cm length of ⅛-in. ID rubber tubing; four disposable weighing dishes (1⅝-in. square); polyethylene transfer pipets with 1-mL calibration marks; 1-mL microburet (Air Tite Norm-Ject Luer Slip 1 × 0.01 mL syringe).

Chemicals: Active dry baker's yeast (Fleischmann's or other brand, available at supermarkets); sodium nitrite, $NaNO_2(s)$; 1.2 M $FeSO_4$/1.8 M H_2SO_4 (freshly prepared); 0.008 M HNO_3; standardized 0.50 M NaOH; 0.01 M $NaNO_2$; 1 M $NaHCO_3$; 0.5% starch indicator (freshly prepared); 1.0 M KI; ~5 M tartaric acid; 3% hydrogen peroxide; 0.1% thymol blue.

▨ NOTES TO INSTRUCTOR ▨

Cylinders of nitrogen monoxide and oxygen can be used to fill the syringes, but NO is available only at relatively low pressure, so a special regulator, com-

patible with the cylinder fitting, is required. The preparation of NO and O_2 in syringes, as described in the experimental procedure, is convenient and relatively safe, but the preparation of NO should be done in the fume hood to assure complete safety.

The procedure for detection of nitrous acid, HNO_2, is semiquantitative but can easily be made more quantitative by using a spectrophotometer to measure the absorbance of the sample. This can be done in two ways. The first method simply uses a spectrophotometer to measure the absorbance at 592 mm of the deep blue triiodide–starch complex, for both the diluted reaction mixture and a sample of known nitrite ion concentration, as described in part 4(c) of the experimental procedure. The second method directly measures the absorbance of nitrite ion itself in basic solution at 353 nm. The molar absorptivity of nitrite ion at this wavelength is approximately 23 L · mol^{-1} · cm^{-1}, so that a 4.0 mM solution of nitrite has an absorbance of about 0.092. (The nitrate ion contributes very little [< 5%] to the total absorbance at this wavelength.) If approximately 3 mL of the aqueous reaction solution is reserved before addition of the thymol blue indicator, the 3 mL of solution can be made basic by the addition of 0.2 mL of 1 M NaOH and the absorbance of the nitrite ion directly measured at 353 nm.

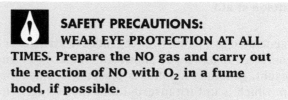

⚠ SAFETY PRECAUTIONS:
WEAR EYE PROTECTION AT ALL TIMES. Prepare the NO gas and carry out the reaction of NO with O_2 in a fume hood, if possible.

☠ WASTE COLLECTION:
The Fe(II)/Fe(III)/H_2SO_4 reaction products produced in making NO(g) should be neutralized with sodium bicarbonate, $NaHCO_3$, and the solid residue placed in a waste collection bottle. All other solutions may be disposed of in the sink, followed by a water rinse.

General procedure for generating gases in syringes

We will study the reactions of gases prepared in 60-mL disposable Luer-tip syringes, a method that minimizes exposure to noxious and toxic gases. Each gas

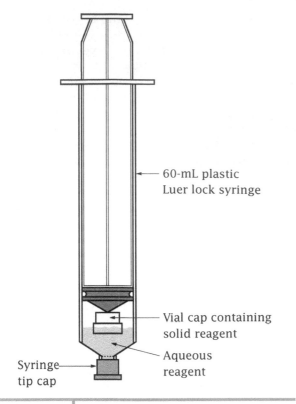

60-mL plastic
Luer lock syringe

Vial cap containing
solid reagent

Syringe
tip cap

Aqueous
reagent

FIGURE 40-1 | **A method for generating gases in a syringe. A vial cap containing a solid reagent is placed in the syringe. A solution of a second reagent is drawn into the syringe and the syringe tip is capped. The syringe is then shaken to mix the two reagents and generate the gas. (After Mattson et al.)**

plunger back and forth. If the plunger movement is stiff and requires a lot of effort, "break in" the syringe by moving the plunger up and down twenty or thirty times. Practice completely removing the plunger, which takes some extra force because of the molded stop designed to prevent the plunger from accidentally coming completely out of the syringe. (Sometimes it also helps to break in a syringe by pulling the plunger completely out and lubricating the plunger tip with a thin film of silicone oil.)

Weighing the Solid Reagent. Weigh 0.050 g of dry active baker's yeast (Fleischmann's or other brand) on a creased sheet of weighing paper, then transfer the yeast into the cavity of a small polyethylene vial cap (use the cap supplied with a 1-dram Titeseal shell vial). The amount of yeast is just about enough to cover the bottom of the vial cap when uniformly distributed.

Loading the Vial Cap into the Syringe. Next, we want to put the vial cap containing the solid reagent into the syringe. This can be done one of two ways, using forceps or using the flotation technique illustrated in Figure 40-2. With either method, begin by completely removing the plunger from the syringe and setting the plunger aside on a clean towel or sheet of paper.

is generated by mixing two reagents in a syringe. The general arrangement is shown in Figure 40-1. One reagent, generally a solid, is placed in a small vial cap, which is in turn inserted into the syringe. The second reagent is prepared as an aqueous solution and is drawn into the syringe from a small plastic weighing dish. Then the syringe is capped and shaken to mix the reagents and generate the gas. If desired, gases that are not too water soluble can be washed by drawing deionized water into the syringe, capping and shaking the syringe, and expelling the water. Repeating this operation will remove nearly all traces of the original reactants.

Because of the small vial cap present in the syringe, not all of the air originally in the syringe is expelled before drawing in the liquid reagent. As a result, the gases prepared contain a small amount of air (about 10% by volume). For many purposes, this is not a significant drawback.

1. Preparing O₂ Gas *Preparing the 60-mL Syringe.* Remove the plastic cap that sometimes covers the tip of new syringes and set it aside. Try to move the

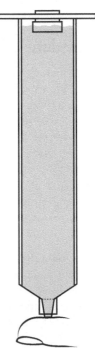

FIGURE 40-2 | **Flotation method for loading the vial cap containing the solid reagent. The plunger is removed from the syringe and the syringe is filled with deionized water; the Luer tip is sealed with a finger. The vial cap containing the solid reagent is floated on the water, and the water is allowed to slowly drain out.**

The vial cap can be loaded using long (10-in.) forceps, grasping the forceps near the top so that you can insert the vial cap all the way to the bottom inside the syringe without your fingers blocking the forceps' entry into the syringe. (If you drop the cap, the solid reagent usually spills out, in which case you must start over.) Then, keeping the syringe upright, with the tip pressed against the bench top, insert the plunger, pushing it snugly against the vial cap. Keep the syringe upright (tip down) in a beaker until you are ready to use it.

The flotation method of putting the vial cap in the syringe works well and requires no forceps. Begin by putting your forefinger on the Luer end of the syringe and filling the syringe nearly to the top with deionized water. Then carefully float the vial cap containing the solid on the water. Controlling the flow with your finger, allow the water to slowly leak out so that the cap rides down to the bottom of the syringe. Perform this operation over the sink, or catch the water in a beaker so you don't spill water all over the bench top. When all of the water has drained out, keep the syringe upright with the tip against the bench top, and insert the plunger, pushing it snugly against the vial cap. Keep the loaded syringe upright in a beaker until you are ready to use it.

Loading the Second Liquid Reagent. Place 6 mL of 3% hydrogen peroxide solution into a small (1⅝-in. square) polyethylene weighing dish. (At room temperature, each mL of 3% H_2O_2 solution produces about 10–11 mL of oxygen gas.) Draw the hydrogen peroxide solution up into the syringe, taking care not to draw in more than a bubble or two of air as you draw in the last of the solution. Cap the syringe with a syringe tip cap. Shake the syringe to mix the yeast and hydrogen peroxide solution. Oxygen is evolved rapidly by catalytic decomposition of the hydrogen peroxide:

$$H_2O_2(aq) \rightarrow H_2O + \frac{1}{2}O_2(g)$$

As the volume of gas nears the full capacity of the syringe, hold the syringe tip up (directed away from your face) and loosen the cap to relieve the pressure. (If the syringe tip is pointed down when you loosen the cap, the solution will spray out, making a mess.) Tighten the cap, shake, and relieve excess pressure again. Continue until the reaction is complete and vent any excess pressure one last time. Turn the syringe tip down and expel the reaction solution into a sink or beaker (it's harmless, being nearly pure water with a small amount of yeast). Keep the syringe capped until you are ready to use the oxygen gas. It isn't necessary to wash the oxygen gas, since the

reaction product is essentially pure water. Label the syringe to identify its contents, making sure that the label does not obscure the volume graduations on the syringe.

2. Preparing NO Gas For generating NO(g) we will use solid sodium nitrite, $NaNO_2$, which reacts with an excess of Fe(II) solution, according to the net ionic reaction

$$NO_2^- + Fe^{2+}(aq) + 2 H^+ \rightarrow$$
$$NO(g) + Fe^{3+}(aq) + H_2O \quad (11)$$

It has been found by trial and error that you can generate 55–60 mL of NO gas by reacting about 0.27 g of solid $NaNO_2$ with 4 mL of a solution that is 1.2 M in $FeSO_4$ and 1.8 M in H_2SO_4. [This is 50% more than the calculated amount of $NaNO_2$, indicating that the reaction is only approximately described by Reaction (11).]

Weigh 0.27 g of solid $NaNO_2$ on a creased sheet of weighing paper and load the $NaNO_2$ into the cavity of a small polyethylene vial cap. Using forceps or the flotation method, load the solid in the vial cap into a 60-mL syringe.

If possible, carry out the next steps in a fume hood. Place 4–5 mL of 1.2 M $FeSO_4$/1.8 M H_2SO_4 solution into a small polyethylene weighing dish. Put the tip of the syringe into the solution and draw up as much of the solution as possible, being careful not to draw in more than a bubble or two of air. Then cap the syringe with a syringe tip cap. In the fume hood, shake the syringe to mix the reagents. The reaction mixture will turn black almost immediately, and NO gas is rapidly evolved. **Caution:** When the reaction (and the plunger) are nearing the end, turn the syringe upside down, tip end up, before removing the cap to relieve the pressure. (Point the tip away from your face in the hood.) If the tip is pointed down when you remove the cap, the liquid reagent will spray out, making a nasty mess. (If this happens, neutralize the spill with sodium bicarbonate [baking soda], then wipe up the spill.)

Washing the NO Gas. After venting excess pressure with the tip up, turn the syringe tip down, remove the tip cap, and expel the black reaction solution into a beaker. (It will be neutralized and collected as waste. Do not pour it down the drain.) If the volume of gas in the syringe is greater than 55 mL, expel some gas (in the fume hood!) until the volume reads 55 mL. In this way, you will have some room to draw in wash water. Place some deionized water in a small beaker and draw up about 5–6 mL of water into the syringe. Immediately cap the syringe and shake it to wash the gas. Expel the water into the waste collection beaker, draw another portion of water into the syringe, and shake again. Finally, expel the water into the waste collection beaker and quickly cap the syringe until you are ready to use the NO gas. Put a label on the syringe identifying its contents as NO.

3. Reacting NO with O$_2$ *If possible, carry out all of the operations in a fume hood.* First, gather together on a piece of paper towel or a clean sheet of paper the materials you will need: the two capped 60-mL syringes containing O$_2$ and NO gas; a clean, dry 10-mL disposable syringe; a short (3-cm) length of $\frac{1}{8}$-in.-ID rubber tubing; and a 100-mL beaker containing 50 ± 1 mL of deionized water. Deionized water is necessary because most tap water contains bicarbonate ion, HCO$_3^-$, which reacts with acids:

$$HCO_3^- + H^+ \rightarrow H_2O + CO_2(g)$$

Therefore, tap water would spoil the titration by consuming an unknown amount of any acid that we later collect in the beaker.

Move the plunger of the 10-mL syringe back and forth until it moves freely, then push the plunger all the way in until it stops. Now you are going to connect this syringe to the NO supply syringe and transfer a 10-mL sample of NO from the 60-mL syringe to the smaller 10-mL syringe. Uncap the 60-mL NO supply syringe, and push one end of the rubber tubing firmly onto the Luer tip. Flush out the tubing by expelling a small volume (2–3 mL) of NO, stopping at the nearest 5-mL division. Then quickly insert the Luer tip of the empty 10-mL syringe into the free end of the tubing and slowly pull back the plunger of the 10-mL syringe by exactly 10 mL, while at the same time pushing in the plunger of the larger NO syringe by exactly 10 mL. At the end, you will have transferred 10 mL of NO, at approximately atmospheric pressure, to the smaller syringe.

Now put the 60-mL O$_2$ syringe beside the 60-mL NO syringe so that the O$_2$ syringe can take its place. When ready to make the transfer, remove the 60-mL NO syringe from the tubing and cap the syringe immediately. Uncap the O$_2$ syringe and immediately connect it to the free end of the rubber tube (with the 10-mL syringe still on the other end). Immediately push on the plunger of the O$_2$ syringe until you have transferred exactly 5 mL of oxygen into the smaller 10-mL syringe containing NO. Do *not* move the plunger of the smaller syringe. Keep it at the 10-mL mark. Then quickly disconnect the O$_2$ syringe from the rubber tubing and cap the syringe. Use your fingers to pinch off the rubber tubing so that no gas will enter or escape. Do you see any evidence of reaction between NO and O$_2$?

Put the free end of the rubber tubing into the previously prepared beaker containing 50 mL of deionized water so that the end of the tubing rests on the bottom of the beaker. Draw a little deionized water into the syringe, taking care to keep the free end of the rubber tubing immersed below the surface of the water. Gently shake the gas/water mixture in the syringe. What happens? Continue shaking and drawing water into the syringe until no more gas is absorbed. At this point, record the final volume of gas in the syringe.

Now expel the solution containing the reaction products into the beaker. Draw a few mL of water into the syringe, then expel the water to rinse out the interior of the syringe. Repeat this a couple of times. At the end, the solution in the beaker will contain all of the NO/O$_2$ reaction products.

Looking at Reaction (2) and recalling that gas volumes combine in exactly the ratios of the stoichiometric equation, we see that Reaction (2) predicts that 10 mL of NO would combine with 5 mL of O$_2$ to form 10 mL of NO$_2$ gas (all at constant atmospheric pressure). That's why we shouldn't have to move the plunger of the 10-mL syringe to accommodate the added volume of O$_2$ gas.

Record Pressure and Temperature. You will need to measure and record the barometric pressure and the room temperature so that you can calculate how many moles of gaseous reactants were used.

4. Analysis of the Reaction Solution

(a) Measurements of pH. If you have a pH meter available, measure the pH of the reaction solution that you collected in the beaker. Rinsing the pH electrode with deionized water between each measurement and blotting the electrode with tissue to remove excess water, also measure the pH of 0.008 M HNO$_3$, then 0.004 M HNO$_3$ solutions. (The latter solution can be prepared by diluting 0.008 M HNO$_3$ with an equal volume of water.) *Save the 0.004 M HNO$_3$ solution to use later in part 4(c) as a reagent blank in the test for nitrite ion.* The HNO$_3$ solutions provide reference pH measurements to check the pH calibration of the meter. The pH difference between the two HNO$_3$ reference solutions should be equal to about log(2) = 0.30 because HNO$_3$ is a strong, completely ionized acid, and the reference solutions differ in concentration by a factor of two.

(b) Titrating the acid. Get a 1-mL microburet (a 1-mL polypropylene syringe equipped with a fine tip cut from a polyetheylene transfer pipet, or whatever form of microburet you usually use). Also get about 6 mL of 0.5 M NaOH in a clean, dry 13 × 100 mm test tube equipped with a size-00 rubber stopper. Fill the syringe with 0.5 M NaOH, taking precautions to remove all bubbles from the syringe (and not to squirt yourself or your neighbor with solution from the syringe). Be sure to record the concentration of the NaOH solution and the initial volume reading of the syringe. Add 5 drops of 0.1% thymol blue indicator to the solution in the beaker. Now add the 0.5 M NaOH solution from the microburet, drop by drop, to your acid solution. At the end point, the solution will turn from yellow to a permanent blue color. (If the end point color drifts back toward yellow, add another drop of 0.5 M NaOH; continue until the solution remains blue for at least 30 sec.) Now read and record the final volume of the added base. *Save the titrated solution for the next test for the presence of nitrite ion.*

(c) Semiquantitative analysis for nitrite ion. In acid solution, nitrous acid oxidizes iodide ion to form iodine (or triiodide ion in the presence of excess iodide ion):

$$2\,HNO_2 + 2\,H^+ + 3\,I^- \rightarrow 2\,NO + 2\,H_2O + I_3^-$$

In the presence of starch indicator, a deep blue starch–triiodide ion complex forms. The test easily detects 1×10^{-4} M nitrous acid, so the reaction solution must be diluted before doing the test.

Dilute the solution that was titrated in part 4(b) by a factor of 50, using a polyethylene transfer pipet with a 1-mL calibration mark to transfer 1 mL of the titration solution to a clean 125-mL Erlenmeyer flask and adding 49 mL of deionized water. Mix this solution thoroughly.

To perform the tests, you will need about 2 mL of each of the following reagent solutions: 1 M NaHCO$_3$, 0.5% starch indicator, 1 M KI (freshly prepared), and 5 M tartaric acid (it's OK if the tartaric acid solution is yellow or light brown).

Then, using a clean polyethylene transfer pipet, put 1 mL of the diluted reaction solution into a 13 × 100 mm test tube and add two drops (0.10 mL) of 1 M NaHCO$_3$, two drops (0.10 mL) of 0.5% starch indicator, two drops (0.10 mL) of 1 M KI, and five drops (0.25 mL) of 5 M tartaric acid. Shake the test tube to mix the solution thoroughly. Allow the solution to stand for 10 min for the blue color to develop fully. While waiting, move ahead to do the control and reagent blank tests.

Control and Reagent Blank Tests for Nitrite Ion. Using the same quantities of the four reagents, perform the same test for nitrite on 1 mL of 1×10^{-4} M sodium nitrite solution (prepared by diluting 1 mL of 0.01 M NaNO$_2$ solution with 99 mL of deionized water and thoroughly mixing). We will call this the control solution because it contains a known amount of nitrite ion.

Finally, perform the same test on 1 mL of a solution containing the 0.004 M HNO$_3$ prepared in part 4(a). We will call this the *reagent blank test*. The purpose of the blank is to show that a solution containing nitrate ion and the reagents does not give a positive test for nitrite ion. It should be colorless. If it shows a significant blue color, most likely the KI solution contains some iodine produced by air oxidation of the iodide ion or there is nitrite ion contamination from some source.

(d) Estimating the total amount of nitrite produced. After allowing the blue color to fully develop by standing for 10 min, compare the intensities of the blue colors of the reaction solution and the control (1×10^{-4} M NaNO$_2$) solution by viewing them against a well-lit white background. (The comparison can be made more quantitative using a spectrophotometer to measure the absorbance at 592 nm.) If the intensities approximately match (or at least appear to be in the same ballpark), you can estimate the total amount of nitrite produced by multiplying by 2500 the amount of nitrite ion in the control. Recall that the total amount of gas (10 mL NO ≅ 0.4 mmol NO) was eventually absorbed in 50 mL of deionized water. Then we took 1 mL of this solution and diluted it again to 50 mL, so we took 1/50th of the reaction solution and diluted it another factor of 1/50 to give a total dilution of 1/2500.

The amount of nitrite ion in the control solution is equal to 1 mL × 1×10^{-4} mol/L = 1×10^{-4} mmol. If the blue colors approximately match, this would also be the amount of nitrite ion in the diluted reaction sample. This amount multiplied by 2500 would then give the total amount of nitrite ion produced by reaction of the gas with water (2500 × 1×10^{-4} mmol = 0.25 mmol of nitrite).

You should have enough NO and O$_2$ left in your syringes to repeat the experiment. If you have time, repeat parts 3 and 4 so that you have two measurements of how much acid and nitrite were formed. Remember, an average of several measurements is always more reliable than a single measurement.

Calculations and data analysis

Using the ideal gas law, determine the quantities (in millimoles) of NO and O$_2$ that were reacted in the syringe. What is the total amount of acid (in units of millimoles) that you would expect to be produced for the given amounts of NO and O$_2$, considering the three reactions summarized in Table 40-1?

In your lab notebook, calculate the expected pH for solutions containing 0.008 M HNO$_3$, 0.004 M HNO$_3$, and 0.008 M HNO$_2$ ($K_a = 7.1 \times 10^{-4}$).

From the results of the titration in part 4(b), calculate the moles of acid formed. Compare this, calculated as a percentage yield, with the theoretical number of moles of acid expected. What is the percentage yield for the formation of acid in your experiment? Is the result consistent with the stoichiometry shown in Table 40-1 for the three net reactions A, B, and C?

Finally, estimate the amount of nitrite produced (in units of millimoles) as described in part 4(d).

CONSIDER THIS

Did you get approximately the expected amount of acid based on the amount of NO gas reacted?

You added 5 mL of O_2 to 10 mL of NO. If these gases reacted as shown in Reaction (2) and if the product NO_2 is completely absorbed by water, we would expect that there would be no gas left after shaking the reaction mixture with water. Was there a measurable amount of gas left? If so, how much? What was the color of the gas? If this gas contained NO_2, what color would it be? Explain how this gas might be NO or O_2 or N_2 or a mixture of two gases (N_2 + NO or N_2 + O_2). Where would N_2 have come from? Is it possible from the results of your experiments to say definitely whether the gas left over in the tube is NO, NO_2, O_2, or N_2, or something else, or a mixture of gases? Can you suggest a way to try to determine the composition of the gas left over? What would be your hypothesis about the identity of the gas?

If NO_2 simply dissolved in the water without reacting, would you expect the solution to have a brown color? Was the solution brown? What do you think is a good hypothesis to explain the color of the solution?

Most chemists would classify Reactions (1) and (2) as oxidation–reduction reactions. For each reaction, specify the oxidation numbers that you would assign to the nitrogen (N), oxygen (O), and hydrogen (H) atoms on both sides of the equation. Based on this (arbitrary) assignment of oxidation numbers, what was oxidized and what was reduced in Reactions (1) and (2)?

DISCUSSION

In your report, include a discussion of the implications of your experimental results for the broad questions:

What human activities lead to the formation of acid rain?

Are nitrogen oxides the only source of the acids in acid rain?

If not, what are the identities and sources of other important atmospheric pollutants that produce acid rain?

Once formed, why doesn't NO just keep building up to higher and higher concentrations in the troposphere? That is, what processes remove nitrogen oxides from the troposphere?

Is it possible from your pH measurements and the semiquantitative estimate of the amount of nitrite produced to determine whether you got exclusively nitric acid, exclusively nitrous acid, or a mixture of nitric and nitrous acids? That is, considering all of the data, which of the three reactions—A, B, or C—is in best agreement with your experimental results?

NOTE The pattern of your pH measurements, when compared with the pattern of calculated pH values for various HNO_3 and HNO_2 solutions, should give you some important clues. The pattern of measurements is more important than the absolute value of the pHs, since the standardization of the pH meter is often in error by as much as ±0.1 pH unit. For this reason, the measured values for 0.008 and 0.004 M HNO_3 may give you a more reliable indicator than the calculated pH values.

Bibliography[1]

Ainscough, E. W.; Brodie, A. M. "Nitric Oxide—Some Old and New Perspectives," *J. Chem. Educ.* **1995,** *72,* 686–692.

Alyea, H. N. "Syringe Gas Generators" (Tested Demonstrations), *J. Chem. Educ.* **1992,** *69,* 65.

Bonner, F. T.; Stedman, G. "The Chemistry of Nitric Oxide and Redox-related Species," in M. Feelisch and J. S. Stamler (eds.), *Methods in Nitric Oxide Research,* John Wiley & Sons, New York, 1996, Chap. 1.

---

[1]The references in the Bibliography use the older name for NO (nitric oxide) rather than the IUPAC approved name (nitrogen monoxide).

Cotton, F. A.; Wilkinson, G.; Murillo, C. A.; Bochmann, M. *Advanced Inorganic Chemistry*, 6th ed., John Wiley & Sons, New York, 1999, pp. 323–333.

Everett, K. G. "The Titration of Air with Nitric Oxide: An Application of Gay-Lussac's Law of Combining Volumes in a General Chemistry Experiment," *J. Chem. Educ.* **1982,** *59,* 802–803.

Hornack, F. M. "Nitric Oxide Leftovers," *J. Chem. Educ.* **1990,** *67,* 496.

Mattson, B.; Anderson, M.; Schwennsen, C. *Chemistry of Gases: A Microscale Approach* (available from Flinn Scientific Inc.). Web site: Center for Microscale Gas Chemistry (http://mattson.creighton.edu/Microscale_Gas_Chemistry.html)

Shaaban, A. F. "The Integrated Rate Equation of the Nitric Oxide–Oxygen Reaction," *J. Chem. Educ.* **1990,** *67,* 869–871.

Atmospheric Chemistry of Nitrogen Monoxide

Smog and Acid Rain

Name _____

Date _____ Section _____

Locker _____ Instructor_____

Observations and Data

1. Preparing O_2 gas

Describe what you see when yeast cells are mixed with hydrogen peroxide in the syringe. Write a balanced equation for the net chemical reaction.

2. Preparing NO gas

Describe what you see when solid sodium nitrite is mixed with iron(II) sulfate in acid solution. Write a balanced equation for the net chemical reaction.

3. Reacting NO with O_2

Describe what you see when oxygen is mixed with nitrogen monoxide, NO. What is the color of the gas produced? Describe what happens next when deionized water is drawn into the syringe. What is the color of the solution produced?

Barometric Pressure and Room Temperature

Room temperature _____ °C

Barometric pressure _____ torr (mm Hg)

Vapor pressure of water at room temperature _____ torr (mm Hg)

Partial pressure of NO and O_2 gases _____ torr (mm Hg)
(after subtracting the vapor pressure of water at room temperature)

4. Analysis of the reaction solution

(a) *pH measurements*

Calculated pH values assume K_a for $HNO_2 = 7.1 \times 10^{-4}$ and all activity coefficients = 1.0

Solution	Measured pH	Calculated pH
NO/O_2 reaction solution		XXXX
0.008 M HNO_3		2.10
0.004 M HNO_3 + 0.004 M HNO_2	XXXX	2.34
0.004 M HNO_3		2.40
0.008 M HNO_2	XXXX	2.69

(b) *Titrating the acid produced*

Molarity of NaOH titrant _____ mol/L

	Trial 1	Trial 2
Initial volume of NaOH titrant	mL	mL
Final volume of NaOH titrant	mL	mL
Net volume of NaOH titrant	mL	mL
Millimoles of acid titrated	mmol	mmol

Name _____   Date _____

(c) *Semiquantitative analysis for nitrite ion*

Describe the nitrite test color comparison of the diluted reaction solution with the control solution $(1 \times 10^{-4}$ M $NaNO_2)$.

Describe the color of the nitrite test on the blank solution.

(d) Estimate the total amount of nitrite produced in units of millimoles.

Calculations and Data Analysis

Using the ideal gas law and assuming room temperature (in kelvins) and the partial pressures of NO and O_2, calculate the millimoles of NO in 10 mL of NO gas and 5 mL of O_2 gas.

Based on the amount (in millimoles) of NO taken, what would you expect to be the total amount (in millimoles) of acid produced?

Calculate the percentage yield of acid $= \dfrac{\text{millimoles acid titrated}}{\text{millimoles NO gas reacted}} \times 100\%$

After shaking the NO/O_2 reaction mixture with water, was there some gas left in the 10-mL syringe? How much? What are your thoughts about what might be the identity and source of this gas?

What was the color of the solution resulting from mixing the NO/O_2 gas mixture with water? What is your hypothesis to explain the color of the solution?

Rewrite Reactions (1) and (2), assigning oxidation numbers to the nitrogen (N), oxygen (O), and hydrogen (H) atoms on both sides of the equations. What species were oxidized and reduced in Reactions (1) and (2)?

Discuss your results, responding to some of the questions posed in the Discussion paragraphs of the Consider This section of the experiment.

Ozone and the Atmosphere

A Web-Based Activity

Purpose

• Use the resources of the Internet to discover information about the chemical and physical properties of ozone.

• Evaluate the role of ozone as both pollutant and protectant in the atmosphere.

• Identify some useful and practical applications of ozone.

NOTES TO INSTRUCTOR

A Web-based activity might be used in several ways, but we typically assign the activity in place of a laboratory experiment, with students allowed about a week to complete the assignment, working at their own pace and on their own schedules.

Introduction

If someone says the word *ozone*, the next word most likely to spring to mind nowadays is *hole*. Most of us have heard about the hole in the ozone layer, but few of us know much about the properties of ozone or even much detail about the ozone hole. After you complete this Web- or Internet-based activity, you should have a much better idea of the properties of ozone and how it affects our lives.

The activity is divided into three parts:

1. Some fundamental chemical and physical properties of ozone.

2. The role of ozone in the atmosphere, particularly in the troposphere (the layer of the atmosphere nearest to the Earth's surface) and the next layer up, called the *stratosphere* (stretching from about 10 km to 50 km above the Earth's surface).

3. Some practical and commercial applications of ozone.

Pre-Lab Activity

Before you log on to the Web, draw Lewis structures for ozone. There are two possible skeletal arrangements for which one could draw reasonable Lewis structures. The three O atoms could be arranged in an equilateral triangle with three equal ideal bond angles of 60°. However, calculations based on quantum mechanics show that this structure has a much higher energy and therefore is unstable with respect to rearrangement to a more stable structure having a central O atom bonded to two other O atoms, the O—O—O skeletal structure.

In principle, the O—O—O skeletal arrangement might be linear (like CO_2) or bent (like SO_2 and OF_2), depending on the total number of valence electrons in the molecule. The most plausible Lewis structures you can draw for the O—O—O skeletal arrangement will conform to the octet rule, have a lone pair on the central O atom, and also have two equivalent resonance structures. How would the presence of a lone pair on the central atom be expected to influence the shape of the O—O—O molecule?

To answer these and other questions, begin by drawing the two equivalent Lewis structures for the O—O—O arrangement and also for hydrogen peroxide, H_2O_2 (H—O—O—H), and molecular dioxygen, O_2. Since oxygen is a Period 2, main Group 16 (or VI) element, a plausible Lewis structure will give each O atom a share of eight electrons, conforming to the octet rule. For each structure, also show the formal charges on the atoms. All of the compounds are neutral molecules (not ions), so the formal charges in any structure should sum to zero. Improbable formal charge distributions (formal charges greater than ±1, adjacent atoms with formal charges of the same sign, etc.) can often be used to rule out implausible Lewis structures.

From their Lewis structures, determine the O—O bond order in O_3, H_2O_2, and O_2. (Revisit Study Assignment B to refresh your understanding of the bond order and how to determine it from the Lewis structures.) For the ozone Lewis structures, obtain the O—O bond order by averaging the two equivalent resonance structures. That is, imagine superimposing the two equivalent resonance structures and taking the O—O bond order as the average of the two superimposed Lewis structures.

Next, classify the ozone molecular structure into one of the thirteen VSEPR classes, AX_mE_n, shown in Table 19-1 of Experiment 19, Models of Molecular Shapes. (Here A represents the central atom, X an atom bonded to A, and E a lone pair of electrons on the central atom, A.) Also recall that in VSEPR theory, the shape of a molecule is assumed to be determined by the mutual repulsion of the electron domains of the central atom, whether they are lone pairs or bonding-electron domains. Further recall that a bonding-electron domain may consist of a single, double, or triple bond (or a bond of fractional bond order), with no distinction being made between an A—X single bond and an A—X bond of higher bond order.

Finally, from the VSEPR class, predict the shape for the ozone molecule and the predicted (ideal) O—O—O bond angle. (For a three-atom molecule with the X—A—X structure, there are only two possible shapes: linear or bent.)

Web Lab Activity

1. Physical and chemical properties of ozone

The most useful approach to finding information on the Web is to use a search engine such as Google (*http://www.google.com*). Google is one of the best search engines in its efficiency in finding relevant links. Enter into the search word box one or more search words such as *ozone,* or *ozone bond angle,* or *ozone bond length,* or *ozone bond energy;* click on the Google *Search* button and you will get a list of links to Web sites giving information about the topic. (In any search of this type, some hits will be useful, others less useful or irrelevant.)

We begin by trying to discover some properties of ozone:

(a) The O—O—O bond angle.

(b) The O—O bond length in ozone.

(c) Some further chemical and physical properties of ozone. For example, does ozone have a dipole moment? Does ozone have oxidizing or reducing properties? Does it have an odor? A color? Is it normally a gas, liquid, or solid at room temperature? How is ozone normally prepared in the laboratory? How does its stability compare to dioxygen, O_2?

2. Ozone in the atmosphere and the ozone hole

In this activity, you will be using the Internet to find information and data about ozone in the troposphere and stratosphere ozone levels. The amount of ozone present in the atmosphere has been studied for many years from many different sites around the globe.

Some of the information you will examine is written text with embedded figures, while some of the data are presented as single-frame pictures. There are even movies depicting the changing ozone levels over extended periods of time. Before beginning your Web exploration, read over the list of questions that you are being asked to answer. The goals of this lab are to learn how to navigate around the Web, to attempt to evaluate whether or not the information posted there is "good," and to discover the answers to the questions posed. You will visit many different sites sponsored by a variety of organizations.

When you find the answers to the questions posed below, be sure to note the Web address where the information appears. If you can, note the agency that has posted the information as well. You will need to include the Web references (a list of useful Web page addresses you visited) at the end of your lab report. Since you will need to write a summary of what you have learned as well as answering the questions posed, you may also want to keep track of any Web pages that contain other interesting information not specifically asked for.

Below are some instructions on how to begin. The words that are in *italics* are either the Web addresses that you need to type into the browser location line at the top of the screen or icons that you need to access with a click of the mouse.

To begin, you will need to gain access to the *World Wide Web* (www) using a Web browser such as Netscape or Internet Explorer. The first site we suggest that you visit is the National Aeronautics and Space Administration (NASA) Web page; you can get there by typing: *http://www.nasa.gov* in the location line. From the opening page go to the main NASA Web site. In the search box near the top of the page, type in *ozone* and click on Search. You will then get to the first of about a dozen pages with more than 50 links to topics related to ozone. Typing words such as *ozone hole* or *Dobson unit* into the search box will bring up a different set of links.

Don't be afraid to click on any link to check out what it contains. You can always click on the *back* button to return to your previous destination. Some of the links actually send you to the site for a different agency, such as the Environmental

Protection Agency (EPA), so keep an eye on the Web addresses.

The NASA Web site is updated periodically, so its appearance and links will change. This should pose no great difficulties, but you may have to use your own ingenuity in navigating and exploring the NASA Web site. Just remember, you can always back out of a page to get to where you started by clicking on the *back* button.

Besides NASA and the EPA (*www.epa.gov/ozone*) there is another U.S. government agency that does atmospheric research, the National Oceanic and Atmospheric Administration (NOAA). Visit the NOAA Aeronomy Laboratory of the NOAA Web site (*www.esrl.noaa.gov/csd*).

Another Web site that contains useful information is posted by the British Antarctic Survey. Their home page can be reached by typing *http://www.antarctica.ac.uk* in the location line.

Cambridge University also has an excellent Web site. Visit *www.atm.ch.cam.ac.uk/tour* and take the ozone hole tour.

The United Nations Environmental Program (UNEP) has produced documents on the environmental effects of ozone depletion. (Try *http://sedac.ciesin.columbia.edu/ozone*)

To find articles about ozone in the *Journal of Chemical Education* (JCE), use the JCE online index. Go to *http://jchemed.chem.wisc.edu* and click on the words "Advanced Search," located just under the search box icon. When the search page appears, type the word *ozone* in the Title search box, leaving the other search option boxes blank; for the Display Results options, choose the Year/Page option and Descending option. When you click on the search button, the search will recover all the articles with ozone in the title, listed in reverse chronological order (latest articles listed first). However, you will be able to print out the full text of the articles only if you are a subscriber or if your library has subscriber privileges. But most university and college libraries subscribe to JCE, so you could look up any article listed in the JCE search.

Make sure to read through the list of questions below before you begin. The questions are not necessarily listed in the order in which you will find the information. The order in which you find the information is unimportant, but you need to try to find answers to all of the questions posed. Do not be afraid to follow any good Web link; you never know what you may find in your quest for knowledge.

The questions in parts 2 and 3 should be answered either individually in the order given or in an essay format, as directed by your instructor. If you use an essay format, the essay should be a maximum of four typed pages. At the end of the questions or essay, include complete references (Web addresses) for all the Web sites you used. Attach your answers to the questions or your essay at the end of the report form.

Part 2 questions

(a) The boundaries of the regions of the atmosphere are defined by the turning points in a plot of altitude versus temperature. Make a rough sketch showing such a plot and indicating the primary regions of the atmosphere (troposphere, stratosphere, mesosphere). Where is most of the ozone in the atmosphere—in the troposphere or stratosphere?

(b) Explain why the effects of ozone in the troposphere are considered "bad," while the effect of ozone in the stratosphere is considered "good."

(c) The amounts of ozone in the total atmospheric column are given in Dobson units. (By atmospheric column we mean a column of air of defined cross-sectional area pointed straight up toward outer space.) Define the Dobson unit. Describe at least one instrument used to measure ozone levels in the total atmospheric column.

(d) How does the concentration of ozone normally vary with altitude, latitude (say, at the equator and the North and South poles), and season? Do ozone holes form over both the North and South poles? At what season(s) of the year do they form? Why do they form during this time over one pole and not the other?

(e) What role does stratospheric ozone play in the levels of UV light that reach the surface of the Earth? Describe (or write) the sequence of photochemical and chemical reactions that are responsible for the formation of ozone in the stratosphere. Describe the chemicals that catalyze ozone depletion and write the chemical reactions that play a role in loss of ozone in the stratosphere.

(f) How has the ozone hole changed since it was discovered? How have global ozone levels changed since the discovery of the hole? What is the current state of the ozone hole? What is the current state of global ozone levels?

(g) List three health hazards that would increase if levels of UV at the Earth's surface were to increase. What steps are countries taking to slow the rate of ozone depletion? What is the Montreal Protocol?

(h) What are some of the effects of ozone in the troposphere on animals and plants?

3. Practical and commercial applications of ozone

To learn something about the useful and beneficial properties of ozone, try using search words such as *ozonation, ozone water purification,* and *ozone air purification.* To get started learning about commercial suppliers of ozone generators, search *ozone generators* on the Web. Also visit the EPA Web site on the evaluation of ozone generators as air cleaners: *www.epa.gov/iaq/pubs/ozonegen.html.*

Briefly describe three practical or commercial applications of ozone and list the Web addresses of the sites that provided useful information about these applications.

■■■■ **CONSIDER THIS** ■■■■

For the ozone molecule, which has the O—O—O bond skeleton, do you think it would be possible for the two O—O bonds to be different, that is, to have different O—O bond lengths or bond energies? Explain your answer.

Use the information you found on the Web for the O—O—O bond angle and O—O bond length in ozone to calculate the distance between the two terminal O atoms (the two O atoms at the ends of the ozone molecule). If the molecule is linear, the distance would simply be the sum of the O—O bond lengths. If ozone has a bent structure, you will need to use some geometry or trigonometry to calculate the distance.

The type of smog found in Los Angeles, California, and other cities of the western United States is often called *photochemical smog* because sunlight plays an important role in its formation. The process begins with nitrogen monoxide, NO, often called nitric oxide, that is produced in any high temperature combustion or flame involving a fuel and air. Beginning with NO, write the sequence of chemical reactions, one of them involving photons of sunlight, that results in the formation of ozone in the troposphere, the layer of the atmosphere nearest the surface of the Earth.

It has been proposed that automobiles should have radiators coated with a catalyst that promotes the decomposition of ozone, so that as they moved along highways they would be purifying the air by removing ozone. Write a balanced net chemical reaction that describes the most likely outcome of the decomposition of ozone.

Ozone in the Atmosphere

A Web-Based Activity

Name _____

Date _____ Section _____

Locker _____ Instructor_____

Pre-Lab Preparation

(a) Given the bond skeletons shown below for ozone, hydrogen peroxide, and dioxygen, finish drawing their Lewis structures. Draw only structures in which the O atoms conform to the octet rule; show any formal charges on the atoms in each structure. Determine from the Lewis structures the O—O bond orders for the three molecules.

Ozone, O_3 (two equivalent resonance structures)	Hydrogen peroxide, H_2O_2	Dioxygen, O_2
O–O–O = O–O–O	H–O–O–H	O–O
O—O bond order: _____	O—O bond order: _____	O—O bond order: _____

(b) Using the Lewis structure for ozone and VSEPR theory, write the AX_mE_n classification for the predicted ozone molecular shape. (Recall that m is the number of atoms bonded to the central A atom and n is the number of lone pairs on the central A atom. (If necessary, revisit Experiment 19 to refresh your understanding of VSEPR theory.)

VSEPR class: _____;   Predicted VSEPR structure: _____ (linear or bent?)

Predicted VSEPR ideal O—O—O bond angle: _____

Web-Based Activity

1. Physical and chemical properties of ozone

Visit the Internet to discover the experimental O—O bond length and O—O—O bond angle in ozone.

(a) O—O bond length: _____ picometers;   O—O—O bond angle: _____ degrees

Web addresses where the information was found: _____

_____

_____

(b) How does the predicted VSEPR ideal O—O—O bond angle compare with the experimental value of the bond angle?

(c) Given the O—O bond lengths in H_2O_2 and O_2 as 141 and 121 pm, respectively, and considering the bond length you found for ozone, complete the table showing the bond lengths and bond orders of the three molecules. What conclusion can you draw about the relation between the O—O bond length and the O—O bond order in these three molecules?

Molecule	O—O bond length (pm)	O—O bond order
H_2O_2 (H—O—O—H)	141	
O_3		
O_2	121	

State your conclusion about the relation between bond order and bond length:

(d) Properties of ozone:

Dipole moment of ozone, if any: _____

Odor of ozone, if any: _____

Color of ozone, if any: _____

Physical state of ozone at room temperature and pressure: _____

Describe the oxidation–reduction properties of ozone.

How is ozone prepared in the laboratory or industrially?

How does the chemical stability of ozone compare with that of dioxygen, O_2?

2. Ozone in the atmosphere and the ozone hole

The following questions in parts 2 and 3 should be answered either individually in the order given or in an essay format, as directed by your instructor. If you use an essay format, the essay should be a maximum of four typed pages. At the end of the questions or essay, include complete references (Web addresses) for all the Web sites you used. Attach your answers to the questions or your essay at the end of the report form.

(a) The boundaries of the regions of the atmosphere are defined by the turning points in a plot of altitude versus temperature. Make a rough sketch showing such a plot and indicating the primary regions of the atmosphere (troposphere, stratosphere, mesosphere). Where is most of the ozone in the atmosphere—in the troposphere or stratosphere?

(b) Explain why the effects of ozone in the troposphere are considered "bad," while the effect of ozone in the stratosphere is considered "good."

(c) The amounts of ozone in the total atmospheric column are given in Dobson units. (By atmospheric column we mean a column of air of defined cross-sectional area pointed straight up toward outer space.) Define the Dobson unit. Describe at least one instrument used to measure ozone levels in the total atmospheric column.

(d) How does the concentration of ozone normally vary with altitude, latitude (say, at the equator and the North and South poles), and season? Do ozone holes form over both the North and South poles? At what season(s) of the year do they form? Why do they form during this time over one pole and not the other?

(e) What role does stratospheric ozone play in the levels of UV light that reach the surface of the Earth? Describe (or write) the sequence of photochemical and chemical reactions that are responsible for the formation of ozone in the stratosphere. Describe the chemicals that catalyze ozone depletion and write the chemical reactions that play a role in loss of ozone in the stratosphere.

(f) How has the ozone hole changed since it was discovered? How have global ozone levels changed since the discovery of the hole? What is the current state of the ozone hole? What is the current state of global ozone levels?

(g) List three health hazards that would increase if levels of UV at the Earth's surface were to increase. What steps are countries taking to slow the rate of ozone depletion? What is the Montreal Protocol?

(h) What are some of the effects of ozone in the troposphere on animals and plants?

3. Practical and commercial applications of ozone

Briefly describe three practical or commercial applications of ozone and list the Web addresses of the sites that provided useful information about these applications.

For the ozone molecule, which has the O—O—O bond skeleton, do you think it is possible for the two O—O bonds to be different, that is, to have different O—O bond lengths or bond energies? Explain your answer.

Use the information you found on the Web for the O—O—O bond angle and the O—O bond length in ozone to calculate the distance between the two terminal O atoms. If the molecule is linear, the distance would simply be the sum of the O—O bond lengths. If ozone has a bent structure, you will need to use some geometry or trigonometry to calculate the distance.

The type of smog found in Los Angeles, California, and other cities of the western United States is often called *photochemical smog* because sunlight plays an important role in its formation. The process begins with nitrogen monoxide, NO, often called nitric oxide, that is produced in any high temperature combustion or flame involving a fuel and air. Beginning with NO, write the sequence of chemical reactions, one of them involving photons of sunlight, that results in the formation of ozone in the troposphere, the layer of the atmosphere nearest the surface of the earth.

It has been proposed that automobiles should have radiators coated with a catalyst that promotes the decomposition of ozone, so that as they moved along highways they would be purifying the air by removing ozone. Write a balanced net chemical reaction that best describes the most likely outcome of the decomposition of ozone.

Ultraviolet Light, Sunscreens, and DNA

Purpose

• Determine the ultraviolet absorption spectra of DNA and sunscreens.

• Develop a better understanding of the interaction of ultraviolet light with DNA.

• Explore the role of sunscreens in protecting us from the effects of exposure to ultraviolet light.

Introduction

The Benefits of Sunshine Sunshine has undeniable physiological and psychological benefits. Vitamin D, a steroid hormone, is synthesized in the body by the action of sunlight on a precursor compound (7-dehydrocholesterol). Vitamin D is necessary in the diet to avoid the bone disease called *rickets*. If we don't get Vitamin D from other sources, we need exposure to sunlight to make it in our bodies.

Some people who live at far-northern or -southern latitudes where the winters are long and dark suffer from a depression disorder called *seasonal affective disorder* (SAD). A treatment for SADness is exposure to light, either artificial sunlight or the real thing. So we can say that sunlight is vital to our feeling of well-being.

Ultraviolet Light and DNA You can get too much of a good thing—sunshine. If you have ever spent a long time in the sun, you may have experienced one of the painful consequences of excessive exposure to ultraviolet (UV) radiation—a nasty sunburn. UV light has been implicated as a cause of skin cancer and has been reported to adversely affect skin structure.

The solar spectrum is shown in Figure 42-1. Solar radiation contains ultraviolet (UV), visible, and infrared light. The UV region extends from about 400 nanometers (nm), at the violet end of the visible spectrum, to 180 nm. UV radiation is often subdivided into three categories: UV-A (320–400 nm), UV-B (290–320 nm), and UV-C (180–290 nm). UV-C, the shortest-wavelength (highest-energy) UV radiation, is the most damaging. Fortunately, it is almost completely absorbed by ozone in the region of the Earth's atmosphere called the *stratosphere*.

The next-highest-energy category of UV radiation, UV-B, causes sunburn and most skin cancer. UV-A is also believed to cause damage to the structure of the skin, and it has been suggested that it is partially responsible for the increase in skin melanoma, a deadly form of cancer. Many sunscreens are more effective at blocking UV-B light than UV-A light.

These undesirable effects of UV radiation result from its absorption by the proteins and nucleic acids of skin cells. Of particular importance is the fact that absorption of UV radiation by deoxyribonucleic acid (DNA), the molecule in which our genetic information is encoded, can cause damage to the molecule, resulting in mutations that lead to cancer.

DNA molecules are the largest known; they consist of repeating units, called *nucleotides*, linked together to form a long chain. Each nucleotide contains one of the four ring-shaped structures shown in Table 42-1 linked to a deoxyribose molecule. The ribose molecules are in turn linked together by phosphate groups to form the long chain of a single strand of DNA. It is the order of these ring-shaped structures along the long double helix chain that encodes the genetic information carried by the DNA, and it is these molecular structures that absorb UV light.

How does UV light damage DNA? An example of a defect produced by absorption of UV light is the joining of two adjacent thymine bases to form a dimer, as shown in Figure 42-2.

Sunscreens We apply sunscreen to protect our skin from the damaging effects of UV radiation. Some active ingredients of sunscreen, such as titanium dioxide and zinc oxide, are called *sunblocks*. They are opaque solids insoluble in water and most solvents, and they work by both blocking and reflecting the UV radiation away from the skin. Other sunscreen ingredients, such as the organic molecules shown in Table 42-2, absorb the incident UV radiation and release the energy they thereby gain as heat (energy at long wavelengths).

Note that the six compounds in Table 42-2 have a structural feature in common—at least one hexagonal C_6 benzene-like ring, with an adjacent C=C or C=O group. The electrons that comprise these bonds are responsible for the absorption of UV light by the

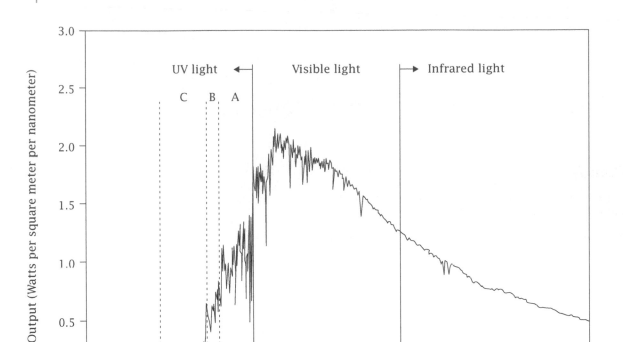

FIGURE 42-1 The wavelength distribution of the solar energy coming from the Sun as observed by satellite above the Earth's atmosphere. Essentially all of the UV-C light (180–290 nm) is absorbed by the normal levels of ozone in the Earth's atmosphere. About half the UV-B light (290–320 nm) penetrates the atmosphere to the surface of the Earth. The solar energy distribution is approximately that of a blackbody at a temperature of 5770 K.

TABLE 42-1 The UV Absorbing Components of DNA

A = ADENINE

T = THYMINE

G = GUANINE

C = CYTOSINE

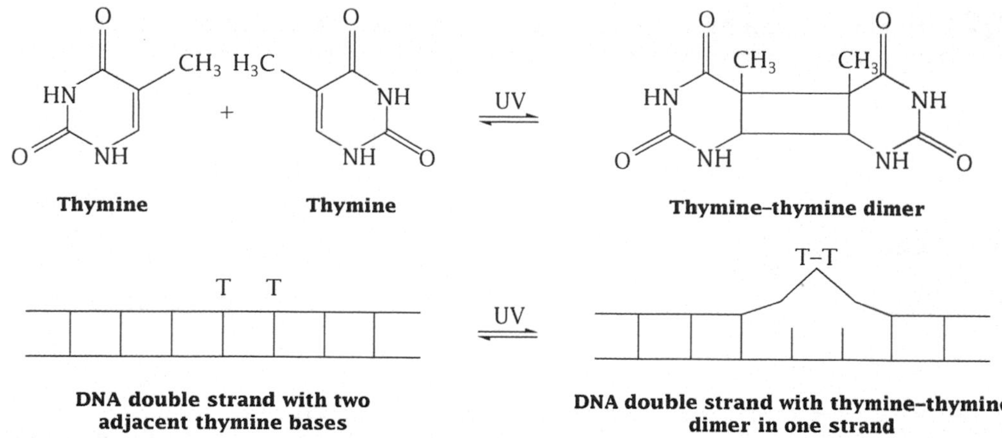

FIGURE 42-2 | **Absorption of UV light can cause the dimerization of two adjacent thymine bases in a strand of DNA, creating a defect.**

compounds. The additional structures attached to the ring cause the compounds to absorb at slightly different wavelengths.

The protective power of sunscreens is indicated by the familiar SPF (sun protection factor) ratings. These SPF ratings are determined by comparing the amount of UV energy necessary to cause a specified effect on protected skin (a mild sunburn), the *minimal erythemal dose* or MED, to the amount necessary to cause the same effect on unprotected skin:

$$\text{SPF} = \frac{\text{MED}_{\text{protected}}}{\text{MED}_{\text{unprotected}}} \qquad (1)$$

The amount of UV energy is usually measured as the length of time the skin must be exposed to the UV light to become sunburned.

There is an approximate relationship between the SPF value of a sunscreen and the fraction of light energy that passes through the layer of sunscreen at a particular wavelength. The fraction of the radiant power of the light that passes through the sample (that is, the fraction transmitted) is called the *transmittance*, defined as the ratio

$$\text{transmittance} = T = \frac{P}{P_0} \qquad (2)$$

where P_0 is the radiant power of the solar radiation incident on the layer of sunscreen covering the skin, and P is the radiant power of the solar radiation that passes through the layer of sunscreen to reach the surface of the skin itself.

The greater the SPF value, the smaller the fraction of solar radiation that reaches the skin, so SPF and transmittance must be inversely related:

$$\frac{1}{\text{fraction of light energy transmitted}} = \frac{1}{T} \cong \text{SPF} \qquad (3)$$

If the fraction of energy transmitted by the sunscreen layer on the skin is equal to 0.5 ($T = 0.5$), then the SPF value of the sunscreen approximately doubles. This means that if it takes 10 min for the unprotected skin to become mildly reddened, it would take about 20 min to redden using a sunscreen of SPF = 2, and 300 min (5 hours) for an SPF value of 30. (However, you would want to use a generous amount of sunscreen and more than one application if you planned to stay out in the sun as long as 300 min.)

The relationship between SPF and transmittance is only approximate because the SPF value corresponds to a biological effect measured over the whole range of wavelengths of the solar spectrum; the transmittance is a physical measurement made at a particular wavelength, which we will take to be the absorption maximum of the UV spectrum of the sunscreen.[1] Moreover, the biological response is not uniform across the range of wavelengths, but rather depends on what is called the *action spectrum* of the

[1]Think of the operation of obtaining the UV spectrum of a sample using a recording spectrophotometer as a process in which you acquire a data point by measuring the transmittance of the sample in a narrow range of wavelengths, typically 2–5 nm wide. This narrow window of wavelengths, called the *spectral bandpass* of the instrument, is then moved along from one end of the wavelength range to the other, acquiring data points as it scans. In most modern spectrophotometers, the transmittance values are converted to absorbance values; these data points are then displayed as an absorbance versus wavelength plot by connecting all the data points with a smooth line.

In another type of spectrophotometer, called a *diode array* spectrophotometer, all of the data points are acquired simultaneously by dispersing the spectrum along an array of light-sensitive diodes.

TABLE 42-2 The Most Common UV Absorbing Ingredients Found in Sunscreens

OXYBENZONE
BENZOPHENONE-3
2-HYDROXY-4-METHOXYBENZOPHENONE

OCTINOXATE
2-ETHYLHEXYL-P-METHOXYCINNAMATE

HOMOSALATE
3,3,5-TRIMETHYLCYCLOHEXYL SALICYLATE

OCTISALATE
2-ETHYLHEXYL SALICYLATE

AVOBENZONE (PARSOL 1789)
BUTYLMETHOXYDIBENZOYLMETHANE

OCTOCRYLENE

skin response, which is different from the absorption spectrum of the sunscreen.

Most modern scanning spectrophotometers display the absorbance versus wavelength for the sample. Absorbance is related to transmittance by the relation

$$\text{absorbance} = A = \text{Log}\left(\frac{P_0}{P}\right) = \text{Log}\left(\frac{1}{T}\right) \quad (4)$$

From Equation (4), we see that we can relate the absorbance of the sunscreen layer on the skin to the SPF value by taking the logarithm of both sides of Equation (3):

$$\log\left(\frac{1}{T}\right)_{\text{skin}} = A_{\text{skin}} \cong \log(\text{SPF}) \quad (5)$$

Now the absorbance, A, in Equation (5) is the absorbance of the sunscreen layer on the skin, which we cannot directly measure. Instead, we will measure the absorbance of sunscreens dissolved in a solvent.

How is the absorbance of the sunscreen layer on the skin related to the absorbance of a solution of sunscreen? We start with the relation known as Beer's law, which states that the absorbance is proportional to a constant times the thickness of the absorbing layer times the concentration of the absorbing molecules:

$$A = \varepsilon\, b\, c \quad (6)$$

where b is the thickness of the absorbing sunscreen and c is the concentration of sunscreen.

Taking the ratio of the solution absorbance to the absorbance of sunscreen on the skin we may write

$$\frac{A_{\text{soln}}}{A_{\text{skin}}} = \frac{(\varepsilon\, b\, c)_{\text{soln}}}{(\varepsilon\, b\, c)_{\text{skin}}} \cong \frac{(b\, c)_{\text{soln}}}{(b\, c)_{\text{skin}}} \quad (7)$$

where we assume that the absorption constants, ε, of the UV-absorbing ingredients are approximately independent of the matrix (inactive ingredients of the sunscreen or solvent) and therefore cancel. So Equation (7) can be rearranged and combined with Equation (5) to give the approximate relation between the absorbance of the sunscreen solution and log(SPF) for the sunscreen:

$$A_{\text{soln}} \cong A_{\text{skin}}\left[\frac{(b\, c)_{\text{soln}}}{(b\, c)_{\text{skin}}}\right] \cong \log(\text{SPF})\left[\frac{(b\, c)_{\text{soln}}}{(b\, c)_{\text{skin}}}\right]$$

$$= k \log(\text{SPF}) \quad (8)$$

where the constant k is equal to the ratio in brackets.

Equation (8) predicts that there will be an approximately linear relationship with slope equal to k between the solution absorbance and log(SPF) of the sunscreens. We will test this relationship by taking the UV spectra of solutions of sunscreens having different SPF values. In making this comparison for several sunscreens, it is important that each sunscreen solution contain the same mass concentration of sunscreen.

Pre-Lab Preparation (Extraction of DNA from Animal or Plant Tissues)

If your instructor so directs, read one or more of the references given in the Bibliography describing simple procedures for extracting DNA from animal or plant tissues. If you are to prepare your own DNA, your instructor will provide in advance of your laboratory a handout summarizing the procedure to be used; the plant or animal material; and the solutions, materials, and equipment required for the extraction procedure.

NOTES TO INSTRUCTOR

If time permits, you may wish to have students extract DNA from an animal or plant tissue, spool the extracted DNA, dissolve it in a buffer solution, and take the UV spectrum of the extracted DNA. If you plan to use this option, it is suggested that you hand out in advance of the scheduled laboratory a one-page summary of the protocol for the DNA extraction. Three sources of DNA extraction procedures are referenced in the Bibliography, differing slightly in the details.

An intermediate option is to provide students with a solution of purified DNA from a commercial source, which they can spool to recover the DNA. (See the Bibliography.)

If you do not plan to have students extract their own DNA, they may be provided with a solution of DNA that they can use to record the UV absorption spectrum.

If only a limited number of scanning UV–visible spectrophotometers are available, you may wish to have students work in small groups of 2–4 students in taking the UV spectra. To save time, it may be practical to take only one spectrum of a solution, then photocopy the original spectrum, dated and initialed by each student in the group, so that each student has copies of all spectra taken.

Experimental Procedure

Special Supplies: Scanning UV–visible spectrophotometers with 1-cm silica cells; at least four commercial sunscreens having a range of SPF values, e.g., SPF = 4, 8, 15, and 30 or 8, 15, 30, and 45.

DNA extraction option: Blenders and centrifuges; liquid detergent (such as Woolite or Ivory, Lemon Joy, or Palmolive dishwashing detergent); ethanol (or 2-propanol).

Chemicals: DNA solution, 40–50 mg/L DNA in Tris-EDTA (TE) buffer. [You can use an inexpensive grade of DNA such as Sigma-Aldrich Catalog No. D3159 DNA (herring sperm), degraded to crude oligonucleotides.]

Tris-EDTA (TE) buffer: 0.01 M Tris/0.001 M Na_2EDTA, adjusted to pH 8 with HCl; [Tris = Tris(hydroxymethyl)aminomethane or 2-amino-2-hydroxymethyl-1,3-propanediol. Contains, per liter of buffer, 1.21 g of Tris (base form), 5.4 mL of 1.00 M HCl, and 0.372 g of $Na_2EDTA \cdot 2H_2O$.]

At least four prepared stock sunscreen solutions with a range of SPF values, each containing 1.00 g sunscreen/100 mL in 2-propanol (isopropanol).

A. The UV absorption spectrum of DNA

1. Extraction of DNA (Optional) If you are to extract DNA from an animal or plant tissue, your instructor will provide in advance a handout summarizing the procedure to be used. Three references listed in the Bibliography (one of them Web-based) describe simple procedures that differ slightly in the details.

Once you have spooled enough DNA to see, describe your DNA in your notebook. Place the tool with the spooled DNA in a small, clean beaker containing 5 mL of Tris-EDTA (TE) buffer and dissolve the DNA in the buffer; then proceed to step 2.

2. Spectrophotometry of DNA With the help of your instructor, use the UV–visible spectrophotometer to record a spectrum for the DNA sample over the range of 220–450 nm. (If you extracted your own DNA and the extract is too concentrated, try diluting it about 4-fold in TE buffer.) Include a copy of this spectrum with your report.

Answer the following questions in your lab report:

(a) What is the approximate wavelength range of UV radiation absorbed by the DNA solution? (Extrapolate the sides of the absorption peak to the wavelength baseline to determine the range.)

(b) What is the wavelength of the absorption maximum—where the absorbance is greatest? Are there any absorbance troughs (absorbance minima)? At what wavelength(s)?

B. UV absorption of sunscreens

1. In your notebook, record the complete brand name, the SPF value, and the active ingredients of each sunscreen provided.

2. Stock solutions of sunscreens (1.0 g/100 mL in 2-propanol) have been prepared for your study. Take a

close look at the sunscreen stock solutions. Are there any undissolved or suspended solids?

Answer the following questions in your lab report:

(a) What differences do you observe in the sunscreen stock solutions? Do any of the solutions stand out as being very different from the rest? A different color? Presence of undissolved solids?

(b) If an undissolved solid is present, could it correspond to an active ingredient listed on the label of the sunscreen? If so, record the name of the material.

3. The stock sunscreen solutions are much too concentrated for UV absorption measurements (the absorbance would be off-scale). Make accurate 1:100 dilutions of each stock sunscreen solution in deionized water by weighing 1.00 ± 0.02 g (~1.0 mL) of the sunscreen solutions in 2-propanol into a 250-mL beaker and adding 99 ± 1 mL of deionized water. The final solutions will contain 0.010 g/100 mL of sunscreen. Mix the solutions thoroughly. With the help of your instructor, use the UV–visible spectrophotometer to record a spectrum for each diluted sunscreen sample over the range of 220–450 nm. Include copies of these spectra with your report.

Answer the following questions in your lab report:

(a) What categories of UV radiation are absorbed by each sunscreen? Which of the sunscreens protect best against both UV-A and UV-B? Explain your answer.

(b) Record the wavelengths of the absorption peaks in the sunscreen spectra. (There may be more than one absorption peak in a spectrum.) Are there any differences among the sunscreens? Do the differences correlate with differences in active ingredients?

(c) How do the *ranges* of the wavelengths absorbed by the sunscreens compare to the range of wavelengths absorbed by the DNA solution?

(d) Determine the absorbance of *each* sunscreen sample at its wavelength of maximum absorbance. Absorbance is proportional to the concentration of the absorbing material. Which sunscreen has the lowest absorbance? Based on the SPF ratings of the sunscreens, would you have expected this result? Explain your reasoning.

(e) When you take the UV spectrum of a solution of a sunscreen, will the SPF value of the sunscreen be accurately reflected by the solution absorbance if the sunscreen contains an active ingredient such as TiO_2 that is an insoluble solid? Explain how the correlation between absorbance and log(SPF) might be affected.

(f) Make a plot of the absorbance of each sunscreen solution (measured at the wavelength of maximum absorbance) versus log(SPF) for the sunscreen. The slope of this plot is predicted from Equation (7) to be approximately equal to k, the ratio of $(b\,c)_{soln}/(b\,c)_{skin}$.

CONSIDER THIS

1. In the marketplace, there is a tendency for makers of sunscreens to get into an SPF ratings game, but the benefits of SPF ratings above 30 are marginal. Australia has no numerical ratings above 30; rather, they use the category 30+, and the U.S. Food and Drug Administration is planning to move to a similar rating scheme.

To understand why protection gains above SPF = 30 are marginal, assume that SPF = $1/T$ as given by Equation (3). The transmittance, T, of a sunscreen layer with SPF = 30 would then be $1/30$ = 0.033. So only about 3.3% of UV-B light (the UV region where sunscreens are most effective) would penetrate the layer of sunscreen. What percentage of UV-B light would penetrate a layer of sunscreen with SPF = 45? What would be the gain in protection in going from SPF 30 to 45?

2. One hypothesis for the effect of UV-A light on skin is that free radicals, produced by absorption of photons of UV light, initiate damage to the skin. The highest energy photon of UV-A light corresponds to a wavelength of 320 nm. Calculate the energy of a mole of photons of wavelength 320 nm in units of kJ/mol. Recall that one photon has an energy given by

$$E = h\nu = \frac{h\,c}{\lambda}$$

where h is Planck's constant, c equals the speed of light, and λ is the wavelength of the light photon. This energy is multiplied by Avogadro's number to obtain the energy of a mole of photons. Consult a table of bond dissociation energies. Are any of the bond dissociation energies comparable to the energy of a mole of photons of wavelength 320 nm? Which ones? (You can exclude bonds involving halogen atoms, which are less likely to be found in skin cells.)

Bibliography

Abney, J. R.; Scalettar, B. A. "Saving Your Students' Skin. Undergraduate Experiments That Probe UV Protection by Sunscreens and Sunglasses," *J. Chem. Educ.* **1998,** *75,* 757–760.

Fujishige, S.; Takizawa, S.; Tsuzuki, K. "A Simple Preparative Method to Evaluate Total UV Protection by Commercial Sunscreens," *J. Chem. Educ.* **2001,** *78,* 1678–1679. Proposes taking UV spectra of sunscreens as thin smears in Visking cellulose tubing instead of dissolving them in solvents.

Kimbrough, D. R. "The Photochemistry of Sunscreens," *J. Chem. Educ.* **1997,** *74,* 51–53. Explains what can happen after absorption of UV light by a sunscreen molecule.

Walters, C.; Keeney, A.; Wigal, C. T.; Johnston, C. R.; Cornelius, R. D. "The Spectrophotometric Analysis and Modeling of Sunscreens," *J. Chem. Educ.* **1997,** *74,* 99–102. Also see the letter by Underwood, G.; MacNeil, J. *J. Chem. Educ.* **2001,** *78,* 453. Recommends using 2-propanol to prepare stock solutions of sunscreen.

Simple DNA Extraction Procedures

Carlson, S. "The Amateur Scientist. Spooling the Stuff of Life," *Scientific American* **1998,** *279,* 96–97 (September 1998 issue).

Genetic Science Learning Center (site last checked 03/26/09): "How to Extract DNA from Anything Living" (split peas recommended) http://gslc.genetics.utah.edu/content/labs/extraction/howto/

"DNA Extraction from (raw) Wheat Germ" http://www.wonderhowto.com/how-to/video/how-to-extract-dna-from-wheat-germ-259776/

Nordell, K. J.; Jackelen, A-M. L.; Condren, S. M.; Lisensky, G. C.; Ellis, A. B. "Liver and Onions: DNA Extraction from Animal and Plant Tissues," *J. Chem. Educ.* **1999,** *76,* 400A–400B (March 1999).

DNA Spooling Educational Kit, Sigma-Aldrich Product No. D8666. The kit provides materials for a hands-on demonstration of DNA spooling (precipitation). The kit costs about $33 (2009 price) and is sufficient for 25 student demonstrations. It contains 25 mL of purified DNA (salmon testis), 1 mg DNA/mL, in Tris-EDTA (TE) buffer, pH 8; 25 mL of 3 M sodium acetate solution; technical bulletin ED-100 describing the procedure. Isopropanol (2-propanol), test tubes, and other simple equipment must be supplied by the user. http://www.sigmaaldrich.com (search for Product D8666).

Ultraviolet Light, Sunscreens, and DNA	Name _____
	Date _____ Section _____
	Locker _____ Instructor_____

A. The UV Absorption Spectrum of DNA

1. Extraction of DNA (optional)

If you isolated DNA from an animal or plant tissue, describe the source of the DNA and attach a copy of the procedure used to extract the DNA.

2. Spectrophotometry of DNA

Attach a copy of the UV absorption spectrum of DNA over the range of 220–450 nm and answer the following questions.

(a) What is the approximate wavelength range of UV radiation absorbed by the DNA solution? (Extrapolate the sides of the absorption peak to the wavelength baseline to determine the range.)

(b) What is the wavelength of the absorption maximum, where the absorbance is greatest? Are there any absorbance troughs (absorbance minima)? At what wavelength(s)?

B. UV Absorption of Sunscreens

1. Describe the sunscreens in the table below.

Code	SPF value	Sunscreen brand name	Active ingredients listed on the label
A			
B			
C			
D			
E			

2. Stock solutions of sunscreens (1.0 g/100 mL in 2-propanol) have been prepared for your study. Take a close look at the sunscreen stock solutions. Are there any undissolved or suspended solids?

Answer the following questions in your lab report:

(a) What differences do you observe in the sunscreen stock solutions? Do any of the solutions stand out as being very different from the rest? A different color? Presence of undissolved solids?

(b) If an undissolved solid is present, could it correspond to an active ingredient listed on the label of the sunscreen? If so, record the name of the material.

3. UV absorption spectra of diluted sunscreen solutions over wavelength range 220–450 nm. Attach copies of the spectra of the sunscreen solutions and answer the following questions in your lab report. Identify the sunscreens by their code letters.

(a) What categories of UV radiation are absorbed by each sunscreen? UV-A, UV-B? Which of the sunscreens protect best against both UV-A and UV-B? Explain your answer.

A:

B:

C:

D:

E:

(b) Record the wavelengths of the absorption peaks in the sunscreen spectra (identified by their code letters). (There may be more than one absorption peak in a spectrum.) Are there any differences among the sunscreens? Do the differences correlate with differences in active ingredients?

A:

B:

C:

D:

E:

(c) How do the *ranges* of the wavelengths absorbed by the sunscreens compare to the range of wavelengths absorbed by the DNA solution?

Name _____   Date _____

(d) Determine the absorbance of *each* sunscreen sample, identified by its code letter, at its wavelength of maximum absorbance. Which sunscreen has the lowest absorbance? Based on the SPF ratings of the sunscreens, would you have expected this result? Explain your reasoning.

A:

B:

C:

D:

E:

(e) When you take the UV spectrum of a solution of a sunscreen, will the SPF value of the sunscreen be accurately reflected by the solution absorbance if the sunscreen contains an active ingredient such as TiO_2 that is an insoluble solid? Explain how the correlation between absorbance and log(SPF) might be affected.

(f) Make a plot of the absorbance of each sunscreen solution (measured at the wavelength of maximum absorbance) versus log(SPF) for the sunscreen. The slope of this plot is predicted from Equation (8) to be approximately equal to k, the ratio of $(b\,c)_{soln}/(b\,c)_{skin}$.

Absorbance versus log(SPF)

Absorbance

log(SPF)

Is there a correlation between the absorbance of the sunscreen solutions, measured at the wavelength of maximum absorbance and log(SPF) of the sunscreens? Is the correlation approximately linear? If, so what is the value of k, the slope of the plot?

CONSIDER THIS

1. In the marketplace, there is a tendency for makers of sunscreens to get into an SPF ratings game, but the benefits of SPF ratings above 30 are marginal. Australia has no numerical ratings above 30; rather, they use the category 30+ and the U.S. Food and Drug Administration is planning to move to a similar rating scheme (see www.fda.gov/consumer/updates/sunscreen082307.html).

To understand why protection gains above SPF = 30 are marginal, assume that SPF = $1/T$ as given by Equation (3). The transmittance, T, of a sunscreen layer with SPF = 30 would then be $1/30 = 0.033$. So only about 3.3% of UV-B light (the UV region where sunscreens are most effective) would penetrate the layer of sunscreen. What percentage of UV-B light would penetrate a layer of sunscreen with SPF = 45? What would be the gain in protection in going from SPF 30 to 45?

2. One hypothesis for the effect of UV-A light on skin is that free radicals, resulting from bond breaking produced by absorption of photons of UV light, initiate damage to the skin. The highest energy photon of UV-A light corresponds to a wavelength of 320 nm. Calculate the energy of a mole of photons of wavelength 320 nm in units of kJ/mol. Recall that one photon has an energy given by

$$E = h\nu = \frac{h\,c}{\lambda}$$

where h is Planck's constant, c equals the speed of light, and λ is the wavelength of the light photon. This energy is multiplied by Avogadro's number to obtain the energy of a mole of photons. Consult a table of bond dissociation energies. Are any of the bond dissociation energies comparable to the energy of a mole of photons of wavelength 320 nm? Which ones? (You can exclude bonds involving halogen atoms, which are less likely to be found in skin cells.)

Natural Radioactivity

The Half-Life of Potassium-40 in Potassium Chloride

Purpose

- Explore the phenomenon of radioactivity using a common natural source, potassium chloride.

- Measure the half-life of potassium-40 in potassium chloride.

Pre-Lab Preparation

Radioactivity and nuclear radiation are usually associated with quite negative contexts: Hiroshima, Chernobyl, and Three Mile Island, for instance. But as you will see in this experiment, it is a common everyday phenomenon and can be found everywhere around (and in) us.

In all of your chemical studies until this discussion of nuclear chemistry, you operated with the *law of mass conservation* and its corollary idea, the conservation of atoms—that is, mass is not created or destroyed, and similarly, atoms are not created or destroyed in a chemical reaction, only rearranged. When radioactivity was discovered by Henri Becquerel in 1896 and further explored by Marie and Pierre Curie, Ernest Rutherford, and Frederick Soddy, the experiments showed that certain nuclei are unstable and disintegrate, giving off particles and electromagnetic radiation.[1]

Types of radioactivity

Rutherford, the Curies, and their colleagues explored the properties of these rays emitted from uranium and the recently discovered polonium and radium. They noted that when the rays were passed through a magnetic field, they split into three components, which Rutherford called *alpha* (α), *beta* (β), and *gamma* (γ). The alpha and beta rays were unlike X rays in that they were deflected by the magnetic field, drastically in the case of beta rays, less so for alpha rays. The gamma rays were unaffected by the magnetic field and were characterized as high-energy X rays, a form of electromagnetic radiation, similar to visible light but with much greater energy. The alpha and beta rays had properties of particles, and by careful measurement of their paths in the magnetic fields, their masses and charges could be determined. Their properties are shown in Table 43-1.

You might remember that Rutherford is remembered for more than his explorations in radioactivity. He went on to utilize the alpha particles of his radium sources to probe the atom and discovered the existence of the atomic nucleus with this tool.

Mass–energy equivalence

A key feature of nuclear reactions that makes this subject so different from the rest of chemistry is the astounding amount of energy released in a nuclear decay. Take the alpha decay of uranium-238 as an example (in the net reaction, the charged alpha particle acquires two electrons to become a helium atom):

$$^{238}_{92}\text{U} \rightarrow \, ^{4}_{2}\alpha \, + \, ^{234}_{90}\text{Th} \quad (1)$$

MASS: 238.05078   4.00260   234.04359

Mass is lost during this process:

change in mass, $\Delta m = (4.00260 + 234.04359)$
$$- 238.05078$$
$$= -0.00459 \text{ amu}$$

---

[1]Wilhelm Roentgen had discovered X rays in 1895, and upon learning of this discovery, Becquerel set about on a search for substances that emit X rays. With a somewhat accidental experiment, he discovered that photographic plates stored in the dark next to samples of uranium became "fogged" by some sort of invisible ray, later identified as nuclear radioactivity.

TABLE 43-1 **Types of Radiation**

Type of Radiation	Alpha Particles (helium nuclei)	Beta Particles (electrons)	Gamma Ray (photons)
Symbol	α	β	γ
Mass (amu)	4	1/2000	0
Charge	+2	−1	0
Speed	Slow	Fast	Very fast (the speed of light)
Ionizing Ability	High	Medium	Very low
Penetrating Power	Low	Medium	High
Stopped by	1 sheet of paper	3 mm of aluminum	50 mm of lead

According to Einstein's *theory of relativity*,[2] this missing mass manifests itself as energy:

$$\Delta E = \Delta m\, c^2$$

$$= (-0.00459\,\text{amu}) \times \frac{1.66 \times 10^{-27}\,\text{kg}}{\text{amu}}$$

$$\times \left(\frac{3.00 \times 10^8\,\text{m}}{\text{s}}\right)^2$$

$$= -\frac{6.86 \times 10^{-13}\,\text{J}}{\text{atom}}$$

$$\times \frac{6.022 \times 10^{23}\,\text{atoms}}{\text{mol}}$$

$$= -4.13 \times 10^{11}\,\frac{\text{J}}{\text{mol}} \qquad (2)$$

Comparing this quantity to the energy of a typical chemical reaction or bond energy (300 kJ/mol = 3×10^5 J/mol) shows that it is a million-fold greater and indicates why radioactivity can be so dangerous. These tremendous energies are carried away from the nuclear disintegration in the kinetic energy of the alpha or beta particle and the photon energy of the gamma ray.

Penetration and shielding

The damage that nuclear radiation can do to biological tissues has mostly to do with the energy and charge of the particles ejected from the radioactive nuclei. As they travel through tissues, charged particles and gamma photons leave behind a trail of ionized atoms and molecules, triggering the breaking of chemical bonds. Photons or particles with energy greater than 10,000 kJ/mol are described as *ionizing*

radiation. This includes ultraviolet, X rays, and cosmic rays as well as alpha, beta, and gamma radiation from nuclear decays.

Alpha particles are helium atom nuclei. They are relatively large and massive compared to beta particles, have a +2 charge, and rapidly lose energy because they interact so strongly with the charged particles in atoms. Collisions with molecules in a few centimeters of air or just a sheet of paper are sufficient to dissipate the energy of an alpha particle; thus, they do not penetrate much beyond the surface of the skin if exposure occurs. Even air can be an efficient shield from alpha radiation. But if alpha-emitting radioisotopes are inhaled, they do great damage to lung tissue, leading to lung cancer.

A beta particle moves much faster than an alpha with similar energy (why?) and thus penetrates more deeply into tissue. Its smaller charge also causes it to interact less strongly with matter. In water or the human body, beta particles travel 1–2 cm before losing their ionizing power, but a 3-mm sheet of aluminum is sufficient to block their travel.

The energetic photons formed in nuclear decays, the gamma rays (also known as cosmic rays if they are formed by the nuclear reactions in the sun and stars), penetrate soft tissues readily. Often a gamma ray can pass right through your body without causing any ionization. Nonetheless, this penetrating power makes them quite dangerous and requires shielding with significant amounts of lead or concrete to provide protection. Figure 43-1 provides a schematic representation of these penetration differences.[3]

---

[2]Einstein's 1905 theory, now verified by experiment, has required an update of the *law of mass conservation* to the *law of mass/energy conservation.*

[3]High-energy neutrons are formed in nuclear reactions such as the fission reactions of a nuclear power plant. With their lack of electrical charge, neutrons are quite penetrating and require significant amounts of lead, concrete, or water shielding (hydrogen atoms are effective neutron absorbers).

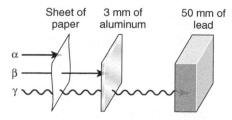

Sheet of paper   3 mm of aluminum   50 mm of lead

α
β
γ

FIGURE 43-1 | **The penetrating power of alpha, beta, and gamma radiation. Alpha and beta particles can be entirely stopped. Gamma radiation is electromagnetic in nature and its intensity declines exponentially as it passes through an absorbing material, so it can be attenuated to an insignificant level but is not entirely stopped.**

Units of radiation measurement

A variety of units are used to measure radioactivity and radiation doses. Table 43-2 summarizes these units and their definitions. The newer SI units are the becquerel (Bq), sievert (Sv), and gray (Gy); the older units are the curie (Ci), rad, roentgen (R), and rem.

Some units are measures of the number of decay events (becquerel and curie); some measure the physical dose or energy deposited by the radiation (the gray, rad, and roentgen); and some attempt to measure the biological dose or amount of biological damage expected (sievert and rem). Because of these different perspectives, it is not always possible to convert from one unit to another without knowing a lot about the context of the specific situation.

Natural radioactivity

You may be used to thinking of radioactivity as a manmade problem created in the last half of the twentieth century. But most of the radiation that you are exposed to comes from natural sources, including those found in your own body.

Where did the elements, including the radioactive ones, come from? The cosmological theory popularly known as the *Big Bang* provides a way to explain the origin of the elements. Calculations based on the theory predict that in the initial stages, hydrogen and helium were formed in about a 3:1 ratio from the

TABLE 43-2 Units of Radioactivity and Radiation Dose

Unit and Abbreviation	Definition	Notes
International System (SI) units of activity and dose		
Becquerel, Bq	1 disintegration per second	Replaces the curie
Gray, Gy	Absorbed or physical dose; 1 joule deposited per kilogram of absorber; 1 gray = 100 rad	Replaces the rad; accounts for the energy of absorbed radiation but not for biological effects
Sievert, Sv	Absorbed dose × quality factor, Q, gives the biological dose; $Q = 1$ for electrons and gamma rays; $Q = 20$ for alpha particles; 1 sievert = 100 rem	Replaces the rem; 3.5 millisieverts is roughly the average dose received in one year's exposure to natural radiations
Old system units of activity and dose		
Curie, Ci	3.7×10^{10} disintegrations per second	Equals the rate for 1 g of radium.
Roentgen, R	2.58×10^{-4} coulombs per kilogram of dry air; exposure to 1 R of gamma radiation gives approximately 1 rem dose	Measures the ionization caused by radiation
Radiation absorbed dose, rad	100 ergs per gram of absorber 1 rad = 0.01 gray	Accounts for the energy of absorbed radiation but not biological effects
Roentgen equivalent, man, rem	Absorbed dose × quality factor; quality factor varies by type of radiation to account for biological effect 1 rem = 0.01 sievert	350 mrem is roughly the average dose received in one year's exposure to natural radiation

energy of the Big Bang, in line with what is observed. As these light elements coalesced into stars, nucleosynthesis began, forming the other elements from hydrogen and helium. Larger stars have shorter lives, which end in a supernova event that disperses the atoms formed in the stars. New stars and planets can be formed when this material again coalesces.

Primordial radioactive atoms with short half-lives have disappeared from Earth,[4] but radioactive atoms with half-lives on the order of a billion years or more still exist on Earth; the energy they produce is sufficient to keep the Earth's mantle at a temperature above 1000 °C, with the thin crust of the continental plates floating on the mantle. Of particular interest is uranium 238, which undergoes a series of decay reactions (8 different alpha emissions and 6 beta decays) ending in stable lead-206. Figure 43-2 shows this decay series.

In the upper portions of the Earth's atmosphere, cosmic-ray neutrons from the sun collide with nitrogen atoms, forming carbon-14 and protons:

$$\,^{14}_{7}\text{N} + \,^{1}_{0}\text{n} \rightarrow \,^{14}_{6}\text{C} + \,^{1}_{1}\text{H} \qquad (3)$$

This reaction produces a steady supply of carbon-14, which is quickly converted to $^{14}\text{CO}_2$ and becomes part of the Earth's carbon cycle as CO_2 in the atmosphere becomes part of the food chain and ultimately part of every living thing, including our bodies. Carbon-14 decays with the emission of a beta particle:

$$\,^{14}_{6}\text{C} \rightarrow \,^{14}_{7}\text{N} + \,^{0}_{-1}\beta \qquad (4)$$

A 70-kg (154-lb) human body contains about 19 ng of carbon-14 and is exposed to about 3,000 disintegrations each second from this source. Your body has to absorb the energy from each of these nuclear decays as well as cope with the damage caused when the carbon atoms turn into nitrogen atoms and change the chemistry of some of the molecules making up your cells.

The potassium-40 that we are studying in this experiment is one of the primordial radioactive elements, those left over from the formation of the Earth. It decays by several different processes (that is, it has a branched decay scheme):

$$\,^{40}_{19}\text{K} \rightarrow \,^{0}_{-1}\beta + \,^{40}_{20}\text{Ca} \quad (\beta^- \text{ or beta decay})$$

$$\,^{40}_{19}\text{K} \rightarrow \,^{0}_{+1}\beta + \,^{40}_{18}\text{Ar} \quad (\beta^+ \text{ or positron decay}) \qquad (5)$$

$$\,^{40}_{19}\text{K} + \,^{0}_{-1}\text{e} \rightarrow \,^{40}_{18}\text{Ar} \quad (\text{electron capture decay})$$

---

[4]*Radioisotope* refers to a radioactive isotope of an element. Some elements have stable isotopes, such as hydrogen, consisting of 99.985% ^{1}H and 0.015% deuterium, ^{2}H, along with traces of radioactive tritium, ^{3}H. Most of the carbon atoms in your body have stable ^{12}C and ^{13}C nuclei but a small fraction are unstable ^{14}C.

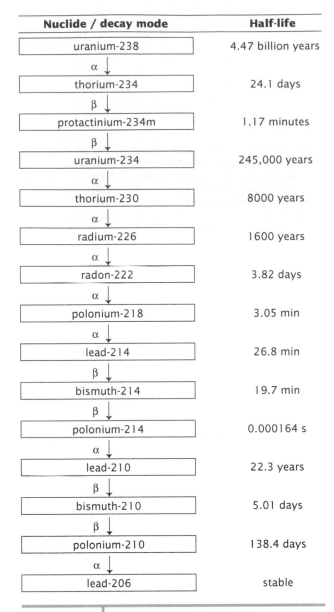

Nuclide / decay mode	Half-life
uranium-238	4.47 billion years
α ↓	
thorium-234	24.1 days
β ↓	
protactinium-234m	1.17 minutes
β ↓	
uranium-234	245,000 years
α ↓	
thorium-230	8000 years
α ↓	
radium-226	1600 years
α ↓	
radon-222	3.82 days
α ↓	
polonium-218	3.05 min
α ↓	
lead-214	26.8 min
β ↓	
bismuth-214	19.7 min
β ↓	
polonium-214	0.000164 s
α ↓	
lead-210	22.3 years
β ↓	
bismuth-210	5.01 days
β ↓	
polonium-210	138.4 days
α ↓	
lead-206	stable

FIGURE 43-2 | **The decay chain for uranium-238. Radon-222 is a gas that penetrates into buildings from the soil. Radon and its short-lived daughter products contribute significantly to the average total annual radiation dose received by the U.S. population. The decay chain ends with the formation of stable lead-206.**

About 88% of the potassium-40 atoms decay by ordinary beta decay to give stable calcium-20 and a combined 12% decay by positron emission or electron capture to form stable argon-40.[5]

---

[5]Electron capture is a rare form of decay in which the nucleus captures an electron from the innermost electron shell, thereby converting a proton into a neutron. It is always accompanied by X-ray emission as electrons drop into the vacant energy state created when an inner electron is captured.

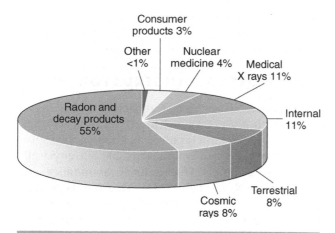

Consumer
products 3%

Other
<1%

Nuclear
medicine 4% Medical
X rays 11%

Radon and
decay products
55%

Internal
11%

Terrestrial
8%

Cosmic
rays 8%

FIGURE 43-3 | **Sources of radiation exposure to the general U.S. population. Exposure to radon and its decay products differs depending on location. Cosmic rays are high-energy gamma rays and energetic particles from extraterrestrial sources like the sun. Total average annual dose to adult U.S. population is about 3600 microsieverts (360 millirem).**

We all are exposed to significant, ongoing amounts of radioactivity from natural sources, typically around 3500 μSv (350 mrem) per year. As you can see from Figure 43-3, more than half of this is the result of exposure to radon and its decay products, some of the intermediate products of the uranium decay series. This is a bigger problem in some areas of the world—particularly regions of India, Brazil, and Sudan that have mineral deposits in the ground with higher levels of radioisotopes. Unavoidable sources of radiation include this exposure to radon and to other products of the uranium decay series (polonium, radium, and thorium).

A smaller but also unavoidable dose comes from extraterrestrial sources, the nuclear reactions that power our sun and the stars. Radioactive decays from the radioisotopes that are naturally in our bodies, such as potassium-40, carbon-14, tritium, and polonium, contribute a dose that is roughly equal to the dose from medical X rays. The small traces of radioisotopes formed from the manufacture of nuclear weapons over the years and the processing of fuel for nuclear power plants comprise a tiny portion of the terrestrial exposure (an estimated 0.1%).

Airline passengers are exposed to additional radiation, since it is our atmosphere that shields us from much of the sun's radioactive emissions. A Los Angeles to New York flight contributes a dose of about 19 μSv (1.9 mrem). But only flight crews who fly for several hours per day on average receive noticeable increases from this source, roughly doubling the average exposure.

Rates of radioactive decay

All radioactive decay processes are governed by a first-order rate law:

number of atom decays per s = decay rate constant × number of radioactive atoms

$$-\frac{dN}{dt} = k \times N \qquad (6)$$

The minus sign means that N is decreasing with time. The integrated form of the differential rate equation is given by

$$\ln\left(\frac{N_t}{N_0}\right) = -kt \quad \text{or} \quad \frac{N_t}{N_0} = e^{-kt} \qquad (7)$$

where N_0 is the number of radioactive nuclei present at the start and N_t the number at some later time, t. The rate constant, k, is related to the half-life of the decaying species by

$$k = \frac{\ln(2)}{t_{1/2}} = \frac{0.693}{t_{1/2}} \qquad (8)$$

These processes are unaffected by temperature and chemical context, so they provide useful atomic clocks for measuring time.

A sample of uranium-238 with a half-life of 4.47×10^9 years (1.41×10^{17} s) has a decay rate expressed as the *specific activity* (number of atoms decaying per second per gram) given by

$$\frac{\text{atom decays}}{\text{s} \times \text{g}} = \frac{\ln(2)}{t_{1/2}} \times \frac{N_A}{M} =$$

$$\frac{0.693}{1.41 \times 10^{17}\,\text{s}} \times \frac{6.02 \times 10^{23}\,\text{atom}}{\text{mol}} \times \frac{1\,\text{mol}}{238\,\text{g}} =$$

12,400 disintegrations/s/g of ^{238}U $\qquad (9)$

where $t_{1/2}$ is the half-life (in seconds), N_A is Avogadro's number (atoms/mol), and M is the molar mass of ^{238}U (g/mol). Note that the shorter the half-life, the greater the specific activity.

For carbon-14 (half-life = 5730 years), the comparable rate is 1.7×10^{11} decays/s/g of ^{14}C. Factoring in the isotopic abundance of carbon-14 in living things ($\sim 1.5 \times 10^{-10}$ %) gives a decay rate of 15 disintegrations/min for each gram of total carbon present.

Radioactive dating methods

Equation (7) provides a way to use radioactive decay processes as atomic clocks for determining the age of objects. In the potassium-40/argon-40 method, the potassium-40 content of a rock sample is measured along with the argon-40 content. The source of the argon-40 is the combined positron and electron capture decay of potassium-40 shown in the Equation (5) decay scheme. The measured amount of ^{40}K in the rock is equal to N_t in Equation (7). The amount of potassium-40 plus argon-40 gives the original potassium-40 content equal to N_0 (assuming no loss of argon-40). Equation (7) can then be solved for the elapsed time, t, since the rock was formed.

Radiation detectors

Various detectors are used to measure radiation. Certain substances emit light when struck by the products of radioactive decay, and the events can be counted by noting the flashes of light. This is referred to as a *scintillation counter.*

The detector that you will use to measure radiation is a neon-gas-filled detector called a *Geiger–Müller (G–M) tube,* shown schematically in Figure 43-4. Radiation passing into the detector through the thin mica window (or through the sides of the tube in the case of gamma or cosmic radiation) knocks electrons out of the gas atoms in the tube; that is, it ionizes the gas atoms. Each time an event occurs ionizing some gas atoms, the detached electrons are accelerated toward the positively charged wire, causing more ionization as they gain energy and collide with other gas atoms. This creates a million-fold or greater amplification of the pulse of detached electrons that

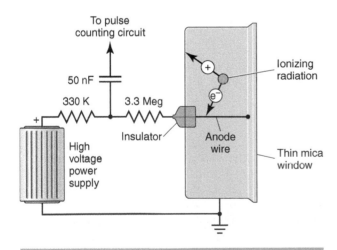

FIGURE 43-4 | **Schematic of a Geiger–Müller (G–M) radiation detector. The tube is filled with a gas such as neon (plus a small amount of a halogen quenching gas). A 500-V positive DC voltage is applied to the insulated anode wire located on the axis of the tube. The end of the tube, closed with a thin window made of mica, is the most sensitive area of the detector. Alpha, beta, or gamma radiation entering the detector ionizes the fill gas, producing a cascade of electrons that are collected on the anode wire as a pulse of electric current.**

are collected on the central wire as a current pulse. These current pulses are conducted to a counting circuit where each event creates one count.

NOTES TO INSTRUCTOR

A substance is classified as radioactive if its activity exceeds 100 Bq/g (1 Bq = 1 atom decay/s). By this definition, KCl is not classified as a radioactive substance. However, it has a measurable activity well above background. The natural abundance of radioactive ^{40}K in potassium chloride is about 0.0118%. Its half-life is sufficiently long (~one billion years) that it may be considered to be a primordial radioisotope, present since the formation of the Earth. A measurement of the approximate rate of decay allows the determination of the half-life using the differential rate equation, Equation (6).

Potassium-40 emits both electrons and positrons (or β^- and β^+ particles). When a thin-end-window Geiger–Müller radiation detector is used, the diameter of the detector should be ~30 mm in order to get satisfactory counting statistics. For example, using the Aware Electronics RM-70 detector (28.6-mm-diameter window), the typical background-corrected activity for a matching (28.5-mm-diameter) thin disk sample of 1.00 g KCl crystals is around 140 counts/min (CPM), about eight times background. The Aware Electronics RM-60 (9.1-mm-diameter window), having a ten times smaller area, gives a count rate that is ten times smaller (around 15 CPM, only twice the background). Because the standard deviation is approximately equal to the square root of the count, the smaller detector gives a factor-of-three larger relative standard deviation, on the order of RSD = 30%. (A larger RSD means poorer statistical precision of the measurement.)

Experimental Procedure

Special Supplies: Nominal 30-mm-diameter thin-end-window G-M tube radiation detector, counter, and sample holder. A satisfactory detector–counter combination is the Aware Electronics RM-70 detector and LCD-60 counting display module (or a computer with a free serial port).[6]

---

[6]We have used with satisfactory results the relatively inexpensive RM-70 radiation detector and LCD-60 counting display module available from Aware Electronics Corp., P.O. Box 4299, Wilmington, DE 19807, tel. 302-655-3800 (www.aw-el.com). The RM-70 also can be directly connected to a computer with a free serial port, drawing its power from the serial port. By using a computer and the software that comes with the RM-70 detector, the LCD-60 display module is not needed.

Sample holder for use with the RM-70 detector: a nominal 1-in. copper pipe coupling (28.5-mm ID × 31-mm OD × 48-mm length), clear plastic food wrap (such as Saran wrap), rubber bands, vinyl electrical tape (7-mil thickness), fine emery cloth (150 grit) or a flat file.

Chemicals: KCl crystals (crystals the size of table salt are satisfactory. The crystals must be free flowing and free of lumps).

Directions for Preparing the Sample Holder: Using emery cloth or a file, round off the sharp outer edge of the copper pipe coupling so it won't cut the plastic wrap. (Don't round the inner edge.) Stretch a single layer of plastic wrap across the rounded end of the coupling, temporarily holding it in place with a rubber band about an inch from the end. Then wrap about five turns of electrical tape tightly around the plastic wrap, keeping the edge of the tape a half-millimeter or so from the end of the coupling. (The few extra turns of tape ensure that the sample holder fits snugly in the cavity of the RM-70 detector, as shown in Figure 43-5, so that it stays centered above the detector window.) When counting samples, the sample holder and plastic wrap "window" are supported by the protective wire mesh of the RM-70 detector (lying on its back with the window of the G-M tube facing upward).

A. Counting background

Obtain a background count by counting for a period of at least 30 min (without the sample holder in place). Divide the total count by the number of minutes in the counting period to express the background count in units of counts per min (CPM). The background count comes from both terrestrial sources (such as the building construction materials) and extraterrestrial sources (such as cosmic rays).

If time permits, you might count background with the empty sample holder in place to see if there is a significant difference. Recall that radioactive decay is a random process so that statistical fluctuations are normal, with the standard deviation being approximately equal to the square root of the number of counts.

If time is limited, your instructor may supply you with the background count for your detector. When counting background, make sure that there are no nearby radioactive sources.

B. Counting samples of potassium chloride of increasing mass

Crease a piece of weighing paper and weigh out 0.70 ± 0.01 g of KCl. Pour the KCl into the sample holder, having one end closed off with plastic wrap. Distribute

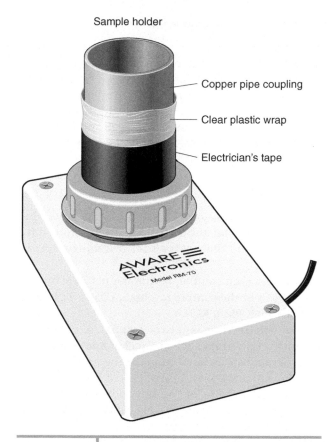

Sample holder

— Copper pipe coupling

— Clear plastic wrap

— Electrician's tape

AWARE Electronics Model RM-70

FIGURE 43-5 | **A Geiger–Müller radiation detector with a homemade sample holder positioned over the thin-end window of the detector. The sample holder is made from a nominal 1-in. copper pipe coupling with one end closed with clear plastic wrap, as described in the text.**

the sample as evenly as possible, by tapping and rotating the holder on the benchtop and/or using a glass rod, to produce a sample in the configuration of a thin disk. Make sure that there are no lumps and that you have evened out the hills and thin spots of the KCl crystals as much as possible.

Place the sample holder with KCl in the cavity of the detector, as shown in Figure 43-5, and count ten 1-min counts and/or one 10-min count. (Make sure there are no other sources of radiation near the detector.) Then, using the same weighing procedure, add increments of 0.30 g, 1.00 g, and 1.00 g KCl (giving totals of 1.00 g, 2.00 g, and 3.00 g). After each addition, distribute the sample evenly and count in the same way as for the first sample.

If you haven't already done so, make sure that you count the background for at least 30 min (without the sample holder in place) or use a previously determined value of the background count supplied by the instructor.

C. A simple shielding experiment

Place about 3 g of KCl in a 50-mL beaker, distributing the KCl evenly over the bottom of the beaker. Rest the beaker on the flange of the radiation detector and make several 1-min counts. What do you observe? How does the count compare to the background count? What conclusions can you draw about the ability of beta particles from ^{40}K to penetrate the glass surrounding the KCl sample?

Data analysis

For each of the four KCl samples, calculate the counts/minute and subtract the background count. We will call this the *net* (or background-corrected) CPM.

Next, for each sample, divide the net CPM by the mass of KCl in the sample to obtain CPM/g KCl. Also calculate log(CPM/g KCl).

You will next make two graphs. A computer spreadsheet program such as Excel or another graphing program can be used to make the graphs if you know how to use the software.

First make a plot of the net counts per minute versus mass of KCl, that is, CPM versus g KCl. There will be four points on the graph, one for each sample you counted. Why does this graph begin to bend over (or show a plateau) where there is no further increase in the number of counts as the mass of KCl increases?

For the second graph, plot log(CPM/g KCl) versus g KCl. Draw the best-fit line through the data points (or add a trendline using an Excel spreadsheet) and extrapolate the line to 0 g KCl, where it intercepts the log(CPM/g KCl) axis. This intercept is the value of log(CPM/g KCl) for an infinitely thin (zero thickness) sample of KCl. From the value of the intercept, calculate, by taking the antilog of the intercept, the extrapolated value of CPM/g KCl corresponding to zero mass of KCl. By using this extrapolation, we can eliminate the effect of self-absorption of beta particles by the KCl sample.

From the extrapolated value of the CPM/g KCl, you can determine the half-life of ^{40}K, knowing the isotopic abundance of ^{40}K and making a correction for the overall counter detection efficiency. This efficiency correction is comprised of two factors, a geometry factor and a shielding factor. About half the beta particles from the sample are heading upward, so they will not enter the detector. And a substantial fraction of the betas are blocked by the wire mesh grid protecting the window or are not able to pass through the window of the detector. By com-

parison with a beta source of known activity, it is estimated that the overall fraction of betas counted from a thin disk sample is about 0.20 (or 20%). Thus, we will adjust the count by applying the estimated efficiency correction factor of 0.20 as follows:

adjusted CPM/g KCL

$$= \frac{\text{extrapolated CPM/g KCl}}{\text{overall beta counting efficiency}}$$

$$= \frac{\text{extrapolated CPM/g KCl}}{0.20} \quad (10)$$

The rate of radioactive decay is described by Equation (6), a first-order rate expression where N is the number of radioactive atoms in the sample and k is the decay constant (units of reciprocal time). We will rewrite Equation (6):

$$-dN/dt = k \times N$$

We are applying this expression to the ^{40}K in the sample whose isotopic abundance is 0.0118%. Let's rewrite the rate expression of the equation in the form

$$-dN(^{40}\text{K})/dt = \text{adjusted CPM/g KCl} = k \times N(^{40}\text{K})$$
$$(11)$$

Let's break down the calculation based on Equation (11) into a step-by-step procedure:

• We first calculate $N(^{40}\text{K})$, the number of ^{40}K atoms in 1 g of KCl, taking into account the isotopic abundance of ^{40}K (0.0118%). (We assume 1 g of KCl in the calculation because we have expressed the adjusted count rate per gram of KCl.)

• Then we calculate from Equation (11) the decay constant, k, which will have units of 1/min. [We know the experimental value for adjusted CPM/g KCl and we calculated $N(^{40}\text{K})$ in the previous step, so the only unknown in Equation (11) is k, the decay constant.]

• Next we calculate $t_{1/2}$ for ^{40}K in units of minutes using Equation (8):

$$k \times t_{1/2} = \ln(2) = 0.693$$

• Finally, we convert $t_{1/2}$ from units of minutes to units of years.

Compare your experimental value for the half-life of potassium-40 to literature values, which may be found in the *CRC Handbook of Chemistry and Physics* or at various Web sites.

CONSIDER THIS

1. Assuming that the human body contains 0.35% by weight of the element potassium, show that a 70-kg human would contain about 0.03 g of potassium-40, and calculate the number of disintegrations per second that would be experienced by this human from this internal source, using Equation (9).

2. Explore the radiation shielding effects of a variety of materials. To accomplish this you will need to count for a length of time long enough to obtain numbers significantly greater than background counts. Sources of alpha, beta, and gamma radiation are available (without license requirements) from Flinn Scientific (Canberra Industries, Inc. is one of their suppliers) and other sources.

3. A point source of radiation is usually governed by the inverse-square law, which states that the intensity of the radiation falls off with the square of the distance between the source and the detector. Test this law with your radiation sources. Does it hold up? Are the shielding effects of air apparent with some types of radiation?

Bibliography

Here are some Web sites that provide historical overviews of the science of radioactivity and biographical sketches of the pioneers in this field (accessed on 03/26/09):

Access Excellence Classic Collection: http://www.accessexcellence.org/AE/AEC/CC

American Institute of Physics: http://www.aip.org/history/curie

Idaho State University presents information on "Radioactivity in Nature": http://www.physics.isu.edu/radinf/natural.htm

Lawrence Berkeley National Laboratory presents the "ABC's of Nuclear Science" and other topics related to radioactivity: http://www.lbl.gov/abc

Natural Radioactivity

The Half-Life of Potassium-40 in Potassium Chloride

Name _____

Date _____ Section _____

Locker _____ Instructor _____

A. Counting background

Count the background for a period of at least 30 min, or your instructor may supply you with a background count for your radiation detector. Make sure that when you count background there are no radioactive sources near the detector.

Counting period (min): _____   Total count: _____   Counts/min: _____

B. Counting samples of potassium chloride of increasing mass

Distribute the KCl over the bottom of the sample holder as evenly as possible. Count each sample for ten 1-min periods or one 10-min period.

Mass of empty sample holder weighed to nearest ±0.001 g: _____

Target mass of KCl sample	0.70 g	1.00 g	2.00 g	3.00 g
Actual mass of sample holder + KCl	g	g	g	g
Actual net mass of KCl sample	g	g	g	g
Total count (10-min period)				
Counting periods	min	min	min	min
Trial 1				
Trial 2				
Trial 3				
Trial 4				
Trial 5				
Trial 6				
Trial 7				
Trial 8				
Trial 9				
Trial 10				
Average of ten counting periods				
Standard deviation				
Net sample counts/min after subtracting background	CPM	CPM	CPM	CPM
Net sample CPM/g KCl				
log(CPM/g KCl)				

1. Make a graph by plotting the net counts per minute for each of the four samples of KCl versus the mass of the KCl sample. There will be four points on the graph, one for each sample of KCl that you counted. Draw a smooth curve through the data points.

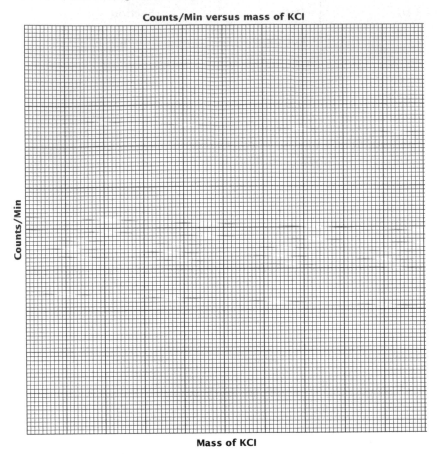

Counts/Min versus mass of KCl

Counts/Min

Mass of KCl

We might expect that as the sample mass of KCl increases, the counts/min would increase in proportion to the sample mass and that the graph would be approximately a straight line. Is this what you observe? What might explain the shape of your graph? *Discuss the shape of your graph and try to explain it.*

To test your hypothesis, you might try adding more KCl (say, a total of 9.00 g of KCl) in the sample holder. Does the net CPM increase by a factor of three compared to the sample containing 3.00 g of KCl? Is this consistent with your hypothesis?

Name _____   Date _____

2. Make a second graph by plotting log(CPM/g KCl) versus mass of KCl. Draw the best fit straight line through the four points on the graph and extend the line to the point where it intercepts the log(CPM/g KCl) axis.

log(CPM/g KCl) versus mass of KCl

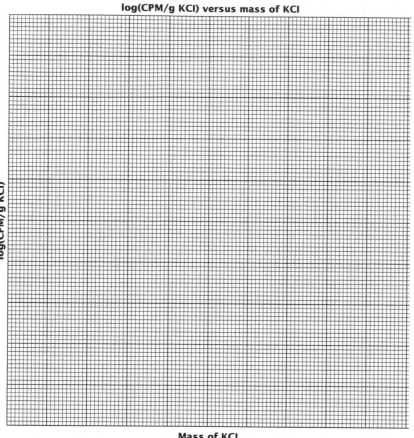

Mass of KCl

Record the value of the intercept on the log(CPM/g KCl) axis: _____

Calculate value of CPM/g KCl as the antilog of the intercept: _____

Calculate the adjusted CPM/g KCl (corrected for the detector efficiency): _____

Using Equation (11), calculate the half-life of potassium-40 in units of years, following the step-by-step procedure outlined in part B. *Show your calculations below and continue on the next page if necessary.*

Compare your results with the literature value for the half-life of potassium-40.

If you repeated your measurements next year, would you expect the values for CPM/g KCl to be significantly different? Why or why not? Can you support your answer with a calculation?

C. A simple shielding experiment

Record the net (background-corrected) counts/min you observed when you counted a 3.0-g sample of KCl in a glass beaker. Compare this with the net CPM you observed with a 3.0-g sample of KCl in the sample holder. What conclusions do you draw about the ability of beta particles from potassium-40 to penetrate a layer of glass? Record your results and write your discussion below:

Uncorrected CPM for 3.0-g sample of KCl in a glass beaker: _____

Net (background-corrected) CPM for 3.0-g sample of KCl in a glass beaker: _____

Explain how you could estimate the thickness of the clear plastic wrap (Saran wrap) covering the bottom of the sample holder using only a ruler, a milligram balance, and a plausible value of the density of the plastic wrap material.

Name _____   Date _____

CONSIDER THIS

1. Assuming that the human body contains 0.35% by weight of the element potassium, show that a 70-kg human would contain about 0.03 g of potassium-40, and calculate the number of disintegrations per second that would be experienced by this human from this internal source, using Equation (9).

2. Explore the radiation shielding effects of a variety of materials. To accomplish this, you will need to count for a length of time long enough to obtain numbers significantly greater than background counts. Sources of alpha, beta, and gamma radiation are available (without license requirements) from Flinn Scientific (Canberra Industries, Inc. is one of their suppliers) and other sources.

3. A point source of radiation is usually governed by the inverse-square law, which states that the intensity of the radiation falls off with the square of the distance between the source and the detector. Test this law with your radiation sources. Does it hold up? Are the shielding effects of air apparent with some types of radiation?

If you explore the experiments suggested in questions 2 and 3, attach a description of the experiments you performed and the conclusions you draw from them.

Tables of Data

TABLE 1 The International System (SI) of Units and Conversion Factors

Basic SI Units

Physical Quantity	Unit	Symbol
Length	meter	m
Mass	kilogram	kg
Time	second	s
Electric current	ampere	A
Temperature	kelvin	K
Amount of substance	mole	mol
Luminous intensity	candela	cd

Common Derived Units

Physical Quantity	Unit	Symbol	Definition
Frequency	hertz	Hz	s^{-1}
Energy	joule	J	$kg \cdot m^2 \cdot s^{-2}$
Force	newton	N	$kg \cdot m \cdot s^{-2} = J \cdot m^{-1}$
Pressure	Pascal	Pa	$kg \cdot m^{-1} \cdot s^{-2} = N \cdot m^{-2}$
Power	watt	W	$kg \cdot m^2 \cdot s^{-3} = J \cdot s^{-1}$
Electric charge	coulomb	C	$A \cdot s$
Electric potential difference	volt	V	$kg \cdot m^2 \cdot s^{-3} \cdot A^{-1} = J \cdot A^{-1} \cdot s^{-1}$

Decimal Fractions and Multiples

Factor	Prefix	Symbol	Factor	Prefix	Symbol
10^{-18}	atto	a	10^{-1}	deci	d
10^{-15}	femto	f	10	deca	da
10^{-12}	pico	p	10^2	hecto	h
10^{-9}	nano	n	10^3	kilo	k
10^{-6}	micro	μ	10^6	mega	M
10^{-3}	milli	m	10^9	giga	G
10^{-2}	centi	c	10^{12}	tera	T

(continued on next page)

TABLE 1 *(continued)*

Common Conversion Factors

LENGTH

1 angstrom unit (Å) $\equiv 10^{-8}$ cm

2.54 cm $\equiv$ 1 inch (in.)

1 m = 39.37 in.

MASS

453.6 grams (g) = 1 pound (lb)

1 kg = 2.205 lb

28.35 g = 1 ounce (oz, avoirdupois)

VOLUME

1 milliliter (mL) $\equiv$ 1 cubic centimeter (cm^3)

(Note that the mL is now defined precisely equal to 1 cm^3.)

1 liter (L) = 1.057 quarts

28.6 mL = 1 fluid ounce

PRESSURE

1 atmosphere (atm) = 1.013×10^5 Pascal (Pa) (N/m^2)

$\qquad$ = 760 torr (mm Hg); pressure of a mercury column

$\qquad$ 760 mm or 29.92 in. high at 0 °C

$\qquad$ = 1.013 bar

$\qquad$ = 14.70 $lb/in.^2$

TEMPERATURE (T)

Absolute zero (K) = -273.15 °C

T (K) = T (°C) + 273.15

Fahrenheit temperature (°F) = 1.8 T (°C) + 32

Celsius temperature (°C) = (T (°F) $-$ 32)/1.8

ENERGY

1 joule $\equiv$ 1 watt-s $\equiv$ 1 kg $\cdot m^2 \cdot s^{-2}$ = 10^7 erg

1 erg $\equiv$ 1 dyne-cm $\equiv$ 1 g $\cdot cm^2 \cdot s^{-2}$

1 calorie = 4.184 joule (J)

1 electron volt/molecule = 23.06 kcal/mol

| TABLE 2 | Fundamental Physical and Mathematical Constants[a] |

Physical Constants

Symbol	Name	Numerical Value
N_A	Avogadro's number	6.0221×10^{23} mol^{-1}
F	Faraday constant	96,485 C per mole of electrons transferred
h	Planck's constant	6.6261×10^{-34} J $\cdot$ s $\cdot$ particle^{-1}
c	Speed of light (*in vacuo*)	2.9979×10^{8} m $\cdot$ s^{-1}
R	The gas constant	0.08206 L $\cdot$ atm $\cdot$ mol^{-1} $\cdot$ K^{-1}
		82.06 mL $\cdot$ atm $\cdot$ mol^{-1} $\cdot$ K^{-1}
		8.3145 J $\cdot$ mol^{-1} $\cdot$ K^{-1}
e	Charge of the electron	1.6022×10^{-19} C
	Volume of 1 mol of an ideal gas	
	at 1 atm, 0 °C	22.41 L
	at 1 atm, 25 °C	24.46 L

Mathematical Constants

$\pi = 3.14159$
$e = 2.71828$
$\ln(x) = 2.303 \log_{10}(x)$

[a]1986 CODATA recommended values from http://physics.nist.gov

| TABLE 3 | Vapor Pressure of Water at Different Temperatures |

Temperature (°C)	Vapor Pressure (torr)	Temperature (°C)	Vapor Pressure (torr)
−10 (ice)	1.0	28	28.3
−5 (ice)	3.0	29	30.0
0	4.6	30	31.8
5	6.5	35	42.2
10	9.2	40	55.3
15	12.8	45	71.9
16	13.6	50	92.5
17	14.5	55	118.0
18	15.5	60	149.4
19	16.5	65	187.5
20	17.5	70	233.7
21	18.6	75	289.1
22	19.8	80	355.1
23	21.1	90	525.8
24	22.4	100	760.0
25	23.8	150	3570.5
26	25.2	200	11659.2
27	26.7		

TABLE 4 Concentration of Desk Acid and Base Solutions

Reagent	Formula	Molarity	Density	Percentage of Solute
Acetic acid, glacial	CH_3COOH	17 M	1.05 g/mL	99.5
Acetic acid, dilute		6	1.04	34
Hydrochloric acid, conc.	HCl	12	1.18	36
Hydrochloric acid, dilute		6	1.10	20
Nitric acid, conc.	HNO_3	16	1.42	72
Nitric acid, dilute		6	1.19	32
Sulfuric acid, conc.	H_2SO_4	18	1.84	96
Sulfuric acid, dilute		3	1.18	25
Ammonia solution, conc. (ammonium hydroxide)	NH_3	15	0.90	58
Ammonia solution, dilute (ammonium hydroxide)		6	0.96	23
Sodium hydroxide, dilute	NaOH	6	1.22	20

TABLE 5 Color Changes and pH Intervals of Some Important Indicators

Name of Indicator	pH Interval	Color Change	Solvent
Methyl violet	0.2–3.0	Yellow to blue-violet	Water
Thymol blue	1.2–2.8	Red to yellow	Water (+NaOH)
Orange IV (Tropeolin OO)	1.3–3.0	Red to yellow	Water
Benzopurpurin 4B	1.2–4.0	Violet to red	20% alcohol
Methyl orange	3.1–4.4	Red to orange to yellow	Water
Bromophenol blue	3.0–4.6	Yellow to blue-violet	Water (+NaOH)
Congo red	3.0–5.0	Blue to red	70% alcohol
Bromocresol green	3.8–5.4	Yellow to blue	Water (+NaOH)
Methyl red	4.4–6.2	Red to yellow	Water (+NaOH)
Chlorophenol red	4.8–6.8	Yellow to red	Water (+NaOH)
Bromocresol purple	5.2–6.8	Yellow to purple	Water (+NaOH)
Litmus	4.5–8.3	Red to blue	Water
Bromothymol blue	6.0–7.6	Yellow to blue	Water (+NaOH)
Phenol red	6.8–8.2	Yellow to red	Water (+NaOH)
Thymol blue	8.0–9.6	Yellow to blue	Water (+NaOH)
Phenolphthalein	8.3–10.0	Colorless to red	70% alcohol
Thymolphthalein	9.3–10.5	Yellow to blue	70% alcohol
Alizarin yellow R	10.0–12.0	Yellow to red	20% alcohol
Indigo carmine	11.4–13.0	Blue to yellow	50% alcohol
Trinitrobenzene	12.0–14.0	Colorless to orange	70% alcohol

TABLE 6 Equilibrium Constants for the Ionization of Acids and Bases (25 °C)

Compound	Ionization Reaction	K_a	pK_a
WATER	$2\,H_2O \rightleftharpoons H_3O^+ + OH^-$ (25 °C)	1.00×10^{-14}	14.00
	(0 °C)	0.11×10^{-14}	14.94
	(60 °C)	9.61×10^{-14}	13.02
WEAK ACIDS			
Acetic	$CH_3COOH + H_2O \rightleftharpoons H_3O^+ + CH_3COO^-$	1.76×10^{-5}	4.75
Boric	$H_3BO_3 + H_2O \rightleftharpoons H_3O^+ + H_2BO_3^-$	6.0×10^{-10}	9.22
Carbonic (CO_2)	$CO_2 + H_2O \rightleftharpoons H_3O^+ + HCO_3^-$	$K_1 = 4.4 \times 10^{-7}$	6.35
	$HCO_3^- + H_2O \rightleftharpoons H_3O^+ + CO_3^{2-}$	$K_2 = 4.7 \times 10^{-11}$	10.33
Chromic	$H_2CrO_4 + H_2O \rightleftharpoons H_3O^+ + HCrO_4^-$	$K_1 = 2 \times 10^{-1}$	0.7
	$HCrO_4^- + H_2O \rightleftharpoons H_3O^+ + CrO_4^{2-}$	$K_2 = 3.2 \times 10^{-7}$	6.50
Formic	$HCOOH + H_2O \rightleftharpoons H_3O^+ + HCOO^-$	1.8×10^{-4}	3.74
Hydrogen cyanide	$HCN + H_2O \rightleftharpoons H_3O^+ + CN^-$	4×10^{-10}	9.4
Hydrofluoric	$HF + H_2O \rightleftharpoons H_3O^+ + F^-$	6.9×10^{-4}	3.16
Hydrogen peroxide	$H_2O_2 + H_2O \rightleftharpoons H_3O^+ + HO_2^-$	2.4×10^{-12}	11.62
Hydrogen sulfate ion	$HSO_4^- + H_2O \rightleftharpoons H_3O^+ + SO_4^{2-}$	$K_2 = 1.2 \times 10^{-2}$	1.92
Hydrogen sulfide	$H_2S + H_2O \rightleftharpoons H_3O^+ + HS^-$	$K_1 = 1.0 \times 10^{-7}$	7.00
	$HS^- + H_2O \rightleftharpoons H_3O^+ + S^{2-}$	$K_2 = 2 \times 10^{-18}$	17.7
Nitrous	$HNO_2 + H_2O \rightleftharpoons H_3O^+ + NO_2^-$	4.5×10^{-4}	3.50
Oxalic	$H_2C_2O_4 + H_2O \rightleftharpoons H_3O^+ + HC_2O_4^-$	$K_1 = 3.8 \times 10^{-2}$	1.42
	$HC_2O_4^- + H_2O \rightleftharpoons H_3O^+ + C_2O_4^{2-}$	$K_2 = 5.0 \times 10^{-5}$	4.30
Phosphoric	$H_3PO_4 + H_2O \rightleftharpoons H_3O^+ + H_2PO_4^-$	$K_1 = 7.1 \times 10^{-3}$	2.15
	$H_2PO_4^- + H_2O \rightleftharpoons H_3O^+ + HPO_4^{2-}$	$K_2 = 6.3 \times 10^{-8}$	7.20
	$HPO_4^{2-} + H_2O \rightleftharpoons H_3O^+ + PO_4^{3-}$	$K_3 = 4.4 \times 10^{-13}$	12.36
Phosphorous	$H_2PO_3 + H_2O \rightleftharpoons H_3O^+ + HPO_3^-$	$K_1 = 1.6 \times 10^{-2}$	1.80
Propionic	$CH_3CH_2COOH + H_2O \rightleftharpoons H_3O^+ + CH_3CH_2COO^-$	1.34×10^{-5}	4.87
Sulfurous ($SO_2 + H_2O$)	$H_2SO_3 + H_2O \rightleftharpoons H_3O^+ + HSO_3^-$	$K_1 = 1.2 \times 10^{-2}$	1.92
	$HSO_3^- + H_2O \rightleftharpoons H_3O^+ + SO_3^{2-}$	$K_2 = 5.6 \times 10^{-8}$	7.25
CATION ACIDS—HYDRATED METAL IONS			
Aluminum ion	$Al(H_2O)_6^{3+} + H_2O \rightleftharpoons H_3O^+ + Al(H_2O)_5OH^{2+}$	1.1×10^{-5}	4.9
Chromium(III) ion	$Cr(H_2O)_6^{3+} + H_2O \rightleftharpoons H_3O^+ + Cr(H_2O)_5OH^{2+}$	1.6×10^{-4}	3.80
Iron(III) ion	$Fe(H_2O)_6^{3+} + H_2O \rightleftharpoons H_3O^+ + Fe(H_2O)_5OH^{2+}$	6.7×10^{-3}	2.17
Zinc ion	$Zn(H_2O)_4^{2+} + H_2O \rightleftharpoons H_3O^+ + Zn(H_2O)_3OH^+$	2.5×10^{-10}	9.60

Compound	Ionization Reaction	K_b	pK_b
WEAK BASES			
Ammonia	$NH_3 + H_2O \rightleftharpoons NH_4^+ + OH^-$	1.8×10^{-5}	4.75
Methylamine	$CH_3NH_2 + H_2O \rightleftharpoons CH_3NH_3^+ + OH^-$	5.0×10^{-4}	3.3
Barium hydroxide	$Ba(OH)_2 \rightleftharpoons BaOH^+ + OH^-$	Strong	
	$BaOH^+ \rightleftharpoons Ba^{2+} + OH^-$	$K_2 = 1.4 \times 10^{-1}$	0.85
Calcium hydroxide	$Ca(OH)_2 \rightleftharpoons CaOH^+ + OH^-$	Strong	
	$CaOH^+ \rightleftharpoons Ca^{2+} + OH^-$	$K_2 = 3.5 \times 10^{-2}$	1.5

TABLE 7 Equilibrium Constants for the Dissociation of Complex Ions, Amphoteric Hydroxides, and Weak Salts ($\approx$25 °C)[a]

Compound	Dissociation Reaction	K	pK
AMMINE (AMMONIA) COMPLEX IONS			
Tetraamminecadmium(II)	$Cd(NH_3)_4^{2+} \rightleftharpoons Cd^{2+} + 4\,NH_3$	2×10^{-7}	6.7
Tetraamminecopper(II)	$Cu(NH_3)_4^{2+} \rightleftharpoons Cu^{2+} + 4\,NH_3$	8×10^{-13}	12.1
Diamminesilver(I)	$Ag(NH_3)_2^{+} \rightleftharpoons Ag^{+} + 2\,NH_3$	6×10^{-8}	7.2
Tetraamminezinc(II)	$Zn(NH_3)_4^{2+} \rightleftharpoons Zn^{2+} + 4\,NH_3$	1×10^{-9}	9.0
HYDROXIDE COMPLEX IONS (AMPHOTERIC HYDROXIDES)			
Tetrahydroxoaluminate	$Al(OH)_4^{-} \rightleftharpoons Al(OH)_3(s) + OH^{-}$	3×10^{-2}	1.5
Tetrahydroxochromate(III)	$Cr(OH)_4^{-} \rightleftharpoons Cr(OH)_3(s) + OH^{-}$	2.5	−0.40
Trihydroxoplumbate(II)	$Pb(OH)_3^{-} \rightleftharpoons Pb(OH)_2(s) + OH^{-}$	2×10^{1}	−1.3
Trihydroxostannate(II)	$Sn(OH)_3^{-} \rightleftharpoons Sn(OH)_2(s) + OH^{-}$	2.6	−0.41
Tetrahydroxozincate	$Zn(OH)_4^{2-} \rightleftharpoons Zn(OH)_2(s) + 2\,OH^{-}$	4×10^{1}	−1.6
CHLORIDE COMPLEX IONS AND WEAKLY IONIZED SALTS			
Dichlorocadmium	$CdCl_2(aq) \rightleftharpoons Cd^{2+} + 2\,Cl^{-}$	2.5×10^{-3}	2.60
Tetrachloroaurate(III) ion	$AuCl_4^{-} \rightleftharpoons Au^{3+} + 4\,Cl^{-}$	5×10^{-22}	21.3
Trichloroiron(III)	$FeCl_3(aq) \rightleftharpoons Fe^{3+} + 3\,Cl^{-}$	8×10^{-2}	1.9
Dichloroiron(III) ion	$FeCl_2^{+}(aq) \rightleftharpoons Fe^{3+} + 2\,Cl^{-}$	8×10^{-3}	2.9
Chloroiron(III) ion	$FeCl^{2+} \rightleftharpoons Fe^{3+} + Cl^{-}$	3.5×10^{-2}	1.46
Mercury(II) chloride	$HgCl_2(aq) \rightleftharpoons HgCl^{+} + Cl^{-}$	$K_1 = 3.3 \times 10^{-7}$	6.48
Chloromercury(II) ion	$HgCl^{+} \rightleftharpoons Hg^{2+} + Cl^{-}$	$K_2 = 1.8 \times 10^{-7}$	6.74
Tetrachloromercurate(II)	$HgCl_4^{2-} \rightleftharpoons Hg^{2+} + 4\,Cl^{-}$	8.5×10^{-16}	15.07
Tin(II) chloride	$SnCl_2(aq) \rightleftharpoons Sn^{2+} + 2\,Cl^{-}$	5.7×10^{-3}	2.24
Tetrachlorostannate(II) ion	$SnCl_4^{2-} \rightleftharpoons Sn^{2+} + 4\,Cl^{-}$	3.3×10^{-2}	1.48
Hexachlorostannate(IV) ion	$SnCl_6^{2-} \rightleftharpoons Sn^{4+} + 6\,Cl^{-}$	10^{-4}	4
Dichloroargentate(I) ion	$AgCl_2^{-} \rightleftharpoons Ag^{+} + 2\,Cl^{-}$	5×10^{-6}	5.3
OTHER COMPLEX IONS AND WEAKLY IONIZED SALTS			
Tetracyanocadmate(II) ion	$Cd(CN)_4^{2-} \rightleftharpoons Cd^{2+} + 4\,CN^{-}$	8×10^{-18}	17.1
Thiocyanatoiron(III) ion	$FeSCN^{2+} \rightleftharpoons Fe^{3+} + SCN^{-}$	1×10^{-3}	3.0
Lead(II) acetate	$Pb(CH_3COO)_2(aq) \rightleftharpoons Pb^{2+} + 2\,CH_3COO^{-}$	1×10^{-4}	4.0
Triacetatoplumbate(II) ion	$Pb(CH_3COO)_3^{-} \rightleftharpoons Pb^{2+} + 3\,CH_3COO^{-}$	2.5×10^{-7}	6.60
Dicyanoargentate(I) ion	$Ag(CN)_2^{-} \rightleftharpoons Ag^{+} + 2\,CN^{-}$	1×10^{-20}	20.0
Dithiosulfatoargentate(I) ion	$Ag(S_2O_3)_2^{3-} \rightleftharpoons Ag^{+} + 2\,S_2O_3^{2-}$	4×10^{-14}	13.4

[a]The formation of complex ions undoubtedly occurs by the step-by-step addition of ligands (the coordinated molecules or anions) as the concentration of ligand increases. It is only in the presence of a large excess of the ligand that the overall cumulative ionization constants given here can be used with any measure of quantitative reliability. Furthermore, the total ionic strength of the solution exerts a major influence on the value of the equilibrium constant. For example, as the total ionic strength increases from zero to 0.5 M, the equilibrium constant for dissociation of the $Fe(SCN)^{2+}$ complex increases from 1×10^{-3} to about 7×10^{-3}.

TABLE 8 **Solubility of Some Common Salts and Hydroxides at Approximately 25 °C**
(The classification is a summary of observed solubility behavior. The division into three solubility groups is arbitrary.)

General Rule, Soluble	Exceptions	
	Sparingly Soluble	Insoluble
Group 1 (alkali metal salts)		
Nitrates (NO_3^-)		
Fluorides (F^-)		MgF_2, CaF_2, SrF_2, BaF_2, PbF_2
Chlorides, bromides	$PbCl_2$, $PbBr_2$	Ag^+, Hg_2^{2+}, PbI_2
Sulfates (SO_4^{2-})[a]	Ca^{2+}, Ag^+, Hg_2^{2+}	Sr^{2+}, Ba^{2+}, Pb^{2+}
Acetates (CH_3COO^-)	Ag^+, Hg_2^{2+}	

General Rule, Insoluble	Exceptions	
	Sparingly Soluble	Soluble
Carbonates (CO_3^{2-})		Group 1 (alkali metals), NH_4^+
Phosphates (PO_4^{3-})[b]		Group 1 (alkali metals), NH_4^+
Sulfides (S^{2-})[c]		Group 1 (alkali metals), NH_4^+
Oxalates ($C_2O_4^{2-}$)		Group 1 (alkali metals), NH_4^+
Hydroxides (OH^-)	Ca^{2+}, Sr^{2+}	Group 1 (alkali metals), NH_4^+

[a]Many hydrogen sulfates such as $Ca(HSO_4)_2$ and $Ba(HSO_4)_2$ are more soluble.
[b]Many hydrogen phosphates are soluble, such as $Mg(H_2PO_4)_2$ and $Ca(H_2PO_4)_2$.
[c]Sulfides of Al^{3+} and Cr^{3+} hydrolyze and precipitate the corresponding hydroxides.

TABLE 9 Solubility Product Constants (18–25 °C)[a]

Compound	K_{sp}	Compound	K_{sp}
ACETATES		**HYDROXIDES** (continued)	
$Ag(CH_3COO)$	4×10^{-3}	*$Mn(OH)_2$	1.7×10^{-13}
		$Pb(OH)_2$	4×10^{-15}
HALIDES AND CYANIDES		$Sn(OH)_2$	10^{-27}
$AgCN$	10^{-16}	$Zn(OH)_2$	5×10^{-17}
*$AgCl$	1.8×10^{-10}		
*$AgBr$	5×10^{-13}	**OXALATES**	
*AgI	8.5×10^{-17}	*$Ag_2C_2O_4$	1.3×10^{-12}
*CaF_2	4.6×10^{-11}	BaC_2O_4	1.5×10^{-8}
$CuCl$	3.2×10^{-7}	CaC_2O_4	1.3×10^{-9}
Hg_2Cl_2	1.1×10^{-18}	MgC_2O_4	8.6×10^{-5}
*Hg_2Br_2	5.5×10^{-23}	SrC_2O_4	5.6×10^{-8}
*Hg_2I_2	1.1×10^{-28}		
MgF_2	8×10^{-8}	**CARBONATES**	
$PbCl_2$	1.6×10^{-5}	Ag_2CO_3	8×10^{-12}
PbI_2	8.3×10^{-9}	$BaCO_3$	1.6×10^{-9}
		$CaCO_3$	4.8×10^{-9}
SULFATES		$CuCO_3$	2.5×10^{-10}
*Ag_2SO_4	6.2×10^{-5}	$FeCO_3$	2×10^{-11}
*$BaSO_4$	1.2×10^{-10}	$MgCO_3$	4×10^{-5}
$CaSO_4$	2.4×10^{-5}	$MnCO_3$	9×10^{-11}
$PbSO_4$	1.3×10^{-8}	$PbCO_3$	1.5×10^{-13}
$SrSO_4$	7.6×10^{-7}	*$SrCO_3$	1.0×10^{-9}
CHROMATES		**SULFIDES**	
Ag_2CrO_4	2×10^{-12}	Ag_2S	10^{-50}
*$BaCrO_4$	2.3×10^{-10}	CdS	10^{-26}
$PbCrO_4$	2×10^{-16}	CoS	10^{-21}
$SrCrO_4$	3.6×10^{-5}	CuS	10^{-36}
		FeS	10^{-17}
HYDROXIDES		HgS	10^{-50}
$Al(OH)_3$	10^{-33}	MnS	10^{-13}
$Ca(OH)_2$	6.5×10^{-6}	NiS	10^{-22}
$Cr(OH)_3$	10^{-30}	PbS	10^{-26}
$Cu(OH)_2$	2×10^{-19}	SnS	10^{-27}
*$Fe(OH)_2$	8.2×10^{-16}	ZnS	10^{-20}
$Fe(OH)_3$	10^{-37}		
$Mg(OH)_2$	9×10^{-12}		

[a]These values are approximate. The solubility is affected by the total ionic strength of the solution, by the temperature, and by the presence of hydronium ion or other ions that may form complex ions or that result in colloidal suspensions. The rate of approach to equilibrium is often slow.

*For these compounds, the molar solubility calculated from the solubility product constant is in reasonable agreement with the experimentally measured solubility. This suggests that these salts completely dissociate into their constituent ions to produce a solution of low ionic strength and that the ions do not react with water to produce hydrolysis side reactions. (See R. W. Clark; J. M. Bonicamp, "The K_{sp}-Solubility Conundrum," *J. Chem. Educ.* 1998, *75*, 1182–1185.)

TABLE 10 Standard Reduction Potentials (25 °C)

The reduction potential convention recommended by the IUPAC is used. The half-cell potentials are those that would be obtained in a hypothetical cell with all species at unit activity when measured with respect to the standard hydrogen (H^+/H_2) half-cell. Values in the table are from W. M. Latimer, *Oxidation Potentials, 2nd ed.*, Prentice-Hall, Englewood Cliffs, NJ, 1952; A. J. Bard; R. Parsons; J. Jordan, *Standard Potentials in Aqueous Solution*, Marcel Dekker, New York, 1985.

Strongest Oxidizing Agents	Half-Reaction	$E°$ (V)	Weakest Reducing Agents
↑ Increasing oxidizing strength	$\frac{1}{2} F_2(g) + H^+ + e^- \rightleftharpoons HF(aq)$	+3.06	Increasing reducing strength ↓
	$\frac{1}{2} F_2(g) + e^- \rightleftharpoons F^-$	+2.87	
	$H_2O_2(aq) + 2\,H^+ + 2\,e^- \rightleftharpoons 2\,H_2O$	+1.77	
	$PbO_2(s) + 4\,H^+ + SO_4^{2-} + 2\,e^- \rightleftharpoons 2\,H_2O + PbSO_4(s)$	+1.685	
	$Ce^{4+} + e^- \rightleftharpoons Ce^{3+}$	+1.61	
	$MnO_4^- + 8\,H^+ + 5\,e^- \rightleftharpoons 4\,H_2O + Mn^{2+}$	+1.51	
	$Mn^{3+} + e^- \rightleftharpoons Mn^{2+}$	+1.51	
	$Au^{3+} + 3\,e^- \rightleftharpoons Au$	+1.50	
	$HClO(aq) + H^+ + 2\,e^- \rightleftharpoons H_2O + Cl^-$	+1.49	
	$ClO_3^- + 6\,H^+ + 5\,e^- \rightleftharpoons 3\,H_2O + \frac{1}{2} Cl_2(g)$	+1.47	
	$Co^{3+} + e^- \rightleftharpoons Co^{2+}$	+1.45[a]	
	$\frac{1}{2} Cl_2(g) + e^- \rightleftharpoons Cl^-$	+1.358	
	$Cr_2O_7^{2-} + 14\,H^+ + 6\,e^- \rightleftharpoons 7\,H_2O + 2\,Cr^{3+}$	+1.33	
	$MnO_2(s) + 4\,H^+ + 2\,e^- \rightleftharpoons 2\,H_2O + Mn^{2+}$	+1.23	
	$O_2(g) + 4\,H^+ + 4\,e^- \rightleftharpoons 2\,H_2O$	+1.229	
	$IO_3^- + 6\,H^+ + 5\,e^- \rightleftharpoons \frac{1}{2} I_2(aq) + 3\,H_2O$	+1.195	
	$\frac{1}{2} Br_2(l) + e^- \rightleftharpoons Br^-$	+1.0652	
	$AuCl_4^- + 3\,e^- \rightleftharpoons 4\,Cl^- + Au$	+1.00	
	$NO_3^- + 4\,H^+ + 3\,e^- \rightleftharpoons 2\,H_2O + NO(g)$	+0.96	
	$2\,Hg^{2+} + 2\,e^- \rightleftharpoons Hg_2^{2+}$	+0.92	
	$ClO^- + H_2O + 2\,e^- \rightleftharpoons 2\,OH^- + Cl^-$	+0.89	
	$HO_2^- + H_2O + 2\,e^- \rightleftharpoons 3\,OH^-$	+0.88	
	$Cu^{2+} + I^- \rightleftharpoons CuI(s)$	+0.861	
	$Hg^{2+} + 2\,e^- \rightleftharpoons Hg(l)$	+0.854	
	$O_2(g) + 4\,H^+ (10^{-7}\,M) + 4\,e^- \rightleftharpoons 2\,H_2O$	+0.815	
	$Ag^+ + e^- \rightleftharpoons Ag$	+0.79991	
	$Hg_2^{2+} + 2\,e^- \rightleftharpoons 2\,Hg(l)$	+0.789	
	$Fe^{3+} + e^- \rightleftharpoons Fe^{2+}$	+0.771	
	$O_2(g) + 2\,H^+ + 2\,e^- \rightleftharpoons H_2O_2(aq)$	+0.682	
	$\frac{1}{2} I_2(aq) + e^- \rightleftharpoons I^-$	+0.621	
	$MnO_4^- + 2\,H_2O + 3\,e^- \rightleftharpoons 4\,OH^- + MnO_2(s)$	+0.60	
	$I_3^- + 2\,e^- \rightleftharpoons 3\,I^-$	+0.536	
	$MnO_2(s) + H_2O + NH_4^+ + e^- \rightleftharpoons NH_3(aq) + Mn(OH)_3(s)$	+0.50	

[a]Newly revised value based on the work of A. L. Rotinjan et al., *Electrochimica Acta* **1974**, *19*, 43.

(continued on next page)

TABLE 10 *(continued)*

Weakest Oxidizing Agents	Half-Reaction	$E°$ (V)	Strongest Reducing Agents
	$O_2(g) + 2\,H_2O + 4\,e^- \rightleftharpoons 4\,OH^-$	+0.401	
	$Ag(NH_3)_2^+ + e^- \rightleftharpoons Ag(s) + 2\,NH_3$	+0.373	
	$Cu^{2+} + 2\,e^- \rightleftharpoons Cu$	+0.337	
	$BiO^+ + 2\,H^+ + 3\,e^- \rightleftharpoons H_2O + Bi$	+0.32	
	$AgCl(s) + e^- \rightleftharpoons Ag + Cl^-$	+0.222	
	$SO_4^{2-} + 4\,H^+ + 2\,e^- \rightleftharpoons H_2O + H_2SO_3(aq)$	+0.17	
	$Sn^{4+} + 2\,e^- \rightleftharpoons Sn^{2+}$	+0.15	
	$S + 2\,H^+ + 2\,e^- \rightleftharpoons H_2S(g)$	+0.141	
	$H^+ + e^- \rightleftharpoons \tfrac{1}{2}\,H_2(g)$	0.000	
	$O_2(g) + H_2O + 2\,e^- \rightleftharpoons OH^- + HO_2^-$	−0.065	
	$Pb^{2+} + 2\,e^- \rightleftharpoons Pb$	−0.126	
	$CrO_4^{2-} + 4\,H_2O + 3\,e^- \rightleftharpoons 5\,OH^- + Cr(OH)_3(s)$	−0.13	
	$Sn^{2+} + 2\,e^- \rightleftharpoons Sn$	−0.136	
	$Ni^{2+} + 2\,e^- \rightleftharpoons Ni$	−0.250	
	$Co^{2+} + 2\,e^- \rightleftharpoons Co$	−0.277	
	$PbSO_4(s) + 2\,e^- \rightleftharpoons SO_4^{2-} + Pb$	−0.356	
	$Cd^{2+} + 2\,e^- \rightleftharpoons Cd$	−0.403	
	$Cr^{3+} + e^- \rightleftharpoons Cr^{2+}$	−0.41	
	$2\,H^+(10^{-7}\,M) + 2\,e^- \rightleftharpoons H_2(g)$	−0.414	
	$Fe^{2+} + 2\,e^- \rightleftharpoons Fe$	−0.44	
	$S + 2\,e^-\,(1\,M\,OH^-) \rightleftharpoons S^{2-}$	−0.45	
	$2\,CO_2(g) + 2\,H^+ + 2\,e^- \rightleftharpoons H_2C_2O_4(aq)$	−0.49	
	$Cr^{3+} + 3\,e^- \rightleftharpoons Cr$	−0.74	
	$Zn^{2+} + 2\,e^- \rightleftharpoons Zn$	−0.763	
	$2\,H_2O + 2\,e^- \rightleftharpoons 2\,OH^- + H_2(g)$	−0.828	
	$SO_4^{2-} + H_2O + 2\,e^- \rightleftharpoons 2\,OH^- + SO_3^{2-}$	−0.93	
	$Mn^{2+} + 2\,e^- \rightleftharpoons Mn$	−1.18	
	$Al^{3+} + 3\,e^- \rightleftharpoons Al$	−1.66	
	$Mg^{2+} + 2\,e^- \rightleftharpoons Mg$	−2.37	
	$Na^+ + e^- \rightleftharpoons Na$	−2.714	
	$Ca^{2+} + 2\,e^- \rightleftharpoons Ca$	−2.87	
	$Sr^{2+} + 2\,e^- \rightleftharpoons Sr$	−2.89	
	$Ba^{2+} + 2\,e^- \rightleftharpoons Ba$	−2.90	
	$Cs^+ + e^- \rightleftharpoons Cs$	−2.92	
	$K^+ + e^- \rightleftharpoons K$	−2.925	
	$Li^+ + e^- \rightleftharpoons Li$	−3.045	

Left margin: Increasing oxidizing strength (upward arrow)

Right margin: Increasing reducing strength (downward arrow)

TABLE 11 International Atomic Masses (based on $C^{12} = 12$ exactly)

Element	Atomic Symbol	Atomic Number	Mass
Actinium	Ac	89	[227][a]
Aluminum	Al	13	26.9815
Americium	Am	95	[243][a]
Antimony	Sb	51	121.76
Argon	Ar	18	39.948
Arsenic	As	33	74.9216
Astatine	At	85	[210][a]
Barium	Ba	56	137.33
Berkelium	Bk	97	[247][a]
Beryllium	Be	4	9.0122
Bismuth	Bi	83	208.980
Bohrium	Bh	107	[267][a]
Boron	B	5	10.811[b]
Bromine	Br	35	79.904[c]
Cadmium	Cd	48	112.41
Calcium	Ca	20	40.08
Californium	Cf	98	[251][a]
Carbon	C	6	12.011[b]
Cerium	Ce	58	140.12
Cesium	Cs	55	132.905
Chlorine	Cl	17	35.453[c]
Chromium	Cr	24	51.996[c]
Cobalt	Co	27	58.9332
Copper	Cu	29	63.546
Curium	Cm	96	[247][a]
Darmstadtium	Ds	110	[271][a]
Dubnium	Db	105	[262][a]
Dysprosium	Dy	66	162.50
Einsteinium	Es	99	[252][a]
Erbium	Er	68	167.26
Europium	Eu	63	151.96
Fermium	Fm	100	[257][a]
Fluorine	F	9	18.9984
Francium	Fr	87	[223][a]
Gadolinium	Gd	64	157.25
Gallium	Ga	31	69.72
Germanium	Ge	32	72.64
Gold	Au	79	196.967
Hafnium	Hf	72	178.49

(continued on next page)

TABLE 11 *(continued)*

Element	Atomic Symbol	Atomic Number	Mass
Hassium	Hs	108	[269][a]
Helium	He	2	4.0026
Holmium	Ho	67	164.930
Hydrogen	H	1	1.0079[b]
Indium	In	49	114.82
Iodine	I	53	126.904
Iridium	Ir	77	192.2
Iron	Fe	26	55.845[c]
Krypton	Kr	36	83.80
Lanthanum	La	57	138.91
Lawrencium	Lw	103	[257][a]
Lead	Pb	82	207.2
Lithium	Li	3	6.941
Lutetium	Lu	71	174.97
Magnesium	Mg	12	24.305
Manganese	Mn	25	54.9380
Meitnerium	Mt	109	[268][a]
Mendelevium	Md	101	[258][a]
Mercury	Hg	80	200.59
Molybdenum	Mo	42	95.94
Neodymium	Nd	60	144.24
Neon	Ne	10	20.180
Neptunium	Np	93	[237][a]
Nickel	Ni	28	58.693
Niobium	Nb	41	92.906
Nitrogen	N	7	14.0067
Nobelium	No	102	[254][a]
Osmium	Os	76	190.2
Oxygen	O	8	15.9994[b]
Palladium	Pd	46	106.4
Phosphorus	P	15	30.9738
Platinum	Pt	78	195.08
Plutonium	Pu	94	[244][a]
Polonium	Po	84	[209][a]
Potassium	K	19	39.098
Praseodymium	Pr	59	140.908
Promethium	Pm	61	[147][a]
Protactinium	Pa	91	[231][a]

(continued on next page)

TABLE 11 (continued)

Element	Atomic Symbol	Atomic Number	Mass
Radium	Ra	88	[226][a]
Radon	Rn	86	[222][a]
Rhenium	Re	75	186.2
Rhodium	Rh	45	102.905
Roentgenium	Rg	111	[283][a]
Rubidium	Rb	37	85.47
Ruthenium	Ru	44	101.07
Rutherfordium	Rf	104	[257][a]
Samarium	Sm	62	150.36
Scandium	Sc	21	44.956
Seaborgium	Sg	106	[266][a]
Selenium	Se	34	78.96
Silicon	Si	14	28.086[b]
Silver	Ag	47	107.868[c]
Sodium	Na	11	22.9898
Strontium	Sr	38	87.62
Sulfur	S	16	32.06[b]
Tantalum	Ta	73	180.948
Technetium	Tc	43	[98][a]
Tellurium	Te	52	127.60
Terbium	Tb	65	158.925
Thallium	Tl	81	204.38
Thorium	Th	90	232.038
Thulium	Tm	69	168.934
Tin	Sn	50	118.71
Titanium	Ti	22	47.867
Tungsten	W	74	183.84
Uranium	U	92	238.03
Vanadium	V	23	50.942
Xenon	Xe	54	131.29
Ytterbium	Yb	70	173.04
Yttrium	Y	39	88.906
Zinc	Zn	30	65.41
Zirconium	Zr	40	91.22

[a]A number in brackets designates the mass number of a selected isotope of the element, usually the one of longest known half-life.

[b]The atomic mass varies because of natural variation in the isotopic composition of the element. The observed ranges are boron, ±0.003; carbon, ±0.00005; hydrogen, ±0.00001; oxygen, ±0.0001; silicon, ±0.001; sulfur, ±0.003.

[c]The atomic mass is believed to have an experimental uncertainty of the following magnitude: bromine ±0.002; chlorine, ±0.001; chromium, ±0.001; iron, ±0.003; silver, ±0.003. For other elements, the last digit given is believed to be reliable to ±0.5.

Atomic masses: Courtesy of the International Union of Pure and Applied Chemistry.

TABLE 12 Twelve Principles of Green Chemistry

1. **Prevention:** It is better to prevent waste than to treat or clean up waste after it has been created.

2. **Atom economy:** Synthetic methods should be designed to maximize the incorporation of all materials used in the process into the final product.

3. **Less hazardous chemical syntheses:** Wherever practicable, synthetic methods should be designed to use and generate substances that possess little or no toxicity to human health and the environment.

4. **Designing safer chemicals:** Chemical products should be designed to effect their desired function while minimizing their toxicity.

5. **Safer solvents and auxiliaries:** The use of auxiliary substances (such as solvents and separation agents) should be made unnecessary wherever possible and innocuous when used.

6. **Design for energy efficiency:** Energy requirements of chemical processes should be recognized for their environmental and economic impacts and should be minimized. If possible, synthetic methods should be conducted at ambient temperature and pressure.

7. **Use renewable feedstocks:** A raw material or feedstock should be renewable rather than depleting whenever technically and economically practicable.

8. **Reduce derivatives:** Unnecessary derivatization (use of blocking groups, protection–deprotection, temporary modification of physical–chemical processes) should be minimized or avoided if possible, because such steps require additional reagents and can generate waste.

9. **Catalysis:** Catalytic reagents (as selective as possible) are superior to stoichiometric reagents.

10. **Design for degradation:** Chemical products should be designed so that at the end of their function they break down into innocuous degradation products and do not persist in the environment.

11. **Real-time analysis for pollution prevention:** Analytical methodologies need to be further developed to allow for real-time, in-process monitoring and control prior to the formation of hazardous substances.

12. **Inherently safer chemistry for accident prevention:** Substances and the form of a substance used in a chemical process should be chosen to minimize the potential for chemical accidents, including releases, explosions, and fires.